Advances in Intelligent Systems and Computing

Volume 1407

The series "Advances in Intelligent Systems and Computing" contains publications on theory, applications, and design methods of Intelligent Systems and Intelligent Computing. Virtually all disciplines such as engineering, natural sciences, computer and information science, ICT, economics, business, e-commerce, environment, healthcare, life science are covered. The list of topics spans all the areas of modern intelligent systems and computing such as: computational intelligence, soft computing including neural networks, fuzzy systems, evolutionary computing and the fusion of these paradigms, social intelligence, ambient intelligence, computational neuroscience, artificial life, virtual worlds and society, cognitive science and systems, Perception and Vision, DNA and immune based systems, self-organizing and adaptive systems, e-Learning and teaching, human-centered and human-centric computing, recommender systems, intelligent control, robotics and mechatronics including human-machine teaming, knowledge-based paradigms, learning paradigms, machine ethics, intelligent data analysis, knowledge management, intelligent agents, intelligent decision making and support, intelligent network security, trust management, interactive entertainment, Web intelligence and multimedia.

The publications within "Advances in Intelligent Systems and Computing" are primarily proceedings of important conferences, symposia and congresses. They cover significant recent developments in the field, both of a foundational and applicable character. An important characteristic feature of the series is the short publication time and world-wide distribution. This permits a rapid and broad dissemination of research results.

Indexed by DBLP, EI Compendex, INSPEC, WTI Frankfurt eG, zbMATH, Japanese Science and Technology Agency (JST).
All books published in the series are submitted for consideration in Web of Science.

More information about this series at http://www.springer.com/series/11156

Vikrant Bhateja · Suresh Chandra Satapathy ·
Carlos M. Travieso-González ·
V. N. Manjunath Aradhya
Editors

Data Engineering and Intelligent Computing

Proceedings of ICICC 2020

Springer

Editors
Vikrant Bhateja
Department of Electronics
and Communication Engineering
Shri Ramswaroop Memorial Group
of Professional Colleges (SRMGPC)
Lucknow, Uttar Pradesh, India

Dr. A.P.J. Abdul Kalam Technical
University
Lucknow, Uttar Pradesh, India

Carlos M. Travieso-González
Department of Signals
and Communications
Institute for Technological Development
Las Palmas de Gran Canaria, Spain

Suresh Chandra Satapathy
School of Computer Engineering
Kalinga Institute of Industrial Technology
Bhubaneswar, India

V. N. Manjunath Aradhya
Department of Computer Applications
JSS Science and Technology University
Mysuru, India

ISSN 2194-5357 ISSN 2194-5365 (electronic)
Advances in Intelligent Systems and Computing
ISBN 978-981-16-0170-5 ISBN 978-981-16-0171-2 (eBook)
https://doi.org/10.1007/978-981-16-0171-2

This Springer imprint is published by the registered company Springer Nature Singapore Pte Ltd.
The registered company address is: 152 Beach Road, #21-01/04 Gateway East, Singapore 189721, Singapore

Conference Organisation Committees

Chief Patrons
Dr. D. Hemachandra Sagar, Chancellor, DSU, Chairman DSI
Dr. D. Premachandra Sagar, Pro Chancellor, DSU, Chairman DSI

Patrons
Shri Galiswamy, Secretary, DSI
Dr. K. N. B. Murthy, Vice Chancellor, DSU
Shri. R. Janardhan, Pro-Vice Chancellor, DSU
Dr. Puttamadappa C., Registrar, DSU

Honorary Chairs
H. P. Khincha, Evangelist, DSU
K. Jairaj, Advisor, DSU

Honorary General Chair
Dr. Hao Ying, Wayne State University, Michigan, USA

General Chair
Dr. S. C. Satapathy, KIIT, Bhubaneshwar, Odisha, India

Publication Chair
Dr. Vikrant Bhateja, SRMGPC, Lucknow, U.P., India
Dr. Carlos M. Travieso-Gonzalez, University of Las Palmas de Gran Canaria (ULPGC), Spain

Organizing Chairs
Dr. M. K. Banga, Dean Research, DSU
Dr. A Srinivas, Dean SoE, DSU
Dr. Sanjay Chitnis, Chairman CSE, DSU

Technical Program Chairs
Dr. V. N. Manjunatha Aradhya, SJCE, Mysore, Karnataka, India
Dr. Wendy Flores-Fuentes, Autonomous University of Baja California, Mexico

Program Chairs
Dr. Rajesh T. M., Dept. of CSE, DSU
Dr. Jasma Balasangameshwara, Dept. of CST, DSU

Organizing Secretaries
Dr. Shaila S. G., Dept. of CSE, DSU
Dr. Girisha G. S., Dept. of CSE, DSU
Dr. Mallanagouda Patil, Dept. of CSE, DSU

Advisory Committee
Dr. Aime' Lay-Ekuakille, University of Salento, Lecce, Italy
Dr. Amira Ashour, Tanta University, Egypt
Dr. Aynur Unal, Standford University, USA
Dr. Bansidhar Majhi, IIIT Kancheepuram, Tamil Nadu, India
Dr. Dilip Kumar Sharma, Vice-Chairman, IEEE U.P. Section
Dr. Ganpati Panda, IIT Bhubaneswar, Odisha, India
Dr. Jagdish Chand Bansal, South Asian University, New Delhi, India
Dr. João Manuel R. S. Tavares, Universidade do Porto (FEUP), Porto, Portugal
Dr. Jyotsana Kumar Mandal, University of Kalyani, West Bengal, India
Dr. K.C. Santosh, University of South Dakota, USA
Dr. Le Hoang Son, Vietnam National University, Hanoi, Vietnam
Dr. Naeem Hanoon, Multimedia University, Cyberjaya, Malaysia
Dr. Nilanjan Dey, TIET, Kolkatta, India
Dr. Noor Zaman, Universiti Tecknologi, PETRONAS, Malaysia
Dr. Roman Senkerik, Tomas Bata University, Zlin, Czech Republic
Dr. Swagatam Das, Indian Statistical Institute, Kolkatta, India
Dr. Siba K. Udgata, University of Hyderabad, Telangana, India
Dr. Shri Nivas Singh, MMMUT, Gorakhpur, U.P., India
Dr. Steven L. Fernandez, The University of Alabama at Birmingham, USA.
Dr. Tai Kang, Nanyang Technological University, Singapore
Dr. Valentina Balas, Aurel Vlaicu University of Arad, Romania

Technical Program Committee
Prof. H R Vishwakarma, VIT, Vellore, India
Prof. Dan Boneh, Stanford University, California, USA
Prof. Alexander Christea, University of Warwick, London UK
Prof. Ahmad Al- Khasawneh, The Hashemite University, Jordan
Dr. Bharat Singh Deora, JRNRV University, India
Prof. Jean Michel Bruel, Department Informatique IUT de Blagnac, Blagnac, France
Prof. Ngai-Man Cheung, University of Technology and Design, Singapore
Prof. Yun-Bae Kim, Sung Kyun Kwan University, South Korea

Prof. Ting-Peng Liang, National Chengchi University Taipei, Taiwan
Prof. Sami Mnasri, IRIT Laboratory Toulouse, France
Prof. Lorne Olfman, Claremont, California, USA
Prof. Brent Waters, University of Texas, Austin, Texas, United States
Prof. Philip Yang, Price Water House Coopers, Beijing, China
Prof. R K Bayal, Rajasthan Technical University, Kota, Rajasthan, India
Prof. Martin Everett, University of Manchester, England
Prof. Feng Jiang, Harbin Institute of Technology, China
Prof. Prasun Sinha, Ohio State University Columbus, Columbus, OH, United States
Dr. Savita Gandhi, Gujarat University, Ahmedabad, India
Prof. Xiaoyi Yu, Chinese Academy of Sciences, Beijing, China
Prof. Gengshen Zhong, Jinan, Shandong, China
Dr. Mukesh Shrimali, Pacific University, Udaipur, India
Prof. Komal Bhatia, YMCA University, Faridabad, Haryana, India
Prof. S. R. Biradar, College of Engineering & Technology, Dharwad, Karnataka
Dr. Soura Dasgupta, SRM University, Chennai, India
Dr Sushil Kumar, Jawaharlal Nehru University, New Delhi, India
Dr. Amioy Kumar, IIT Delhi, India

Preface

This book is a collection of high-quality peer-reviewed research papers presented at the '4th International Conference on Intelligent Computing and Communication (ICICC-2020)' held at the School of Engineering (SoE), Dayananda Sagar University (DSU), Bengaluru, Karnataka, India, during 18–20 September 2020.

After the success of the past three editions of ICICC held in 2016 at the University of Kalyani, West Bengal, India, in 2017 at MAEER's MIT College of Engineering, Pune, India, and in 2019 at SoE, DSU, Bengaluru, India, the 4th edition of ICCC was again organized by SoE, DSU, Bengaluru, India. All papers of the past ICICC editions are published by Springer AISC series. Presently, ICICC-2020 provided a platform for academicians, researchers, scientists, professionals and students to share their knowledge and expertise in the diverse domain of intelligent computing and communication.

ICICC-2020 had received a number of submissions from the field of ICT, intelligent computing and its prospective applications in different spheres of engineering. The papers received have undergone a rigorous peer-review process with the help of the technical programme committee members of the conference from various parts of the country as well as abroad. The review process has been crucial with the minimum of two to three reviews each along with due checks on similarity and content overlap. This conference has featured seven exclusive theme-based special sessions as enlisted below along with the main track and hence witnessed more than 400 submissions in total.

S1: Application of Artificial Intelligence in Mobile Apps: A Machine and Deep Learning Techniques Perspective
S2: Deep Learning for Intelligent Systems
S3: Computer Vision for Multimedia Security and Forensics
S4: Emerging Trends in Machine Learning Methodologies and Applications in Artificial Intelligence
S5: Artificial Intelligence and Nature-Inspired Algorithms
S6: Recent Innovations in Artificial Intelligence Research and Computing Technologies
S7: Artificial Intelligence and Machine Learning Applications (AIML)

Out of the above pool, only quality papers were finally presented in 10 parallel tracks/day (of virtual presentation sessions chaired by esteemed professors from premier institutes of the country) and compiled in this volume for publication. This volume consists of these selected papers focussing on the theme of 'Data Engineering and Intelligent Computing'.

The conference featured many distinguished keynote addresses by eminent speakers like Mr. Upendra M Kulkarni (Vice President and G M, Firmware and Software (Non-volatile Memory Solutions Group) at Intel Corporation, California, USA, who delivered a lecture on the Role of High Performance and High-Density Storage in Intelligent Computing; Mr. Andrew Haschka (APAC Technology Practice Lead, Application Modernization, Google Cloud at Google, Australia), who delivered a lecture on Building on Application Platform; Dr. Dinakar Sitaram (Chairperson of the Cloud Computing Innovation Council of India working group, ex-CTO HP System Technology, Software Division, India), who delivered a lecture on Speech Recognition for Indian and Low Resource Languages; Mr. Swaminathan Venkataraman (Worldwide Product and Solution Architect, Communications and Media Solutions (CMS), Hewlett Packard Enterprise, India), who delivered a lecture on Cloud Native 5G.

These keynote lectures/talks embraced a huge toll of audience of students, faculties, budding researchers as well as delegates. The editors thank the General Chair, TPC Chair and the Organizing Chair of the conference for providing valuable guidelines and inspirations to overcome various difficulties in the process of organizing this conference. The editors also thank SoE, DSU, Bengaluru, for their whole-hearted support in organizing this conference.

The editorial board take this opportunity to thank the authors of all the submitted papers for their hard work, adherence to the deadlines and patience during the review process. The quality of a refereed volume depends mainly on the expertise and dedication of the reviewers. We are indebted to the TPC members who not only produced excellent reviews but also did these in short time frames.

Lucknow, India
Bhubaneswar, India
Las Palmas de Gran Canaria, Spain
Mysuru, India
October 2020

Dr. Vikrant Bhateja
Dr. Suresh Chandra Satapathy
Dr. Carlos M. Travieso-González
Dr. V. N. Manjunath Aradhya

The original version of the book was revised: Book volume number has been corrected as 1407. The correction to the book is available at https://doi.org/10.1007/978-981-16-0171-2_61

Contents

About the Editors

Vikrant Bhateja is Associate Professor, Department of Electronics and Communication Engineering, Shri Ramswaroop Memorial Group of Professional Colleges (SRMGPC), Lucknow, and also Head (Academics and Quality Control) in the same college. He completed his doctorate in Bio-Medical Imaging and Signal Processing and has a total academic teaching experience of 16 years with around 125 publications in reputed international conferences, journals and online book chapter contributions. His areas of research include digital image and video processing, computer vision, medical imaging, machine learning, pattern analysis and recognition. Dr. Vikrant has edited 15 proceeding books/editorial volumes with Springer Nature. He is Editor-in-Chief of *IGI Global—International Journal of Natural Computing and Research* (IJNCR). He is also Associate Editor in *International Journal of Synthetic Emotions* (IJSE) and *International Journal of Ambient Computing and Intelligence* (IJACI) under IGI Global press. He is Guest Editor of Special Issues in reputed Scopus/SCIE indexed journals under Springer Nature: *Evolutionary Intelligence* and *Arabian Journal of Science and Engineering*.

Suresh Chandra Satapathy is a Ph.D. in Computer Science, currently working as Professor and at KIIT (Deemed to be University), Bhubaneshwar, Odisha, India. He held the position of the National Chairman Div-V (Educational and Research) of Computer Society of India and is also Senior Member of IEEE. He has been instrumental in organizing more than 20 International Conferences in India as Organizing Chair and edited more than 30 Book Volumes from Springer LNCS, AISC, LNEE and SIST Series as Corresponding Editor. He is quite active in research in the areas of swarm intelligence, machine learning and data mining. He has developed a new optimization algorithm known as Social Group Optimization (SGO) published in Springer Journal. He has delivered number of keynote address and tutorials in his areas of expertise in various events in India. He has more than 100 publications in reputed journals and conference proceedings. Dr. Suresh is in Editorial Board of IGI Global, Inderscience, Growing Science journals and also Guest Editor for *Arabian Journal of Science and Engineering* published by Springer.

Carlos M. Travieso-González received the M.Sc. degree in 1997 in Telecommunication Engineering at Polytechnic University of Catalonia (UPC), Spain, and Ph.D. degree in 2002 at the University of Las Palmas de Gran Canaria (ULPGC-Spain). He is Full Professor on Signal Processing and Pattern Recognition and Head of Signals and Communications Department at ULPGC, teaching from 2001 on subjects on signal processing and learning theory. His research lines are biometrics, biomedical signals and images, data mining, classification system, signal and image processing, machine learning, and environmental intelligence. He has researched in 50 International and Spanish Research Projects, some of them as Head Researcher. He has 440 papers published in international journals and conferences. He has published 7 patents in Spanish Patent and Trademark Office. He is Evaluator of project proposals for European Union (H2020), Medical Research Council (MRC—UK), Spanish Government (ANECA—Spain), Research National Agency (ANR—France), DAAD (Germany), and other Institutions. He has been General Chair in 16 international conferences, mainly sponsored by IEEE and ACM. He is Associate Editor in CIN and Entropy (Q2-WoS journals). He has been awarded in "Catedra Telefonica" Awards in editions 2017, 2018 and 2019 on Knowledge Transfer Modality.

V. N. Manjunath Aradhya is currently working as Associate Professor in the Department of MCA, Sri Jayachamarajendra College of Engineering, Mysuru. He received the M.S. and Ph.D. degrees in Computer Science from the University of Mysore, Mysuru, India, in 2004 and 2007, respectively. He is a recipient of "Young Indian Research Scientist" from Italian Ministry of Education, University and Research, Italy, during 2009–2010. His professional recognition includes as Technical Editor for *Journal of Convergence Information Technology* (JCIT), Editor Board in *Journal of Intelligent Systems*, a reviewer for *IEEE Transactions on System, Man and Cybernetics—Part B*, *Pattern Recognition* (PR), *Pattern Recognition Letters* (PRL), etc. Also he is member of program committees of over 45 international conferences. His research interest includes pattern recognition, image processing, document image analysis, computer vision, machine intelligence, applications of linear algebra for the solution of engineering problems, biclustering of gene expression data and web data analysis and understanding. He has published more than 135 research papers in reputed international journals, conference proceedings and edited books.

Heuristic Approach to Detect Pectoral Muscle and Nipple in Mammogram for Computer-Aided Diagnosis

Vibha Bafna Bora, Ashwin Kothari, and Avinash Keskar

Abstract In computer-aided diagnosis algorithms for mammograms, many preprocessing steps like pectoral muscle detection and the coordinates of the nipple are done automatically before the actual detection and diagnosis process begin. Normally, in medio-lateral oblique (MLO) views, the pectoral muscle should be seen to the tip of the nipple or below. Consequently, straight line estimate of pectoral muscle is parallel to the tangent drawn at nipple. Considering this heuristic anatomical relation, in the proposed work, hybrid combination of textured Hough transform with distant surround suppression was used to detect straight or curved pectoral muscle on MLO view followed by orthogonal Radon transform to detect the nipple position. The proposed algorithm was tested on 266 MLO views from miniMIAS and 74CC views from local hospital. The pectoral muscle was detected with accuracy of 97%, and detected nipple position was acceptable in 93.67% of mammograms with average error of 5.755 mm with location annotated by a radiologist.

Keywords Pectoral muscle detection · Nipple detection · Computer aided diagnosis · Hough transform · Radon transform · Distant surround suppression

1 Introduction

Mammograms are X-ray images of the breast with low dose and are used for screening asymptomatic women for early finding of breast cancer. Breast cancer is one of the wide spread diseases in women, and a lot of effort has been put in avoidance through

V. B. Bora (✉)
G H Raisoni College of Engineering, Nagpur, India
e-mail: vibha.bora@raisoni.net

A. Kothari · A. Keskar
Visvesvaraya National Institute of Technology, Nagpur, India
e-mail: ashwinkothari@ece.vnit.ac.in

A. Keskar
e-mail: avinashkeskar@ece.vnit.ac.in

V. Bhateja et al. (eds.), *Data Engineering and Intelligent Computing*,
Advances in Intelligent Systems and Computing 1407,
https://doi.org/10.1007/978-981-16-0171-2_1

different types of screening processes. Such screening will necessarily generate a huge number of mammograms which must be sighted and read by a limited number of professional radiologists. Hence, usage of computer-assisted interpretation of the results can reduce the workload on radiologists [1]. These systems are usually used as a second adviser, with the final conclusion regarding the presence of a cancer left to the radiologist.

Pectoral muscle and nipple location are normally the key points for mammogram registration in computer-aided diagnosis (CAD). Removal of pectoral muscle is also vital in following cases:

1. Pectoral muscle must be removed completely in an automated breast tissue density quantification method, otherwise, in case of fatty tissues, it will cause an over evaluation of the breast density. On the contrary, mammogram density may be underestimated, if the triangular pectoral region accidentally includes some of the dense glandular area adjacent to the pectoral region [2].
2. Due to resemblance between mammographic glandular parenchyma and pectoral muscles, higher number of false positives occurs in mass detection method [3].

Similarly, location of nipple also plays crucial role in following cases:

1. Nipple location is one of the significant landmarks for measuring possible asymmetry between the right and left breast of a patient in computer-aided detection (CAD) analysis [4].
2. Radiologists normally pay more attention to the nipple while examination, as cancer can arise in glandular tissues which converge onto nipple [5, 6].
3. Coordinates of nipple are important in finding the architectural distortion of the mammogram [7].
4. In CAD, nipple is an alignment pivot [8].

Proper detection of pectoral muscle and nipple is being so important that these have been a main support to the researchers working on CAD. The methods [9–12] in which muscle was assumed to be straight give moderate accuracy since majority of the muscle follows curvature. As reported by Kinoshita et al. [12], images with straight line edges had enhanced performance with and inferior performance with other types of edges of the pectoral muscle region. The method by Ferrari et al. [13], based on Gabor transform, was reported to be very time-consuming, and only 84 miniMIAS images were taken which excluded very challenging images like mdb053l, mdb061l, mdb179l, mdb240r, etc. The performance of Kwok et al. [14] was visually reviewed by a five point index only, but no pixel-wise quantitative consideration was performed. Also method Raba et al. [15] results into excess of segmentation for mammograms when the dense glandular tissues are close to the pectoral muscle and the performance was only visually evaluated. In texture field method [16], quantitative assessment was performed but for mammogram with BI-RADS type-4; the breast that contains glandular and fibrous tissue greater than 75% and skin folded in upper part covers 20–30% of the breast region near to muscle as a pectoral muscle. Thus, the registration of pectoral muscle is a challenging task due to several factors like overlapping of muscle by fibro glandular tissue, muscle periphery masked by artifacts such as sticky tapes and axillary fold due to ill positioning.

The method [6, 8] uses the breast boundary technique to obtain nipple location and neither of the two methods reported whether the nipple was in silhouette on the mammogram. In [5] also, breast boundary was extracted, and the performance of the method was evaluated using 24 images; a very limited data set with 16 medio-lateral (MLO) views and eight cranio-caudal (CC) views. As reported in [12], the nipple is detected only if nipple is located on breast surface as bright regions. Thus, nipple registration is also a challenging task because of the nipple visibility or invisibility in mammogram, variations in image quality and patient positioning.

The aim of the proposed work is to develop a robust method for pectoral muscle detection in MLO view which must consider overlapping of muscle with dense tissues as well as curvature followed by nipple detection.

In the proposed method, textured Hough transform with distant surround suppression technique is developed to delineate straight or curved pectoral muscle accurately. Further, using the slope of pectoral muscle, two orthogonal Radon transforms are taken to register nipple (visible or invisible) on mammogram.

2 Materials and Methods

2.1 Database

The database consists of total 340 mammograms, 266 are MLO views from a miniMIAS database [17], and 74 are CC views locally procured from (Central India Cancer Research Institute (CICRI), Nagpur, Maharashtra, India). The original image of miniMIAS database, digitized at 50 micron pixel edge, has been condensed to 200 micron pixel edge and padded so that single image is 1024×1024 pixels. CICRI mammograms were digitized with GE Senographe600T scanner with a pixel edge of 100 micron. Each original image is thus 2370×1770 pixels and a gray value resolution of 12 bit. These images are down sampled, without loss of significant information to 1185×885 pixels with a resolution of 200 micron pixel and a gray value resolution of 8 bit. The proposed method was executed on MATLAB ver. R2016.

2.2 Artifact Removal, Orientation Correction and Enhancement

In scanned or digital mammograms labels, tags and scanning artifacts were present, which make detection of key point more challenging. Artifacts were removed using morphological process and connected component algorithm [18]. There is no orientation standard in mammography, and as reported by Mustra et al. [19] compared to unknown position in the image, known orientation makes detection of muscle somewhat easier, so muscles are correctly oriented with the help of heuristic approach [20]. Hence, each mammogram had pectoral muscle in the top left corner before muscle and nipple detection algorithm is applied. Next breast background is cropped and contrast enhanced [18, 21].

2.3 *Pectoral Muscle Detection*

The pectoral region is the high intensity triangular section in the top left corner of the mammogram. Let the mammogram image be indicated by $I(x, y)$. The top left corner of the image is the origin (1, 1), where X is defined to be the vertical axis and Y to be the horizontal axis. M: no. of rows and N: no. of columns. Hough transform can be used to approximate the muscle edge with straight lines [18]. But in many mammograms, pectoral muscles are heavily textured and occluded by dense tissues. Hence, in proposed work, instead of applying Hough transform to an intensity mammogram, it was applied to textured plot of mammogram to obtain straight line approximation of muscle edge accurately.

2.3.1 Textured Hough Transform, Region of Interest and Straight Line Approximation

The textured plot of mammogram (TPM) can be plotted using texture feature such as local entropy, phase gradient and gray intensities [22]. Local entropy is a statistical measure of randomness in the subimage. It achieves maximum value when all elements of the image are maximally random and minimum for constant image. As muscle boundary is characterized by high intensity change, hence in TPM, muscle edge will have high values of local entropies. Since orientation correction is done, muscle is always located in upper left corner. Hence geometrically, it is always situated in first quadrant from 30° to 90°. The phase gradient at each pixel location (x, y) in the mammogram is obtained using first derivative Sobel operator [18]. Hence, TPM is the set of those pixels having orientation in the range of 30°–90°, local entropies greater than mean entropy of the mammogram and gray intensities greater than 0.5 of $I_{\max}$, where $I_{\max}$ is maximum intensity present in mammogram. Since muscle is high intensity region, pixels present in muscle are always greater than 0.5 of $I_{\max}$.

During image acquisition, depending on anatomy and position of patient, half of the breast region or near about of it could be occupied by the pectoral muscle [13]. Hence, from the TPM, initial rectangle shaped region of interest (ROI) is considered for straight line approximation of pectoral edge and is represented as.

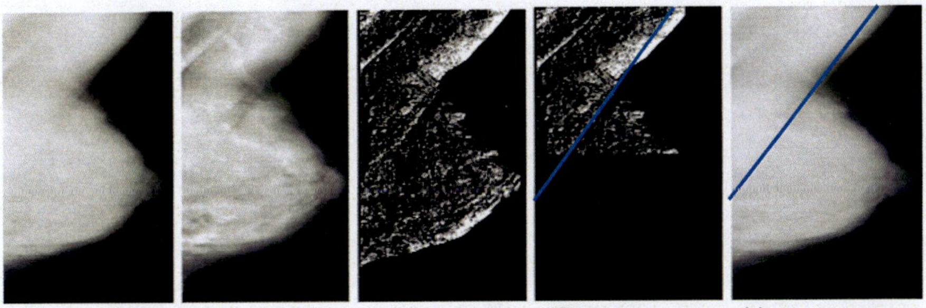

Fig. 1 Straight line approximation of pectoral muscle for mdb053l. **a** Original images mdb053l, **b** mammogram after contrast enhancement, **c** textured plot of mammogram; TPM, **d** region of interest and approximated straight edge in blue color using Hough transform, **e** approximated straight pectoral edge overlapped on original mammogram

$$\text{ROI} = \{\text{TPM}(x, y) : 1 \le x \le M/2 \text{ and } 1 \le y \le 3 * N/4\} \tag{1}$$

To the ROI of Eq. (1), Hough transform is applied to obtain strongest edge [18, 22]. If the strongest straight line intersects top and left edge of the mammogram, then it is a valid straight line, otherwise ROI is tuned by about 20%.

Figure 1a–e shows the entire process of straight line approximation of pectoral muscle for the case mdb053l.

2.3.2 Pectoral Edge Refinement Using Distant Surround Suppression

The anatomical shape of the pectoral muscle is either straight, concave or convex or mixture of both. Using Hough transform on TPM, pectoral edge was approximated as a straight line PQ as shown in Fig. 2a. Let the coordinates of the approximated straight pectoral boundary be P (1, P_{wt}) and Q (P_{ht}, 1), where P_{wt} and P_{ht} are the distance from the upper left corner of the mammogram to the top and bottom edge of the approximate pectoral muscle, respectively. To obtain the actual curvature of the muscle, the region near the straight edge is considered by shifting straight line toward left and right by value Px from the approximated edge PQ as shown in Fig. 2a. The left straight boundary AB is plotted from $A(1, P_{\text{wt}}\text{-}Px)$ to $B(P_{\text{ht}}\text{-}Px, 1)$, whereas right straight boundary CD is plotted from $C(1, P_{\text{wt}} + Px)$ to $D(P_{\text{ht}} + Px, 1)$, and the selected textured plot of mammogram (STPM), i.e., APCDQBA for mdb053l is shown in Fig. 2a. Further STPM is $n \times n$ (here 5×5 empirical value taken) block averaged so that the pixels which are least involved in pectoral edge representation are removed as seen in Fig. 2b. The averaged STPM is obtained as

$$\begin{aligned} &\text{Avg\_STPM}[x : x + (n-1), y : y + (n-1)] \\ &= \frac{\sum_{s=0}^{(n-1)} \sum_{t=0}^{(n-1)} \text{ROI}(x + s, y + t)}{n^2} \end{aligned} \tag{2}$$

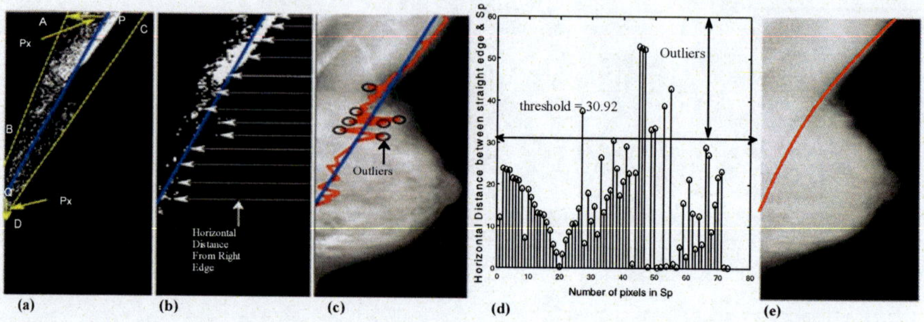

Fig. 2 Pectoral muscle detection using distant surround suppression. **a** Selected textured plot of mammogram (STPM) for mdb053l, **b** 5 × 5 block averaged STPM, scanned horizontally to obtain S_P; a set of first nonzero pixels from right edge of mammogram. **c** S_p plotted as a rough pectoral edge; black circle is indicated as some outliers. **d** Absolute distance plot between pixels in S_p and their horizontal distance between straight pectoral edge and S_P. **e** Smoothened pectoral muscle using cubical polynomial on original mammogram

where $x = 1, n, 2n, \ldots, M/2$ and $y = 1, n, 2n, \ldots, N/4$. Blocks with average value of 0.51 or higher are considered as 1 (white) otherwise to 0 (Black). The size of *Px* is optimally chosen to wrap pectoral boundary variation for vastly concave or convex shaped muscle. If the smaller value of Px is chosen, then STPM will miss pixels which are significant on pectoral edge to give the accurate shape. Contrary if larger value Px is chosen, then unwanted nearby pixels may get perceived to form imprecise shape of the muscle. In this work, the experimental value of $Px = 80$ pixels is taken.

In next step, binary Avg_STPM image is scanned horizontally from right vertical edge to pick the first nonzero pixel as shown in Fig. 2b. As a result, a string of pixels position (S_P) is obtained and considered as rough pectoral edge as shown in Fig. 2c, out of which majority of pixels lies on the pectoral edge. The outliers in S_P may be due to heavy textured pattern nearby muscle edge or due to axillary folds. To remove the outliers, distant surround suppression (DSS) technique is applied. In DSS technique, horizontal distance between approximated straight pectoral edge *PQ* and S_P is calculated as follows

The Cartesian slope intercept form of approximated straight pectoral edge is

$$X = m_y Y + b \tag{3}$$

where *m* is the slope of the line PQ with respect to horizontal *y*-axis and *b* as the *x* intercept. The two end points are $P\,(1, P_{wt})$ and $Q(P_{ht}, 1)$, and then, the slope m_y and *x* intercept *b* is given as

$$m_y = \frac{(1 - P_{ht})}{(P_{wt} - 1)} \tag{4}$$

$$b = 1 - m_y P_{wt} \tag{5}$$

The horizontal coordinate of points satisfying the straight pectoral muscle is obtained using the equation

$$Y(y) = \frac{X(x) - b}{m_y} \tag{6}$$

Then, the horizontal distance between points on PQ and S_P is

$$\mathrm{dst}(x) = |Y(y) - S_P(y)| \tag{7}$$

where S_P is the set of position of first nonzero pixels from right edge of mammogram when Avg_STPM is scanned horizontally and the plot is drawn in Fig. 2d.

Now to remove outliers, empirical threshold is calculated as

$$\mathrm{threshold} = \mathrm{med(dst)} + \mathrm{IQR(dst)} \tag{8}$$

where med is the median value and IQR is the interquartile range of dst.

If the horizontal distance dst(x) for pixel of S_P is greater than threshold then it will be treated as outliers as shown in Fig. 2c and removed from S_P. The filtered S_P is further smoothened using cubical polynomial and is plotted on original mammograms shown in Fig. 2e. (The coordinates of point P and Q are concatenated with S_P before polynomial modeling).

2.4 Nipple Detection Using Orthogonal Radon Transform

Mathematically, Radon transform of mammogram $I(x, y)$ is represented [18, 23] as

$$I(\rho, \phi) = \int_{-\infty}^{\infty} \int_{-\infty}^{\infty} I(x, y)\delta(x \cos\phi + y \sin\phi - \rho)\mathrm{d}x\mathrm{d}y$$
$$-\infty < \rho < \infty, 0 \le \phi \le \pi \tag{9}$$

where $\rho = x \sin\phi + y \cos\phi$ is the straight line equation in polar form. If orthogonal Radon transform $g(\rho, \phi)$ and $g(\rho, \phi + 90°)$ is drawn, then the intersection of two Radon projections at some specific location has the ability to detect the position of nipple on mammogram. In the proposed method of nipple detection, in MLO view, the first step is to obtain the slope of the pectoral muscle m_x with respect to vertical x-axis, and angle of pectoral muscle with respect to vertical x-axis is given by

$$m_x = \frac{(P_{wt} - 1)}{(1 - P_{ht})} \tag{10a}$$

$$\phi = \tan^{-1}(m_x) \tag{10b}$$

Then, the two Radon transforms are taken at an angle equal to ϕ and $\phi + 90°$. The first Radon transform taken at the angle ϕ gives the location A (Loc(A)) and is marked as the last nonzero intensity on the projection as shown in Fig. 3a, b. Mathematically denoted as

$$\text{Loc}(A) = \{\text{lst}(g(x \sin\phi + y \cos\phi, \phi)) | \forall g(\rho, \phi) \sim 0\} \tag{11}$$

where lst indicates the last element of the set having nonzero value on the projection. The tangent at point A is parallel to the approximated straight pectoral muscle edge. The equation of line passing Loc(A) $[x_1, y_1]$ is

$$(x - x_1) = m_a(y - y_1) \tag{12}$$

where $m_a = m_y$ (refer Eq. 4) is slope of the pectoral edge with respect to horizontal y-axis.

The second projection taken at $\phi + 90°$ gives the location of point B which is the highest peak point on the radon projection as shown in Fig. 3c, d.

$$\begin{aligned}\text{Loc}(B) = \{&g(x \sin((\phi + 90) + y \cos(\phi + 90), (\phi \\ &+ 90)) | \forall g(\rho, \phi + 90°) = \max(g(\rho, \phi + 90°)\}\end{aligned} \tag{13}$$

The equation of line passing from Loc(B) $[x_2, y_2]$ is

$$(x - x_2) = m_b(y - y_2) \text{ where } m_b = \tan(\tan^{-1}(m_y) + 90°) \tag{14}$$

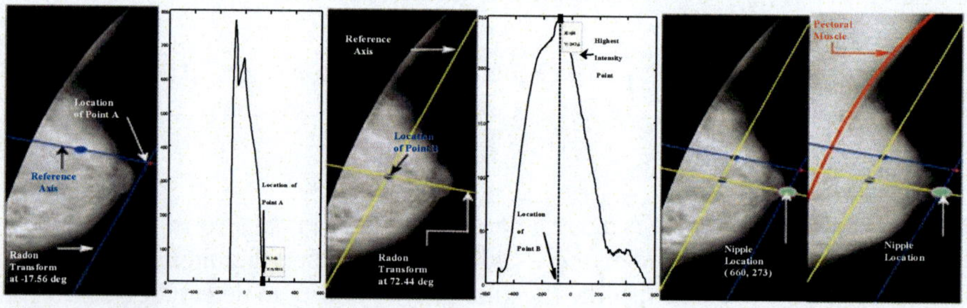

(a) (b) (c) (d) (e) (f)

Fig. 3 Nipple detection on mdb0531 with pectoral muscle delineated. **a** Radon transform at angle $\phi = -17.56°$ indicating location of Point A, **b** radon projection at angle ϕ indicating the last nonzero intensity as location of Point A. **c** radon transform at angle ϕ + 90° (i.e., 72.44°) indicating Point B on the reference axis. **d** Radon projection at ϕ + 90° indicating Point B as the highest intensity point. **e** Nipple position located on intersection of two orthogonal Radon transforms. **f** Registration of pectoral muscle and nipple position on original mammogram

The intersection of line drawn from location A and B is the location of nipple as shown in Fig. 3e, f. Similarly, for CC view, Radon transform is applied at angle 0° and 90° to find the location of the nipple, which is in majority of the cases at pinnacle of the breast.

2.5 *Performance Evaluation*

2.5.1 Performance Metrics for Pectoral Muscle

The accuracy of the proposed technique was estimated through comparison of segmented pectoral "mask" $P = (p_1, p_2, \ldots, p_x)$ with its corresponding "gold standard" $Q = (\mathrm{q}_1, \mathrm{q}_2, \ldots, q_y)$ created by experienced radiologist using quantitative measures. For each MLO view mammogram, the accuracy of proposed method is evaluated by three performance metrics and qualitatively by pectoral score index (PSI).

1. **Pixel Percentage of False Positive (FP):**

A pixel outside the reference region but inside the acquired pectoral edge. Mathematically,

$$\text{Pixel Percentage of FP} = \frac{|P \cup Q| - |Q|}{|Q|} \times 100\% \tag{15}$$

where $P \in$ attained pectoral muscle region, $Q \in$ gold standard pectoral muscle.

2. **Pixel Percentage of False Negative (FN):**

A pixel inside the reference region but outside the attained pectoral edge. Mathematically,

$$\text{Pixel Percentage of FN} = \frac{|P \cup Q| - |P|}{|Q|} \times 100\% \tag{16}$$

3. **Hausdorff Distance (HD):**

The Hausdorff distance is defined by as

$$\mathrm{HD}(P, Q) = \max(h(P, Q), h(Q, P)) \tag{17}$$

where $h(P, Q) = \max_{p \in P} \min_{q \in Q} \| p - q \|$ and $\| . \|$ is the Euclidean distance.

4. **Pectoral Score Index (PSI):**

Radiologist qualitatively determines accuracy of pectoral muscle using four point pectoral score index (PSI), as follows. If FP and FN > 15%: Not acceptable (1), if FP and FN < 15%: Tolerable (2), if 5% < FP and FN < 10%: Acceptable (3) and if FP and FN < 5%: Accurate (4). An index of 3 or more indicates a sufficient muscle registration. But pectoral muscle registration with tolerable limit can be used for nipple registration.

2.5.2 Performance Metrics for Nipple Registration

With respect to the visibility of the nipple as a "gold standard", radiologists interpret all the images in the database to evaluate the performance of the nipple detection algorithm. Let $N_m(x, y)$ be the result of automated nipple detection method and $N_r(c, d)$ be nipple location annotated by radiologist. The accuracy of algorithm is quantitatively evaluated by calculating the chess board distance between N_m and N_r.

$$C_{\text{dist}}(N_m, N_r) = \max(|x - c|, |y - d|) \tag{18}$$

Radiologist determines accuracy of nipple registration using three point nipple score index (NSI). If C_{dist}> 10 mm: Not Acceptable (1), if 5 mm < C_{dist}< 10 mm: Acceptable (2) and if C_{dist}< 5 mm: Accurate (3).

3 Results and Discussion

3.1 Accuracy of Pectoral Muscle and Nipple Detection on MLO and CC Views

The method for the detection of pectoral muscle and nipple was applied to 266 MLO views from miniMIAS which includes many challenging images and 74 CC views from CICRI database. The algorithm seems to be very effective for pectoral muscle overlapped by dense tissues or muscles having very low distinction with breast tissues. For mammograms having large and small pectoral muscle as shown in Fig. 4a–e, the proposed method gives accurate registration of pectoral muscle and nipple except in mdb031l, and nipple is not detected correctly (distance between automated nipple and annotated nipple, C_{dist}> 12 mm). As many reported methods are unable to segment mammograms having muscle with heavy texture and edge obscured by highly dense breast tissues, for example, mdb062r (see Fig. 4b), but the proposed method can adequately segment it with PSI of 4. Here, the pectoral muscle is estimated with FN = 6.3% and FP = 2.8%. As mentioned by Kwok et al. [14], no muscle segmentation was produced for mdb179l due to fuzzy pectoral edge. With proposed textured-based surround suppression method as shown in Fig. 4d, pectoral muscle is registered with FP = 7.82% and FN = 10.37% and can be used

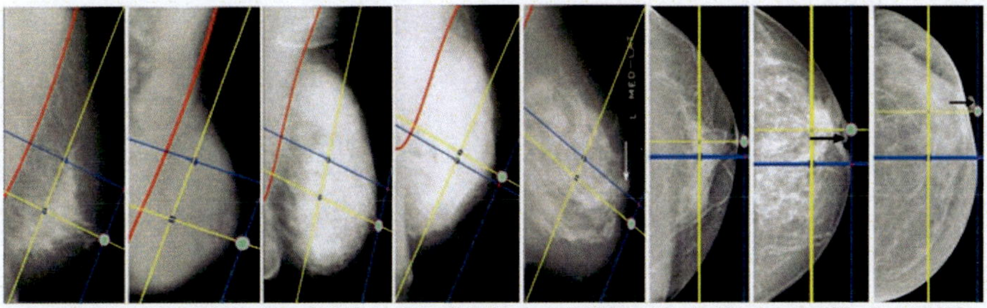

(a) mdb233l [4,4] (b) mdb062r [4,4] (c) mdb240r [3,4] (d) mdb179l [2,4] (e) mdb031l [4,1] (f) cicri01 [3] [g] cicri02 [2] (h) cicri03 [2]

Fig. 4 Results of pectoral muscle and nipple detection on original MLO views from the miniMIAS and nipple detection on CC views from CICRI database. Red line indicates the muscle delineated using proposed method. Nipple detected automatically using proposed method is indicated by circle (light green face color). The qualitative indices are shown as [PSI, NSI] for MLO and [NSI] for CC views

to calculate the nipple position. In Hough [11] and Gabor [13], edge segmentation is unacceptable if FP and FN > 10%, while in Radon [12], it is unacceptable if FP and FN > 15%. For the proposed method, pectoral muscle detection is unacceptable if FP and FN > 10%.

The three performance metrics for pectoral muscle detection namely HD, FP and FN are tabulated in Table 1. From Table 1, proposed method outperforms Hough [11], Kinoshita et al. [12], Gabor [13] method, all of which were implemented on the same miniMIAS platform with different number of images. Table 2 shows range of chess board distance C_{dist} in terms of NSI, accuracy and average error for proposed method of nipple registration on both the database and is compared with Kinoshita et al. [12].

The proposed method fails to register the vertical pectoral muscle. Because a valid strongest straight line should intersect top and left edge of the oriented mammogram, the vertical pectoral muscle cannot satisfy the validation condition. For example, in mdb151l.

4 Conclusion

Precise detection of pectoral muscle and nipple registration on mammograms is demanding because of muscle texture, dense tissue overlap on muscle edge, image quality and nipple projections. A new accurate texture-based distance surround suppression method for automatic detection of the pectoral muscle on the MLO views and nipple registration using orthogonal Radon transform is proposed on the both the views. In two stage pectoral registration technique, textured Hough transform was applied to obtain straight line approximation of muscle followed by distant surround suppression and polynomial modeling to refine it to a curve. Without breast

Table 1 Comparison of proposed method with three reported methods

Author/methods	Database/(No. of images)	HD (in mm)	FP (%)	FN (%)	Images with (FP and FN) < 5%	Images with 5% < (FP&FN) < 10%	Images with 10% < (FP&FN) < 15%	Images with (FP and FN) > 15%	Accuracy (%)
Proposed/Textu-redHough + distant surround suppression	miniMIAS (266)	3.22 ± 1.66	1.54 ± 3.86	3.7 ± 3.88	214 (80.45%)	44 (16.55%)	6 (2.25%)	2 (0.75%)	97
Ferrari et al. [11] /Hough	miniMIAS (84)	7.08 ± 5.26	1.98 ± 6.09	25.19 ± 19.14	11.90%	9.52%	78.57%	NA	21.42
Ferrari et al. [13] /Gabor	miniMIAS (84)	3.84 ± 1.73	0.58 ± 4.11	5.77 ± 4.83	53.57%	26.19%	20.24%	NA	79.76
Kinoshita et al. [12] /radon	Local (540)	12.45 ± 22.96	8.99 ± 38.72	9.13 ± 11.87	28.9%	40.7%		30.4%	69.6

Table 2 Range of chess board distance C_{DIST} in terms of NSI, accuracy and average error and comparison

Author/method	Database/(No. of images)	Nipple presence	Accurate (3) C_{dist}< 5 mm	Acceptable (2) 5 mm < C_{dist}< 10 mm	Not acceptable (1) C_{dist}> 10 mm	Accuracy (%)	Average error (mm)
Proposed/orthogonal radon	miniMIAS (266 MLO views)	Visible(114) Invisible(152)	67(58.77%) 73(48.02%)	40(35.08%) 68(44.73%)	07(6.14%) 11(7.23%)	93.85 92.75 93.67	6.98 5.755
	CICRI (74 CC views)	Visible (54) Invisible (20)	38(70.37%) 09(45.00%)	15(27.77%) 09(45.00%)	01(1.85%) 02(10.00%)	98.14 90.00	4.53
Kinoshita et al. [12]/radon	Local hospital (1080, MLO + CC)	NA	585(54.2%)	271(25.1%)	224(20.7%)	79.30	7.4

boundary extraction, nipple is extracted using two Radon transform at ϕ and $\phi + 90°$, where ϕ is the angle of pectoral muscle with respect to vertical axis. The method was tested on 266 MLO views from miniMIAS and 74 CC views from CICRI database and a radiologist to assess the registration results. The results demonstrate that pectoral muscle was registered with the accuracy of 97% and nipple with overall accuracy of 93.67% and average distance of 5.755 mm.

Acknowledgements The authors would like to thanks Dr. Suchitra Mehta, Central India Cancer Research Institute (CICRI), Nagpur, India, for providing Mammography Images and also for assessing and giving ground truth results.

References

1. K. Doi, Computer aided diagnosis in medical imaging: historical review, current status & future potential. Comput. Med. Imaging Graph. **31**(4-5), 198–211 (2007)
2. P.K. Saha, J.K. Udupa, E.F. Conant, D.P. Chakraborty, D. Sullivan, Breast tissue density quantification via digitized mammograms. IEEE Trans. Med. Image **20**, 792–803 (2001)
3. R. Gupta, P.E. Undrill, The use of texture analysis to delineate suspicious masses in mammography. Phys. Med. Biol. **40**, 835–855 (1995)
4. V. Andolina, S. Lille, K. Willison, *Mammographic Imaging: A Practical Guide* (Williams & Wilkins, Lippincott, 2001)
5. R. Chandrasekhar, Y. Attikiouzel, A simple method for automatically locating the nipple on mammograms. IEEE Trans. Med. Imaging **16**(5), 483–494 (1997)
6. A.J. Mendez, P.G. Tahoces, M.J. Lado, M. Souto, J.L. Correa, J.J. Vidal, Automatic detection of breast border and nipple in digital mammograms. Comput. Methods Prog. Biomed **49**, 253–262 (1996)
7. R.M. Rangayyan, S. Banik, J. Chakraborty, S. Mukhopadhyay, J.E. Desautels, Measures of divergence of oriented patterns for the detection of architectural distortion in prior mammograms. Int. J. Comput. Assist. Radiol. Surg. **8**(4), 527–545 (2013)
8. F. Yin, M. Giger, K. Doi, C. Vyborny, R. Schmidt, Computerized detection of masses in digital mammograms: automated alignment of breast images and its effect on bilateral-subtraction technique. Med. Phys. **21**, 445–452 (1994)
9. N. Karssemeijer, Automated classification of parenchymal patterns in mammograms. Phys. Med. Biol. **43**, 365–378 (1998)
10. S.R. Aylward, B.M. Hemminger, E.D. Pisano, Mixture modeling for digital mammogram display and analysis. Dig. Mammogr. Comput. Imaging Vis. Ser. **13**, 305–312 (1998)
11. R.J. Ferrari, R.M. Rangayyan, J.E.L. Desautels, A.F. Frère, Segmentation of mammograms: identification of the skin boundary, pectoral muscle, and fibroglandular disc, in *Proceedings of the 5th International Workshop on Digital Mammography* (2000), pp. 573–579
12. S.K. Kinoshita, P.M. Azevedo-Marques, R.R. PereiraJr, J.A.H. Rodrigues, R.M. Rangayyan, Radon-domain detection of the nipple and the pectoral muscle in mammograms. J. Digit. Imaging **21**, 37–49 (2008)
13. R.J. Ferrari, R.M. Rangayyan, J.E.L. Desautels, R.A. Borges, A.F. Frère, Automatic identification of the pectoral muscle in mammograms. IEEE Trans. Med. Imaging **23**, 232–245 (2004)
14. S. Kwok, R. Chandrasekhar, Y. Attikiouzel, M. Rickard, Automatic pectoral muscle segmentation on mediolateral oblique view mammograms. IEEE Trans. Med. Imaging **23**(9), 1129–1140 (2004)

15. D. Raba, A. Oliver, J. Marti, M. Peracaula, J. Espunya, Breast segmentation with pectoral muscle suppression on digital mammograms, in *Pattern Recognition and Image Analysis*, LNCS; 3523 (Springer, Berlin, 2005), pp. 471–478
16. C. Zhou, J. Wei, H. Chan, C. Paramagul, L. Hadjiiski, B. Sahiner, et al., Computerized image analysis: texture-field orientation method for pectoral muscle identification on MLO-view mammograms. Med. Phys. **37**(5), 2289–2299 (2010)
17. J. Suckling, J. Parker, D. Dance, S. Astley, I. Hutt, C. Boggis et al., The mammographic image analysis society digital mammogram database. Exerpta Med. Int. Congr. Ser. **1069**, 375–378 (1994)
18. A.K. Jain, *Fundamentals of Digital Image Processing*. Prentice Hall Information and System Sciences Series NJ, USA: Prentice Hall (1989)
19. M. Mustra, M. Grgic, Robust automatic breast and pectoral muscle segmentation from scanned mammograms. Sig. Process. **93**(10), 2817–2827 (2013)
20. V. Bora, A.G. Kothari, A.G. Keskar, A Heuristic Approach for Automated Mammogram Orientation in Computer Aided Diagnosis. IEEE-ICETET -7, Japan, pp. 56–61 (2015)
21. Vikrant Bhateja, Mukul Misra, Shabana Urooj, Aimé Lay-Ekuakille, A robust polynomial filtering framework for mammographic image enhancement from biomedical sensors. IEEE Sens. J. **13**(11), 4147–4156 (2013)
22. Vibha Bora, A.G. Kothari, A.G. Keskar, Robust automatic pectoral muscle segmentation from mammograms using texture gradient and Euclidean distance regression. J. Dig. Imaging **29**, 115–125 (2016)
23. R. Rangayyan, W.A. Rolston, Directional image analysis with the Hough and Radon transforms. J. Indian Inst. Sci. **78**(1), 3–16 (1998)

Design of LSTM–CNN with Feature Map Merge for Crime Scene Detection in CCTV Footage

E. Thirumagal and K. Saruladha

Abstract Nowadays, circuited televisions are installed in many places, and it is been monitored by the authorities to detect the crimes. The authorities carelessness may prone to miss the important crimes that are been recorded. There are two existing architectures for detecting crimes and human activities such as 3D CNN and two-stream CNN. The drawbacks of the existing techniques are that it is unable to capture the complete local features and the accuracy is less. In order to overcome these problems, this paper proposes long short-term memory based convolutional neural network with feature map merge block to capture the complete features of the scene. The proposed architecture detects the weapons such as guns and knives, robberies, and if anybody is fighting. The experiments are conducted by using 500 crime scene videos collected from Internet. The experimental results shows that proposed architecture yields 90% of accuracy. Further, the results are compared with existing architectures.

Keywords CCTV · Convolutional neural network · Crime detection · Long short-term memory · Surveillance

1 Introduction

In recent years, the deep neural networks has become widespread in various research areas such as natural language processing, image processing, and recognition of human activities. Deep neural networks has become researcher's favorite, as it yields better prediction results and extracts complex features and its classification efficiency

E. Thirumagal (✉) · K. Saruladha
Department of Computer Science and Engineering, Pondicherry Engineering College, Puducherry, India

K. Saruladha
e-mail: charuladha@pec.edu

E. Thirumagal
REVA University, Bengaluru, India

V. Bhateja et al. (eds.), *Data Engineering and Intelligent Computing*, Advances in Intelligent Systems and Computing 1407,
https://doi.org/10.1007/978-981-16-0171-2_2

in various research fields. The rapid growth of deep neural network techniques has made the difficult task of analyzing the images and videos easy. The closed-circuit television video (CCTV) surveillance is been globally installed to detect crimes. But this CCTV monitoring involves human to monitor it to detect the crimes. Monitoring of many CCTV's at the same time might tend to missing of the crime scenes by human. Hence, an automated system is needed to detect the crimes in the CCTV footages and to alert the human.

For detecting crime and the human activity tracing, there are two main existing architectures such as two-stream CNN and 3D CNN. The two-stream CNN uses spatial and temporal convnets. The temporal convnets uses optical flow trajectory stacking. The drawback of two-stream CNN is that the spatial pooling technique used with it are not taking the trajectories in account. This may lead to the problem of missing important features from the scene. The drawback of the 3D CNN is that the placing or movement of camera. It uses more resources such as time, and hardware.

With the motivation of surplus amount of CCTV's installed, increased crime rate, and the applications of deep neural networks, the LSTM–CNN with feature merge is been proposed to detect the crimes in CCTV footage. The main contributions of proposed LSTM–CNN with feature fusion are (i) to capture the spatial features of the CCTV footage using CNN with the feature map merge block, (ii) to capture the temporal features of the CCTV footage using LSTM, (iii) to fuse the spatial and temporal features using feature fusion block, (iv) to classify the crime scene using the soft-max classifier, and (v) to improve the accuracy of crime detection compared to existing two-stream CNN and 3D CNN.

The main advantages of the proposed LSTM–CNN with feature map merge is that with limited resources, all the important local features of the scene can be captured. So that the crime scenes will not be missed. The accuracy is more when compared to the existing architectures. The paper is organized as follows: Sect. 2 describes the related works. The CNN, LSTM, feature map merge technique used are described in Sect. 3. Sect. 4 discuss the experimental results obtained. Sect. 5 describes the conclusion.

2 Related Works

The convolutional neural networks are employed in various applications, as it yields better results. The paper [1] proposed two-stream CNN which uses two ResNet CNN for capturing the spatial and the temporal features. It also uses LSTM module to track the human activity. The drawback is that it cannot capture the complete local features. Training two ResNets and LSTM is time consuming and complex. The paper [2, 3] proposed a 3D CNN to detect the fights recorded in the CCTV's and localize it. The advantage is that it uses the optical flows to capture the motion of the human which demonstrated positive. The drawback is that the most of the spatial features captured has not demonstrated positive in many cases. The paper [4] used VGG-19 for detecting the crime, and it produced a better accuracy. The paper [5] proposed a deeper VGG-19 architecture for recognizing human behavioral activities which is done by soft-max supervision. The paper [6] uses the 3D CNN for capturing the spatial and temporal features of the CCTV footage. The 3D CNN is used for detecting the abnormal driving of vehicles by humans which has the yielded the accuracy of 86%. The paper [7] proposed two-stream CNN which uses two fully connected convolutional networks for capturing the spatial and the temporal features. The pre-trained AlexNet CNN architecture was used. For finding the abnormal scenes, the high-dimensional features maps are encoded, and the patterns are found using the recurrent binary codes. The advantage is that it yields both spatial and temporal consistency. The drawback is the complexity. The encoding measures must be optimized to reduce the complexity. The paper [8, 9] proposed a two-stream CNN architecture which uses the darknet CNN architecture. The two stream indicates the spatial [10] and the temporal features. The spatial features are detected by YOLO detector, and the temporal features are obtained by the SORT [11] or DeepSORT tracking system with deep appearance feature. Hence in this paper, CNN–LSTM with feature map merge is proposed to detect the crimes in the CCTV footages and also to increase the accuracy and speed of recognition. The next section discuss the methodologies used to capture the spatial and the temporal features and the techniques used for merging them.

3 Methodology

This section discusses the various techniques used in the proposed LSTM–CNN with feature map merge. The proposed LSTM–CNN with feature map merge architecture is shown in Fig. 1.

3.1 Pre-processing

The input surveillance video is first been loaded. The video must be pre-processed [12], i.e., the videos must be resized and also decreased the frame rate to be 30

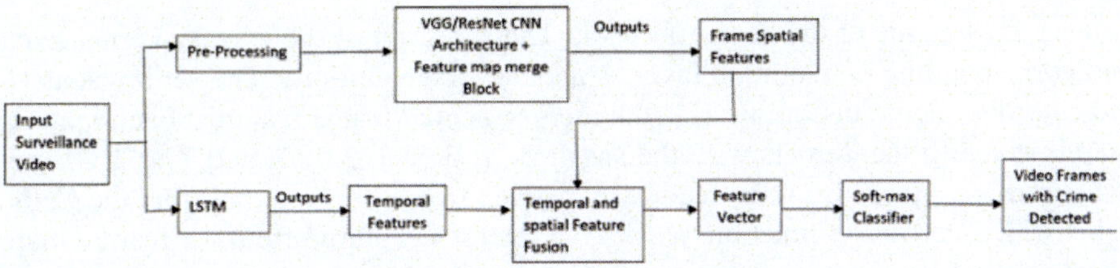

Fig. 1 Proposed LSTM–CNN with feature map merge architecture

frames per second. The video is the divided into smaller clips and sent to CNN. The pre-processed frames are fed as input to the CNN architectures.

3.2 CNN Architectures

The CNN architecture takes the pre-processed frames as an input. The spatial features can be captured either using VGG or ResNet [13] architecture.

3.3 Proposed Feature Map Merge Block

There are various feature map merging techniques available such as multiscale dense block, deep residual learning, etc. In all the existing feature merge techniques, the drawback is that the feature maps are not completely utilized. As a result, some important features in the scene can be missed. Hence, in this paper, a feature map merge block is proposed for merging all the local features of the scene. This feature map merge block is concatenated with the CNN architecture which is been used. The feature maps obtained from the CNN after convolutions are given as input to the feature map merge block. The feature map merge block is shown in Fig. 2. The feature maps from the convolutional layers of VGG/ResNet is given by C_n. The C_n is fed as an input to the feature map merge block. There are four kernels with the size $1 \times 1, 3 \times 3, 5 \times 5$, and 7×7 for detecting the feature maps on different scales.

The ReLU activation function is applied over the merge operation. The feature map merge is given by,

$S_1 = \text{ReLu}(w^1_{7\times7} * C_n)$	$R_1 = \text{ReLu}(w^1_{5\times5} * C_n)$
$Q_1 = \text{ReLu}(w^1_{3\times3} * C_n)$	$P_1 = \text{ReLu}(w^1_{1\times1} * C_n)$
$S_2 = \text{ReLu}(w^2_{7\times7} * [P_1, Q_1, R_1, S_1, C_n])$	$R_2 = \text{ReLu}(w^2_{5\times5} * [P_1, Q_1, R_1, S_1, C_n])$
$Q_2 = \text{ReLu}(w^2_{3\times3} * [P_1, Q_1, R_1, S_1, C_n])$	$P_2 = \text{ReLu}(w^2_{1\times1} * [P_1, Q_1, R_1, S_1, C_n])$
$F = \text{ReLu}(w^3_{1\times1} * [P_1, Q_1, R_{1,}, S_1, P_2, Q_2, R_2, S_2, C_n])$	

w is the weights of the neural network. The subscript of the w is the kernel size, and corresponding convolution layer. * indicates convolutions. The set brackets [] indicates the merge operation. F is the merged spatial frame features. Each feature map is given to the kernels with the size 1×1, 3×3, 5×5, and 7×7. $S1$ are the features captured from the feature map given to the 7×7 kernel from the CNN using ReLU activation function. $R1$ are the features captured from the feature map given to the 5×5 kernel from the CNN using ReLU activation function. $Q1$ are the features captured from the feature map given to the 3×3 kernel from the CNN using ReLU activation function. $P1$ are the features captured from the feature map

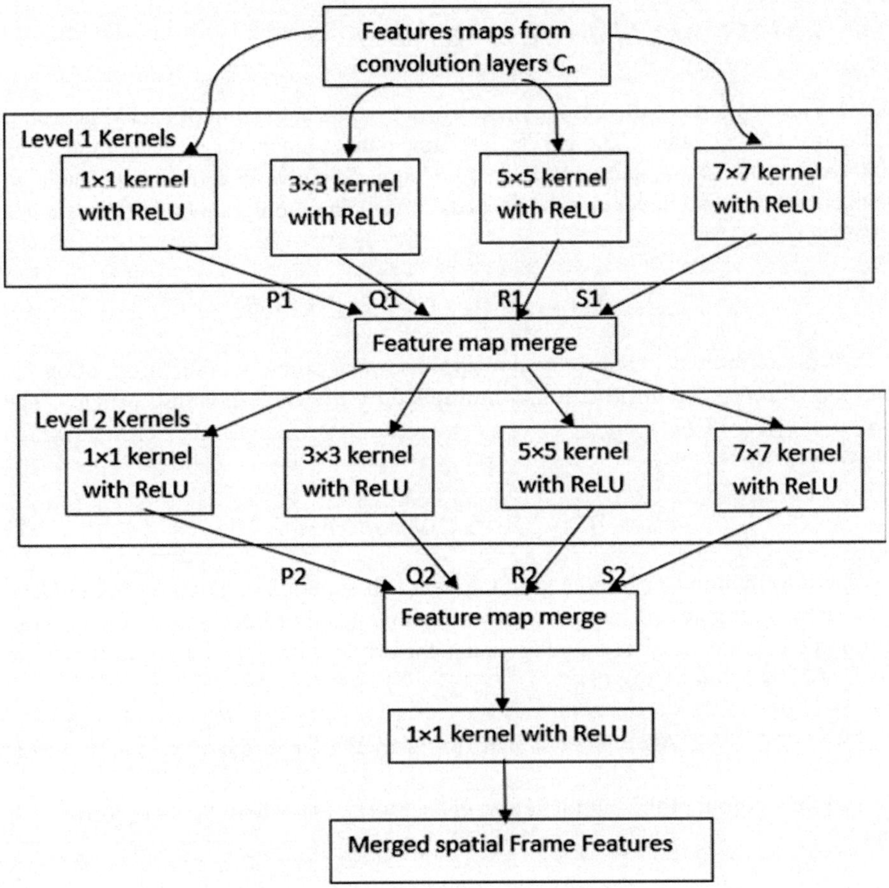

Fig. 2 Feature map merge block

given to the 1 × 1 kernel from the CNN using ReLU activation function. $S2$ are the features captured by merging P_1, Q_1, R_1, S_1, C_n, and ReLU is applied over it. $R2$ are the features captured by merging P_1, Q_1, R_1, S_1, C_n, and ReLU is applied over it. $Q2$ are the features captured by merging P_1, Q_1, R_1, S_1, C_n, and ReLU is applied over it. $P2$ are the features captured by merging P_1, Q_1, R_1, S_1, C_n, and ReLU is applied over it. F is the resultant spatial frame features which merges all the features such as P_1, Q_1, $R_{1,}$, $S_{1,}$ P_2, Q_2, R_2, S_2, C_n. Hence, all the local spatial features of the scene will be captured.

3.4 LSTM (Long Short-Term Memory)

The LSTM neuron is called as the memory block. The activation of LSTM neuron is called as the 'cell state'. The LSTM has three gates such as forgot gate, input gate, and output gate. These gates are made up of sigmoid and tanh activations. The input gate consists of gate layer and the proposal layer. The input gate layer (IL) is given by Eq. (1).

$$\mathrm{IL} = \sigma(x_{t-1}) \times W_{\mathrm{IL}-1} + x_t \times W_{\mathrm{IL}} + B) \tag{1}$$

σ is the sigmoid activation function applied to the previous state information $x_{t\text{-}1}$ and the current state information x_t multiplied by the corresponding weights. The proposal layer (PL) is used for giving a new candidate value for the cell state which is given by

$$\mathrm{PL} = \sigma(x_{t-1} \times W_{\mathrm{PL}-1} + x_t \times W_{\mathrm{PL}} + B) \tag{2}$$

The information to be updated in the next cell state NS is given by **NS = IL + PL**. The output gate will decide which are all the desired information to be passed through the cell state. The output gate has the affine layer. The output of the affine layer (AL) is given by Eq. (3),

$$\mathrm{AL} = \sigma(x_{t-1} \times W_{\mathrm{AL}-1} + x_t \times W_{\mathrm{AL}} + B) \tag{3}$$

The final output of the output gate is given by Eq. (4), where S_t is the current cell state.

$$O_{\mathrm{t}} = \mathrm{AL} \times \tanh(S_{\mathrm{t}}) \tag{4}$$

The CCTV footage is given as an input to the LSTM. The LSTM [14–17] is used to capture the temporal sequence and to track the activity of the object from the scene. LSTM will remember the important activities of the object from the scene for short time.

3.5 Spatial and Temporal Feature Fusion

The spatial and the temporal features [18] are given as input to the feature fusion block. The feature fusion block is shown in Fig. 3. The spatial feature vector extracted from CNN is given by f_{s}. The temporal feature vector extracted from LSTM is given by f_{t}. The probability that the feature is spatial frame features is given by Eq. (5).

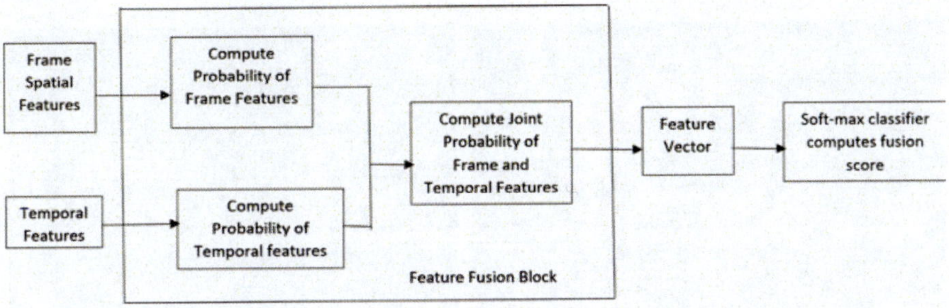

Fig. 3 Feature fusion block

$$p(i|f_s) = \frac{\exp(\theta_s^i . f_s)}{\sum_{i'=1}^{n} \exp(\theta_s^{i'} . f_s)} \quad (5)$$

The probability that the feature is temporal feature is given by Eq. (6).

$$p(i|f_t) = \frac{\exp(\theta_t^i . f_t)}{\sum_{i'=1}^{n} \exp(\theta_t^{i'} . f_t)} \quad (6)$$

The fusion score is computed by taking the joint probability of the spatial and the temporal features which is given by Eq. (7).

$$\text{Fusion score} = p(i|f_s, f_t) = \frac{p(f_s, f_t|i)p(i)}{\sum_{i'=1}^{n} p(f_s, f_t|i')p(i')} \quad (7)$$

In above equations, f_s—spatial feature vector, f_t = temporal feature vector, θ_s—weights and bias of CNN, and θ_t—weights and bias of LSTM. After fusing the spatial and the temporal features, the fused features are given as input to the soft-max classifier. The soft-max classifier will classify the crimes from the scene. The Adam optimizer is used for training the networks. The next section discusses the experimental results obtained and the analysis of it.

4 Experimental Results and Discussion

The experiments were conducted by the 500 CCTV footages collected from Internet. This paper uses CNN + feature map merge block (proposed) and LSTM. The implementation is been done using pycharm and tensorflow [19]. The training and testing is been done using one Nvidia Titan X Pascal GPU. The CNN used for conducting experiments is VGG-16/ResNet for collecting the spatial features. The soft-max classifier is been used to classify the crime scene. Figure 4 shows the detection of weapons and the fight scene in the video footage. Figure 5 shows the loss graphs of

Fig. 4 Detection of weapons and fight in the scene

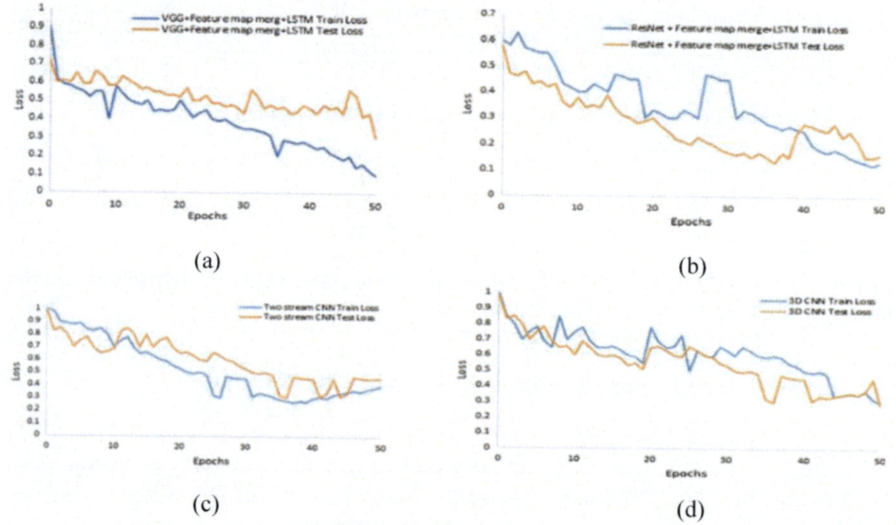

Fig. 5 Loss graphs of (**a**) proposed LSTM–CNN(VGG) with feature map merge (**b**) proposed LSTM–CNN(ResNet) with feature map merge (**c**) existing two-stream CNN, and (**d**) existing 3D CNN

two-stream CNN, 3D CNN, and the LSTM–CNN(VGG/ResNet) with feature map merge obtained during training and testing phase. The loss graph shows that the loss at the 50th epoch is nearly 0.2 for LSTM–CNN(VGG/ResNet) with feature map merge. The loss at the 50th epoch for two-stream CNN is 0.5. The loss at the 50th epoch for 3D CNN is 0.4. Hence, the loss of LSTM–CNN(VGG/ResNet) with feature map merge is less when compared to two-stream CNN and 3D CNN. The quantitative results such as precision, recall, F1-score, and accuracy obtained by LSTM–CNN are compared with the existing two-stream CNN and 3D CNN as given in Table 1. The accuracy obtained by the proposed LSTM–CNN with feature map merge is 90% which is more compared to the existing architectures. The precision, recall, and F1-score obtained by the LSTM–CNN with the feature map merge are 0.956, 0.898, and 0.915, respectively.

Table 1 Experimental analysis of CNN architectures

Crime detection architectures	Precision	Recall	F1 score	Accuracy (%)
Existing two_stream CNN	0.894	0.833	0.86	85
Existing 3D CNN	0.921	0.852	0.88	87
Proposed LSTM–CNN (VGG) + Feature map merge	0.936	0.864	0.89	88
Proposed LSTM–CNN (ResNet) + Feature map merge	0.956	0.898	0.91	90

5 Conclusion

The existing two-stream CNN and 3D CNN architectures for crime scene detection suffer from drawback such as it could not capture the complete local features of the scene and the complexity is more as the high-dimensional features are encoded. In order to avoid these drawbacks, this paper proposed a LSTM–CNN with the feature map merge architecture. The spatial features of the CCTV footage are captured using CNN with the feature map merge block. The temporal features of the CCTV footage are captured using LSTM. Then, the spatial and temporal features using feature fusion block and the crime scenes are classified using soft-max layer. The quantitative results obtained by the LSTM–CNN with the feature map merge is good when compared to the existing two-stream CNN and 3D CNN. The accuracy obtained LSTM–CNN with the feature map merge is 90%. Hence, this proposed LSTM–CNN with feature map can be used for crime detection in the CCTV footages. With this detection, the security people can be altered. In future for detecting the human activity, PosNets can be used.

References

1. W. Ye, J. Cheng, F. Yang, Y. Xu, Two-stream convolutional network for improving activity recognition using convolutional long short-term memory networks. IEEE Access **7**, 67772–67780 (2019)
2. M. Perez, A.C. Kot, A. Rocha, Detection of real-world fights in surveillance videos, ICASSP 2019—*2019 IEEE International Conference on Acoustics, Speech and Signal Processing (ICASSP)*, Brighton, United Kingdom, 2019, pp. 2662–2666
3. G. Zhu, L. Zhang, P. Shen, J. Song, Multimodal gesture recognition using 3-D convolution and convolutional LSTM. IEEE Access **5**, 4517–4524 (2017)
4. U.V. Navalgund, P.K., Crime intention detection system using deep learning, *2018 International Conference on Circuits and Systems in Digital Enterprise Technology (ICCSDET)*, Kottayam, India, 2018, pp. 1–6
5. T.H.N. Le, Y. Zheng, C. Zhu, K. Luu, M. Savvides, Multiple scale faster-RCNN approach to driver's cell-phone usage and hands on steering wheel detection, *2016 IEEE Conference on Computer Vision and Pattern Recognition Workshops (CVPRW)*, Las Vegas, NV, 2016, pp. 46–53

6. Y. Hu, M. Lu, X. Lu, Spatial-temporal fusion convolutional neural network for simulated driving behavior recognition, *2018 15th International Conference on Control, Automation, Robotics and Vision (ICARCV)*, Singapore, 2018, pp. 1271–1277
7. H. Wei, Y. Xiao, R. Li and X. Liu, Crowd abnormal detection using two-stream fully convolutional neural networks, *2018 10th International Conference on Measuring Technology and Mechatronics Automation (ICMTMA)*, Changsha, 2018, pp. 332–336
8. X. Li, M. Xie, Y. Zhang, G. Ding, W. Tong, Dual attention convolutional network for action recognition. IET Image Proc. **14**(6), 1059–1065 (2020)
9. K. Simonyan, A. Zisserman, Two-stream convolutional networks for action recognition in videos, in *Proceedings of the 27th International Conference on Neural Information Processing Systems—Volume 1*, NIPS '14, 2014, pp. 568–576
10. S. Ren, K. He, R. Girshick, J. Sun, Faster r-cnn: towards real-time object detection with region proposal networks, in *Proceedings of the 28th International Conference on Neural Information Processing Systems—Volume 1*, NIPS '15, 2015, pp. 91–99
11. A. Bewley, Z. Ge, L. Ott, F. Ramos, B. Upcroft, Simple online and realtime tracking, in *Image processing (ICIP), 2016 IEEE International Conference on IEEE*, 2016, pp. 3464–3468
12. D. Moreira, S. Avila, M. Perez, D. Moraes, V. Testoni, E. Valle, A. Rocha, Temporal robust features for violence detection, in *IEEE Winter Conference on Applications of Computer Vision (WACV)*, 2017, 1–9
13. BRAC university on Crime scene prediction by detecting threatening objects using convolutional neural network, submitted on, 13th April 2017
14. A.A. Aljarrah, A.H. Ali, Human activity recognition using PCA and BiLSTM recurrent neural networks, *2019 2nd International Conference on Engineering Technology and Its Applications (IICETA)*, Al-Najef, Iraq, 2019, pp. 156–160
15. F.C. Heilbron, J.C. Niebles, B. Ghanem, Fast temporal activity proposals for efficient detection of human actions in untrimmed videos, in *IEEE Conference on Computer Vision and Pattern Recognition (CVPR)*, 2016, 1914–1923
16. K. Zhou, Y. Zhu, Y. Zhao, A spatio-temporal deep architecture for surveillance event detection based on ConvLSTM, in *Proceedings VCIP*, Saint Petersburg, FL, USA, 2017, pp. 1–4
17. L. Zhang, G. Zhu, L. Mei, P. Shen, S.A.A. Shah, M. Bennamoun, Attention in convolutional LSTM for gesture recognition, in *Proceedings NIPS*, 2018, pp. 1953–1962
18. Y. Jia, E. Shelhamer, J. Donahue, S. Karayev, J. Long, R. Girshick, S. Guadarrama, T. Darrell, Caffe: convolutional architecture for fast feature embedding, *MM 2014—Proceedings of the 2014 ACM Conference on Multimedia*, 2014
19. R. Al-Rfou, Theano: a python framework for fast computation of mathematical expressions. arXiv preprint arXiv: 1605.02688, 2016

Malware Family Classification Model Using Convolutional Neural Network

T. Gayathri and M. S. Vijaya

Abstract Malware is dangerous for system and network users. Malware identification is an essential task in effectively detecting and preventing the computer system from self-infection, protecting it from potential data loss and system compromise. Commonly, there are 25 malware families exist. Traditional malware detection and anti-virus systems fail to classify the new variants of unknown malware into their corresponding families with the development of malicious code engineering, and it is possible to understand the malware variants and their features for new malware samples that carry variability and polymorphism. The detection methods can rarely detect such variants, but it is important in the cybersecurity field to investigate and detect large-scale malware samples more efficiently. In this paper, an accurate malware family classification model using a convolutional neural network technique is proposed. Malware family recognition is formulated as a multi-classification task, and an accurate solution is obtained by training convolutional neural network with images of malware executable files. Ten families of malware have been considered here for building the models. The image dataset with 2000 instances is applied to a convolutional neural network to build the classifier. The experimental results, based on a dataset of ten classes of malware families and 2000 malware images trained model, provide an accuracy of over 95% in discriminating from malware families. The techniques provide better results for classifying malware into families.

Keywords Convolutional neural network · Deep neural network · Malware · Malware family classification

---

T. Gayathri (✉) · M. S. Vijaya
PSGR Krishnammal College for Women, Coimbatore 641004, India

M. S. Vijaya
e-mail: msvijaya@psgrkc.ac.in

V. Bhateja et al. (eds.), *Data Engineering and Intelligent Computing*,
Advances in Intelligent Systems and Computing 1407,
https://doi.org/10.1007/978-981-16-0171-2_3

1 Introduction

Malware or malicious code is any program or file that is harmful to a system and network user. Malware is any piece of software code that was created with the intent of damaging devices, sealing sensitive data, and commonly causing a mess. These are proposed to access computer systems and network resources, interrupt computer operations, and bring together private information without taking the permission of the system's owner. This malicious software can perform heterogeneity of functions such as encrypting and destroy data, hijacking core computing functions and accessing user system activity without their permission.

Web usage mining is the detection of meaningful patterns from data generated by client–server transactions on one or more web localities. Several web transactions automatically make the data that gets gathered in server access logs, refer logs, agent logs, client side's cookies, user profile, metadata, page attribute, page content and site structure. Web services enable the detection of malware with an enormous partner's data. But even a massive malware database does not assure detection of recent ones. Most anti-malware software products, such as Kaspersky, Symantec, and McAfee, basically use the signature-based method to recognize threats.

The damage affected by a virus that corrupts a computer or a corporate network can be different from an irrelevant increase in outgoing traffic to the complete network breakdown or the loss of hypercritical data. A virus on a commercial network can be considered a force mature, and the damage affected by it is being equal to the loss associated with the network downtime essential for disinfection.

Malware comes in a wide range of variations like virus, worm, trojan horse, ransomware, backdoor, zeus, key loggers, adware, and ten malware families, such as Allaple, CryptoLocker, Agent, Trojan generic, WannaCry, AutoRun, Dialplatform, Alueron.gen, Simda, and zbot, are taken into account for implementation.

2 Literature Survey

Several research works have been carried out currently using deep learning. Various features such as static and dynamic have been used to build models. In few cases, signature-based methods and image-processing methods have been used. Based on the study of various kinds of literatures available on malware family classification, a brief report is presented in this section about the developments in the respective area in the last several years.

Zarras et al. [1] developed malware collected from the dataset as the input of Sandbox. The neural network comprises two parts, a convolutional part and recurrent part. In recurrent layer, LSTM cells were utilized to model the consequential sequence and sort the significance based on mean pooling. Finally, dropout and a softmax layer were used to prevent overfitting to occur in the output. As per the experimental result, it shows that the combination of convolutional network and LSTM gave better accuracy (89.4%) compared to feed-forward network (79.5%) and convolutional network (90.2%).

Karthikeyan et al. [2] were the first to explore the use of byte plot visualization for automatic malware classification. They converted all the malware samples to grayscale byte plot representations and extracted texture-based features from the malware image. The dataset consisted of 9,456 malware samples belonging to 25 different classes, collected from the Anubis system. They used the global image-based features to train a K-Nearest Neighbour model, with Euclidean distance as distance measure, to classify malware samples into their respective classes, and an accuracy of 97.18% was obtained.

David et al. [3] propose a method for a signature generation that does not depend on a particular part of malware, so it can be resistant to code modifier. They used a total of 1800 samples, which are 300 samples for each category. During the experiment, they used 1,200 samples for training and 600 samples for testing. The number of features generated is 30, and the accuracy is 98.6%. The input noise was also set as 0.2 and learning rate as 0.001.

In the existing work, the classifications were performed by training malware executable files or images or PE (portable executable) files. The features based on texture were extracted from executable files that assisted in recognizing the malwares or in classification of malware. In a few cases, malware image datasets were used, and the corresponding binary values were trained to build classifiers. In few cases, an array of user-defined features corresponding to PE files was used in building the model, but the results were not promising. This motivated to carry out research work for building the malware family classification model through powerful deep learning architecture, namely convolutional neural network which is designed specifically for image classification.

3 Malware Family Classification Model

The problem of malware family identification is formulated as multiclass image classification task and solved using contemporary convolutional neural network technique. The methodology of proposed malware family classification includes four different stages. In the first stage, malware family corpus development is performed wherein the virus executable files corresponding to ten families of malware are collected. In the second stage, the executable files are converted into images. Finally, the malware family classification model is built by implementing convolutional neural network with malware image as input. The methodology of the malware family classification model is shown in Fig. 1.

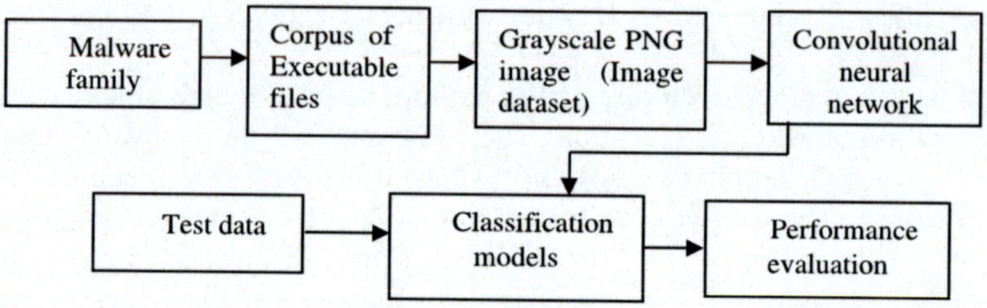

Fig. 1 Framework of malware family classification model

3.1 *Corpus Development and Dataset Creation*

Initially, a corpus consisting of 2000 malware executable files related to ten types of malware family is developed. Here Allaple, CryptoLocker, Agent, Trojan generic, WannaCry, AutoRun, Dialplatform, Alueron.gen, Simda, and zbot are ten different types of malware family considered. These viruses corresponding to ten malware families are collected from the das malwark, virustotal, and a malwr.com malware samples. These portals have to collect malware using honeypots, and also users around the world submit files over them for analysis and sharing malware samples. A set of 200 files of Allaple malware, 200 files of CryptoLocker, 200 files of Agent family, 200 files of Trojan generic, 200 files of WannaCry, 200 files of AutoRun, 200 files of Dialplatform, 200 files of Alueron.gen, 200 files of Simda, and 200 files of zbot malware family has been collected, and a malware family corpus of size 2000 is developed. Malware executable file in hexadecimal view is shown in Fig. 2.

Malware coders modify small parts of the original source code to create a new variant. Images can capture small variation yet retain the global structure. Images

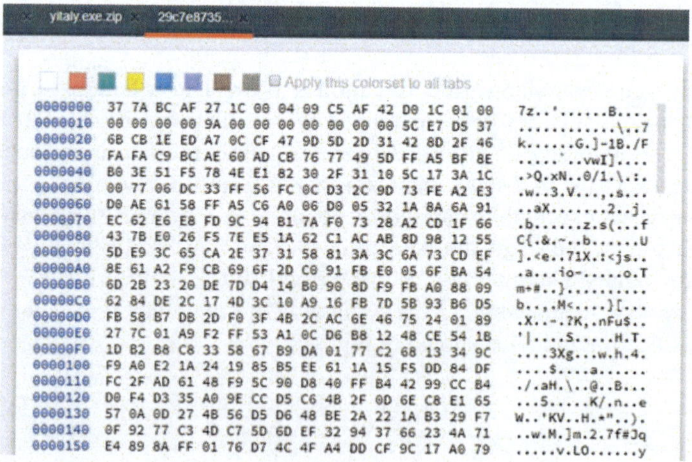

Fig. 2 Malware executable files in hexadecimal view

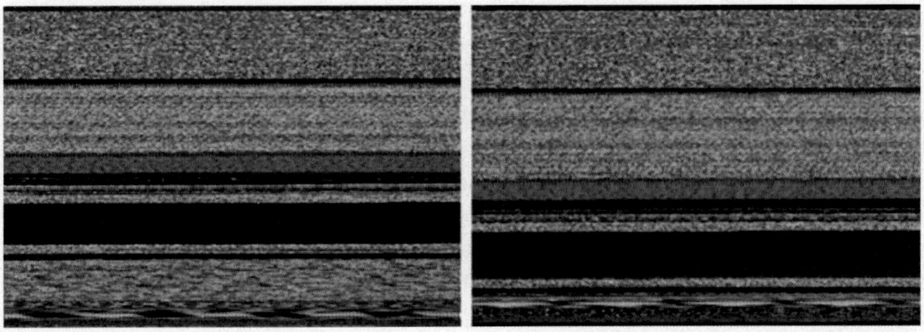

Fig. 3 Image of zbot exe file

can give more information about the structure of the malware. The malware files in the corpus are converted into a grayscale Portable Network Graphics image of size 64 × 64 using Python code. Conversion of malware executable file to image sample is shown in Fig. 3. The image dataset with 2000 instances is created.

3.2 Model Building

Malware family classification model is built with the above dataset as input, and a deep learning architecture, namely convolutional neural network, is employed to build the malware family classification model. CNN is consisting of neurons with learnable weights and biases. CNNs mainly contain three components, namely convolutional layer, pooling layer, and fully connected layer. These convolutional layers apply a certain number of convolution operations to the image in sequence. Typically, these filters extract edge, color, and shape information from the input image. The output of this layer is typically forwarded to a nonlinear function.

This pooling layer is responsible for reducing spatial resolution of the input layer. The data produced from convolution layers so that processing time can be reduced, and hence computational resources can handle the scale of the data. This fully connected layer performs classification on the output generated from convolution layers and pooling layers. Every neuron in this layer is connected to every neuron present in the previous layer. This type of layer is typically followed by a dropout layer that improves the generalization capability of the model by preventing overfitting. Various hyperparameters, such as learning rate, epochs, batch size, activation function, and optimizers, are defined to improve the efficiency of learning and to build the accurate malware family classification model.

4 Experiments and Results

Ten types of malware families are taken into account for developing the CNN-based image classifier, and hence, malware recognition problem turns out to be multiclassification. In these experiments, the sequential model is developed to build CNN classifier. All the images were of different sizes and had to be converted into 64 by 64 pixels before they are used as input to the model. The input layers have been given 2000 images in sequential model. The output layer is of neurons corresponding to the ten families of malwares available in the dataset used. The activation functions, namely ReLU with adam optimizers, are used in this work. A softmax function is used in output layer, and ReLU function is used for hidden layer. Adam optimizer used with value of epochs is 80, and batch size is 10 to increase the prediction accuracy. The functioning of CNN-based malware family classification method is determined based on classification metrics such as precision, recall, F-measure, and accuracy. It has been observed that the prediction accuracy of 95% is achieved by CNN classifier. Class-wise performance of CNN model is shown in Tables 1 and 2.

Table 1 Performance results of convolutional neural network classifier by class

Class labels	Precision	Recall	F-measure
0	0.98	0.96	0.97
1	0.99	0.92	0.98
2	0.99	0.90	0.99
3	0.78	0.79	0.78
4	1.00	0.95	0.97
5	0.65	0.76	0.70
6	0.97	0.99	0.98
7	0.99	0.99	0.99
8	0.72	0.79	0.75
9	0.94	0.92	0.95
Accuracy	0.96	0.94	0.95

Table 2 Results of convolutional neural network model

Classification	Accuracy	Precision	Recall	F-measure
Convolutional neural network (%)	95	96	94	95

4.1 Comparison with Deep Neural Network

The performance of the CNN-based malware detection model is compared with DNN-based malware detection model developed in our earlier work which was trained with image features. These image features were nothing but a vector of 8-bit unsigned integers between 0 and 255 derived from the binary sequence of ones and zeros. Each image was resized to 28 × 28 and represented in binary form. Width of the image was set according to the file size. Height of the image varies depending on the file size. This vector was then reshaped into a two-dimensional array of unsigned integers. The two-dimensional feature matrix of size 28 × 28 amounting to 784 feature values was derived to form the feature vector. The dimensionality reduction technique, namely principal component analysis (PCA), was used to reduce the size of the feature vectors. This dimensionality reduction method efficiently represents interesting parts of an array as a compact feature vector of size 1840. A malware family dataset with 2000 instances of dimension 1840 was developed and used as input to train the deep neural network. Class-wise performance of DNN model is shown in Tables 3 and 4.

The results of CNN classifier are compared with the implementation results of DNN-based malware family classifier. It is discovered that convolutional neural network has achieved highest prediction accuracy of 95% than the DNN model. CNN classifier obtained a precision of 96, recall value of 94 with F-measure of

Table 3 Performance results of deep neural network classifier by class

Class labels	Precision	Recall	F-measure
0	1.00	1.00	1.00
1	0.97	0.95	0.96
2	1.00	0.96	0.98
3	0.78	0.64	0.68
4	0.48	0.58	0.53
5	1.00	1.00	1.00
6	0.58	0.68	0.60
7	0.78	0.77	0.77
8	0.80	0.81	0.80
9	0.75	0.83	0.78
Accuracy	0.80	0.86	0.81

Table 4 Results of deep neural network model

Classification	Accuracy	Precision	Recall	F-measure
Deep neural network	87%	0.80	0.86	0.81

Table 5 Comparative results of CNN and DNN classification algorithms

Performance evaluation	Classifiers	
	Deep neural network	Convolutional neural network
Precision	80	96
Recall	86	94
F-measure	81	95
Accuracy	87	95

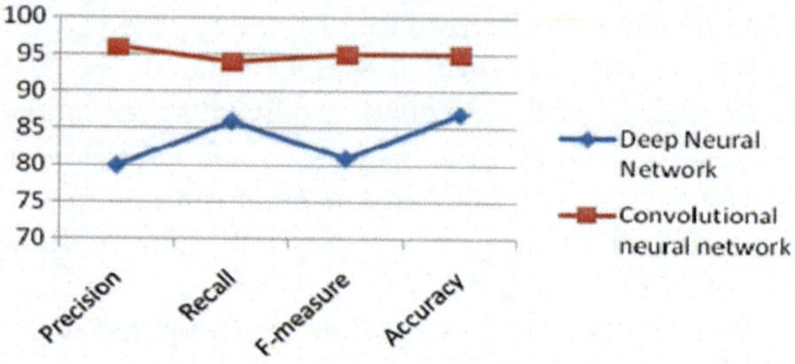

Fig. 4 Comparison of classifiers based on various evaluation metrics

95, and accuracy 95%. The comparative predictive performances of malware family classifiers are shown in Table 5 and Fig. 4.

In case of the deep neural network, the model takes user-defined features as input, and they perform the classification task individually, whereas CNN performs feature learning and training through convolution and pooling layers to recognize the type of executable file and performs classification task simultaneously. The convolutional neural network model has capability to extract high-level features from the images. The weights of the neurons are self-adjusted through the layers so the error rate is minimized and the model can perform accurate prediction. The result of convolutional neural-network-based classification model is attained by maximizing hidden layers and number of epochs in sequential model. The performance of the convolutional neural network classifier is validated using its measures with high accuracy and least error rate for malware family classification.

5 Conclusion

This paper demonstrated the implementation of convolutional neural-network-based malware family classification model built with the aim to predict the malware family through CNN classifier using malware family image dataset. CNN method automatically learns high-level features from the images and learns the pattern of malware family. The executable malware files are collected from various malware resources, and malware family corpus has been developed. The performance of the deep neural network and convolutional neural-network-based classifiers is evaluated in terms of accuracy, recall, precision, F-measure, and the results are compared. The CNN-based malware family classification model has shown high classification accuracy compared with DNN classifier. Finally, it is concluded that convolutional neural-network-based classification method achieves accurate prediction in classifying the malware among its families.

References

1. B. Kolosnjaji, A. Zarras, G. Webster, C. Eckert, Deep learning for classification of malware system call sequences, in *Australasian Joint Conference on Artificial Intelligence*. Springer, 2016, 137–149
2. L. Nataraj, S. Karthikeyan, G. Jacob, B. Manjunath, Malware images: visualization and automatic classification, *Proceedings of the 8th International Symposium on Visualization for Cyber Security*, Article No. 4, 2011
3. N.S. Netanyahu, O.E. David Deepsign, Deep learning for automatic malware signature generation and classification in Neural Networks (IJCNN), *2015 International Joint Conference on. IEEE*, 2015, pp. 1–8
4. R. Tian, L. Batten, R. Islam, S. Versteeg, An automated classification system based on the strings of trojan and virus families, *Proceedings of the 4th International Conference on Malicious and Unwanted Software*, (Montréal, 13 October 2009)
5. K. Kim et al., Network intrusion detection using deep learning, Springer Cyber Security Systems and Networks Springer Nature Singapore Pte Ltd
6. B. Cakir, E.. Dogdu, Malware classification using deep learning methods on ACM SE '18: *ACM SE '18: Southeast Conference*, Richmond, KY, USA. ACM, New York, NY, USA, p. 5
7. M. Bailey, J. Oberheide, J. Andersen, Z.M. Mao, F. Jahanian, J. Nazario, Automated classification and analysis of internet malware, *in Proceedings of the 10th Symposium on Recent Advances in Intrusion Detection*
8. R. Islam, R. Tian, L. Battenb, S. Versteeg, Classification of malware based on integrated static and dynamic features. J. Netw. Comput. Appl. **36**, 646–556 (2013)
9. N. Bhodia, P. Prajapati, F.D. Troia, M. Stamp, Transfer learning for image-based malware classification in ICISSP 2019, Czech Republic, 2019
10. M. Kalash, M. Rochan, N. Mohammed, N. D.B. Bruce, Y. Wang, Malware classification with deep convolutional neural networks, 2018 9th IFIP NTMS, 2018

Malware and Benign Detection Using Convolutional Neural Network

V. Krithika and M. S. Vijaya

Abstract Malware is a virus program written to enter a system and damage or alter files and data. A computer virus is more vulnerable as it makes changes or deletes files and can enter into a system as an attachment of images or audio and video files. It also infects the system through the downloads on the Internet. This paper proposes an intelligent learning model to recognize malware using convolutional neural network approach. Here, malware detection identification models through binary classifier, and two approaches of deep neural network classification and convolutional neural network are used. A database consisting of 600 exe files of which 300 malware files and 300 benign files have been considered here for building the malware and benign detection model. All the exe files are converted into images files. The contemporary CNN-based binary classification which takes the grayscale images as input. The convolutional neural networks-based classification model proves accuracy of 93% in discriminate from malware and benign files. The convolutional neural network-based malware detection model has higher performance when compared with deep neural network classification model trained with GIST features of image.

Keywords Malware detection · Convolutional neural network · Binary classification · Deep neural network

1 Introduction

Malware is a various type of program that is installed without user permission on computer to gather user data, their system, or browsing habits tracks everything without user knowledge and send it to other user. It also can download different malware programs from the Web site and install it on the system. Malware works like adware but is frequently a separate program that is installed unknowingly when

V. Krithika (✉) · M. S. Vijaya
PSGR Krishnammal College for Women, Coimbatore 641004, India

M. S. Vijaya
e-mail: msvijaya@psgrkc.com

V. Bhateja et al. (eds.), *Data Engineering and Intelligent Computing*,
Advances in Intelligent Systems and Computing 1407,
https://doi.org/10.1007/978-981-16-0171-2_4

installing a different freeware-type program. Malware detection model helps to recognize malware when an executable file with a virus code is downloaded. Malware detection is type of cybercrimes in which the attacker uses a spyware to install key logging software, thereby allowing an illegal access to the computer. Malware bypass is the usual authentication used to access a computer. The use of the spyware is to allow the cybercrimes future access to the system even if the organizations fix the original vulnerability used to attack the system.

Malware detection contains mechanisms to spot and defend against damage from viruses, worms, Trojan horses, spyware, and alternative sorts of malicious code. Cybersecurity consists of security mechanisms that try to provide solution to cyberattacks. Many cyberthreats are suitable to malware intrusions. Much about network intrusion is heared. What happens here is that intruder attempts to tap into the networks and acquires the information that is being transmitted. This intruder may be human intruder or automatic malicious software set up by humans. The cyberthreat analysis and intrusion detection are quite similar. Web mining techniques are also used for protection of user system or logging information against such cybercrimes as hacking and virus spreading. Clustering and classification techniques of Web mining can reveal identifies of cybercrimes [1].

2 Literature Survey

The background study is carried out to grasp the nature of the work. The proposed work is identified based on literature survey on malware detection that is presented below.

Lee et al. [2] proposed a model the malware programs by using a machine learning method. A behavioral profile was created based on information recorded regarding sample interaction with method resources like registry keys, write files, and network activities. The parallel between two profiles was calculated and then by applying k-medoids, different samples are grouped into different clusters. They use hybrid techniques which include both static and dynamic features simultaneously for improved malware detection and classification.

Anderson et al. [3] proposed a model, in which different data sources—the static binary, the disassembled binary file, a dynamic training trace or system call trace, and a file in sequence feature vector—were used. Then, several kernel learnings were implemented to find a weighted combination of the data source and support vector machine classifier was used to classify the dataset into malware and benign.

In the existing work, the classification was performed by using learning models. The features based on static were extracted from samples that assisted in recognizing the malwares or in classification of malware. In few cases, malware detection datasets were used and the corresponding static binary values were trained to build the model. But the results were not promising. Hence, its proposed work is carried out by training the features of malware and benign samples of exe files and binary classification models through deep neural network and convolutional neural network which are designed specifically for image classification are built.

3 Methodology of Malware Detection Model

The methodology of proposed malware and benign identification model comprises three different stages. The first stage is the creation of corpus consisting of executable files related to malware and benign files collected from various malware resources and windows applications. In the second stage, all the exe files are converted into images which can be directly fed to CNN. In the third stage, the malware and benign detection model is built by implementing convolutional neural network. The methodology of the malware detection model is shown in Fig. 1.

3.1 *Corpus Development*

Initially, a corpus consisting of malware exe files and benign exe files related to windows files of malware detection is developed. These virus corresponding malware are collected 300 files from the TekDefense and Das Malwerk malware samples. Benign are collected 300 files from the windows application benign samples.

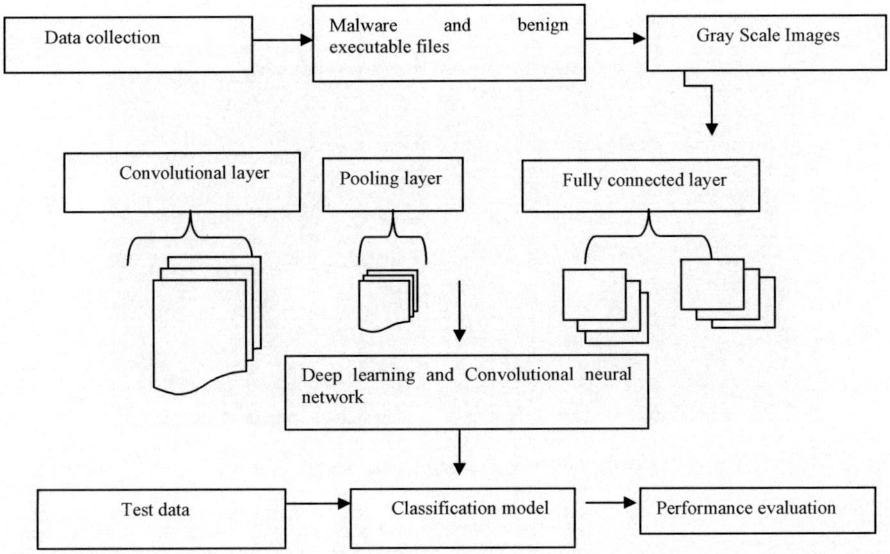

Fig. 1 Methodology of malware and benign classification model

Malware and benign have been collected and a malware detection corpus of size 600 is developed. A sample malware exe file in hexa view is shown in Fig. 2.

Each malware and benign files in the corpus is converted into a grayscale image of size 64 × 64 using Python code. The sample images of virus exe file and benign exe file are shown in Fig. 3 (A) and (B), respectively.

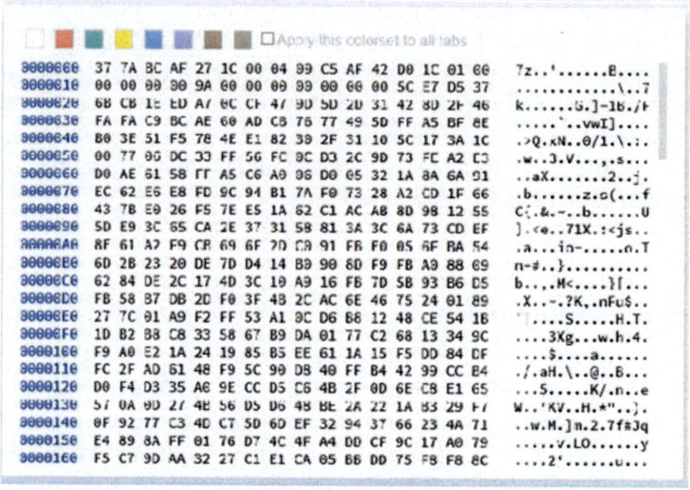

Fig. 2 Sample malware file in hexa view

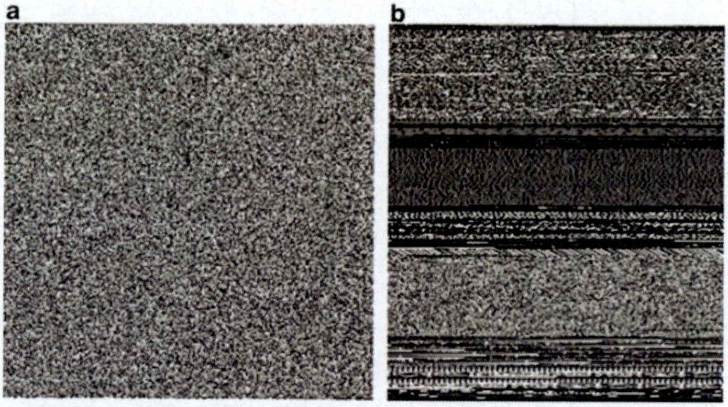

Fig. 3 (**A**) Image of virus exe file (**B**) Image of benign exe file

3.2 Model Building

The convolutional neural network is used as it is reliable and it can be applied to mainly image classification use for feature extraction. The CNN is a neural network where the connectivity pattern between neurons is inspired by the structure of visual cortex and that has proved great value in the analysis of visual imagery. The number of parameter of the CNN uses the weight sharing and the convolutional kernel mechanism. To build a model, CNN is formed in three layers, called the convolutional layer, pooling layer, and fully connected layer.

The convolutional layers refer to extracting features from the input image and multiply the values in the filter with the original pixel values. Convolutional layer will determine the output of neurons which are connecting to local region of the input through the calculation of the scalar product between their weights and the region connects to the input volume. The pooling layer operation reduces dimensionality of the given input, further reducing the number of parameters within that activation. The fully connected layer is a classic layer with perceptrons and an activation softmax in the output layer. It be also suggested that ReLU may be used between these layers, as to improve performance [4].

The output layers have been given malware belongs to. Cross-entropy loss function that is frequently used for binary classifier is used in this work. Initially all the images are converted into 64 X 64 grayscale images before they are used as input to the model. Here, the malware and benign image are easily identified by pattern learning through convolutional, pooling, and subsampling layer, and the malware detection model is developed by training the self-learned features in convolutional neural network for binary classification.

4 Experiments and Results

The malware detection uses CNN algorithm that is used to training and testing the model. In this process, three hidden layers are used, and each layer has its own parameter. Here, the sequential model is developed to build CNN classifier. All the images are converted into 64 by 64 grayscale images before they are used as input to the model. The input layers have been given 600 images in sequential model. The output layers have been specified as 2 class labels where the class label 1 is assigned for malware file and the class label 0 is assigned for benign file. Adam optimizer is used with value of epochs 75 and batch size 10 is used. The activation function softmax with dropout 0.1 and learning rate within the range 0.1, 0.3 used. The functioning of CNN-based malware classification methods attains maximum precision, recall, F-measure, and accuracy. It has been observed that the prediction accuracy of 93% is achieved by CNN classifier. The classwise results of CNN model is shown in Table 1. Its measure values of CNN model with precision 90%, recall 89%, and F-measure 89% are obtained. The precision has high performance than other

Table 1 Performance results of convolutional neural network classifier by class

Class labels	Precision	Recall	F-measure
0	83	80	81
1	85	85	85
Result	90	89	89

Table 2 Results of convolutional neural network-based model

Classification	Accuracy	Precision	Recall	F-measure
Convolutional neural network	93	90	89	89

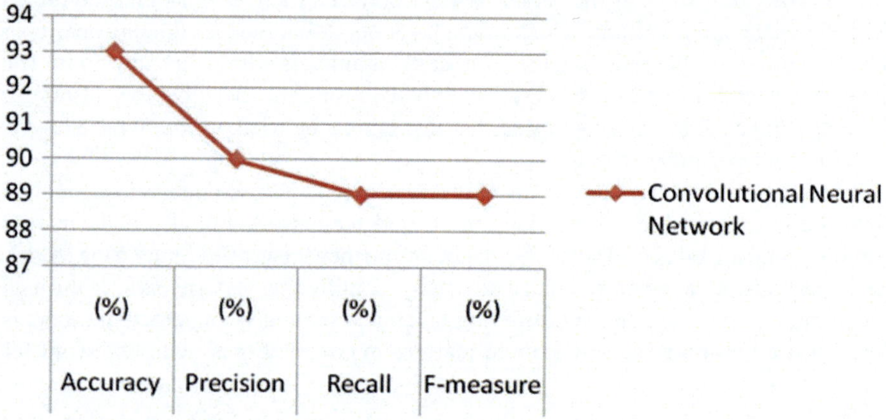

Fig. 4 Results of CNN-based malware detection model

values such as recall and F-measure. The CNN classification method is evaluated by malware detection, and results for each malware and benign concerning different metrics are shown in Table 2 and illustrated in Fig. 4.

4.1 Comparison with Deep Neural Network

In these experiments, CNN classifier is compared with the DNN-based malware detection model trained with GIST features. The input layers have been given attributes, and the hidden layers have been given instances in the sequential model. The sequential model is developed to build DNN classifier. The output layers have been specified as 2 class labels where the class label 1 is assigned for malware file and the class label 0 is assigned for benign file. The activation functions namely ReLU and Adam optimizers are used in this work. Adam optimizer used with the

Table 3 Performance results of deep neural network classifier by class

Class labels	Precision	Recall	F-measure
0	0.64	0.64	1.00
1	0.70	0.72	0.94
Result	0.64	0.66	0.91

Table 4 Results of deep neural network-based model

Classification	Accuracy	Precision	Recall	F-measure
Deep neural network	81	64	66	91

value of epochs is 85, and batch size which is 10 is used. The functioning of DNN-based malware classification methods attains maximum precision, recall, F-measure, and accuracy. It has been observed that the prediction accuracy of 81% is achieved by DNN classifier. The classwise results of DNN model are shown in Table 3. Its measure values of CNN model with precision 64%, recall 66%, and F-measure 91% are obtained. The F-measure has high performance than other values such as precision and recall. The DNN classification method is evaluated by malware detection, and results for each malware and benign concerning different metrics are shown in Table 4.

The performances of both binary classifiers built using convolutional neural networks and deep neural network are compared against their performance metrics. From the comparative study, it is detected that the convolutional neural networks-based malware detection model produces effective results than deep neural network-based binary classifier for distinguishing malware and benign files. By examining the results, it is observed that the convolutional neural networks-based classification model proves high accuracy of 93% than deep neural network model. The results of comparative analysis of CNN and DNN classifiers are shown in Table 5 and illustrated in Fig. 5.

From above comparison of CNN and DNN classifiers, CNN model is achieved high results for distinguishing malware and benign files accurately. The trained CNN classification methods attain maximum precision, recall, and F-measure when evaluated with DNN methods. The DNN algorithm produced the accuracy of 81% although CNN classification algorithm gives high prediction accuracy of 93%.

Table 5 Comparative analysis of CNN and DNN

Classification	Precision	Recall	F-measure	Accuracy
Convolutional neural network	90	89	89	93
Deep neural network	64	66	91	81

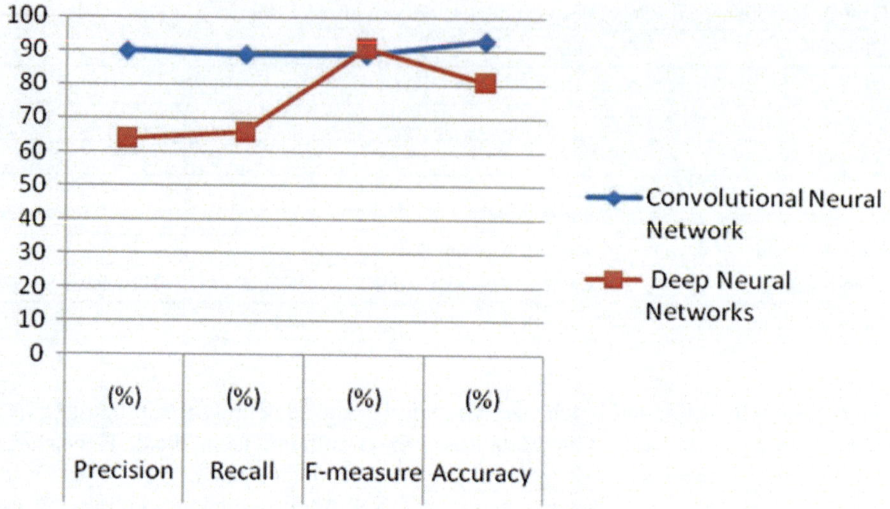

Fig. 5 Comparative analysis of binary classifiers CNN and DNN

5 Conclusion

This paper demonstrates the model of malware and benign detection task as binary classifier and implementation of CNN. The corpus with 600 instances of malware and benign files is created and used for building binary classifier and two approaches of DNN and CNN are used. The performance of both deep neural network and convolutional neural network-based malware identification models is evaluated and analyzed with respect to various metrics like precision, recall, F-measure, and accuracy. Though DNN algorithms are effective in malware detection, CNN-based malware detection model has higher performance when compared with DNN classification model. Finally, CNN model achieved high results for distinguishing malware and benign files accurately. In future, the efficiency of the models can be improved using alternate deep architectures like recurrent neural network and deep Boltzmann machine with different feature set.

References

1. B. Thuraisingham, L. Khan, M.M. Masud, K.W. Hamlen, Data Mining For Security Applications. *IEEE/IFIP International Conference on Embedded And Ubiquitous Computing*, 2010
2. T. Lee, J.J. Mody, Behavioral classification. *Proceedings of the European Institute for Computer Antivirus Research Conference* (EICAR'06), 2006
3. B. Anderson, C. Storlie, T. Lane, Improving malware classification bridging the static/dynamic gap. *Proceedings of 5th ACM Workshop on Security and Artificial Intelligence (AISec)*, 2012
4. K. O'Shea, R. Nash, An introduction to convolutional neural networks, Department of Computer Science, Aberystwyth University, Ceredigion, SY23 3DB, 2015

5. M. Siddiqui, M.c. Wang, J. Lee, A survey of data mining techniques for malware detection using file features. In *Proceedings of the 46th Annual Southeast Regional Conference* on Xx. 2008. Acm
6. M.E. Boujnouni, M. Jedra, N. Zahid, New malware detection framework based on N-grams and support vector domain description, *11th International Conference On Information Assurance And Security (Ias)*, pp 123–128
7. C.I. Fan, H.W. Hsiao, C.H. Chou, Y.F. Tseng, Malware detection systems based on api log data mining, *IEEE 39th Annual Computer Software And Applications Conference*, 2015, pp. 255–260

An Evaluation and Validation of Contemporary Approaches in Scene Text Extraction and Recognition

M. J. Roopa and K. Mahantesh

Abstract In recent decades, Scene Text Extraction and Recognition (STER) has remarkably attracted many researchers from the area of image processing and computer vision. Most of the methods discussed in this paper are subjected to evaluation and validation based on different feature properties such as size, orientation, color information, edges, texture property, connected components, and geometry of scene text. Combining different features like CNN and region-based methods together have gained importance and increase in performance of STER. Critical analysis of several state-of-the-art techniques along with benchmark datasets has been carried out with standard testing procedures and evaluation metrics in order to rationalize the decision of selecting an efficient algorithm for better and fast scene text extraction and recognition. Findings of this paper are discussed exclusively by specifying open area for future research in the field of STER system.

Keywords Text localization · Scene text · Text extraction · Recognition · Validation · Evaluation

1 Introduction

STER in a natural image is a popular research area in the field of image processing and pattern recognition, which tries to build a device with capacity to read text on its own that is visually embedded in complex backgrounds. For the most part, pictures will have variations in text dimension, style, shading, color, orientation, alignment so forth with background which is complex in nature, added noise, with low resolution and less contrast. These issues need to be addressed in an efficient way while automating the process of extraction of text from the image and recognize them correctly. Although the recognition of text involved in many applications, to

M. J. Roopa (✉)
Department of CSE, SJB Institute of Technology, Bengaluru, India

K. Mahantesh
Department of ECE, SJB Institute of Technology, Bengaluru, India

V. Bhateja et al. (eds.), *Data Engineering and Intelligent Computing*,
Advances in Intelligent Systems and Computing 1407,
https://doi.org/10.1007/978-981-16-0171-2_5

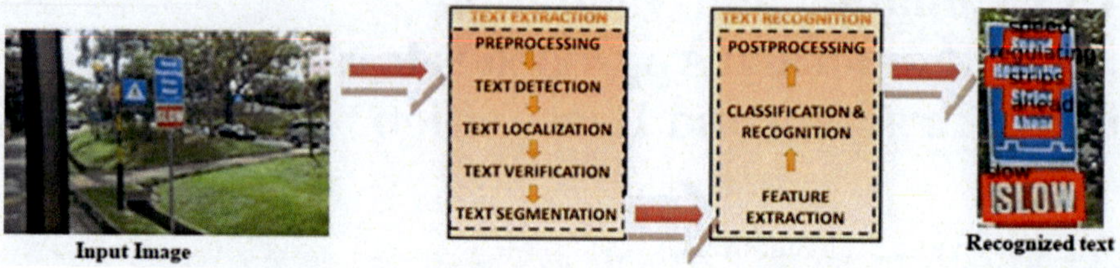

Fig. 1 Architecture of STER system

fetch text content in an image, the primary goal is to identify text means checking for the presence of texts, and if present, then it needs to be localized and recognized. The procedure of STER includes majorly extraction and recognition of the text that are present in natural images [1]. The general architecture of STER is provided in Fig. 1. Extraction of the text location can be achieved by expanding this phase into preprocessing, detection, localization, verification, and segmentation. Similarly, text recognition can be achieved by involving feature extraction, classification, and post-processing processes. Preprocessing is done as fundamental task on input images for better enhancement. Searching out the presence of text components in scene images is nothing but process of detection. After existence of text, they need to localize position of the text in that image. In segmentation step, given localized text, it needs to segmented line by line then by each character out of segmented line. Feature extraction calculates the characteristics of character. Role of classification involves the database comparison. This comparison results in recognition of text objects.

Further sections of this paper are coming up in the following way: Sect. 2 brief about trends and research avenues, methodologies used so far, challenges involved, recent advancement, bench mark datasets, and performance evaluation metrics; Sect. 3 summarizes about findings obtained and finally presents conclusion in Sect. 4.

2 Trends and Research Avenues

In this section, some of the contemporary approaches in STER with its crucial research objectives are identified. Research objectives for STER can be viewed as: the methods in text extraction, text recognition, and end-to-end recognition; challenges, recent advancements, benchmarking datasets, and its performance evaluation. Brief explanations of the aforementioned techniques can be found in the following sections.

2.1 *Text Extraction, Text Recognition, and End-to-End Recognition Methods*

Text exists around us in different forms like display posters, advertisement boards, shop and house names, and so on which is trivial in image content and gives useful information of the scenario.

Finding the exact position of the text and being enable recognize is an important task. In this process of extraction and recognition, there occurs many noise or unwanted contents (see Table 1) which can be a big threat. All these anomalies make it hard to process data inside the image when compared to classical image processing. Plenty of algorithms/methods have been introduced to address above said issues from the research community such as ICDAR, RRC, etc. Some of the authors contributed for text extraction and used previous methods for recognitions. Some authors worked only on recognition part and used previous methods for extraction. Some authors focused on end-to-end system. Some of these methods have been briefly discussed in the following section.

Text extraction methods: Text extraction means extracting location of the text from the image. These methods can be categorized as edge-based, region-based, texture-based, and hybrid methods. Following researchers adopted one of these methods and achieved better results.

Table 1 Scene text challenges and its types

Challenges involved in scene text	Types	Description
Inference factors during image acquisition(C_1)	Noise(C_{11}), blur(C_{12}), distortion(C_{13}) low resolution(C_{14})	Leads to poor performance in text detection from natural scene images
Environmental affects(C_2)	Complex background(C_{21}) Uneven lightening (C_{22})	Many natural objects look like text in natural scene image scenario. Due to such complexity of background, it always produces either confusion or errors during recognition It might cause partial text when image is captured. Because of this performance of the recognition system will decrease
Diversification of text(C_3)	Different font(C_{31}) size(C_{32}), color(C_{33}) orientation(C_{34}) multilingual(C_{35})	Generally, scene images are with diversity of text, various aspect ratio, different text orientation and having more than one language in single image. This makes complexity for the researcher to build single unified model which is suitable for all

In images, dominations of text edges were retained by Contourlet transforms to obtain accurate text localization regions [2]. By combining GLCM features with SVM classifier, potentiality of the text part or non-text part was tested. Hybrid technique combines edge detection and Connected Component (CC) methods for detecting and localizing of text in image [3]. Another hybrid approach was proposed by joining texture method and connected component-based method [4]. Word Fence Detection Network (WFDN) was proposed to ensure that individual word detection and word localization achieved by using weighted softmax loss function [5].

Text recognition methods: To localize and recognize real-time scene text images, combination of HOG transformation and stroke algorithm was used [6]. Multiple different models were utilized to approach the problem of identify character from images. This method integrates both localization and segmentation via the use of convolutional neural networks. Basic HoG/HSV-based SVM was used to get a decent accuracy [7]. Authors of [8] used existing method for extraction of text and contributed only in recognition phase. They designed color channel specific feature extraction method then applied multi-label Support Vector Machine (SVM) classifier for selecting the better color channel. This resulted in best recognition rate.

End-to-end methods: An end-to-end word recognition [6, 9] achieved greater accuracy using CNN. Histogram of Gradient (HOG) feature descriptor was used to detect text regions in scene images. To segment connected components, local binarization was applied. Removed non-text regions using basic features like height and width normalized ratio and compactness. Finally, texts were recognized by distance metric. Morphological operation helped in partitioning the image and grouping of CC to find out text region in images [10].

One of the most important applications of STER system is mobile app, where AmI [11] is one of such application proposed as alternative tool to outdoor localization system when GPS of the mobile fails to find location of the mobile.

2.2 *Challenges in STER*

This survey made us to summarize all challenges involved in the scene text extraction and recognition processes, and it is shown in Table 1. It has been noticed that very few percentage of work has been done on diversification of text, i.e., multi-oriented texts [12].

2.3 *Recent Advancements in STER*

It is observed that recent survey papers on text detection were extremely good. However, this section focuses on recent advancements and left over papers novel approaches which were efficient enough to handle some of the challenges shown in Table 1, and it is summarized in Table 2 of this section. Table 2 concludes ICDAR

Table 2 Analysis of challenges addressed in state-of-the-art scene text datasets

Dataset used	Year	C_1				C_2		C_3				
		C_{11}	C_{12}	C_{13}	C_{14}	C_{21}	C_{22}	C_{31}	C_{32}	C_{33}	C_{34}	C_{35}
ICDAR 2011—[18], [10]	2017	√	X	X	X	X	√	√	√	√	X	X
ICDAR 2013—[18], [19], [10], [18]	2017	√	X	X	X	X	X	X	X	X	X	X
MSRA-TD500—[16], [19], [6]	2017	√	X	X	X	√	√	√	√	√	√	√
COCO-text—[21], [18], [6], [17]	2017, 2018	√	√	√	√	√	√	√	√	√	√	X
SVT—[18], [16], [13], [7], [14]	2018	√	√	√	√	√	√	√	√	√	X	X
ICDAR 2015—[1], [19], [18], [6], [17], [15]	2017,2018,2019	√	√	√	√	√	√	√	√	√	√	X

2015 and COCO-text are the major datasets which include more challenges of Table 1 and hence attracts STER researchers.

It is observed that MSER is one of the stable algorithms in finding out text regions due to its properties like invariant to size, orientation, and language [1]. Gaussian low-pass filter used to enhance image because it always results in positive value to get better image. Enhanced image is given as input to the region detector (MSER). It was able to generate stable regions by varying threshold value (Δ) from 1 to 20. This output, collection of regions then only candidate text regions are selected by applying two level filtering. Bounding box is created for each character and overlapped boxes grouped as one word using graph concept. Finally, text is localized. This method was tested on 50 randomly selected ICDAR2015 images. Results proved that how an image enhancement technique impact on increasing accuracy of the system. It is possible to improve results by bringing neural network into enhancement technique or by considering different color channels.

Authors make use of basic edge and connected component information in [13–15, 11], machine learning approach in [11, 16, 17], neural network approach and deep neural network in [6, 9, 12, 18–20] papers and achieved better results. Recent advancements shown tremendous result on benchmark datasets such as SVT, ICDAR 2015, MSRA-TD500, and COCO-text, which are summarized in Tables 3, 4, 5, and 6.

Table 3 Performance of reviewed algorithms on SVT dataset

Method	Text localization results			Accuracy (%)
	P (%)	R (%)	F (%)	
Candidate text + least-square SVM [16]	80.7	81.9	81.2	–
LUV channel + MSER + CNN [13]	85.6	77.9	81.4	89.5

Table 4 Performance of reviewed algorithms on ICDAR 2015 (Incidental dataset)

Method	Text localization results		
	P (%)	R (%)	F (%)
CNN + FCN [19] (ResNet-50)	89	73	80
Candidate text + least-square SVM [16]	77.9	74.1	75.9
CNN [22]	74	52	61
Enhanced MSER [1]	54.6	64.3	59.1
Color prior guided MSER +SWT + CNN [15]	42.1	55.7	48.9

Table 5 Performance of reviewed algorithms on MSRA-TD500 dataset

Method	Text Localization Results		
	P (%)	R (%)	F (%)
CNN + FCN [19]	85	70	76
X Zhou et al. [6]	87	67	76
Candidate text + least-square SVM [16]	82.4	82.2	81.1

Table 6 Performance of reviewed algorithms on coco-text dataset

Method	Text localization results		
	P (%)	R (%)	F (%)
CNN [22]	68.42	28.33	40.07
	End–to-end text recognition Results		
CNN [18]	31.4	16.7	21.8

2.4 *Benchmarking Datasets*

Communities like Document International Analysis and Recognition (DIAR), Robust Reading Competition (RRC), etc. are constantly encouraging the authors to upload their datasets which were used to train and/or evaluate their algorithm. To make competition healthier and collaboration to be better, they encourage data like ground truth, software, tools, etc., to easily represent results. Texts type can be classified as printed or handwritten. There are several categories in datasets some of them are scene text, machine-printed documents, and many more. This paper mainly focuses on scene text due to its challenges. Dataset means collection of data with its information. Experimental procedure involves dataset division into training and testing group. In this section, datasets discussed along with its metrics are benchmark. Topic-wise datasets can be obtained [http://www.iapr-tc11.org/mediawiki/index.php/Datasets_List], and we can upload our own datasets into TC-11 online resources.

MSRA-TD500 dataset: MSRA text detection 500 database (MSRA-TD 500) is a publicly available collection of natural images; it consists of 500 annotated images (red: unsuccessful cases, green: successful cases) which can be split into 300 train and 200 test images (http://www.iapr-tc11.org/mediawiki/index.php?) title = MSRA_Text_Detection_500_Database_(MSRA—TD500). Images are acquired in both indoor (office and mall) and outdoor (street) environment with challenges such as multi-oriented and multilingual characters; resolution varies between 1296 × 864 and 1920 × 1280 (Fig. 2).

Street View Text (SVT): A Google street view data collection (http://tc11.cvc.uab.es/datasets/SVT_1) shows greater variability and lesser resolution which posses a hard challenge that ICDAR 2013. Four key features are (a) there are 100 images for training while 250 images for testing, (b) Most images are horizontally oriented,

Fig. 2 Sample images of dataset. (**a**) MSRA-TD500, (**b**) Street view (**c**) ICDAR2015-incidental text dataset, and (**d**) COCO-text

(c) each image is given with a lexicon having 50 words, (d) Labeling is not done for all the images. Thus, the dataset can be used only for word detection evaluation.

ICDAR 2015: The International Conference on Document Analysis and Recognition 2015 (ICDAR 2015 Incidental Text http://tc11.cvc.uab.es/datasets/SVT_1) robust reading competition has 1000 images for train set and 500 images for test set. All these are of low resolutions which are captured by Google glasses with different orientation. Bounding box annotations are given for words and three lexicons (a) strong lexicon which can provide 100 words for each image in test set; (b) weak lexicon which provides 100 words for whole test set; (c) generic with 90,000 words.

COCO-Text: A complex day-to-day scene image dataset called MS COCO (COCO—Common Objects in Context https://vision.cornell.edu/se3/coco-text-2/) has a wide variety of collection. It is the largest collection which has different orientations in it. This dataset has 239,506 labeled text regions in over 63,686 images which are way greater than RRC 2015—Challenge 4 which has 7,548 text regions in over 1500 images. Structure of COCO-text dataset is done on three levels: Text localization, recognition of trimmed word, and complete recognition means end-to-end.

2.5 Performance Analysis

Metrics: To describe the efficiency of STER model, evaluation metrics are used. In STER, precision and recall are the important metrics used. True Positive (TP) means texts correctly classified as texts, True Negative (TN) means non-texts classified as non-texts, False Positive (FP) means non-text classified as text, and False Negative (FN) means text classified as non-text. Precision means correctly classified texts out of total selected texts. Recall means correctly classified texts out of actual texts.

Analysis: Performance analysis of contemporary methods summarized in Table 2 made use of these metrics for evaluation. The results are show in Tables 3, 4, 5, and 6. This paper summarizes text localization results but not recognition results this is because, as recognition methods accuracy majorly depends on localization/detection of text.

3 Findings

This section lists findings of the observation made in Sect. 2 and highlights guidelines to the researchers:

- It is observed in [1], bringing neural network into preprocessing step then combing with traditional method such as region detector would result in good accuracy. Hence, researcher contribution is in demand in the field of preprocessing.
- In this paper, methods presented by many researchers were compared for the performance analysis (Tables 3, 4, 5, and 6). This analysis implies future focus can be made on deep learning models, regularization and non-linearity for optimization, domination sets to localize multi-oriented texts optimally and efficient machine learning technique like unsupervised learning to achieve satisfactory performance.
- This review concludes that there is no single unified model for different language. Analysis made in Sects. 2.2, 2.3, 2.4, and 2.5 directs researchers to concentrate more challenges like multilingual texts, text with different orientation, distorted texts, end-to-end recognition, and post-processing as there is less work has been done.
- When it comes to deployment of STER system in real-time applications, time taken to recognize text is the major issue. Researcher can contribute in this area.

4 Conclusion

Extraction of scene texts and its recognition is noteworthy in most of the applications and hence attracted many researchers from the field of image processing, pattern recognition, computer vision, and machine learning. This paper provides the evaluation and validation of scene text extraction and recognition methods. Initially, elemental requirement for review of STER system has been addressed. Trends and research avenues were listed to carry out this work effectively. This paper follows well-ordered analysis to summarize past works on STER system. Methods applied on STER were discussed to conclude progress of those methods with respect to challenges. For this progress analysis, scene text datasets used are MSRA-TD500, SVT, ICDAR 2015, and COCO-text. At the end, findings of this paper followed by conclusion were presented by analyzing all challenges, methods, and their progress.

References

1. M.J. Roopa, K. Mahantesh, An impact of frequency domain filtering technique on text localization method useful for text reading from scene images, *2019 4th International Conference (ICEECCOT)*, Mysuru, India, 2019, pp. 37–43
2. S.V. Seeri, J.D. Pujari, P.S. Hiremath, Text localization and character extraction in natural scene images using contourlet transform and SVM classifier, I.J. Image, Graph. Signal Process., 36–42 (2016)
3. S.A. Ali, A.T. Hashim, Wavelet transform based technique for text image localization. Karbala Int. J. Mod. Sci., 138–144 (2016)
4. T. Kumuda, L. Basavaraj, Hybrid approach to extract text in natural scene images. Int. J. Comput. Appl. **6**(4), 1614–1618 (2016)
5. A. Polzounov, A. Ablavatski, S. Escalera, S. Lu, J. Cai, Wordfence: text detection in natural images with border awareness, in *IEEE International Conference on Image Processing*, 2017, pp. 1222–1226
6. X. Zhou, C. Yao, H. Wen, Y. Wang, S. Zhou, W. He, J. Liang, EAST: an efficient and accurate scene text detector, in *IEEE Conference on Computer Vision and Pattern Recognition, Proceedings—30th IEEE Conference on Computer Vision and Pattern Recognition*, 2017, pp. 2642–2651
7. P. He, W. Huang, Y. Qiao, C.C. Loy, X. Tang, Reading scene text in deep convolutional sequences, *AAAI Conference on Artificial Intelligence*, arXiv:1506.04395, 2016
8. A.K. Bhunia, G. Kumar, P.P. Roy, R. Balasubramanian, U. Pal, Text recognition in scene image and video frame using color channel selection. Multimed. Tools Appl. **77**(7), 8551–8578 (2018)
9. M. Liao, B. Shi, X. Bai, X. Wang, W. Liu, TextBoxes: a fast text Detector with a single deep neural network, *31st AAAI Conference on Artificial Intelligence*, 2016, pp. 4161–4167
10. V.V. Rampurkar, S.K. Shah, G.J. Chhajed, S.K. Biswash, An approach towards text detection from complex images using morphological techniques, in *2nd International Conference on Inventive Systems and Control*, 2018, pp. 969–973
11. T. Im, D. Coelho, K. Mueller, P. De, Smartphone based approximate localization using user highlighted texts from images. Pervasive Mob. Comput. **46**, 1–17 (2018)
12. Z. Zhang, C. Zhang, W. Shen, C. Yao, W. Liu, X. Bai, Multioriented text detection with fully convolutional networks, in *Proc. IEEE conf. Computer Vision and Pattern Recognition*, 2016, pp. 4159–4167
13. G.J. Ansari, J.H. Shah, M. Yasmin, M. Sharif, S.L. Fernandes, A novel machine learning approach for scene text extraction. Futur. Gener. Comput. Syst. **87**, 328–340 (2018)
14. B. Shi, X. Wang, P. Lyu, C. Yao, X. Bai, Robust scene text recognition with automatic rectification, *Proceedings of the IEEE Computer Society Conference on Computer Vision and Pattern Recognition*, 2016, pp. 4168–4176
15. X. Zhang, X. Gao, C. Tian, Text detection in natural scene images based on color prior guided MSER. Neurocomputing, 61–71 (2018)

16. L.M. Francis, N. Sreenath, TEDLESS—text detection using least-square SVM from natural scene. J. King Saud University—Computer Inf. Sci., 287–299 (2020)
17. M. Liao, B. Shi, X. Bai, TextBoxes ++: a single-shot oriented scene text detector, in IEEE Trans. Image Process. **27**, 3676–3690 (2018)
18. M. Busta, L. Neumann, J. Matas, Deep textspotter: an end-to-end trainable scene text localization and recognition framework, in *IEEE International Conference on Computer Vision, 2017-October*, 2017, pp. 2223–2231
19. W. He, X.Y. Zhang, F. Yin, C.L. Liu, Multi-oriented and multi-lingual scene text detection with direct regression. IEEE Trans. Image Process. **27**(11), 5406–5419 (2018)
20. B. Shi, X. Bai, S. Belongie, Detecting oriented text in natural images by linking segments, in *Proc. IEEE conf. Computer Vision and Pattern Recognition*, 2017, pp. 2550–2558
21. V. Andreas, M. Tomas, N. Lukas, M. Jiri, B. Serge, COCO-text: dataset and benchmark for text detection and recognition in natural images, arXiv:1601.07140, 2016
22. Z. Tian, W. Huang, T. He, P. He, Y. Qiao, Detecting text in natural image with connectionist text proposal network, in *Proc. European Conference on Computer Vision*, 2016, pp. 56–72

Sentiment Analysis for Movie Ratings Using Deep Learning

Vijay Verma, Anurag Gupta, and Simran Kumawat

Abstract With the rapid development of network and technology, an enormous amount of data is produced and shared every day across the Internet over Facebook, Google, Twitter, IMDB, shopping sites, and online review sites. This growing availability of opinion-rich resources leads to a new technology that is concerned with what other people think and their emotions. This textual data may have rich information with various labels such as sentiment, age, country, gender, and so forth. Using these labels may bring benefits to some industrial fields such as product quality and market research. The technology used to analyze people's sentiments, opinions, evaluations, attitudes, and emotions from their written review is known as opinion mining and sentiment analysis. This work introduces a recurrent neural network (RNN) and long short-term memory (LSTM)-based method to predict movie ratings based on movie reviews. Sentiment analysis is a natural language processing approach to identify the emotional tone behind the opinion text from negative to a positive level. This technique uses fine-grained sentiment analysis that provides a more precise level of polarity by breaking it from very negative to very positive on the scale of the five-star range. A user-friendly web application has been made that provides users to write a review and predict sentiments. Maximum accuracy of around 75% has been achieved on this fine-grained system as text classification is very challenging when it comes to sarcasm, double negation, use of emojis, tone, and subjectivity.

Keywords Sentiment analysis · LSTM · Movie review · RNN

V. Verma (✉) · A. Gupta
Computer Engineer Department, National Institute of Technology Kurukshetra, Kurukshetra, India

S. Kumawat
Computer Engineer Department, Indian Institute of Information Technology Sonepat, Sonipat, India

V. Bhateja et al. (eds.), *Data Engineering and Intelligent Computing*,
Advances in Intelligent Systems and Computing 1407,
https://doi.org/10.1007/978-981-16-0171-2_6

1 Introduction

In recent years, rapid development of e-commerce platforms and online social platforms has changed people's lifestyles. More and more Internet users turn from simple information viewers to information producers such as commenting on movies, products, and expressing views on politics and news. This comment information of the users is using as an essential reference for buying products, insight for business, and social marketing strategies. In the same way, sentiment analysis can be highly exercisable in political campaigns, or the platform which concerns public opinion. A lot of data with a diverse variety is generated across the Internet through many microblogging [1] (Facebook or Twitter) and online review Web sites. For a better understanding of the data, new technology has emerged, i.e., sentiment analysis. The task of sentiment analysis typically involves taking the text like a sentence, a review, a comment, or an entire document and returning a "score" that measures how positive or negative the review is. There are mainly four types [2] of sentiment analysis which includes fine-grained sentiment, emotion detection sentiment analysis, aspect-based sentiment analysis, and intent analysis. Recently, deep learning techniques have been proved effective in the natural language processing field, including sentiment analysis. These techniques do not need specific hand-picked words for classification; they learn themselves with training data. These words are represented in high-dimensional space where words with similar meanings are nearby vectors. This report is focusing on a deep learning technique of sentiment analysis, where the LSTM layer proved efficient. A Web site having IMDB features has been created, where users can watch reviews, and further, these reviews are used for rating prediction for that movie.

1.1 Motivation

The motivation behind choosing this topic is that there are a lot of challenges associated with sentiment analysis of reviews such as sarcasm, irony, subjectivity, tone, context, emojis, and how to define neutral. When it comes to sarcasm and irony, people generally communicate their negative sentiments using positive words and hashtags, which can be difficult for our model to understand it. We are experimenting with different approaches with combinations to other procedures such as convolutional neural network, recurrent neural network, gated recurrent unit, and long short-term memory to improve the accuracy and efficiency.

Opinion mining (sentiment analysis) is the process of extraction of a user's attitude from the text. Sentiment analysis for review is very challenging. It is not about the overall review but also about some specific aspect of that movie. So, it is important to understand that the review points explicitly toward which aspects. It helps in the marketing department for the improvement of their product. For ensuring customers' needs, sentiment analysis plays an essential role by providing information about which feature they liked, which feature they did not like, and, based on that, make improvements to their product. Also, sentiment analysis helps in finding out political campaigns. Since the data is of diverse variety and significant in amount, it can be used for marketing strategies, business, and political purposes using sentiment analysis techniques. The applications and use cases of sentiment analysis are as follows:

A. Social Media Monitoring
B. Brand Monitoring
C. Customer Service
D. Market Research
E. Voice of Customers.

2 Related Work

There has been a significant amount of study and research on sentiment analysis and opinion mining throughout the last two decades. The importance of text data had increased since when many organizations started using the reviews and feedback given by the people to improve their functionality and also with the growth of various social media platforms, which helps people to express their views on different topics. Since the increased demand of sentiment analysis has attracted many researchers and organizations, various techniques had been used for sentiment analysis for reviews. Some of them are discussed here. Word processing is becoming very important for natural language processing. Now word embedding is used to perform word preprocessing. The expression vector first came into the picture in 2003 when a neural probabilistic language model was proposed by Bengio [3] for binary language model construction through neural networks. Collobert [4] in 2008 uses word vectors for preprocessing techniques like entity recognition and role labels. Later in 2013, a research team at Google created [5] Word2vec to convert simple words into word vectors. Word vectors help the preprocessing work in natural language processing. One of the first approaches is an analysis based on sentiment dictionary. This method uses either an open-source dictionary or the extended [6] dictionary, and new sentiment rules according to the problem are used to evaluate the values of the sentiment weights. There are many open-source sentiment dictionaries like Hownet [6], the Chinese and the WordNet [7], and the English open-source sentiment dictionaries, respectively. This approach uses a sentiment analyzer to analyze the sentiment.

This method uses a sentiment classification based on sentiment dictionaries. For a dataset that has an apparent sentiment tendency, this approach results in high accuracy. But for the dataset, which has a very vague sentiment tendency, this method will turn inefficient. Also, for specific fields, this method turns inefficient because, for all particular areas, the dictionary generation is not possible. Another technique uses support vector machine (SVM) [8] or Naive Bayes classifier. In this approach first, the text is preprocessed. Preprocessing includes removing the relevant links, images, hyperlinks, etc., in text. Then feature selection is performed. Then this is fed to the SVM classifier, which projects it to a higher dimension space for classifying the reviews. The advantage of the above approach is SVM's potential to handle significant features that produce a better text classification, and also most of the text is linearly separable. The disadvantage is that it does not produce better results for the large dataset. Another approach [9] uses a method by checking the polarity of the words. In this approach, first, the text is preprocessed by removing the hyperlink and other unnecessary tags. After that, the words are divided into categories like adverbs and adjectives. After that, the overall polarity of the text is devised using a predefined word sentiment dictionary. The overall sentiment of the sentence is devised from the average of the sentiment of the words. This would give us the overall sentiment of the review. The disadvantage of the presented approach is that the method fails on the touches of sarcasm and also did not provide good results for hybrid reviews.

3 Proposed Approach

The proposed approach will work on the prediction of the polarity of sentiment from the movie reviews. We have taken a dataset from the "Rotten Tomatoes" movie Web site, where the sentiment of the review is given on a scale of 1–5. As neural networks deal with numbers, we have to convert movie reviews in the form of numbers. It is part of the tokenization in TensorFlow. We have created a Web site where users can watch reviews, write a review, and the star rating is given after predicting the sentiment of the review. This approach has the following steps.

3.1 Text Preprocessing

Text preprocessing includes removal of the extra unusable information from the text and further uses it to improve result efficiency. After the removal of stop words, punctuations, additional sparse words, and HTML tags, we have left with more semantic information. As the neural networks deal with numbers quickly, we are going to convert each word into their corresponding unique number. This part can be by tokenizer approaches. The stream of words is now turned into sentences of numbers. In the below technique, we need a similar input size to define the architecture of the model. However, reviews may vary on their length, so to overcome this, we use padding (Fig. 1).

3.2 Model Training

We use a sequential model to train our training dataset. We tried a combination of neural network layers to ensure more efficiency. The output of text preprocessing will become the input for the embedded layer that results in the 2D array. We will

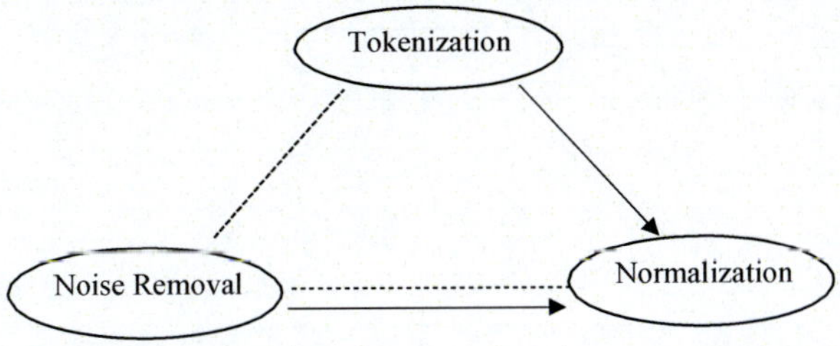

Fig. 1 Text data processing framework

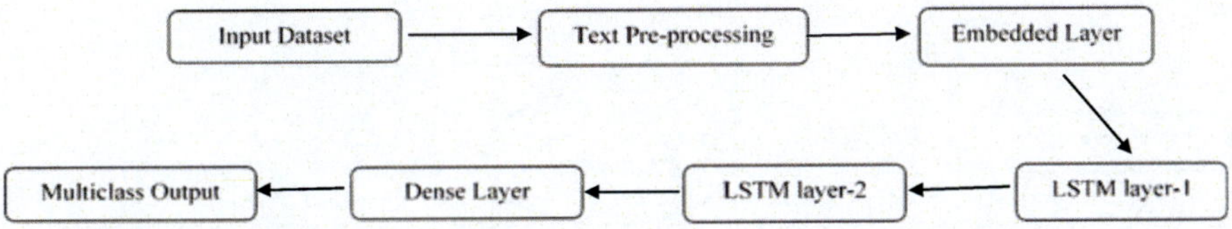

Fig. 2 Model architecture

use two layers of LSTM. This data is the input for the first LSTM layer, with some neurons dropped to avoid overfitting. The result from the previous LSTM layer is going to be the input for the next LSTM layer. Then the output will be given to the fully connected layer and then to the final layer with five classes such as negative, somewhat negative, neutral, positive, and somewhat positive. The last layer's output shows the positiveness of the comment regarding content. Then from this percentage, we calculate the rating that users want to give. Also, with the help of a trained model of the presented approach, we can analyze the sentiment of comments at run time (Fig. 2).

3.3 *Taking Review from User Interface*

After training the model, we stored the model's value and its weights for further usage. This pre-trained model is used to predict the rating of user input reviews. For real-time interaction with the model, we had created a Web site that will allow users to write reviews for their favorite movies. These reviews have passed to the model at runtime, where these reviews are given a score from 1 to 5. These scores are used cumulatively for the overall rating of the movie. The overall rating of the movie is displayed after the user enters its review. The review and the overall rating have stored in the SQL database. Storing the data helps us retrieve the latest rating from the database, and the user can see the comments of other users (Fig. 3).

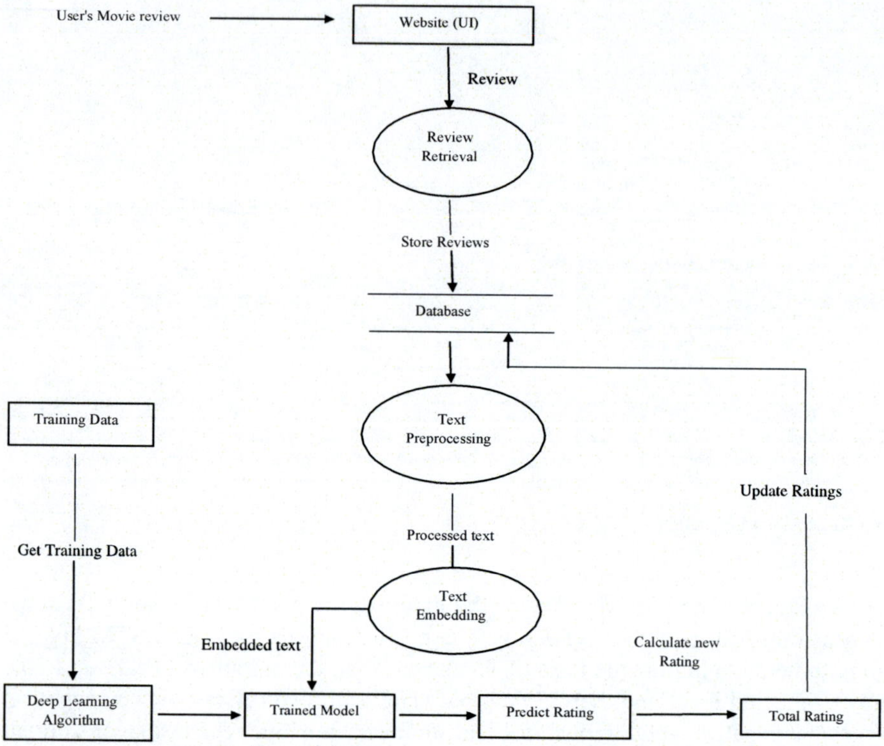

Fig. 3 Flow of data of the system

4 Result

While training the model and other processes, we have calculated a few important features for evaluating the accuracy of the model. For improving the accuracy, we have taken different configurations of neural networks. We also used the CNN layers above the LSTM layer, but their accuracy is low. For improving the accuracy, we try to optimize the hyperparameter of the model. By optimizing the hyperparameter to the optimum values and by making the neural network layers at optimizing position, the training accuracy of the above model after running over 40 epochs increased from 50% to around 75%.

In Table 1, we show the performance of the two-layer LSTM model, which achieves the 75% accuracy after running for 40 epochs. Figures 4 and 5 show the graph between training–validation accuracy and training–validation loss, respectively.

Table 1 Model accuracy

Epochs	Accuracy	Loss
10	0.7281	0.6421
20	0.7391	0.6212
30	0.7447	0.6094
40	0.7491	0.6022

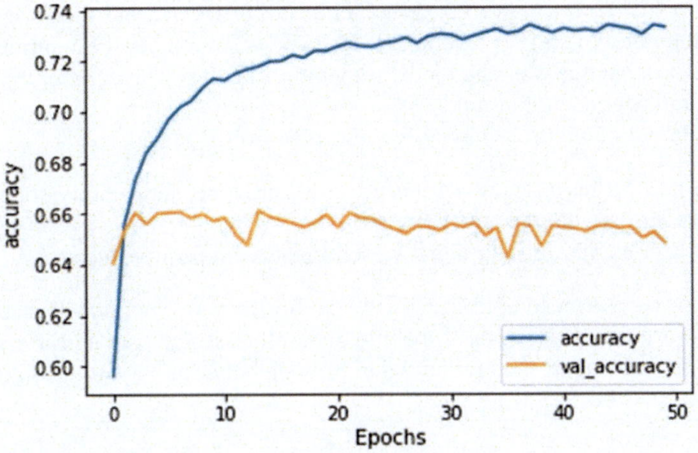

Fig. 4 Graph between training and validation accuracy

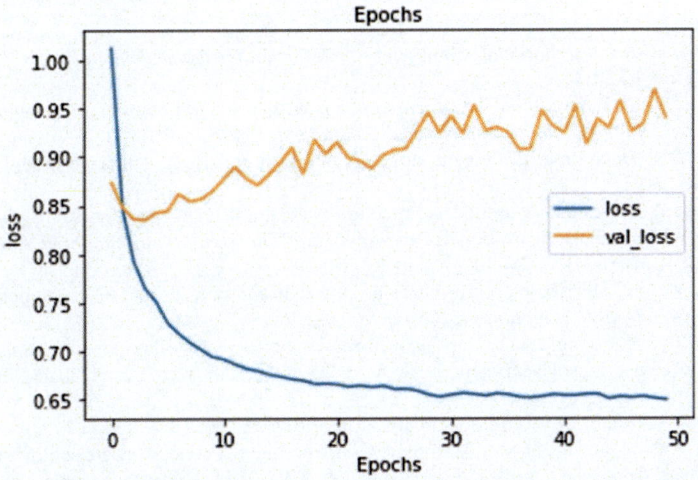

Fig. 5 Graph between training and validation lost

5 Conclusion and Future Plan

In this paper, we introduce a sentiment analysis approach to classify the movie review from a negative to positive on a scale of 1–5. In this work, we first refine the reviews by removing unnecessary details and noise (like HTML tags). Then these text reviews are re-refined by removing punctuations. Then these reviews are fed into the neural network. In the neural network, we use embedding layers, and then we use long short-term memory (LSTM) layers. LSTM helps to remember the sentiment of the previous words; in this way, the model remembers the last words. After training, we get an overall accuracy of around 75%.

The following limitations have been observed of this model:

- when a user writes a review with irony, sarcasm, tone, subjectivity, tone
- when a user uses double negations and
- when a user expresses their negative sentiment with positive words.

In future, we can try to optimize our model for the abovementioned cases and try to improve efficiency. Nowadays, people generally use sarcasm and irony, so it is important to understand such reviews and provide better movie ratings to users.

References

1. W. Maharani, Microblogging sentiment analysis with lexical based and machine learning approaches, *2013 International Conference of Information and Communication Technology (ICoICT)*, IEEE, 2013
2. Cäcilia, Z. et al., Fine-grained sentiment analysis with structural features, *Proceedings of 5th International Joint Conference on Natural Language Processing*, 2011
3. Y. Bengio, R. Ducharme, P. Vincent, et al., A neural probabilistic language model. J. Mach. Learn. Res. **3**(Feb), 1137–1155 (2003)
4. R. Collobert, J. Weston, A unified architecture for natural language processing: deep neural networks with multitasks learning, *Proceedings of the 25th International Conference on Machine Learning, ACM*, 160–167 (2008)
5. T. Mikolov et al., Distributed representations of words and phrases and their compositionality. *Advances in neural information processing systems*, 2013
6. W. Du, S. Tan, X. Cheng, et al., Adapting information bottleneck method for automatic construction of domain-oriented sentiment lexicon. ACM Int. Conf. Web Search Data Min., 2010, 111–120
7. Y.Y. Zhao, B. Qin, T. Liu, Sentiment analysis. J. Softw. **21**(08), 1834–1848 (2010)
8. Y. Lu, M. Castellanos, U. Dayal, Automatic Construction of a context-aware sentiment lexicon. *International Conference on World Wide Web*, (WWW), 2011, Hyderabad, India
9. N. Zainuddin, A. Selamat, Sentiment analysis using a support vector machine. *2014 International Conference on Computer, Communications, and Control Technology (I4CT)*, IEEE, 2014

Subject Noncontingent EEG-Based Emotion Detection Using Deep Learning Algorithm

Vaishali M. Joshi and Rajesh B. Ghongade

Abstract This paper proposes electroencephalography (EEG)-based emotion recognition subject noncontingent approach using 32, 3, 2, and single channel. The subject noncontingent experiment uses a benchmark training model of another person for analysis of emotions. Work is performed on an openly available benchmark DEAP database. This experiment is conducted using features of differential entropy, Hjorth parameters, and power spectral density (PSD). For training purposes, deep learning classifier bidirectional long short-term memory (BiLSTM) network is used. Results established that despite using few channels, recognition accuracy does not vary to a large extent. We achieved almost the same accuracy level by selecting less number of optimal frontal channels.

Keywords EEG · Emotions · Deep learning · BiLSTM

1 Introduction

In our everyday routine, professional challenges and stressful lifestyle generate huge strain on the brain. This generates various emotions such as anxiety, depression, sadness, stress [1]. The type and extent of emotions are also depending upon age, professions, and mental ability to react to the situations. Therefore, it becomes necessary to have these emotions analyzed to keep the mind in full control which helps to handle various emotional situations in a better way to avoid health complications. Many portable devices of EEG analysis are available in the market. However, it is necessary to develop the algorithm and design the process which can give almost equal and accurate results compared against the higher channel density machines. Data is analyzed in two ways, namely subject contingent and subject noncontingent. The performance of the subject contingent approach is better than subject noncontingent. It is challenging to maintain the individual's EEG database for the varying situations, and therefore, it necessary to generate a subject independent benchmark

V. M. Joshi (✉) · R. B. Ghongade
Electronics Engineering, Bharati Vidyapeeth (Deemed to be University) College of Engineering, Pune 411043, Maharashtra, India

V. Bhateja et al. (eds.), *Data Engineering and Intelligent Computing*,
Advances in Intelligent Systems and Computing 1407,
https://doi.org/10.1007/978-981-16-0171-2_7

training model and use it as a comparator for analysis. In view of this paper, it focused on subject noncontingent emotion analysis using deep learning algorithm.

2 Related Work

Various researchers are working across the globe to work out the most accurate method by using combinations of feature extraction and training modules and developing the most accurate predictive emotion recognition model [2, 3]. Li et al. [4] worked on two databases 'Database for Emotion Analysis using Physiological Signals' (DEAP) and 'SJTU Emotion EEG Dataset' (SEED). They achieved an accuracy of 59.06% and 83.33% using DEAP and SEED database respectively. Their approach used statistical feature and support vector machine (SVM) for positive and negative emotions. R. M. Mehmood.et al. [5] achieved an accuracy of 76.6% using the Hjorth parameter feature extractor to train a deep neural network. They generated own database for research work. Fang Wang et al. [6] used data augmentation approach and deep learning. For research work, they have used the SEED database and achieved 75% accuracy. John Atkinson et al. [7] used Hjorth parameters and fractal dimension for training SVM. They have achieved 62.33% and 60.33% accuracy for arousal and valance. Pandey et al. [8] proposed use of wavelet coefficient extractor and multilayer perceptron neural network classifier on DEAP database to achieve recognition accuracy of 58.5%. Joshi et al. [9] used differential entropy as feature and BiLSTM classifier and achieved 74% accuracy using DEAP database. Joshi et al. [10] in another work used SEED database and achieved 74.91% emotion recognition accuracy using differential entropy and BiLSTM classifier. Yucel Cimtay et al. [11] used convolutional neural network to train model and median filter to remove false detections. They used the SEED database and analyzed two and three class emotions and achieved accuracy of 86.56% and 78.34%, respectively. On DEAP database, they have achieved recognition accuracy of 72.81%. Luka Kraljević [12] proposed new features using linear predictive coefficients and trained model using SVM and achieved 60.33% and 62.67% accuracy for valance and arousal using DEAP database.

Deep learning [13–15] is the current buzz word and used as the latest trend in machine learning. Different deep learning architectures are introduced in various literatures. The models with deep architectures attain excellent results against old methods.

3 Proposed Methods and Materials

This paper aims EEG-based emotional analysis using subject noncontingent approach. From DEAP database, data has been segregated and analyzed for 32, 3, 2, and single channel. While selecting a lower number of channels care has been

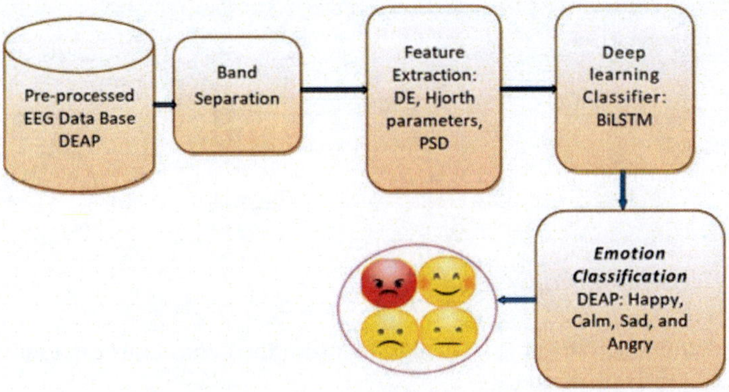

Fig. 1 Block diagram of proposed work

taken that selected channels falls from frontal area and is important for analysis. Figure 1 shows a block diagram of the implemented process. There are mainly four phases, namely database collection, band selection, feature extraction, and emotion classification.

3.1 EEG Signal Database

In the proposed work, openly available emotion database of DEAP [16] is employed. DEAP EEG database generated for 32 subjects using 40 channels EEG machine systems. Thirty-two channels were EEG and remaining were peripherals. Subjects asked to watch 40 videos of 1-minute duration. Emotions were scaled on valence, arousal coordinates. Russell valence arousal scale was used for modeling. Participants were asked to fill up self-assessment manikin developed by Bradley and Lang. DEAP database was collected using a 10–20 international electrode placement system at a sampling rate of 512 Hz. For analysis, the collected data was downsampled to 128 Hz and bandpass filtered in the range from 4 to 45 Hz.

3.2 Feature Extraction Method

Features are extracted by using three different methods known as differential entropy (DE), Hjorth parameters, and power spectral density (PSD). Features are extracted in various bands. Used frequency band details are as below.

Theta: (4 ~ 8 Hz), Alpha: (8 ~ 14 Hz), Beta: (14 ~ 30 Hz), Gamma: (30 ~ 44 Hz).

Table 1 Hjorth parameters

Activity	Mobility	Complexity
Activity = variance$(u(\mathrm{t}))$ u (t) is a signal	Mobility $= \sqrt{\frac{\text{variance}(\overline{u}(t))}{\text{variance}(u(t))}}$ $\overline{u}(t) = \frac{du}{dt}$	Complexity $= \frac{\text{Mobility}(\overline{u}(t))}{\text{Mobility}(u(t))}$

3.2.1 Differential Entropy

Differential entropy (DE) [17] is used to measure the continuous random variable's entropy. The DE is illustrated as,

$$\mathrm{DE} = \frac{1}{2}\log_{\mathrm{e}} 2\pi e\sigma^2$$

where σ^2 is variance, and π and e are constants.

3.2.2 Hjorth Parameters

Hjorth parameters [18] are activity, mobility, and complexity as given in Table 1, which are often used in EEG signal analysis.

3.2.3 Power Spectral Density

The power spectral density (PSD) is an average energy distribution per unit time over different frequency bands [19].

3.3 Classification Method

BiLSTM classifiers is used to examine the performance of the feature extraction methods.

Long Short-Term Memory (LSTM):

Generally, long sequences of EEG signal data are not remembered by conventional recurrent neural network (RNN). RNN also has a limitation of vanishing gradient. To overcome this subclass of RNN known as long short-term memory network (LSTM) [20] is introduced.

LSTM can very efficiently extract the temporal and spatial features from the EEG signal. Figure 2 shows BiLSTM architecture, containing a forward and a reverse LSTM network. Three control units, namely input gate, output gate, and forget gate, are present in LSTM cell. Bidirectional LSTM network has the ability to remember

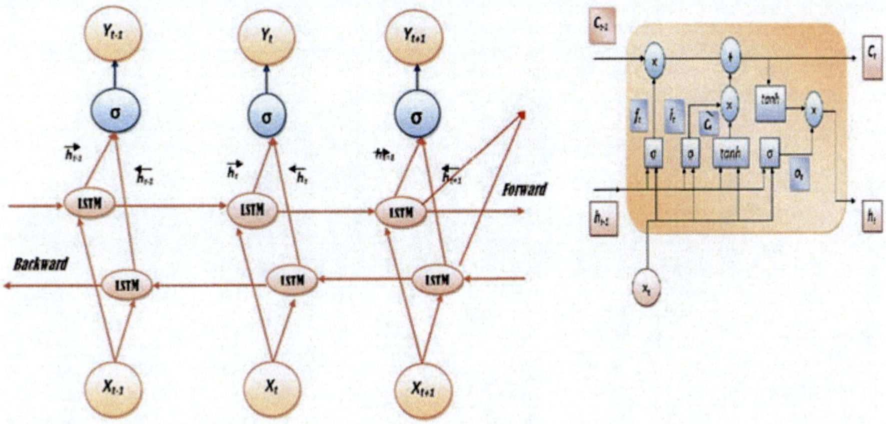

Fig. 2 BiLSTM architecture

more information as compared to unidirectional networks. BiLSTM network learns spatial and temporal information which is effective for emotion recognition.

4 Result and Discussion

4.1 Hyper-Parameter

The hyper-parameters used for BiLSTM are:

Gradient Decay Factor: 0.9000, squared gradient decay factor: 0.9990, initial learning rate: 0.001, hidden unit: 200, and optimizer: Adam (adaptive moment estimation).

4.2 Experimental Result

In the proposed work, testing data and training data are from different subject, i.e., subject noncontingent. In this work features of DE, Hjorth parameters, and PSD of 25 subjects are used to train the BiLSTM classifier and the other five subject's data are used to test performance. Figures 3, 4, and 5 show performance using DE, Hjorth parameters, and PSD.

From Figs. 3, 4, and 5, it can be seen that differential entropy attains the highest average accuracy among all feature extractors. From result, it is observed that for 32 channels the average recognition accuracy is 77%, for single channel (FP1) is

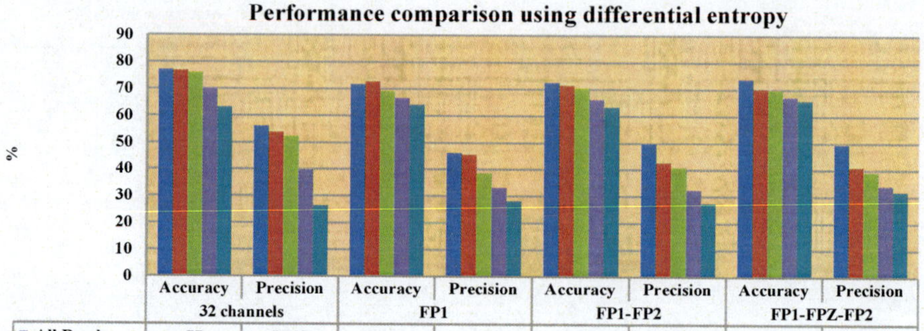

	Accuracy (32 channels)	Precision (32 channels)	Accuracy (FP1)	Precision (FP1)	Accuracy (FP1-FP2)	Precision (FP1-FP2)	Accuracy (FP1-FPZ-FP2)	Precision (FP1-FPZ-FP2)
All Band	77	56.07	71.75	46.14	72.5	49.87	73.75	49.44
Theta Band	76.75	53.9	72.75	45.42	71.5	42.68	70.25	40.98
Alpha Band	76	52.36	69.25	38.67	70.5	40.95	69.75	39.17
Beta Band	70	39.67	66.75	33.3	66.25	32.54	67.25	34.05
Gamma Band	63.25	26.38	64.25	28.27	63.5	27.53	66	31.87

Fig. 3 Performance comparison using differential entropy

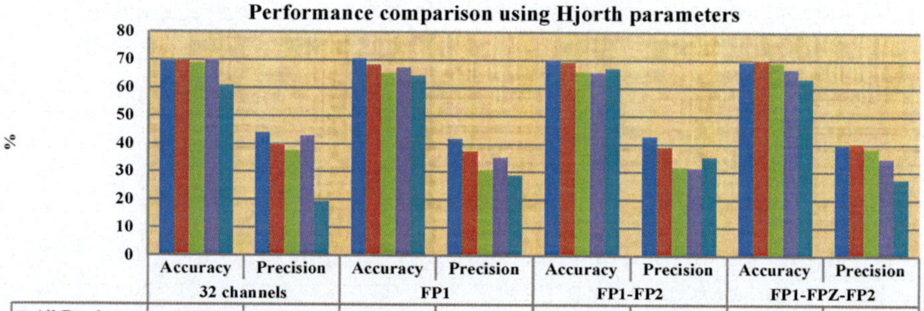

	Accuracy (32 channels)	Precision (32 channels)	Accuracy (FP1)	Precision (FP1)	Accuracy (FP1-FP2)	Precision (FP1-FP2)	Accuracy (FP1-FPZ-FP2)	Precision (FP1-FPZ-FP2)
All Band	69.75	44.13	70.75	41.99	70.25	42.85	69.5	39.69
Theta Band	70	39.94	68.5	37.38	69.5	39.13	70.25	40.48
Alpha Band	69	37.71	65.5	30.71	66	31.77	69.25	38.55
Beta Band	70.25	42.97	67.5	35.27	65.75	31.4	67	34.86
Gamma Band	61	19.33	64.5	28.85	67.25	35.39	63.5	27.46

Fig. 4 Performance comparison using Hjorth parameters

71.75%, for two channels (FP1-FP2) is 70.25%, and for three channels (FP1-FZ-FP2) the average recognition accuracy is 73.75%. From the obtained results, it can be clear that for a less number of channels recognition rate is almost equal to the complete multi-channel system. The result obtained by the proposed method is also compared with that of the other research in this field and is given in Table 2.

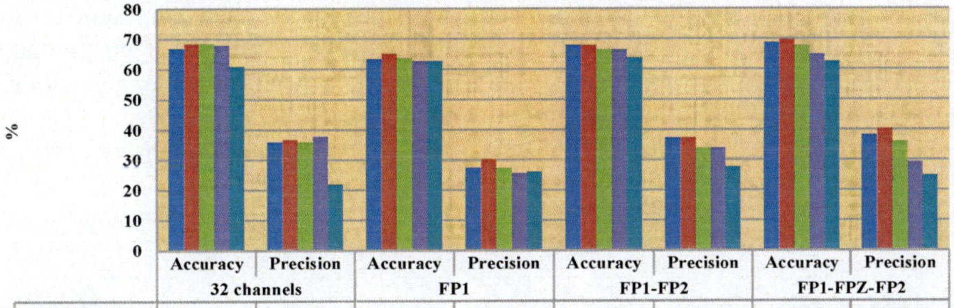

	32 channels		FP1		FP1-FP2		FP1-FPZ-FP2	
	Accuracy	Precision	Accuracy	Precision	Accuracy	Precision	Accuracy	Precision
All Band	67	36.13	63.5	27.5	68	37.41	68.75	38.31
Theta Band	68.5	36.89	65.25	30.36	68	37.41	69.75	40.43
Alpha Band	68.5	36.1	63.75	27.39	66.5	34.07	67.75	36.21
Beta Band	68	37.9	62.75	25.65	66.5	34.07	65	29.26
Gamma Band	61	22	62.75	26.11	63.75	27.77	62.5	24.86

Fig. 5 Performance comparison using PSD

Table 2 Performance comparison with related method using DEAP database

Sr. No.	Article	Method	Name of class	Average accuracy (%)
1.	Yucel Cimtay et al. 2020 [11]	Convolutional neural Network(CNN)	Positive, negative	72.81
2.	Pandey et al. 2019 [21]	Variational mode decomposition (VMD), Deep Neural Network (DNN)	Valance, Arousal	62.50, 61.25
3.	Lan et al. 2019 [22]	Differential Entropy, Domain Adaption technique	Valance	48.93
4.	**Proposed**	**Differential entropy, BiLSTM**	**Happy, Calm, Sad Angry**	**82, 71.5, 74.5, 80**

5 Conclusion

This work proves that, for subject noncontingent approach differential entropy extractor with BiLSTM network is the most effective combination. This combination generates almost equal emotion recognition accuracy even with less number of channels. BiLSTM network learns spatial and temporal information which is the main key factor for improving the performance of the emotion recognition system. Currently, there are many single, two, and three-channel devices available in the market. With this work, it is established that for emotion analysis, we do not need heavily invested multi-channel machines. However, we can achieve the same result using portable devices. Individuals can monitor emotions while on the job and take corrective actions on own while facing emotionally real-time challenging situation. It also establishes that patients can monitor emotion at home and analyze. In future, other deep learning classifiers like CNN, DNN will be used and suitability of these classifiers for emotion recognition will be examined.

References

1. S.A. Hosseini, M.A. Khalilzadeh, Emotional stress recognition system using EEG and psychophysiological signals: Using new labelling process of EEG signals in emotional stress state, in *2010 international conference on biomedical engineering and computer science*, 2010, pp. 1–6
2. P. Vergallo, A. Lay-Ekuakille, S. Urooj, V. Bhateja, Spatial filtering to detect brain sources from EEG measurements, in *2014 IEEE International Symposium on Medical Measurements and Applications (MeMeA)*, 2014, pp. 1–5
3. A. Mishra, V. Bhateja, A. Gupta, A. Mishra, S., C. Satapathy, Feature fusion and classification of EEG/EOG signals, in *Soft Computing and Signal Processing*, ed, Springer, 2019, pp. 793–799
4. X. Li, D. Song, P. Zhang, Y. Zhang, Y. Hou, B. Hu, Exploring EEG features in cross-subject emotion recognition. Front. Neurosci. **12**, 162 (2018)
5. R.M. Mehmood, R. Du, H.J. Lee, Optimal feature selection and deep learning ensembles method for emotion recognition from human brain EEG sensors. Ieee Access **5**, 14797–14806 (2017)
6. F. Wang, S.-h. Zhong, J. Peng, J. Jiang, Y. Liu, Data augmentation for eeg-based emotion recognition with deep convolutional neural networks, in *International Conference on Multimedia Modeling*, 2018, pp. 82–93
7. J. Atkinson, D. Campos, Improving BCI-based emotion recognition by combining EEG feature selection and kernel classifiers. Expert Syst. Appl. **47**, 35–41 (2016)
8. P. Pandey, K. Seeja, Emotional state recognition with eeg signals using subject independent approach, in *Data Science and Big Data Analytics*, ed, Springer, 2019, pp. 117–124
9. V.M. Joshi, R.B. Ghongade, Optimal number of electrode selection for EEG based emotion recognition using linear formulation of differential entropy. Biomed. Pharmacol. J. **13** (2020)
10. V.M. Joshi, R.B. Ghongade, Emotion detection with single channel EEG signal using deep learning algorithm. Int. J. Recent. Technol. Eng. **8** (2020)
11. Y. Cimtay, E. Ekmekcioglu, Investigating the use of pretrained convolutional neural network on cross-subject and cross-dataset EEG emotion recognition. Sensors **20**, 2034 (2020)
12. L. Kraljević, M. Russo, M. Sikora, Emotion classification using linear predictive features on wavelet-decomposed EEG data, in *2017 26th IEEE International Symposium on Robot and Human Interactive Communication (RO-MAN)*, 2017, pp. 653–657
13. L. Wang, W. Huang, Z. Yang, C. Zhang, Temporal-spatial-frequency depth extraction of brain-computer interface based on mental tasks. Biomed. Signal Process. Control **58**, 101845 (2020)
14. V. Majhi, A. Saikia, A. Datta, A. Sinha, S. Paul, Comprehensive review on deep learning for neuronal disorders: applications of deep learning. Int. J. Nat. Comput. Res. (IJNCR) **9**, 27–44 (2020)
15. V.M. Joshi, R.B. Ghongade, IDEA: Intellect database for emotion analysis using EEG signal. J. King Saud Univ.-Comput. Inf. Sci. (2020)
16. S. Koelstra, C. Muhl, M. Soleymani, J.-S. Lee, A. Yazdani, T. Ebrahimi et al., Deap: a database for emotion analysis; using physiological signals. IEEE Trans. Affect. Comput. **3**, 18–31 (2011)
17. R.-N. Duan, J.-Y. Zhu, B.-L. Lu, Differential entropy feature for EEG-based emotion classification, in *2013 6th International IEEE/EMBS Conference on Neural Engineering (NER)*, 2013, pp. 81–84

18. B. Hjorth, EEG analysis based on time domain properties. Electroencephalogr. Clin. Neurophysiol. **29**, 306–310 (1970)
19. T. Song, W. Zheng, C. Lu, Y. Zong, X. Zhang, Z. Cui, MPED: a multi-modal physiological emotion database for discrete emotion recognition. IEEE Access **7**, 12177–12191 (2019)
20. S. Hochreiter, J. Schmidhuber, LSTM can solve hard long time lag problems, in *Advances in neural information processing systems*, 1997, pp. 473–479
21. P. Pandey, K. Seeja, Subject independent emotion recognition from EEG using VMD and deep learning, J. King Saud Univ.-Comput. Inf. Sci. (2019)
22. Z. Lan, O. Sourina, L. Wang, R. Scherer, G.R. Müller-Putz, Domain adaptation techniques for EEG-based emotion recognition: a comparative study on two public datasets. IEEE Trans. Cogn. Dev. Syst. **11**, 85–94 (2018)

Performance Analysis of Dijkstra's and the A-Star Algorithm on an Obstacle Map

N. G. Sathvik and Sharan Patil

Abstract Today and in future, there is a need to fulfill the requirement of obtaining the shortest path between two points in a graph. These shortest paths are majorly to be found in the presence of obstacles; these obstacles can be even a ball that is present in a play area or it can be houses that are present on a map which act like as a blockade through which the object cannot be passed and a deviation needs to be taken to get to the destination. The selection of the right algorithm is necessary to find the shortest path in such cases. In this work, there is an implementation of Dijkstra's and A-star search algorithms with two heuristic functions for A-star and test which of these performed better and can be better implemented in real-life scenarios. This paper also compares the Dijkstra's and the A-star algorithm for their performance based on the number of operations on different obstacle maps.

Keywords Dijkstra's algorithm · A-star algorithm · Heuristics · Euclidean distance · Manhattan distance · Obstacle map · Path searching

1 Introduction

The shortest path problem is a problem of finding the shortest path or route from a starting point to a final destination. Generally, to represent the shortest path problem, a graph is used. A graph is a mathematical abstract object, which contains sets of vertices and edges. Edges connect pairs of vertices. Along the edges of a graph, it is possible to walk by moving from one vertex to other vertices.

There is a need to fulfill the finding of the shortest path algorithm to reduce the amount of resources required to move from one location to another in a given space or for an object/person to move from one place to another in the quickest way possible. With the development of obstacle detection, there is also a need for computing the shortest path between the two locations in the graph. These calculations of the shortest path sometimes need to be done in real time.

N. G. Sathvik (✉) · S. Patil
Department of Computer Science and Engineering, School of Engineering, Dayananda Sagar University, Bengaluru, Karnataka, India

V. Bhateja et al. (eds.), *Data Engineering and Intelligent Computing*,
Advances in Intelligent Systems and Computing 1407,
https://doi.org/10.1007/978-981-16-0171-2_8

As an example, let us assume that there are essential goods that need to be moved from one location to another location in an event of emergency, and the transportation team needs to find out the shortest path to move these goods, there are many possible path options that the team has, but they want the shortest path possible the shortest path for this problem can be solved in a short amount of time, but if there comes a situation where these goods need to be transported in a very large area with several number of obstacles such as buildings or uneven land with several numbers of possible paths. This would lead to the necessity of finding the shortest path to be computed in the least time possible which can be very computationally intensive. This leads to urgency to implement the algorithm that would be taking the least amount of time to solve large maps with obstacles. The Dijkstra's and A-star algorithms are classical algorithms that can be used to find the shortest path from the destination to the source with A-star just making intelligent guesses using heuristics. So, the results mentioned in this paper are going to justify, which among the two algorithms, including both the heuristics for A-star algorithm but limited to a two-dimensional plane, the obstacles here are also 2-dimensional figures like a square, circle or even some complex figures like a polygon, accordingly, the best algorithms can be chosen for finding the shortest path in a given scenario.

2 Related Works

The works that were done in the domain with respect to the algorithms have helped us to get the proper and the required information for the comparison that was desired. Gong and Liu [1] have taken the heuristic function and showed how the math works for the A-star search algorithm and then demonstrated the different A-star algorithms along with its heuristics. The heuristic functions used in the above-stated paper were implemented to get the desired working of the A-star algorithms. Similarly, the related work has been done by Sharma and Pal [2] where they have found out the total time taken to find the shortest path in the presence and the absence of an obstacle for A-star algorithm. In addition to these papers, Geeks for geeks was also used for help with the implementation of the heuristics. Alashoury and Amarif [3] have shown that how the Dijkstra's algorithm produces the best results with the increase in the number of nodes in the graph when the graph is undirected. Magzhan and Jani [4] have also included the extension of the Dijkstra's algorithm to work on graphs that includes the negative edges as well, which is the Bellman–Ford algorithm and the Floyd–Warshall algorithm and evaluated these algorithms for the shortest possible path and showed how their genetic algorithm is advantageous for finding out the optimal solution. Papers/articles on the practical applications [5–8] have helped us to see and get the possible outcomes and what can be expected and how choosing the best shortest path algorithm is going to have an impact on the time that can be saved to move around from the starting point to the destination. In the case study done by Pimploi Tirastittam and Phutthiwat on the Bangkok metropolitan area using the Dijkstra's algorithm [9] and used a GUI-based system to get the evaluation results using the mean and standard deviation to find the best bus that is available and why public transport system should be more flexible to save time and users must be able to select busses with more number of options and how graph theory and Dijkstra's algorithm helped them to obtain the most efficient route. Similar work has been done by Goyal et al. [10] where they have compared the performance of Dijkstra's and A-star algorithm by choosing a single vertical line as an obstacle and proved that Dijkstra's is slower than A-star without taking multiple obstacles and showed valid efficiency results. With reference to these papers and the implementation, the combined results of the performance of the Dijkstra's and the A-star search algorithm and the exact movement of these algorithms visually using pygame are presented in this paper.

3 Setup and Used Methodology

3.1 Accept Inputs

As an initial step, the input for the starting and the goal coordinates are taken from the user [11]. The accepted coordinates are stored in a numpy array as init_node ande and the goal in the form $\{x, y\}$ (Fig. 1).

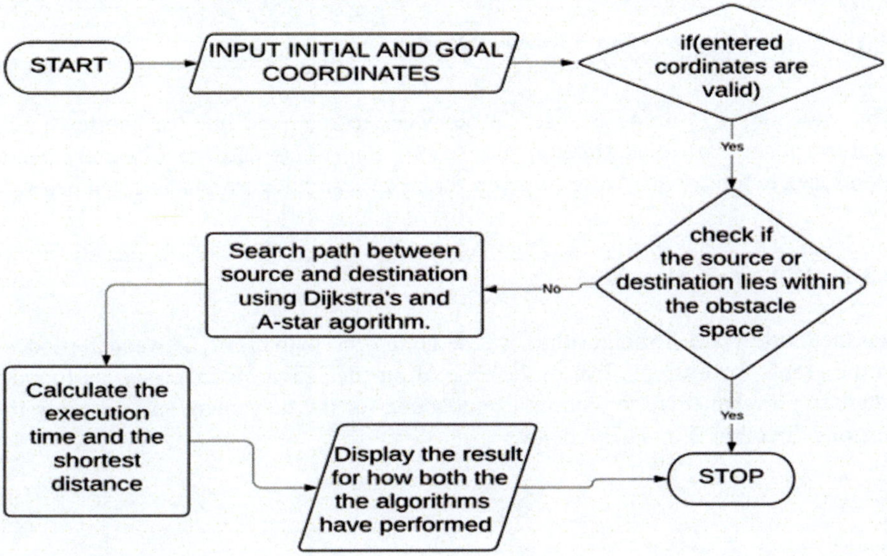

Fig. 1 Basic system design working

Fig. 2 Obstacle map

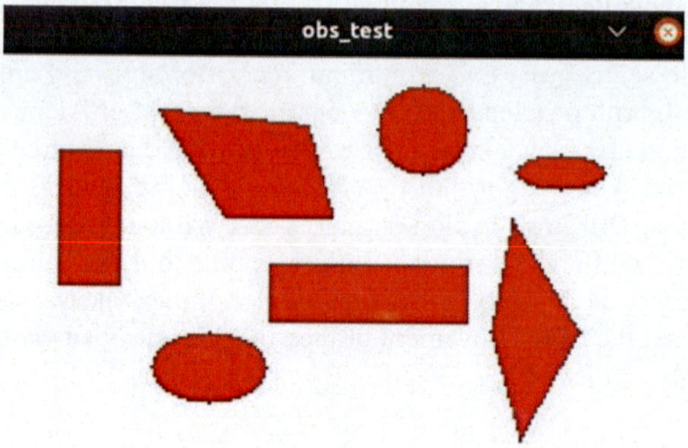

3.2 *Condition Check*

The entered coordinates are made to go through two condition checks where first it is checked if the points are valid or not, i.e., if the entered points are within the defined obstacle maps area (Fig. 2), and then, the next condition check is to see if the entered points are on the obstacle map, and the searching process is going to continue if and only if these required conditions are satisfied.

3.3 *Searching for the Shortest Path*

The searching algorithms are used in our work to scan and find the shortest path to move from start to destination; the search algorithms used have already been mentioned before, so the basic function for these search algorithms is given below.

3.3.1 Dijkstra's Algorithm

It is the shortest path first algorithm, where the shortest path [3, 12] between the nodes in the graph, for example like in the case of our designed obstacle map, is found. The basic implementation with the pseudocode for the movement and scanning in just one direction is as shown below.

```
#goes up
nd,cost=up(ndx)
    if (nd[1]>=0 and obstacle_space(nd[0],nd[1],r)!=1):
        if nd not in visited_node:
            xl=range(0,len(c_nd))
            xl=xl[::-1]
            check=0
            for cku in xl:
                if(nd == c_nd[cku]):
                    check=1
                    if(cst[cku]>=(cst[x]+cost)):
                        p_nd[cku]=ndx
                        cst[cku]=round((cst[x]+cost),1)
                        break
            if (check!=1):
                p_nd.append(ndx)
                c_nd.append(nd)
                  cst.append(round((cost+cst[x]),1))

def up(current_node):
    next_node=[0,0]
    next_node[0]=current_node[0]
    next_node[1]=current_node[1]-1
    cost=1
     return next_node,cost
```

This is just the pseudocode for the movement in the north direction; similarly, the movement and destination scanning can be done. What happens here is that the first 'if' condition checks if the cell is valid or not, then it is verified if the destination node is the same as the current nodes successor; if this is also true, then the parent of the destination node is stored as *ndx* (node *x*) and then checks if the next node lies on the closed obstacle. If this is true then the next 'if' condition is not executed and the total cost is going to be calculated, and then if this node is visited, then it is going to be stored in the visited_node numpy array, and then, calculations are done to find the shortest path from the starting vertex to the ending vertex by taking the visited array as our new map, from which shortest path can be found very easily.

The evaluation function of this algorithm [1] can be given by

$$D = f_1(i) = g_1(i) \quad (1)$$

where i is the current position of the algorithm while searching and $g(i)$ denotes the path-cost function from the start to the ith position.

3.3.2 A-Star Algorithm

A-star search algorithm is a smarter version of the Dijkstra's algorithm as it uses heuristics to make a smart guess of where the next point is going to be instead of scanning in all the possible directions like the Dijkstra's search algorithm [13]. The two approximation heuristic functions that have been used to find the shortest path are (1) Manhattan distance and (2) Euclidean distance

1. Manhattan distance is the sum of the absolute values of the differences in the goals and current cell's *x*- and *y*-coordinates [14]; it can be given by

$$\sum_{i} |a_i - b_i| \tag{2}$$

2. Euclidean distance is the distance between two points in a Euclidean space; the Euclidean distance formula is given by the polynomial formula [14, 15]

$$\sqrt{\sum_{i=1}^{n} (\boldsymbol{a}_i - \boldsymbol{b}_i)^2} \tag{3}$$

Here, a_i and b_i are the coordinates of the current cell and the destination cell, respectively, [in (2, 3)]. The two heuristic functions given above have been implemented to get two different implementations for A-star search algorithm.

The evaluation function for these two algorithms [15] can be given by

$$A^*(1) = f_2(i) = g_2(i) + h_2(i) \tag{4}$$

$$A^*(2) = f_3(i) = g_3(i) + h_3(i) \tag{5}$$

where h_2 is the Manhattan distance function as given by (2) and h_3 is the Euclidean distance function as given by (3); so, here by seeing Eqs. (1, 4, 5), it is clear that Dijkstra's is just a search algorithm with $h = 0$ because of which it scans for the destination in all the directions of the map hence taking more time, whereas in the case of A-star algorithm, because of the presence of the heuristics, the search process is going to take place only in the direction of the destination as shown in Fig. 4 (Figs. 3 and 5).

3.4 Calculations

So this is the final step of our implementation; the total time taken by both these algorithms is calculated along with the total number of operations that it has performed to get to the destination and then display the comparison results in the form of a bar graph as shown in Fig. 6, so that the best performing algorithm can be chosen to be used in practical applications.

Fig. 3 Area scanned and shortest path for Dijkstra's search algorithm

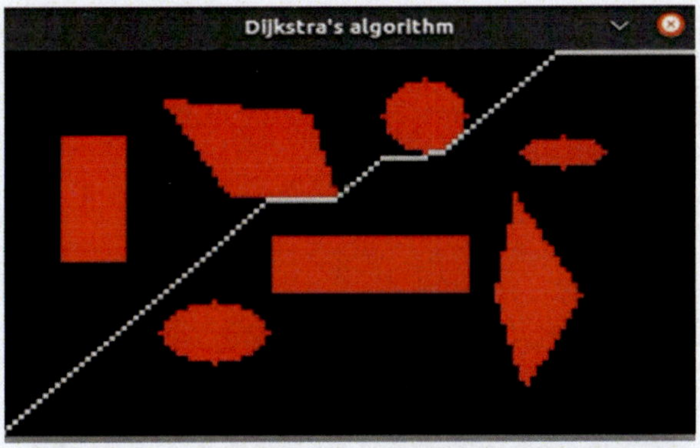

Fig. 4 Area scanned and shortest path for A-star algorithm using Euclidean distance

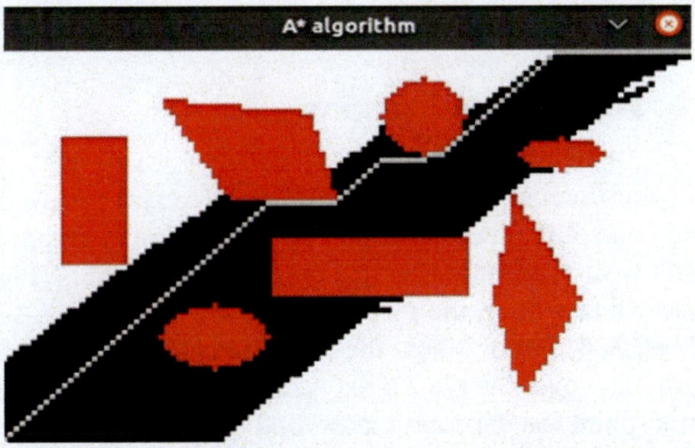

Fig. 5 Area scanned and shortest path for A-star algorithm using Manhattan distance

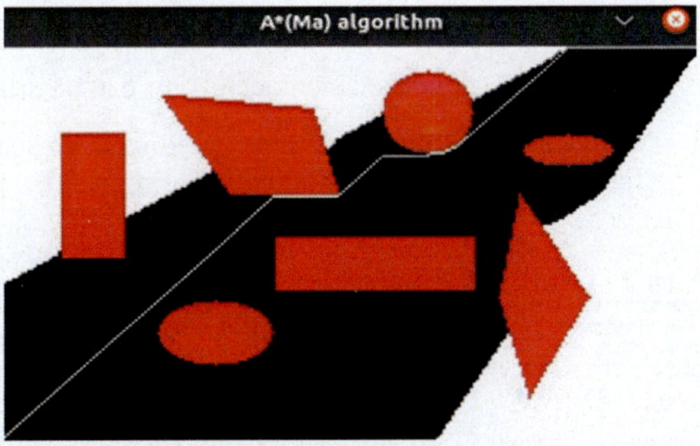

Remark 1 The results shown above are the final comparisons of the obtained time complexities so it can be clearly seen here that the Dijkstra's algorithm is consuming almost double the time taken by both the A-star algorithms.

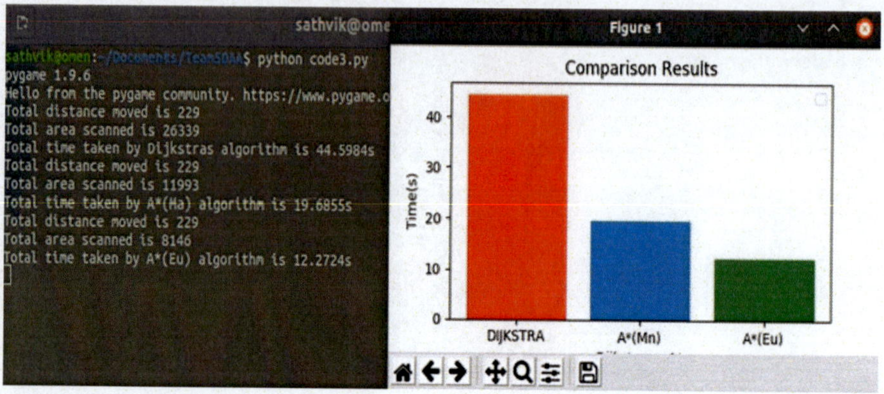

Fig. 6 Comparison results of all three algorithms for a random test case

4 Results and Comparisons

Five different obstacle maps of the same size are used in this comparison analysis, and one of them is shown in Fig. 2. The size of this designed obstacle map is 230 × 144 with a resolution set to 1 pixel. The red spaces (Fig. 1) denote the obstacles, and the white area in the map is where the algorithm's search area. The black area in Figs. 3, 4, 5 shows how the algorithm is scanning for the destination node, and the gray line over the black area is the shortest distance from source to destination. The start point has the coordinates (0, 0), and destination is (230,144) in all the test cases used.

Remark 2 N is the total number of nodes that are present on the obstacle map.

From Table 1, the following conclusions can be drawn:

1. The A-star algorithm which uses the Euclidean distance as the heuristic function has outperformed the other two algorithms. This table helps us to validate that

Table 1 Time taken and operations performed by the algorithms

Algorithms	Dijkstra's		A-star(A*(1))		A-star(A*(2))	
time complexity	$O(N^2)$		$O(N \log N)$		$O(N \log N)$	
Graph	Time (s)	Operations	Time (s)	Operations	Time (s)	Operations
1	45.214	26,339	19.773	11,993	12.476	8146
2	66.712	33,494	32.855	16,995	20.70	12,217
3	45.952	27,659	20.520	13,223	13.829	9263
4	52.216	29,552	23.581	14,233	16.345	10,205
5	53.849	30,268	25.661	15,174	18.274	11,219

with the help of the Euclidean distance as the heuristic, more time can be saved and it is going to perform much better than the other two algorithms.

2. Using Manhattan distance as the heuristic will be useful but it will always be outperformed by the A-star algorithm which uses Euclidean distance as the heuristic, it can be seen that the average time taken and the number of operations performed by $A^*(2)$ is much better than $A^*(1)$ so $A^*(1)$ has no advantage over $A^*(2)$.

5 Conclusion

Through comparison analysis, this paper compared the performance of the three algorithms with different heuristics in some randomly generated obstacle maps. The obstacle maps through the results show that in general when given a fixed start point and an end point the A-star algorithm is a better performer than the Dijkstra's algorithm. It is also true that with varying the heuristics for the A-star algorithm or by using different heuristic functions, the results obtained are different as seen in the table. The A-star algorithm using the Euclidean distance $A^*(2)$ as a heuristic has outperformed $A^*(1)$, so $A^*(2)$ is automatically a better choice for finding the shortest path in an obstacle map. However, it has to be kept in mind that the results produced by the A-star algorithm are not always the shortest but surely is the quickest. The results may vary entirely while being used for practical applications such as to find the shortest path in an obstacle map like Google Maps where transportation networks [16] are used to go from the source to destination in the shortest time possible. [10] does not give any graphical scanning visualizations but the implementation done in this paper does, this work has been done using libraries such as numpy and pygame, the libraries available in python, which led us to show the visualization which cannot be done easily using C++, this has not only allowed us to show the visualizations of the scanning process and route representation but also shows the side by side comparison with a bar graph (Fig. 6).

In future work, the best performing A-star algorithm $A^*(2)$ will be used to make automatic or strong bots in a racing game [6] where the bot has the shortest and the quickest path that it can go from the start to the finish point and win the race. This algorithm as mentioned in [1] can also be used in a rescue task in a building or in a maze so that many lives can be saved in a short duration of time by sending a mobile robot [5, 7, 8] instead of manually sending in a human to search.

References

1. D. Gong, X. Liu, A comparative study of A-star algorithm for search and rescue in perfect maze, in *International Conference on Electric Information and Control Engineering* (2011)
2. S.K. Sharma, B.L. Pal, Shortest path searching for road network using A* algorithm. Int. J. Comput. Sci. Mob. Comput. **4**(7), 513–522 (2015)

3. I. Alashoury, M. Amarif, An effect of using Astorage medium in Dijkstra algorithm performance for ideal implicit path cost. Int. J. Comput. Sci. Inf. Technol. (IJCSIT) **11**(4) (2019)
4. K. Magzhan, H.M. Jani, A review and evaluations of shortest path algorithms. Int. J. Sci. Technol. Res. **2**(6) (2013)
5. W.W. Maw, W.P. Ei, Shortest path finding and tracking system based on Dijkstra's algorithm for mobile robot. Int. J. Sci. Eng. Technol. Res. (IJSETR) **6**(11), (2017). ISSN: 2278-7798
6. X. Cui, H. Shi, A*-based path finding in modern computer games. IJCSNS Int. J. Comput. Sci. Network Secur. **11**(1) (2011)
7. Ė. František Ducho, A. Babinec, M. Kajan, P. BeĖoa, M. Florek, T. Fico, L. Jurišica, Path planning with modified A star algorithm for a mobile robot. Model. Mech. Mechatron. Syst. MMaMS (2014)
8. B. Uppalancha, Kranthi, Optimizing the robot's path using Dijkstra algorithm. Int. J. Innov. Res. Sci. Eng. Technol. (An ISO 3297: 2007 Certified Organization) **4**(6) (2015)
9. P. Tirastittam, P. Waiyawuththanapoom, Public transport planning system by Dijkstra algorithm: case study Bangkok metropolitan area
10. A. Goyal, P. Mogha, R. Luthra, N. Sangwan, Path finding: A* Or Dijkstra's? IJITE **02**(01) (2014). ISSN: 2321-1776
11. S.B. Mansuri, S. Kumar, Comparative analysis of path finding algorithms. IOSR J. Comput. Eng. (IOSR-JCE) **20**(5), 38–45 (2018). e-ISSN: 2278-0661, p-ISSN: 2278-8727
12. R.K. Arjun, P. Reddy, Shama, M. Yamuna, Research on the optimization of Dijkstra's algorithm and its applications. Int. J. Sci. Technol. Manage. **04**(01) (2015)
13. D.S. Yershov, S.M. LaValle, Simplicial Dijkstra and A* algorithms for optimal feedback planning, in *2011 IEEE/RSJ International Conference on Intelligent Robots and Systems*, 25–30 Sept 2011
14. GeeksForGeeks contents on A-Star Algorithm, https://www.geeksforgeeks.org/a-search-algorithm/
15. A. Ghaffari, An energy efficient routing protocol for wireless sensor networks using A-star algorithm. J. Appl. Res. Technol.
16. N.F. Darquah, J. Annan, K. Baah Gyamfi, Shortest path algorithm for transportation networks, case study: Kumasi metropolitan assembly. Am. Res. J. **2**(8). ISSN 2304-7151

Automated Microaneurysms Detection in Fundus Images for Early Diagnosis of Diabetic Retinopathy

M. L. Jadhav, M. Z. Shaikh, and V. M. Sardar

Abstract The detection of microaneurysms at an early stage plays an imperative role to reduce the severity of diabetic retinopathy (DR) disease. As microaneurysm is a common sign of diabetic retinopathy, which appears as a red lesion in the fundus image. This paper deals with the classification of microaneurysms and non-microaneurysms. The proposed work follows four steps, specifically pre-processing, candidate extraction, extraction of texture features from candidate region using Gabor filter and finally classification based on extracted features. Support vector machine (SVM) has been used for classification of microaneurysms (MAs). The proposed approach has been tested on Indian diabetic retinopathy image dataset (IDRiD). The reported result of accuracy is 80.85%, specificity is 78.76% and sensitivity 76.75% highlights the effectiveness of the presented technique.

Keywords Diabetic Retinopathy (DR) · Microaneurysms (MAs) · Classifier · Support Vector Machine (SVM) · Gabor filter · Fundus image

1 Introduction

Diabetic retinopathy (DR) is an outcome of damage to small blood vessels due to diabetes. It is a major cause of blindness as reported in [1]. As per WHO estimations, the 19% of the world's diabetic population is Indian, and by the year 2030 nearly 80 million, people from India will be suffering from diabetes. The patients with diabetic retinopathy are 25 times more prone to blindness as compared to non-diabetic person [2].

M. L. Jadhav (✉)
Department of E&TC Engineering, MET's Institute of Engineering, Nashik (M.S.), India

M. Z. Shaikh
Shree Bhagubhai Mafatlal Polytechnic, Mumbai (M.S.), India

V. M. Sardar
Department of E&Tc Engineering, J.S. College of Engineering, Pune (M.S.), India

V. Bhateja et al. (eds.), *Data Engineering and Intelligent Computing*,
Advances in Intelligent Systems and Computing 1407,
https://doi.org/10.1007/978-981-16-0171-2_9

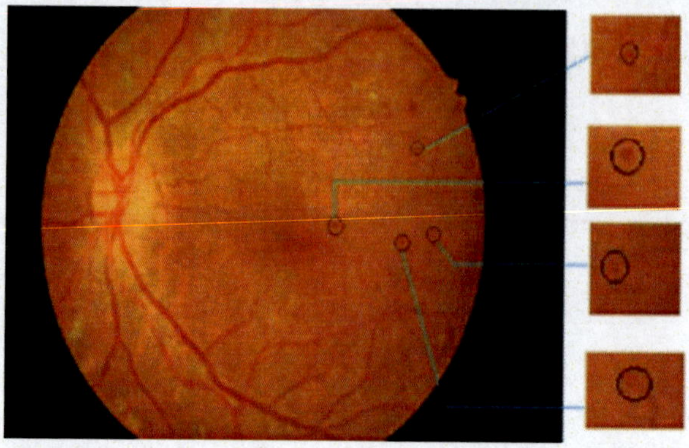

Fig. 1 Fundus image with MAs

Thus, to overcome the problem of diabetic retinopathy, regular eye screening is essential. Capturing the fundus image is a non-invasive technique for detecting the retinal lesions such as microaneurysms, haemorrhages and exudates as a result of diabetic retinopathy.

The introductory sign of DR is the presence of microaneurysm in the retinal image. Microaneurysms occur due to swelling of tiny vessels which appears as sag structures [3]. The increase in the level of blood sugar leads to swelling and blocking of vessels causing red dots, having a diameter less than 125 μm. These dots appear on capillaries and have an isolated pattern; capillaries are not visible in the retinal image.

Thus, detection of dots at the primary stage can be helpful for proper treatment to stop or delay the progression of the disease. The fundus image representing microaneurysms is illustrated in Fig. 1.

The paper is organized as follows: The related literature on MA detection using different approaches has been demonstrated in Sect. 2. Section 3 describes a methodology for pre-processing, including green plane extraction, denoising and histogram equalization followed by segmentation using morphological operations and extended maxima transform. The texture feature extraction from the candidate extracted image is done using Gabor filter, and further classification is done using SVM classifier to classify MA and non-MA candidates. Experimental results based on the proposed technique are illustrated in Sect. 4. The conclusion based on experimentation is stated in Sect. 5.

2 Related Literature

Different techniques for microaneurysms detection have been demonstrated for proper diagnosis of diabetic retinopathy in the literature. An automatic detection of microaneurysms and haemorrhages based on moat operator for 30 images has been carried out and obtained sensitivity and specificity of 77.5 and 88.7%, respectively [3]. Morphological top hat algorithms for detecting MAs have been developed by [4]. They evaluated the unique performance of various feature vectors by using receiver operating characteristics (ROC) based on different kernel functions of SVM. A method of diameter closing and automatic thresholding for feature extraction is proposed by [5]; features were classified as MA and non-MA candidates using the kernel density with variable bandwidth. The sensitivity obtained was 88.5% and per image with the average number of 2.13 false positives.

A random forest algorithm for classification of lesions and non-lesions was extracted using regional minimas with sufficient contrast and intensity-based dynamic shape features [6].The global score for the FROC curve was achieved was approximately 0.393 over 143 fundus images of 2210 diameter. The classification of microaneurysms is done using Markov chain Monte Carlo algorithm for features extracted using histogram-based feature [7]. A hybrid classifier combining a Gaussian mixture model, SVM and extension of multi-mediod-based modelling approach was developed by [8] to get better accuracy. Feature vector was prepared from shape, intensity, colour and statistics by extracting it from possible MA regions.

The authors in [9] have proposed classification of MA and non-MA candidates detected with curvelet transform using rule-based classifier; the algorithm was applied to 336 images retinopathy online challenge database (ROC) and achieved the sensitivity of 48.21%. The authors in [10] have proposed 10 layer CNN for segmentation of DR features such as exudates, haemorrhages and MAs. The novel approach for MA detection by eliminating vessels and calculating the gradient vectors to reflect the characteristic of DR lesions is proposed in [11]; they have calculated seven features to discriminate true MAs and non-MAs. In [12], the blob descriptors and local-scale estimation were used to extract the scale adopted features from the blob regions.

3 Methodology

The proposed approach is divided into four major stages, as described below. Figure 2 demonstrates the schematic of the proposed approach. The colour fundus images from IDRiD dataset of 4288 × 2848 size has been used for the experimentation [13]. This dataset consists of 81(54 training + 27 testing) images with microaneurysms (MA) ground truths.

The dataset was developed for grading diabetic retinopathy with grading ground truths, which comprise images with all types of lesions along with normal images.

3.1 Pre-Processing

The green plane is extracted from the colour fundus image, as it provides good contrast and useful details in the image. The median filter removes the artefacts

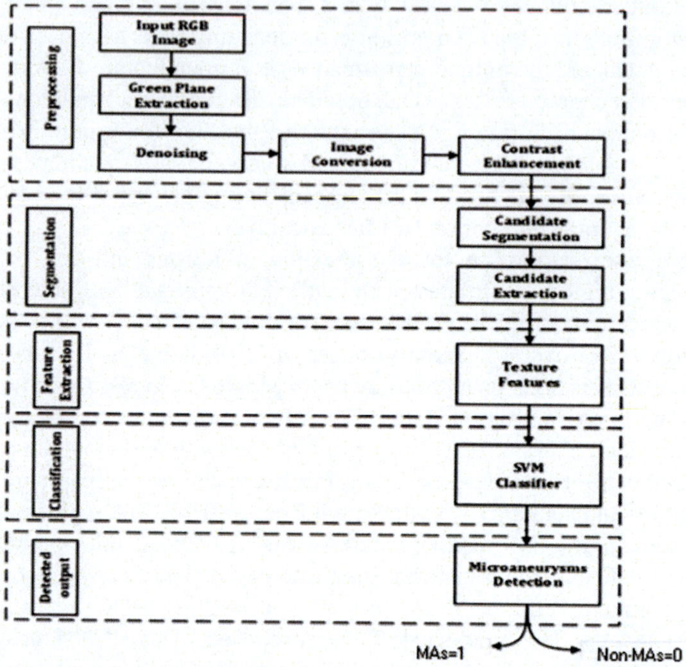

Fig. 2 Schematic representation of the proposed system for detection of microaneurysms

present in the image. The filtered image is inverted to make MAs more clearly visible. The contrast limited adaptive histogram equalization (CLAHE) [14] with clip limit of '0.015' is further applied to the inverted image to enhance the contrast of each small region to maintain the high frequency contents.

3.2 Segmentation

Segmentation plays a key role in the complete detection process. It removes those objects which are not similar to MA. However, if MAs are mistakenly removed at this stage, they can't be recovered back. As MAs appear as a bright spot in the pre-processed image, with at least one regional maximum [15], the pixels with local maxima can be contemplated as MA candidates.

The extended maxima transform is applied on an inverted image, to suppress all maximas whose depth is lower or equal to the threshold h. It can extract the local maxima related to target objects in the image [15]. Here, the value of h is considered 15 empirically. It is attained by performing reconstruction using dilation of I from $(I - h)$, as shown in Eq. 1.

$$H_{\max}(I) = R_I^{\partial}(I - h) \tag{1}$$

where $I is$ the inverted version of the green channel image and h is the threshold value. An extended maximum is the regional maximum of H-maxima transform.

$$E_{\max} = R_{\max}[H_{\max}(I)] \tag{2}$$

The output image highlights the regional maximum calculated using extended maxima transform with selected threshold, i.e. image Ie_1.

The resultant image Ie_1 consists of parts of blood vessels and haemorrhages along with candidate pixels. By using area opening function, the pixels having an area less than 150 are removed to obtain image Ie_2

Ie_2 image will consist of only the areas of pixels greater than 150, i.e. vessels part, fovea and haemorrhages. It is subtracted from extended maxima transformed image to obtain image Ie_3.

$$Ie_3 = Ie_1 - Ie_2 \tag{3}$$

Ie_3 shows only the areas less than 150 pixels. This includes short vessels also. It is removed by calculating the eccentricity [0–0.9]. The eccentricity value ranges from 0 (circle) to 1 (line segment).

This returns the binary image Ie_4 consisting of only small circular dots. The binary image is dilated with the help of disc-type structuring element to include the part of MA which may be missed in an earlier operation.

$$Ie_5 = Ie_4 \oplus B_1 \tag{4}$$

where B_1 is disc shape structuring element of size 5.

The output image Ie_5 consists of microaneurysms as well as some artefacts; to get rid of artefacts, the pixels with zero perimeters are eliminated being non-MA parts. The locations are found out for the candidate pixels from the binary image.

3.3 Feature Extraction

The texture features are calculated for each candidate region using Gabor filter. As a result, the false detected candidates are reduced, which improved the detection accuracy.

3.4 Gabor Filter

The Gabor filter can be finely tuned to frequencies as well as orientation selectiveness [16]. It is having certain optimal localization property in spatial and frequency domain. These filters behave like band-pass filter and appear as a sinusoidal modulated by Gaussian. Gabor filter gives a prominent response at the edges and at the points where the texture changes [16]. It is given by the equation

$$g(x, y) = \frac{f^2}{\pi\gamma\eta}\exp\left(-\frac{x'^2 + \gamma^2 y'^2}{2\sigma^2}\right)\exp(j2\pi f x' + \varphi) \quad (5)$$

$$where x' = x cos\theta + y sin\theta$$

$$y' = x sin\theta + y cos\theta$$

where f is the frequency of the sinusoidal factor, θ is the orientation, φ-phase offset, σ standard deviation; γ is the spatial aspect ratio.

The Gabor array is generated of the size 39 × 39 with 5 scales and 8 orientations without any downsampling. The values of γ and η are chosen $\sqrt{2}$ as an optimum value. It plays important role to decide the correct classification as increase in these values beyond $\sqrt{2}$ reduces the correct classification. However, the change in values of scale and orientation does not have significant effect on correct classification. The Gabor features are calculated for each candidate region in the candidate extracted binary image for a given size of Gabor array. This has helped in enhancing the true MAs in the image.

The ground truth images of microaneurysms consist of MAs as red dots, so the red plane is extracted from the ground truth image. The region of the ground truth containing microaneurysms with their indices is stored in an array. The intersection of indices of the candidate from the extracted image is performed with indices of actual ground truth. The result of the intersection is stored with a label. If the result of the intersection is logic '1', it is labelled as 'MA' else labelled as 'other'. The SVM is trained for 54 images with their Gabor features and labels (MA or other).

3.5 Image Classification

Support vector machine is used for binary (0 and 1) classification efficiently, and the hyperplane is formed based on training dataset to set the standards for classification [17, 18]. In the earlier stage, the true MA regions are marked as '1' and non-MA regions are marked as 'other'. The linear kernel functions are used for mapping the predictor data. So SVM is trained for a total of 54 images with their ground truths.

For the input image, the Gabor features are calculated for the extracted candidates. Every single feature vector is compared with the trained dataset. If it matches with predefined label 'other', it is considered as a false candidate; otherwise, it is treated as 'MA' candidate.

Likewise, for every feature vector, the connected components which are falsely detected are made zero to generate resultant array. The intersection of the indices of the resultant array is calculated with the indices of its ground truth array. If the result of intersection is zero, it is treated as false positive. If the intersection result is one, it is considered as true positive.

4 Experimental Results

The proposed technique of microaneurysms detection is applied on the IDRiD dataset; the dataset is provided with microaneurysms ground truths. The experimentation is carried out with MATLAB 17 platform. The processing time for each image is 3.3 s. The proposed technique goes through the following stages:

image pre-processing, candidate extraction using extended maxima transform and morphological operations, texture feature extraction of candidate region using Gabor filter, classification of MA and non-MA candidates using SVM.

The experimental result of microaneurysms detection is demonstrated in Fig. 3 which includes green plane extraction, histogram equalization of inverted image, removal of vessels and large area parts from image and candidate extraction using texture features.

The sensitivity, specificity and accuracy are calculated for the detected microaneurysms as follows:

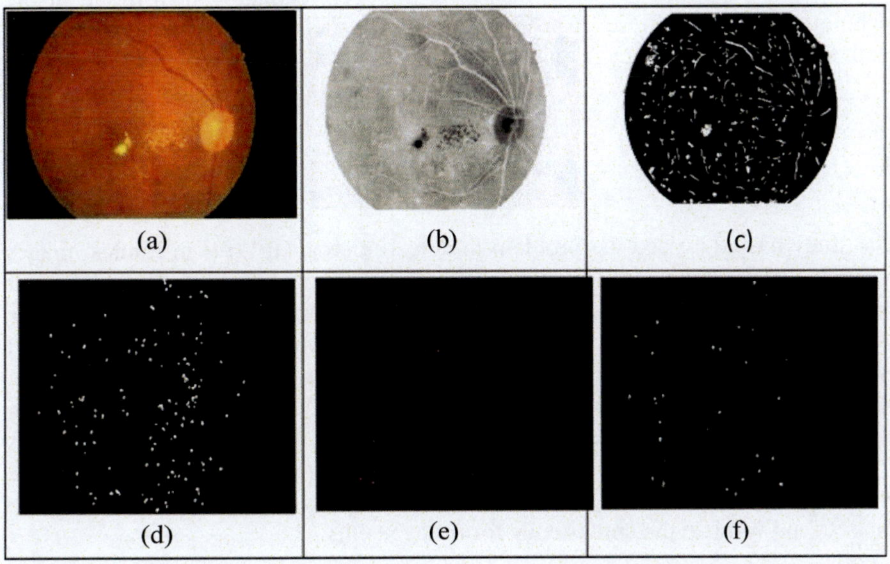

Fig. 3 (**a**) Original image (**b**) histogram equalized inverted image (**c**) image after extended maxima transformation(Ie_1) (**d**) extracted candidates with small circular dots (Ie_4) (**e**) ground truth of microaneurysms (**f**) detected microaneurysms

Table 1 Results of microaneurysms detection for sample images

Authors	Methodology	Sensitivity	Specificity	Accuracy
Sayyad Ali [9]	Curvelet transform	0.482	–	–
Tan J.H. [10]	Convolutional neural network	0.461	–	–
Dai [11]	Gradient vector analysis and class imbalance classification	0.691	–	–
Adal [12]	Local scale estimators and blob descriptors	0.646	–	–
Proposed method	**Morphological operations and texture features**	**0.767**	**0.787**	**0.808**

$$\text{sensitivity} = \frac{(\text{True positive})}{(\text{True positive} + \text{False negative})} \quad (6)$$

$$\text{specificity} = \frac{(\text{True negative})}{(\text{True negative} + \text{False positive})} \quad (7)$$

$$\text{Accuracy} = \frac{(\text{True positive} + \text{True negative})}{(\text{True positive} + \text{True negative} + \text{False positive} + \text{False negative})} \quad (8)$$

Table 1 shows accuracy, sensitivity and specificity based on experimentation.

The result Table shows comparison of proposed technique with the available techniques in the literature. It proves better in sensitivity as compared to existing methods. And the accuracy is on to be high, as 80.85%.

5 Conclusion

The study has addressed the problem to detect microaneurysms in fundus images which are of great concern to reduce the severity of diabetic retinopathy as a result of early detection. The undertaking presents the scheme based on texture feature extraction using Gabor filter from candidate region segmented using morphological operations. The technique can be used as computer-aided diagnosis (CAD), to assist ophthalmologist for diagnosis of diabetic retinopathy. The proposed technique using texture features-SVM reported accuracy of 80.85% on IDRID dataset. The robustness of system is also analysed the other relevant parameters, namely specificity and sensitivity, and found to be optimum. The future scope is to extend the work for large datasets and validate the consistency for of the results.

References

1. B. Antal, A. Hajdu, An ensemble-based system for microaneurysm detection and diabetic retinopathy grading, IEEE transaction on Biomedical Engineering **59**(6), 1720–1726 (June 2012)
2. N. Gupta, R. Gupta, Diabetic retinopathy—an update. JIMSA Jan.—Mar. 28 2015
3. J.F. Sinthanayothin, T.H. Boyce, H.K. Williamson, E. Cook, S. Mensah, D. Lal, Usher, automated detection of diabetic retinopathy on digital fundus images. Diabet. Med. **19**(2), 105–112 (2002)
4. L. Xu, S. Luo, Optimal algorithm for automatic detection of microaneurysms based on receiver operating characteristic curve. J. Biomed. Opt. **15**(6), 065004 (November/December 2010)
5. T. Walter, P. Massin, A. Erginay, R. Ordonez, C. Jeulin a, J.-C. Klein, Automatic detection of microaneurysms in color fundus images. Med. Image Anal. **11**, 555–566 (2007)
6. L. S´eoud, T.´eeFaucon, T. Hurtut, J. Chelbi, F. Cheriet, J.M.P. Langlois, Automatic detection of microaneurysms and haemorrhages in fundus images using dynamic shape features, 2014 *IEEE 11th International Symposium on Biomedical Imaging* (ISBI 2014)
7. M. Tamilarasi, K. Duraiswamy, Automatic detection of microaneurysms using microstructure and wavelet methods. Sadhana **40**, Part 4, June 2015, pp. 1185–1203. @ Indian Academy of Sciences
8. M. UsmanAkram, S. Khalid, S.A. Khan, Identification and classification of microaneurysms for early detection of diabetic retinopathy. Pattern Recognit. **46**, 107–116 (2013)
9. S.A.A. Shah, A. Laude, I. Faye, T.B. Tang, Automated microaneurysm detection in diabetic retinopathy using curve let transform. J. Biomed. Opt. **21**(10), 101404 (October 2016)
10. J.H. Tan, H. Fujita, S. Sivaprasad, S.V. Bhandary, A.K. Rao, K.C. Chua, U.R. Acharya, Automated segmentation of exudates, haemorrhages, microaneurysms using single convolutional neural network. Inf. Sci. **420**, 66–76 (2017)
11. B. Dai, X. Wu, W. Bu, Retinal micro aneurysms detectionus ingradient vector analysis and class imbalance classification. PLoS ONE **11**, e0161556 (2016)
12. K.M. Adal, D. Sidibé, S. Ali, E. Chaum, T.P. Karnowski, F. Mériaudeau, Automated detection of microaneurysms using scale-adapted blob analysis and semi-supervised learning. Comput. Methods Progr. Biomed. **114**, 1–10 (2014)
13. P. Porwal, S. Pachade, R. Kamble, M. Kokare, G. Deshmukh, V. Sahasrabuddhe, F. Meriaudea, Indian Diabetic Retinopathy Image Dataset (IDRiD): a database for diabetic retinopathy screening research. Dataset: https://doi.org/10.21227/h25w98, Dataset License: CC-BY 4.0, Data 2018, 3, 25; https://doi.org/10.3390/data3030025, www.mdpi/journal/data
14. S.M. Pizer et al., Adaptive histogram equalization and its variations. Comput. Vis., Graph., Image Process. **39**(3), 355–368 (Sep. 1987)
15. Y. Qin, W. Wang, W. Liu, N. Yuan, Extended-maxima transform watershed segmentation algorithm for touching corn kernels, Hindawi Publishing Corporation, Advances in Mechanical Engineering, Article ID 268046 (2013)
16. M. Haghighat, S. Zonouz, M. Abdel-Mottaleb, Identification using encrypted biometrics, Computer Analysis of Images and Patterns, Springer Berlin Heidelberg, 2013, pp. 440–448
17. Haralick features-based classification of mammograms using SVM. Information systems design and intelligent applications. Springer, Singapore, 2018, 787–795
18. J. Sushitha Susan, D. Aju. A comparative objective assessment on mesh-based and SVM-based 3D reconstruction of MRI brain. Int. J. Nat. Comput. Res. (IJNCR) **8**(3), 41–54 (2019)

Plant Health Report Through Advanced Convolution Neural Network Methodology

Amith Kiran Kumar, G. H. Dinesh Babu, and P. Rajeswari

Abstract This paper proposes an ideal approach to detect the micro (B, Cl, Cu and Fe), macronutrients (nitrogen phosphorus and potassium) deficiency and disease detection in plants for effective development of the plant health monitoring system for farm fields. It is possible to obtain the pattern of a specific disease and the characteristics of a deficiency using the Convolution Neural Network (CNN) algorithm, which employs approximately 128 filters; this enables to identify 128 specific features hence providing accuracy of 92%, using the dataset of 50 thousand diseased plants from plant village and 1.8 thousand nutrient deficient plants from other sources were analysed to provide greater accuracy in detection, by using large dataset it has improved the quality and accuracy of the disease detection in plants using Keras and TensorFlow for image processing. By doing so, the mobile device is capable of identifying the disease with higher efficiency and is able to suggest the measures that farmers can take to avoid the pest infection and diseases that have been identified in their plants, to grow a healthy plant for high yield. The disease detection is done using the classifier present in the cloud, and hence, it gets a regular update with little change in the hardware so as to reduce the investment cost.

Keywords Plant nutrient deficiency · Plant disease detection · Neural network · Plant disease

1 Introduction

Agriculture is an important part of the world economy. Essentially, the world's population's exponential growth means food scarcity, and the stresses of the environment will become global issues in the following decades. Plants display some unusual symptoms when they are infected with deficiencies in potassium (K), phosphorus

A. K. Kumar · G. H. Dinesh Babu (✉) · P. Rajeswari
Dayananda Sagar Institutions, Bangalore, Karnataka, India

P. Rajeswari
e-mail: rajeswari-tce@dayanandasagar.edu

V. Bhateja et al. (eds.), *Data Engineering and Intelligent Computing*,
Advances in Intelligent Systems and Computing 1407,
https://doi.org/10.1007/978-981-16-0171-2_10

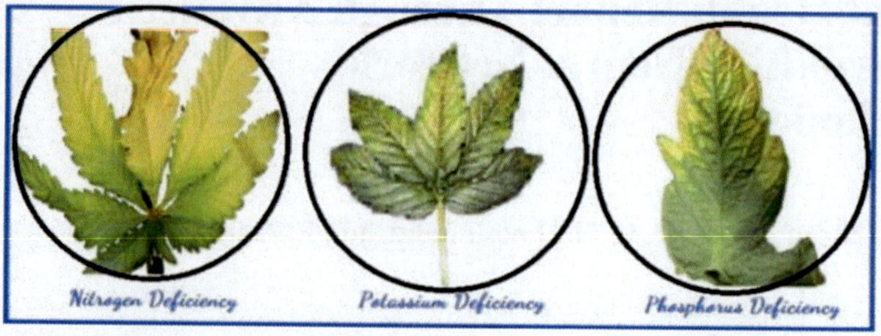

Fig. 1 Defining characteristics of NPK deficient leaf

(P), nitrogen (N), and mostly, these symptoms develop several rapid morphological diagnoses throughout the agricultural field. Morphological diagnoses are quite difficult, and often human perception can still go wrong, involving a great deal of experience in the domain of botany, understanding the pattern of venation, colour pigmentation, as well as other symptoms, is hard for humans, yet the use of convolutional neural networks has made it possible to assess the health of plants with little to no attempt to consider the complexity of the farm teams.

Plants from NPK deficiencies typically exhibit many symptoms. Old leaves as well as occasionally all the leaves turn pale green and chlorophyll deficient around the edges under N deficiency. In K deficiency, deep-green plants mostly with a brownish-yellow margin of leaves or deep-brownish spots will occur on only the margins at first of ageing leaves. Ageing leaves often change their colour from dark yellow to light-brownish [1]. The structure and the colour of the leaves and the sheath that therefore suggest the health and nutritional status of the plants are always closely linked to the nutritional quality (Fig. 1).

Lately, the assessment of plant nutritional status has also been found by hyper-spectral properties that are calculated by using a device namely hyper-spectrometer, which tests the canopy and leaf reflectance [2–9]. Just focusing on only the hyper-spectral characteristics makes it almost impossible to create a uniform standard for functional research to assess the diet status.

All of the plants need an adequate supply of macronutrients to ensure stable growth, and nitrogen (N) is a resource widely found in high demand. Plant nitrogen-deficiency may occur when high carbon organic matter, such as cow dung, is applied to the soil. Soil species use some nitrogen to disintegrate carbon sources, making nitrogen-deficient to the plants [2]. This is considered 'robbing' nitrogen soil. All plants are susceptible to this disorder with the exception of legumes which fix nitrogen to the soil.

Deficiency of phosphorus is a plant disease related to inadequate phosphorus availability. Deficiency of phosphorus can be regulated by the application of phosphorus sources such as rock phosphate, bone meal, compost, and phosphate fertilizers [3].

Potassium mostly also participates in certain biological processes such as synthesis of protein, photosynthesis, activation of certain enzymes, transportation of photo-solute phloem to source tissues, and preservation of anion and cation balance in cytosol and vacuole [4].

In precision farming, controllers, sensor networks, static stations, drones, and handheld robots are used for identification. It is important for such devices to just be able to interpret and draw conclusions from gathered data, such as a specialist in fields or also in greenhouses, in addition to being able to do precision farming [5].

2 Motivation

This project is influenced by a single chip computer used for identification and classification of plant diseases. The detection of nutrient deficiencies, leaf diseases, rodent damages and leaf damage is important. Such adverse effects alter the outward appearance of the plant such that cameras can distinct the source of the damage from the images. For this case, you would need a mobile device and a normal red, green, and blue webcam in disease identification using a device/robot. Mostly because disease identification is performed on just the computer, the mobile system needs to be sufficiently powerful in doing computational work. There are also additional features for mobile devices, such as vehicle control, autonomous navigation, and image processing. Enforcement of the deep learning algorithms emerged to develop through the multidisciplinary field. This research was effective in mobile computing technologies, using the PlantVillage dataset [8] for Deep Learning. In their study, they have mostly used all of the datasets from PlantVillage and tested them on AlexNet as well as in GoogleNet[11]. The AlexNet model of 227.9 MB and the GoogleNet models of 51.3 MB all of which have a relatively wide spectrum for smartphone applications. The scale of the MobileNet model is 2.9 Mb, and the accuracy is still very much the same.

3 Methodology

Each section addresses the different image processing techniques, algorithms, and strategies used to recognize different nutritional imbalances and diseases in plants by analysing the structure and colour of the leaves.

3.1 Classification Method on Nutrient Deficiency

For plant nutritional deficiency, the lateral convolution sequence [[01110], [01110], [01110], …, [01110]] is being used to determine the midrib of the leaf, and then, the next convolution sequence is used to determine the area of effect based on 64 distinct

Table 1 Macronutrients deficiency

Plant micronutrient deficiency	
Symptoms	Source
Fresh leaves are twisted or shaped like hooks. The tip which is growing could die and in tomatoes causes blossom end rot, escarole and celery black and or brown heart and also promotes tip burns in cabbages	The compounds consist of 'calcium.' as well as gypsum
Older leaves at the bottom (plants) may turn yellow. Leftover leaves are mostly lighter in green colour. Stems can as well as become yellow and spindly. Growth is declining	The compounds consist of: 'nitrate,' 'ammonium', or 'urea.' As well as manure
Slower growth and occasionally the outer edges of leaves turn into light yellow. Sometimes, there might be dark spots in the yellow colour in new growth	The compound's with a 'magnesium,'
Relatively small leaves which can get a shade of a variety of colours like red, purple or combination of both. Mainly, the edges of leaves look charred, and fewer times, older ones can turn black which drastically degrade production of seeds or fruits	The compounds consist of 'phosphate or bone.' As well as greensand

convolution techniques, comparing the area wherein the leaf colour has changed at midrib, and hence, the nutrient deficiency is observed.

For some deficiency of micronutrients where leaf curl characterizes the deficiency, the ventricular pattern is detected using vertical and horizontal convolution [010]. We describe the symptoms using the predetermined data.

Here, in Table 1 [9], some characteristics of a few deficiencies in macronutrients are shown.

3.2 *Classification Method on Disease Detection*

Following are some examples of how diseases are detected in some plants and what features are extracted from them.

3.2.1 Fungal Disease Identification Using Image Processing

This research is concentrated predominantly on recognizing diseases caused by a fungus, with visual symptoms in the previous stage [6]. The research was categorized according to the quality of the data found between edible and profit-oriented crops.

3.2.2 Citrus Disease—Cassification and Diagnosis

This section consists of two subsections:

1. Citrus plant-spots of the bruise diagnosis.
2. Classification of predetermined diseases.

Diagnosis of lesions in a spot in fruits is performed via the process of weighted segmentation. The slice of the classification contains cankers, spots which are black in colour, melanose, scab, and many more. The biggest downside to this experiment is to construct a set of data with a limited no. of samples, meaning not much precision has been achieved. They as well as proposed developing a deep learning system for a larger dataset of citrus.

3.2.3 Detection of Vine Disease

In the context of vine cultivation, the biggest challenge facing us is to monitor the expansion of any minor disease across vineyard cultivated lands. Kerkech et al. [8] suggested a co-evolutionary neural network to classify the disease by extracting the colour specimens.

Similarly, around 37 different common diseases in seven variety of plants have been identified and classified.

4 Procedure

See Fig. 2.

4.1 Image Acquisition

Once the image is received, they will be forwarded based on the criteria for image pre-processing level.

The calculation as well as carried out in agricultural data is to measure, since robustness, duration and the size of fruit and crops for the study purpose. Suppose images obtained from the given classifier are not appropriate for a pre-processor. Then, the image enhancement supports this step.

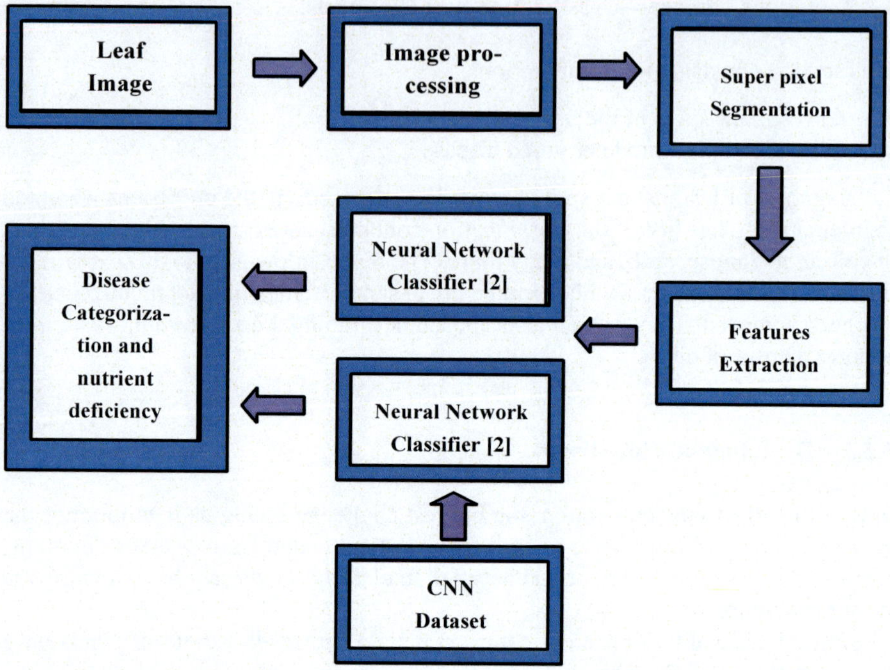

Fig. 2 Block diagram of methodology

4.2 *Image Pre-processing*

Pre-processing of images is a common term used in earlier stages to process the raw image. This involves mostly fine quality of the images, suppressing the unwanted external noise and enhancing some other functionality for further pre-processing. Throughout the processing of agricultural photographs, the pictures of a crop will be collected for processing from either the satellite or the drone pictures.

4.3 *Super Pixel Segmentation*

Segmentation of the images is known as the partitioning of the digital processed image into several segments. The important main purpose of dividing the image into the segment is to finally transform an image representation into something that facilitates a decent study. Segmentation is mainly used primarily for classifying objects (in this case, leaves, fruits, and vegetables), and margins are the recorded image. That will always result in a collection of segments covering the entire image. It has a broad variety of uses, such as the medical image analysis, recognition tasks, object detection, and finally, traffic signals, in the image processing.

4.4 Feature Extraction

The extraction of features is commonly used in the fields of image pre-processing, machine fundamental learning, and for recognition of the pattern. This will begin with extracting the best-extracted features from a raw unknown data, for the purpose of computation. Some of the data analytics packages include dimension reduction and extraction of functions. In the case of the crops, characteristics are extracted and can be the colour of the taken leaf, and shapes and the thickness of the given fruit. The most versatile methods of extraction of features are deformable and parameterised shapes, active contours.

4.5 Convolutional Neural Networks Classification

Image classification takes the form of an image input array of 225 * 225 * 3 (each image may have different sizes, but here, the convolution model is designed for 225 * 225-pixel image with RGB code) and a class output (plant disease or nutritional deficiency) or a class likelihood that best describes the picture. After the image is obtained, various processes can be implemented to make the specific operation depending on the given requirements. The calculation is as well as carried out in agricultural data to measure robustness, duration, and the measurement of the leaf or the crops to study. If the obtained images to the network classifier are not the best appropriate to process, then the image enhancement supports this step.

4.6 Dataset Model Collection

The PlantVillage dataset contains an additional 50 thousand image array of diseased crop leaf, and some 1800 images of nutritional imbalance leaves are obtained from other sources. This paper explains how to process using Keras' sequential model to shape the neural network. Sequential model is used to build simple models with a shone linear stack of layers below, which is a simple temple to generate dataset examples.

4.7 Output

Finally, we get the disease name and nutrient imbalance of that particular plant.

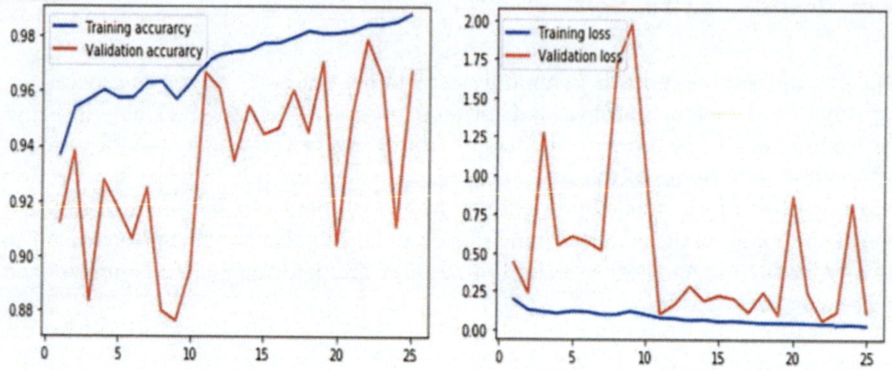

Fig. 3 Training and validation graph output

5 Results

In order to visualize the result, an algorithm in Python was developed to depict the default input image and convert it to (256, 256, 3) array colour space, and enhancements are made. Then, the convolution neural network is used to classify these characteristics, and the classification is shown in the label.

First, NPK deficiency is analysed, and then, the plant disease is identified. Output is then presented to the end user (Fig. 3).

6 Conclusions and Future Works

The paper takes inference from a lot of other journals that were reviewed, which gives an insight on automatic crop nutrient abnormality recognition, disease detection, categorization, and crop evaluation. Given that smart agriculture is the future, further research works can be developed in this direction. These works should not only identify the symptoms, but they should also diagnose diseases at initial stages, henceforth suggesting preventive measures. Along with advancements, integrating big data analytics in future can yield accurate results. This survey has helped to define different techniques of image processing which could be carried out for different purposes in further research in the field of agriculture.

References

1. D.L. Armstrong, Nutrient deficiency symptoms in rice. Better Crops Int. **16**(Special Supplement: 23–25) (2002)
2. J. Jacobi, Identification of fungal infection and nitrogen deficiency in wheat crop using remote sensing, in *2006 IEEE International Symposium on Geoscience and Remote Sensing* (2006)
3. S. Abel, C.A. Ticconi, C.A. Delatorre, Phosphate sensing in higher plants, in*2008 IEEE International Journal* (2008)
4. M. Masrie, M.S. Rosman, R. Sam, Z. Janin, Detection of nitrogen, phosphorus, and potassium (NPK) nutrients of soil using optical transducer, in *IEEE 4th International Conference on Smart Instrumentation, Measurement and Application (ICSIMA)* (2017)
5. H. Durmus, E.O. Gunes, M. Kirci, Disease detection on the leaves of the tomato plants by using deep learning, in *2017 6th International Conference on Agro-Geoinformatics* (2017)
6. J.D. Pujari, R. Yakkundimath, A.S. Byadgi, Image processing based detection of fungal diseases in plants. Procedia Comput. Sci. **46**(Icict 2014), 1802–1808 (2015)
7. M. Sharif, M.A. Khan, Z. Iqbal, M.F. Azam, M.I.U. Lali, M.Y. Javed, Detection and classification of citrus diseases in agriculture based on optimized weighted segmentation and feature selection. Comput. Electron. Agric. **150**, 220–234 (2018)
8. M. Kerkech, A. Hafiane, R. Canals, Deep learning approach with colorimetric spaces and vegetation indices for vine diseases detection in UAV images. Comput. Electron. Agric. **155**, 237–243 (2018)
9. https://www.gardeningknowhow.com/edible/vegetables

A Systematic Review on Approaches to Recognize Emotions Using Electroencephalography (EEG) Signals

Megha D. Bengalur and Aditya K. Saxena

Abstract Emotions play a very significant role in the daily life of the individuals, in decision-making processes and in the understanding of the surrounding world. The emotional state of the human has become a priority due to the growing interest of the research community in determining the emotional interactions between humans and computers. There are multiple measures to achieve the same, and one such promising tool is electroencephalography (EEG). EEG is an electrophysiological monitoring method to record electrical activity of the brain which is noninvasive, with the electrodes placed on the scalp to measure the real-time changes in voltage caused by brain's electrical activity, providing good temporal resolution. This paper presents a neurophysiological survey of research performed on emotions (basically valence, arousal and dominance) using EEG signals. The main focus is to compare the conventional methods carried out on the aspects of recognition process like the number and type of the subjects preferred, the features extracted and the classifiers chosen. The review summarizes a set of current practices and performance outcomes providing practical suggestions that research scholars can follow to achieve validated and high-quality results.

Keywords Electroencephalogram (EEG) · Emotion recognition · Valance · Arousal · Dominance

1 Introduction

Emotions are basically biological states related to the nervous system play an important role in the everyday life of humans in decision-making, human-to-human/computer interaction, perception and intelligence. These are associated with

M. D. Bengalur (✉) · A. K. Saxena
Department of Computer Science & Engineering, Presidency University, Bangalore, India
e-mail: megha@presidencyuniversity.in

A. K. Saxena
e-mail: adityasaxena@presidencyuniversity.in

V. Bhateja et al. (eds.), *Data Engineering and Intelligent Computing*,
Advances in Intelligent Systems and Computing 1407,
https://doi.org/10.1007/978-981-16-0171-2_11

thoughts, feelings, behavioral responses and a degree of pleasure or displeasure. The relationship between cognition and emotion has absorbed the researchers to work on an interesting field called human–computer interaction. However, in the past two decades, a substantial amount of work has been online on the interdependence of the two.

Physiological signals assist to get better understanding of the participants' responses during the time of observation corresponding to multichannel recordings from the nervous systems. Signals commonly used in measuring emotions involve galvanic skin response (GSR), which rises linearly with a participant's level of arousal. Electromyography (EMG) which is frequency of muscle tension correlates to negative valence emotions. Heart rate (HR), giving increased signals with negatively valenced emotions like fear, lastly respiration rate (RR) which measures how deep and fast the breath is, becomes irregular with highly aroused emotions like anger. Measurements recorded from brain also enable the observation of the emotions felt [1].

Neuroimaging techniques like functional magnetic resonance imaging (fMRI), electroencephalography (EEG) and positron emission tomography (PET) are also used in the study of emotions. EEG requiring some training directly measures the brain activities compared to fMRI and PET. Though EEG has poor spatial resolution requiring several electrodes to be placed at different locations on the head, it provides good temporal resolution, to study the phase changes responding to emotions. Further, EEGs are noninvasive, inexpensive and fast because of which they become the researcher's choice to study the brain's responses to emotional stimuli [2]. Nowadays, varieties of EEG sets based on wearability, portability, ease-of-use and cost are available in the market, hence making it accessible for EEG-based emotion recognition in different areas like health care, E-Commerce, human–computer interactions, etc. [3, 4].

2 Review Methodology

The survey started by the search on Web sites apt for research like Google Scholar, Mendeley and IEEE Xplore as "*Emotion recognition from EEG signals using video/audio stimuli*". The resulting set of papers obtained was filtered based on the quality of papers considering their publications and the number of citations gained. The papers with greater citations number than the median of citations for that year were selected. As smaller citation number is not a worthy point to rate a paper not good, we also analyzed the novelty of work and paper content to maximum extent.

3 Emotion Recognition Using EEG

Emotions: A research on emotions shows that there are absolutely two categories of emotions called valence and arousal. Valence is a degree to which humans feel good or bad, and arousal, the degree indicating calmness or excitement. On this basis, the emotions like anger and fear are termed as both negative (low valence), highly excited (high arousal) states.

To recognize emotions using EEG signals, we perform the step by step procedure as displayed in Fig. 1. The subject/participant is exposed to a stimulus like video/ an audio clips with deep emotions. This induces an electrical activity in the form of signals in the subject's brain due to voltage changes. The signals containing multiple artifacts are removed. From the filtered signals, features are extracted and classified using different machine learning algorithms and signal processing methods [5].

3.1 Stimulus

An evoked potential parallels to an electrical signal recorded after the stimulus is presented. There exists basically two types: auditory evoked potentials (AEP), visual evoked potentials (VEP) [6]. AEP are generated by an audio/visual stimulus accessed through earphones and VEP's by light flashes or pattern changes on the monitor screen.

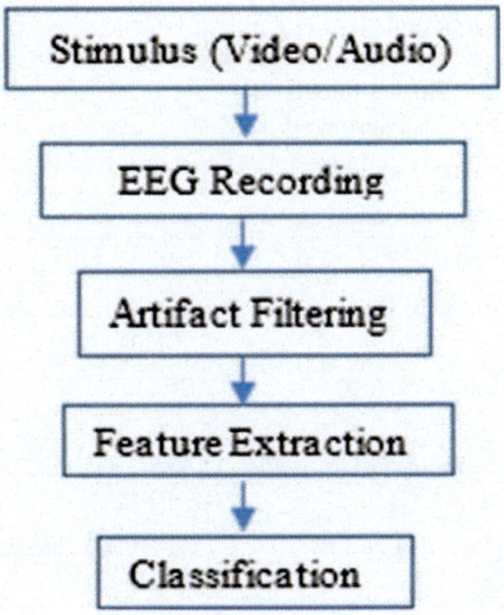

Fig. 1 Process of emotion recognition using EEG

3.1.1 Subject

Table 1 presents the analysis performed based on the type and durations of the stimulus used, the total number of subjects, the gender, and the emotions to be recognized. It is seen that majority of the papers have not used statistically maximum number

Table 1 State-of-art Stimulus

References	Audio/video stimulus	Male–female subjects	Emotions
Koelstra et al. [9]	Music Videos (2 min)	6 (–/–)	Valence, arousal, and like/dislike
Liu and Sourina [10]	Video (30 s)	4 (–/–)	Valence, arousal
Soleymani et al. [11]	IADS (30 s)	12 (3/9)	Valence, arousal, dominance
Soleymani et al. [12]	Video (1 to 2 min)	30 (17/13)	Valence, arousal
Koelstra and Patras [13]	MAHNOB HCI (35 to 117 s)	12 (3/9)	Valence, arousal
Liu et al. [14]	MAHNOB HCI (35 to 117 s)	24 (–/–)	Valence, arousal, control
Liu and Sourina [15]	IAPS (7 s)	7 (–/–)	Valence, arousal
Yoon and Chung [16], Hatamikia et al. [17], Jie et al. [18], Atkinson and Campos [19], Kumar et al. [20], Mohammadi et al. [21], Yin et al. [22], Yin et al. [23], Mohammadi et al. [24], Alhagry et al. [25]	DEAP (60 s)	32 (16/16)	Valence, arousal, dominance, liking, familiarity
Jiang et al. [26]	CAPS (–)	10 (–/–)	Valence
Thammasan et al. [27]	Music (73 to 147 s)	15 (0/15)	Valence, Arousal
Yadava et al. [28]	personal dataset	40(25/15)	Likes, dislikes
Liu et al.[29]	personal dataset (60 s video)	30(–/–)	Valence, Arousal
Lan et al. [30]	SEED	15	Positive neutral negative
Santamaria-Granados et al. [31]	AMIGOS dataset, DEEP	– , 60 s video	Valence, Arousal,
Correa et al. [32]	AMIGOS dataset, DEEP	– , 60 s video	Valence, arousal, personality traits, mood and social context
Nakisa et al. [33]	DEAP, MANHOB AND PERSONAL	–	–
Mert and Akan [34]	USING BIOPIC disposable electrodes	26 (13/13)	Happy, unhappy

of participants to provide a better level of validity with 50% of the papers consider less than 15 subjects each, and only about 30% using at least 30 subjects. Gender of the participants also plays an important role in distinguishing emotions when considered equally. It is also observed that some papers have not given priority on gender while many consider unequal number of male/female participants leading to reduced accuracy on online tests. From Table 1, the average duration of the stimulus is observed from 7 seconds, varying between 30 to 60 seconds involving the dataset like (DEAP) [7], the Mahnob HCI dataset [8] and seconds to 2 minutes in case of SEED dataset.

From Table 1, the average duration of the stimulus is observed from 7 s, varying between 30 and 60 involving the dataset like (DEAP) [9], the Mahnob HCI dataset [10] and seconds to 2 min in case of SEED dataset.

3.2 EEG Recordings

Recordings of the EEG signals vary with the types of the equipment used, no. of electrodes present in the equipment and the placement positions of the electrodes over the subjects scalp. The listed features play an important role in determining accuracy of the emotions captured. But, ideally, the number of electrodes should be reduced. Table 2 highlights the different types of clinical devices used to record the brain's electrical activities. Majority of the research are found to using Biosemi Active Two [35] with varied sampling rates, while others refer Emotiv [32] wireless headset and EEG module from Neuroscan, Inc [36]. Among these devices, the most portable found easy to use is the Emotiv wireless headset. There are many works that have not provided the information about the devices used.

3.3 Artifact Filtering

Artifacts in the EEG signals occur due to many reasons like eye blinks, head movement of the subject, muscular contractions, disturbance of the equipment connections, so on. Some of the works manually removed certain data due to artifacts associated with recording. Methods such as blind source separation (BSS) and independent component analysis (ICA) were used in the removal of eye movements, blinks, muscle, heart and line noise. We observe that around 50% of the works use band pass filters. Some works re-referenced the electrodes using filters such as the common average reference (CAR) (20)%, Laplacian or average mean reference (AMR) (15%). Also, around 15% of the works focused on independent component analysis (ICA). It is found that bandpass filters provide a better result. Table 3 shows the different filtering methods used.

Table 2 EEG machines for recordings and the features considered

Reference	Equipment with frequency	Electrode location
Soleymani et al. [11]	Biosemi Active Two (1024 Hz)	IS(32)
Liu and Sourina [10]	Emotiv (2048 Hz)	AF3, F7, F3, FC5, T7, P7 1, O2, P8, T8, FC6, F4, F8, F4(14)
Huang et al. [37]	32Ch EEG Device	IS(32)
Pantic et al.[12]	Biosemi Active Two (256 Hz)	
Koelstra et al. [9], Liu and Sourina [15]	Biosemi Active Two (512 Hz)	IS(32)
Liu et al. [14], Jiang et al. [26]	EEG NeuroScan (500 Hz)	N/A(64)
Thammasan et al. [27]	Waveguard EEG (250 Hz)	FP1, FP2, F3, F4, F7, F8, Fz, C3, C4, T3, T4, Pz (12)
Koelstra and Patras [13]	Biosemi Active Two (256 Hz)	CP6, Cz, FC2, Oz, CP1, T7, C4, FC6, PO4, Cz, CP6, CP2, T8, F8 (14)
Yoon and Chung [16]	Biosemi Active Two (512 Hz)	FP1, AF3, F3, F7, FC5, FC1, C3, T7, CP5, CP1, P3, P7, PO3, O1, Oz, Pz, FP2, AF4, Fz, F4, F8, FC6, FC2, Cz, C4, T8, CP6, CP2, P4, P8, PO4, O2 (32) + 61 virtual channels
Hatamikia et al. [17]	Biosemi Active Two (512 Hz)	FP1, AF3, F3, F7, FC5, FC1, C3, T7, CP5, CP1, P3, P7, PO3, O1, Oz, Pz, FP2, AF4, Fz, F4, F8, FC6, FC2, Cz, C4, T8, CP6, CP2, P4, P8, PO4, O2 (32)
Jie et al. [18]	Biosemi Active Two (512 Hz)	IS(40)
Atkinson and Campos [19]	Biosemi Active Two (512 Hz)	N/A(14)
Kumar et al. [20]	Biosemi Active Two (512 Hz)	FP1, FP2 (2)
Mohammadi et al. [21]	Biosemi Active Two (512 Hz)	F3-F4, F7-F8, FC1-FC2, FC5-FC6, FP1-FP2 (5)
Thammasan et al. [27]	Waveguard EEG (250 Hz)	FP1, FP2, F3, F4, F7, F8, Fz, C3, C4, T3, T4, Pz (12)

(continued)

Table 2 (continued)

Reference	Equipment with frequency	Electrode location
Yin et al. [22]	Biosemi Active Two	Fp1, AF3, F3, F7, FC5, FC1, C3, T7, CP5, CP1, P3, P7, PO3, O1, Oz, Pz, Fp2, AF4, Fz, F4, F8, FC6, FC2, Cz, C4, T8, CP6, CP2, P4, P8, PO4, and O2
Yadava et al. [28]	Biosemi Active Two	AF3, AF4, F3, F4, F7, F8, P7, P8, O1, O2, T7, T8
Yin et al. [23]	Biosemi Active Two	Fp1, AF3, F3, F7, FC5, FC1, C3, T7, CP5, CP1, P3, P7, PO3, O1, Oz, Pz, Fp2, AF4, Fz, F4, F8, FC6, FC2, Cz, C4, T8, CP6, CP2, P4, P8, PO4, and O2
Liu et al. [29]	Biosemi Active Two	AF3-AF4, F3-F4, F7-F8, FC5-FC6, T7-T8, P7-P8, and O1-O2
Mohammadi et al. [24]	Biosemi Active Two	F3– F4, F7–F8, FC1–FC2, FC5–FC6 and FP1–FP2
Alhagry et al. [25], Tripathi et al. [38], Kwon et al. [39], Moon et al. [35], Yang et al. [36]	Biosemi Active Two	N/A(32)
Zhuang et al. [40]	Biosemi Active Two	FP1, FP2, F7, F8, T7, T8, P7, and P8
Lan et al. [30]	Biosemi Active Two	AF3, AF4, C3, C4, P1, CP2, CP5, CP6,Cz, F3, F4, F7, F8, FC1, FC2, FC5, FC6, Fp1, Fp2, Fz, O1, O2, Oz, P3, P4,P7, P8, PO3, PO4, Pz, T7, and T8
	For SEED Dataset	AF3, AF4, C1, C2, C3, C4, C5, C6, CB1,CB2, CP1, CP2, CP3, CP4, CP5, CP6, CPZ, CZ, F1, F2, F3, F4, F5, F6, F7,F8, FC1, FC2, FC3, FC4, FC5, FC6, FCZ, FP1, FP2, FPZ, FT7, FT8, FZ,O1, O2, OZ, P1, P2, P3, P4, P5, P6, P7, P8, PO3, PO4, PO5, PO6, PO7, PO8, POZ, PZ, T7, T8, TP7, and TP8

3.4 Feature Extraction

3.4.1 EEG Features

Most of the authors do not disclose the types of EEG features used. From our survey, we find that around (89.4%) used mainly the alpha, beta, delta, theta and gamma bands (13.7%), prefer to use only alpha, beta, theta and gamma (13.7%), alpha and

Table 3 Artifact filtering used across the research works

Reference	Artifacts filtering
Koelstra et al. [9]	CAR applied to EEG data; high pass filter with a cutoff of 0.5 Hz; BSS to filter EOG and EMG; band pass filter with a bandwidth of 0.5–35 Hz; downsampling to 256 Hz
Liu and Sourina [10]	CAR applied to EEG data; band pass frequency filter with a bandwidth of 4–45 Hz; BSS; downsampling to 512 Hz (DEAP dataset) Band pass filter with a bandwidth of 0.2–45 Hz; notch filters at 50 Hz and 60 Hz; downsampling to 128 Hz (own dataset)
Soleymani et al. [11]	Downsampling to 256 Hz; band pass frequency filter with a bandwidth of 4–45 Hz; CAR applied to EEG data
Soleymani et al. [12]	CAR applied to EEG data
Liu et al. [14]	N/A
Liu and Sourina [15], Yoon and Chung [16], Hatamikia et al. [17], Jie et al. [18], Kumar et al. [20]	Band pass frequency filter with a bandwidth of 4-45 Hz; BSS; downsampling to 128 Hz
Atkinson and Campos [19]	Band pass frequency filter with a bandwidth of 4–45 Hz; downsampling to 128 Hz; automatic removal of ocular artifacts
Jiang et al. [26]	Band pass frequency filter with a bandwidth of 0.1–60 Hz; notch filter at 50 Hz; downsampling to 250 Hz; filter data with wavelet of order 6.8; manually artifact rejection
Mohammadi et al. [21]	AMR and normalization
Thammasan et al. [27]	Band pass frequency filter with a bandwidth of 0.5–60 Hz; notch filter at 60 Hz; ICA to remove EOG Artifact Filtering: Average Mean Reference (AMR), Blind Source Separation (BSS), Common Average Reference (CAR), Electromyography
Yin et al. [22]	Band pass Butterworth filter, Independent Component Analysis
Yadava et al. [28]	
Yin et al. [23]	Band pass filter, Independent Component Analysis, Stacked Autoencoder
Liu et al. [29]	Band pass filter, Independent Component Analysis
Mohammadi et al. [24]	Average Mean Reference
Alhagry et al. [25]	High pass Filter
Kwon et al. [39]	Band pass Filter, wavelet transform

(continued)

Table 3 (continued)

Reference	Artifacts filtering
Li et al. [41]	Band pass filter between 0.3 and 70 Hz

beta, alpha, beta and gamma (7.8%), delta, theta, alpha and beta (3.92%), alpha, beta, gamma (3.92%), among other combinations.

3.4.2 Methods

The feature extraction process can be handled using various methods (for further information please see [42, 7]). Table 4 highlights the various feature extraction methods that are employed to extract the features from the acquired EEG signals. The review shows there were around 30 different methods used. The most used feature extraction methods were the Fourier transform (FT) such as the short-time Fourier transform (STFT) or discrete Fourier transform (DFT), power spectral density (PSD), statistical, wavelet transform (WT), entropy such as the sample entropy (SE) or wavelet entropy (WE), differential entropy (DE), approximate entropy (AE), higher order crossings (HOC), common spatial patterns (CSP), fractal dimensions and asymmetry index (AI).

3.5 Classification

The analysis of Table 4 showcases that about 50% of the work uses support vector machine SVM classifier alone or SVM in combination with other kernels such as Radial Basis Function (RBF) (18%). Other forms of SVM such as multi-class support vector machine (ML-SVM), adaptive SVM, or least squares support vector machine (LS-SVM), were used in 8% of these works. Most of the works using SVM do not specify the kernel used. The k-nearest neighbors (kNN) was chosen by almost 12.5% of the work, again many of the works do not specify value of k, but by observation, most of the works choose $k = 2$ to 8. Linear discriminant analysis (LDA) is selected by 3.125%. Lastly, the Naive Bayes (NB) is used by 6.35%.

4 Conclusion

In this paper, the analysis of emotion recognition methods using EEG signal is presented novel approaches. To achieve statistically good and meaningful results, at least 30 subjects are recommendable and having a balance between the genders is a good practice. The number of electrodes used (and the equipment) has a leading role

Table 4 Feature extraction, classification and result analysis

Ref	Feature extraction	Classification	Accuracy
Koelstra et al. [9]	PSD, and CSP	SVM Linear	58.8% valence; 55.7% arousal; 49.4% like/dislike
Huang et al.[37]	ASP, CSP, and FBCSP	NB	>80.00% arousal; >65.00% valence
Liu and Sourina [10]	HFD, and Statistical	SVM	DEAP dataset as benchmark: 76.51% arousal/dominance; 50.80% valence
Soleymani et al. [11]	PSD, and AI	SVM RBF	62.10% arousal; 50.50% valence
Soleymani et al. [12]	SPA	SVM RBF	52.4% arousal; 57.0% valence
Koelstra and Patras [13]	FBCSP	LR	71.30%
Liu et al. [14]	DFT	Adaptive SVM	73.57% arousal; 73.42% valence
Liu and Sourina [15]	Statistical and HFD	SVM Polynomial	Using only 11 subjects: 63.04% arousal/dominance; 51.49% for valence
Yoon and Chung [16]	FFT	NB	2 classes: 70.9% valence, 70.1% arousal; 3 classes: 55.4% valence, 55.2% arousal
Hatamikia et al. [17]	AR with Burg method	kNN	2 classes: 72.33% valence, 74.20% arousal; 3 classes: 61.10% valence, 65.16% arousal
Jie et al. [18]	SE	SVM	User-Dep: 80.43% LVHA-HVHA, 71.17% LVLA-LVHA; User-Ind: 79.11% LVHA-HVHA, 64.47% LVLA-LVHA
Atkinson and Campos [19]	MRMRM and Statistical	SVM RBF	2 classes: 73.06% arousal; 73.14% valence; 3 classes: 60.7% arousal; 62.33% valence

(continued)

Table 4 (continued)

Ref	Feature extraction	Classification	Accuracy
Jiang et al. [26]	SIM algorithm	SVM	75% two levels of extremely negative; 79.55% moderately negative and neutral
Kumar et al. [20]	HOS	LS-SVM RBF	60.9% low arousal, 68.8% high arousal, 59.4% low valence, 62.5% high valence
Mohammadi et al. [21]	DWT	kNN (k = 3)	84.05% arousal, 86.75% valence
Thammasan et al. [27]	HFD, PSD, DWT	DBN	88.24% valence, 82.59% arousal
Ali et al. [4]	DWT, WE, and Statistical	SVM RBF	Valance: 84%, Arousal: 85%
Yin et al. [22]	T-RFE, FFT	LLSSVM	Valence-78%, Arousal-75% Dominance-78% Liking-80%
Yadava et al. [28]	DFT	HMM	70%
Yin et al. [23]		MSMAE	
Liu et al. [29]	STFT	SVM with RBF kernal	Valence:92.26% Arousal: 86.63%
Mohammadi et al. [24]	DWT	SVM, KNN	Valence:84.05% Arousal: 86.75%
Alhagry et al. [25]	LSTM & DROPOUT LAYERS		VALENCE: 85% AROUSAL: 85% LIKING: 87%
Tripathi et al. [38]	–	DNN, CNN	–
Zhuang et al. [40]	Empirical Mode Decomposition (EMD)	SVM	VALANCE: 69.10 AROUSAL: 71.05 GOOD COMPARED TO OTHERS
Lan et al. [30]	Differential Entropy (DE)	–	7–13%> [13]–[16]
Kwon et al. [39]	CNN		73.40%
Yang et al. [36]	PSD and DE	CNN, RNN	Valence: 90.80 Arousal: 91.3
Li et al. [41]	PSD and DE	Integrated MS STM, semi- MS-STM,	Multi-source STM-88.92%, semi-Multi-source STM-91.31%
Song et al. [43]	Novel Attention-long short-term memory (A-LSTM)	SVM, KNN	

(continued)

Table 4 (continued)

Ref	Feature extraction	Classification	Accuracy
Nakisa et al. [33]	EC ALGORITHM	PNN	DEAP: 67.474, 3.389% MAHNOB: 96.97, 1.893% PERSONAL DATASET: 65.043028, 3.195%

determining extent of features to process. Available datasets like DEAP with Active Two (512 Hz) equipment are found to be used by many of the works. Choice and accuracy of feature extraction method differ and are found to be dependent of the EEG features also. Analysis shows short-time Fourier transform (STFT) or discrete Fourier transform (DFT), power spectral density (PSD) provide good features. We find a family of classifiers, and majority of the works used more than one classifier. Among these, the most commonly used are Bayesian, decision trees, support vector machines, etc. It becomes complex to do the comparisons between them. Hence, conclusion can be drawn based on the combinations of EEG features considered, feature extraction method and classification algorithm chosen.

References

1. H. Gunes, B. Schuller, M. Pantic, R. Cowie, Emotion representation, analysis and synthesis in continuous space: a survey, in *IEEE FG*, pp. 827–834 (2011)
2. C.P. Niemic, Studies of emotion: a theoretical and empirical review of psychophysiological studies of emotion. J. Undergraduate Res., 15–18 (2002)
3. C. Hondrou, G. Caridakis, Affective, natural interaction using EEG: sensors, application and future directions, in *SETN*, pp. 331–338 (2012)
4. M. Ali, A. H. Mosa, F. Al Machot, K. Kyamakya, EEG-based emotion recognition approach for e-healthcare applications, in *ICUFN*, pp. 946–950 (2016)
5. D.O. Bos, EEG-based emotion recognition: the influence of visual and auditory stimuli. Capita Selecta Paper (2006)
6. P. Walsh, N. Kane, S. Butler, The clinical role of evoked potentials. J. Neurol. Neurosurgery Psychiatry **76**(suppl 2), ii16–ii22 (2005)
7. R. Jenke, A. Peer, M. Buss, Feature extraction and selection for emotion recognition from EEG. IEEE Trans. Affect. Comput. **5**(3), 327–339 (2014)
8. J. Li, et al., Multisource transfer learning for cross-subject EEG emotion recognition. IEEE Trans. Cybern. (2019)
9. S. Koelstra, C. Muhl, M. Soleymani, J.-S. Lee, A. Yazdani, T. Ebrahimi, T. Pun, A. Nijholt, I. Patras, DEAP: a database for emotion analysis using physiological signals. IEEE Trans. Affect. Comput. **3**(1), 18–31 (2012)
10. M. Soleymani, J. Lichtenauer, T. Pun, M. Pantic, A Multimodal database for affect recognition and implicit tagging. IEEE Trans. Affect. Comput. **3**(1), 42–55 (2012)
11. S. Koelstra, A. Yazdani, M. Soleymani, C.M. uhl, J.-S. Lee, A. Nijholt, T. Pun, T. Ebrahimi, I. Patras, Single trial classification of EEG and peripheral physiological signals for recognition of emotions induced by music videos, in *BI*, pp. 89–100 (2010)

12. Y. Liu, O. Sourina, EEG-based valence level recognition for real-time applications, in *Cyberworlds*, pp. 53–60 (2012)
13. M. Soleymani, M. Pantic, T. Pun, Multimodal emotion recognition in response to videos. IEEE Trans. Affect. Comput. **3**(2), 211–223 (2012)
14. S. Koelstra, I. Patras, Fusion of facial expressions and EEG for implicit affective tagging. Image Vis. Comput. **31**(2), 164–174 (2013)
15. Y.H. Liu, C.T. Wu, Y.H. Kao, Y.T. Chen, Single-trial EEGbased emotion recognition using kernel Eigen-emotion pattern and adaptive support vector machine, in *IEEE EMBS* (2013)
16. Y. Liu, O. Sourina, Real-time subject-dependent EEG-based emotion recognition algorithm. Trans. Comput. Sci. **XXIII**, 199–223 (2013)
17. H.J. Yoon, S.Y. Chung, EEG-based emotion estimation using Bayesian weighted-log-posterior function and perceptron convergence algorithm. Comput. Biol. Med. **43**(12), 2230–2237 (2013)
18. S. Hatamikia, K. Maghooli, A.M. Nasrabadi, The emotion recognition system based on autoregressive model and sequential forward feature selection of electroencephalogram signals. J. Med. Signals Sens. **4**(3) (2014)
19. X. Jie, R. Cao, L. Li, Emotion recognition based on the sample entropy of EEG. Bio-Med. Mater. Eng. **24**(1), 1185–1192 (2014)
20. N. Kumar, K. Khaund, S.M. Hazarika, Bispectral analysis of EEG for emotion recognition. Procedia Comput. Sci. **84**, 31–35 (2016)
21. Z. Mohammadi, J. Frounchi, M. Amiri, Wavelet-based emotion recognition system using EEG signal. Neur. Comput. Appl., 1–6 (2016)
22. Z. Mohammadi, J. Frounchi, M. Amiri, Wavelet-based emotion recognition system using EEG signal. Neural Comput. Appl. **28**(8), 1985–1990 (2016). https://doi.org/10.1007/s00521-015-2149-8
23. J. Jiang, Y. Zeng, L. Tong, C. Zhang, B. Yan, Single-trial ERP detecting for emotion recognition, in *SNPD*, pp. 105–108 (2016)
24. M. Yadava, P. Kumar, R. Saini, P.P. Roy, D. Prosad Dogra, Analysis of EEG signals and its application to neuromarketing. Multimedia Tools Appl. **76**(18), 19087–19111 (2017). https://doi.org/10.1007/s11042-017-4580-6
25. Y.-J. Liu, M. Yu, G. Zhao, J. Song, Y. Ge, Y. Shi, Real-time movie-induced discrete emotion recognition from EEG signals. IEEE Trans. Affective Comput., 1–1 (2017).https://doi.org/10.1109/taffc.2017.2660485
26. J. Atkinson, D. Campos, Improving BCI-based emotion recognition by combining EEG feature selection and kernel classifiers. Expert Syst. Appl. **47**, 35–41 (2016)
27. Z. Yin, Y. Wang, L. Liu, W. Zhang, J. Zhang, Cross-Subject EEG feature selection for emotion recognition using transfer recursive feature elimination. Front. Neurorobot., 11(2017). https://doi.org/10.3389/fnbot.2017.00019
28. S. Alhagry, A.A. Fahmy, R.A. El-Khoribi, Emotion recognition based on EEG using LSTM recurrent neural network. Emotion **8**(10), 355–358 (2017)
29. N. Thammasan, K.I. Fukui, M. Numao, Application of deep belief networks in EEG-based dynamic music-emotion recognition, in *IJCNN*, pp. 881–888 (2016)
30. J.A.M. Correa, et al., Amigos: a dataset for affect, personality and mood research on individuals and groups. IEEE Trans. Affective Comput. (2018)
31. A. Mert, A. Akan, Emotion recognition from EEG signals by using multivariate empirical mode decomposition. Pattern Anal. Appl. **21**(1), 81–89 (2018)
32. https://emotiv.com
33. D. Huang, C. Guan, K. K. Ang, H. Zhang, Y. Pan, Asymmetric spatial pattern for EEG-based emotion detection, in *IJCNN*, pp. 1–7 (2012)
34. S. Tripathi, et al., Using deep and convolutional neural networks for accurate emotion classification on DEAP dataset, in *Twenty-Ninth IAAI Conference* (2017)
35. https://www.biosemi.com/products.htm
36. https://www.cephalon.eu/products/erp-ep/neuroscan-recording-systems/
37. Z. Yin, M. Zhao, Y. Wang, J. Yang, J. Zhang, Recognition of emotions using multimodal physiological signals and an ensemble deep learning model. Comput. Methods Programs Biomed. **140**, 93–110 (2017). https://doi.org/10.1016/j.cmpb.2016.12.005

38. Z. Lan, et al., Domain adaptation techniques for EEG-based emotion recognition: a comparative study on two public datasets. IEEE Trans. Cogn. Dev. Syst. **11**(1), 85–94 (2018)
39. B. Nakisa, et al., Evolutionary computation algorithms for feature selection of EEG-based emotion recognition using mobile sensors. Expert Syst. Appl. **93**, 143–155 (2018)
40. L. Santamaria-Granados, et al., Using deep convolutional neural network for emotion detection on a physiological signals dataset (AMIGOS). IEEE Access **7**, 57–67 (2018)
41. E.S. Dan-Glauser, K.R. Scherer, The Geneva affective picture database (GAPED): a new 730-picture database focusing on valence and normative significance. Behav. Res. Methods, 468–477 (2011)
42. M.-K. Kim, M. Kim, E. Oh, S.-P. Kim, A review on the computational methods for emotional state estimation from the human EEG. Comput. Math. Methods Med. (2013)
43. M.M. Bradley, P.J. Lang, International affective digitized sounds (IADS): stimuli, instruction manual and affective ratings, The Center for Research in Psychophysiology, University of Florida, Tech. Rep. (1999)

Pediatric Skeletal Age Assessment Using Deep Learning Proceedings

Shrinivas D. Desai and Prajwal T. Kadiyavar

Abstract Skeletal age assessment (SAA) is one of the techniques to verify the claimed age and actual age. It is also performed to evaluate whether skeletal age is normal or delayed compared to the patient's chronological age (CA). A delayed or advanced bone age can indicate growth disorders. It is generally performed by using either the Greulich & Pyle (G&P) method or the Tanner–Whitehouse (TW) method. However, inter- and intra-observer differences occur that could be resolved by developing an automatic system. Three years back (in 2017) Radiological Society of North America (RSNA) challenged the data science community to leverage computer vision to predict SA more efficiently and accurately than existing methods. The primary goal of this research work is to use convolutional neural networks to build a model to predict skeletal age with high accuracy given RSNA's provided dataset. The secondary goal is to leverage deep learning visualization techniques for better interpretation of our results. Overall, our proposed model achieved a competitive MAE of 7.61 months on the test set provided by Radiological Society of North America (RSNA).

Keywords Skeletal Age · Chronological Age · InceptionV3 · RSNA · MAE · Digital X-Ray · Left hand

1 Introduction

In recent years, artificial intelligence (AI) has been applied more and more within the medical field which has led to some novel insights. Noticeable examples of these are, e.g., IBM Watson identifying cancer and heart diseases, tele-robotic surgery performed with the da Vinci surgical system and mental health chat-bots such as Woebot. AI has tremendous applications in the area of face recognition [1], fingerprint recognition [2], head pose estimation [3] and medical applications such as diagnosing

S. D. Desai (✉) · P. T. Kadiyavar
School of Computer Science & Engineering, KLE Technological University, Hubballi 580024, Karnataka, India
e-mail: sd_desai@kletech.ac.in

V. Bhateja et al. (eds.), *Data Engineering and Intelligent Computing*,
Advances in Intelligent Systems and Computing 1407,
https://doi.org/10.1007/978-981-16-0171-2_12

psychological disorders [4] and also other domain applications such as sign language detection [5]. In this section, an overview to research work, motivation, literature survey, problem statement and objectives are presented. In fall 2017, the RSNA challenged the AI community to leverage computer vision and CNNs to predict SA. They posed their challenge and accompanying hand radiograph dataset via Kaggle, an online platform for predictive modeling and analytics competitions. RSNA aims to yield an accurate model from the competition that leverages technological innovation to promote excellence in patient care.

This paper focuses on applying AI in the radiology department with the goal being the assessment of skeletal bone age from digital X-ray images. SAA is performed to evaluate advancement of bone and whether SA is delayed compared to the patients CA [6]. A delayed or advanced skeletal age can indicate pediatric disorders. In that case, serial measurements can help to assess if a treatment is succeeding.

In this research work, we propose deep convolutional neural network to data provided by RSNA to predict skeletal age with high accuracy. Our final model architecture is influenced by the published models of successful Kaggle competitors in conjunction with additional tuning and alteration by us. We employed visualization techniques to gain a deeper understanding into how convolutional neural networks tackle the skeletal age prediction challenge.

Our goal is to accurately and efficiently predict SA given the hand radiograph dataset. Our model takes in as input an arbitrary-sized black-and-white hand radiograph, paired with an indicator of male–female gender. We then use a convolutional neural network, comprised of an Inception V3 network reading the hand radiograph and a fully connected network reading the gender, to output a predicted numerical skeletal age of the radiograph in months. We selected mean absolute error (MAE), calculated as the mean of the absolute values of the difference between the model estimates and those of the reference standard bone age, as our loss criterion, as instructed by RSNA's challenge. Our baseline goal is to achieve MAE of 0.65 years (7.2 months), outperforming Bone-expert, the only automated SA assessment tool on the market at present.

1.1 Motivation

The skeletal age analysis (BAA) can complement diagnosis of rate of growth of a child's skeletal being maturing, and this can help doctors to come out with reports commenting on eligibility in sport context or permission during legal and law suits. SA is effective in the treatment of children with various endocrine growth disorders because a noteworthy incongruity between the skeletal age (SA) and the actual age of a child indicates deviation in skeletal development. In developing countries, 65% of birth records are not registered up to the age of five years. Manual methods of SA are laborious and prone to expert's variability. This is a motivation for developing automated methods of SA.

1.2 Objectives

The aim of skeletal age assessment is to evaluate the age of a child using radiographic images so that doctors can diagnose pediatric disorders. Therefore, a higher accuracy of bone age is very important which can be achieved by analyzing specific regions of a child's left hand x-ray. Our objective is to develop automated pediatric skeletal age analysis model using left hand X-ray images, compare the performance of proposed model with that of domain experts and that of existing automated models (RSNA) and to make the model robust.

2 Literature Survey

Following literatures which are recently reported for bone age assessment are considered for review purpose. Literature is reviewed from the perspective of proposed method, dataset used, programming platform, results achieved, etc., and special remarks are presented for each reported methods.

Pan et al. [7] proposed U-net architecture for SAA, by using RSNA dataset with deep active learning (AL) to select unlabeled data samples with sufficient information intentionally. Result recorded is MAE of 8.08 for all mixed radiograph test. To summarize, author has employed six off-the-shell deep convolutional neural networks (CNNs) with pre-trained weights on ImageNet. Iglovikov et al. [8] proposed VGG architecture for SAA, by using Stanford bone age dataset on Keras, TensorFlow to compare two types of CNNs: regression and classification. Result recorded is MAE of 8.48 for all mixed radiograph test. To summarize, carpal bones or metacarpals and proximal phalanges posed 15–20% increase in error as compared to the whole hand assessment. Larson et al. [9] proposed deep learning-based method for SAA, by using X-ray data from local hospital with 14,036 clinical hand radiographs. Programming was performed using Python OpenCV and TensorFlow. Computing for 6–8 h to train on a single K80 graphics processing unit (Nvidia, Santa Clara, Calif) is carried out. A result shows that Mean RMS and MAD of 0.63 and 0.50 years have been recorded. Spampinato et al. [10] proposed deep learning-based method using pre-trained CNN (OverFeat, GoogLeNet and OxfordNet) for SAA. 1391 X-ray left hand scans (age up to 18 years old) were employed for experiment. Results recorded are mean absolute error (years):OxfordNet = 0.83 GoogLeNet = 0.82 and BoNet = 0.79. The author concluded that SAA is possible even with limited data with acceptable MAE score. Lee et al. [11] proposed fully automated deep learning system for SAA, and dataset was collected during 2005 to 2015 pertaining to left hand of X-ray of children's from local hospital. Accuracy of 98.56% within 2 years and 92.29% within 1 year for the female and male cohorts was achieved.

From the above survey, it is observed that there is need for design and development of deep CNN-based model which can predict bone age with MAE less than 0.65 years using dataset provided by RSNA challenge.

3 Proposed Methodology

Figure 1 presents the proposed model, which is Inception V3 network for skeletal age assessment. From the literature, it is observed that many researchers have tried with popular networks such as Inception V4, ResNet 152, DenseNet and InceptionV3_SE [12–16]. This is one of influential factor for exploring Inception V3 in this research work. Keras supports this network and is computationally feasible for experiment. Network is customized by extracting the layer after the final concatenation layer from the Inception V3 network, flattened it and concatenated it with gender information with binary information (0 for female or 1 for male) and fed it through a 32-neuron densely connected layer. The concatenated layer is fed to two additional 1000-neuron densely connected layers with 'ReLU' activation before the final single-output linear layer.

The motivation for this design stems from the relative contribution of each input (pixels and gender) into the final decision. At our concatenation layer, the pixels contribute 100,384 inputs, while the gender contributes 32. This ratio is selected, because it will avoid the bias in network too significantly due to gender input and provides the ability to impact the overall prediction. More learnable parameters to adjust during training are provided by two dense layers. This also allows reasoning out the relationship between the pixel and gender information.

3.1 Dataset Description

The network is trained and tested on a dataset made available by Kaggle RSNA bone age analysis challenge 2017. The total dataset contains 12,611 training DR images and 200 testing DR images of patients between the ages of 1–19. The images are in.png format that contains image data as well as.csv file which contains information about the patient that is image id, gender and actual age.

The dataset contains few images with improper orientation, limited ROI (metacarpal and phalanx of the thumb) and has around 5% right hand DR images.

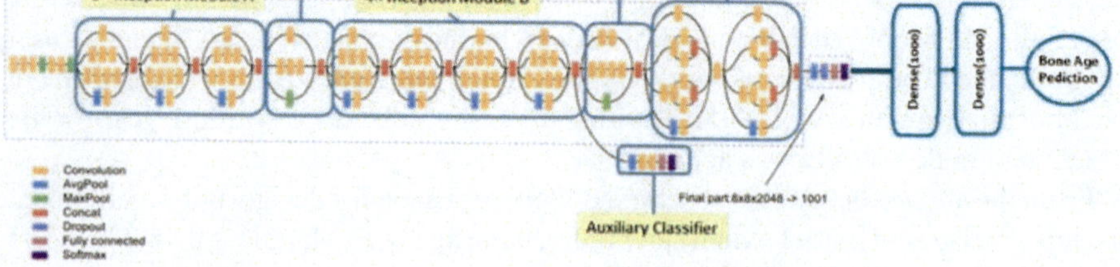

Fig. 1 Proposed model

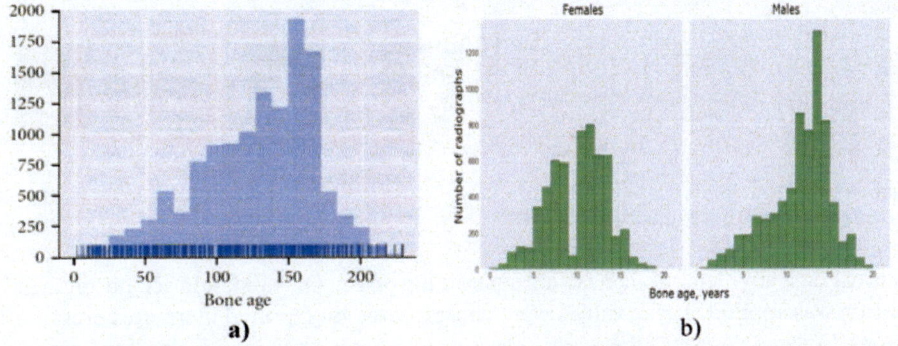

Fig. 2 **a** Dataset distribution as per age in moths. **b** Gender wise distribution

Table 1 Splitting of dataset

Total dataset (X)	Training 0.75 * X	Validation 0.75 *0.25 *X	Testing 0.25 * 0.25 * X
12,611	9458	2364	789

The pixel data of the images is varying in size, usually around 1514 × 2044. The network requires images of the same size, and since hardware puts restrictions on the data size, all images are scaled to 128 × 128.

Figure 2 presents data distribution as per age in moths as well as gender-based age distribution.

3.1.1 Splitting the Dataset

The dataset is split into training, validation and testing set as mentioned in Table 1.

3.2 Preprocessing

Key factor in skeletal age assessment is images with clear features of wrist and carpal bones. In the dataset provided by RSNA, these features are not prominent in 20% of the dataset. Hence, preprocessing is very much needed in this case. Following preprocessing as presented in Fig. 3 is applied on all the images.

Images of size 500 × 500 pixel as dataset, gender information and labels were loaded into memory during runtime. Data augmentation on the entire dataset is being carried out using a Keras ImageDataGenerator. Rotation range of 20°, horizontal/vertical translation up to 20%, zoom up to 20% and a horizontal flip is applied

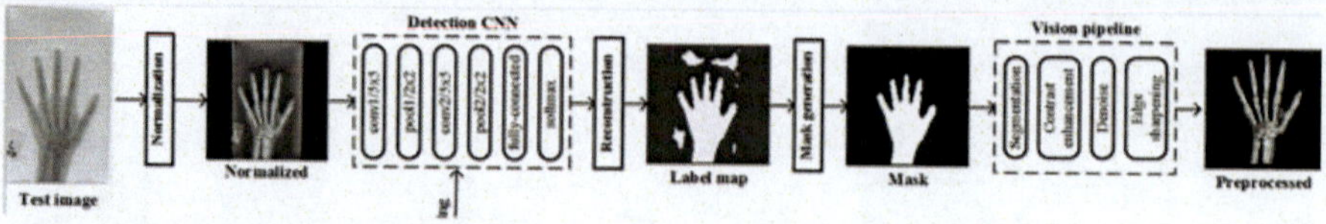

Fig. 3 Preprocessing of image

during data augmentation phase during training phase. On validation set, no augmentation was applied. These values were chosen based on reported literatures presentations. Figure 4a presents the preprocessed digital X-ray images with clear background and enhanced features.

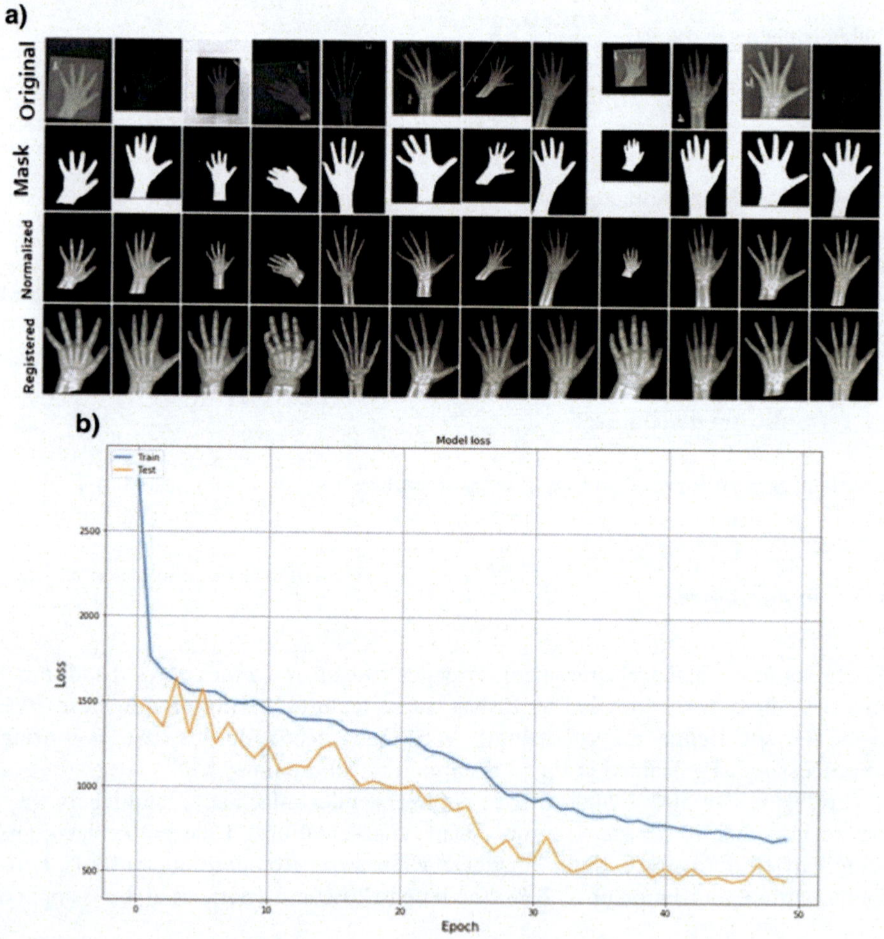

Fig. 4 **a** Images after Preprocessing. **b** Training and testing loss

Without augmentation, at a resolution of 250 × 250 pixels, MAE is recorded in the range of 8–9 months. Real-time augmentation drastically multiplies the dataset which improves learning, and also, each of the transformations selected will improve generalizability. By this, the network will learn features which are intrinsic to the patient rather than the imaging technique. Finally, model is trained with a minibatch size of 16 for 100 epochs (approximately 20 h) with the ADAM optimizer attempting to minimize mean absolute error of the output. Figure 4b presents loss during training and testing. It is observed that the model is getting well trained as loss is gradually decreasing over epochs.

4 Results and Discussions

In order to test the trained model, test dataset is fed to model for skeletal age assessment. Figure 5 presents the actual and predicted skeletal age. It is observed that the predicted values lie very close to the ground truth and as the performance suggests the average deviation from true value is around 7.61 months.

Occlusion method is used to generate attention maps to find which part of an image is locally significant for fine-grained classification. The colors indicate the strength for a location to be activated for bone age assessment (blue: low; yellow: strong). From Fig. 6, it is clear that the most prominent feature for skeletal age assessment is given more weightage during training.

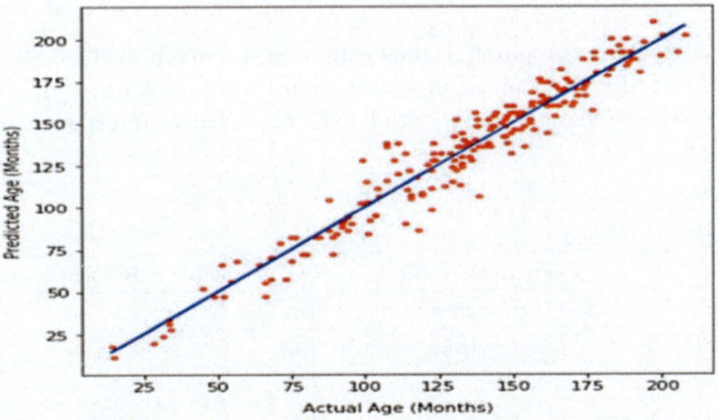

Fig. 5 Testing (predicted age vs. actual age)

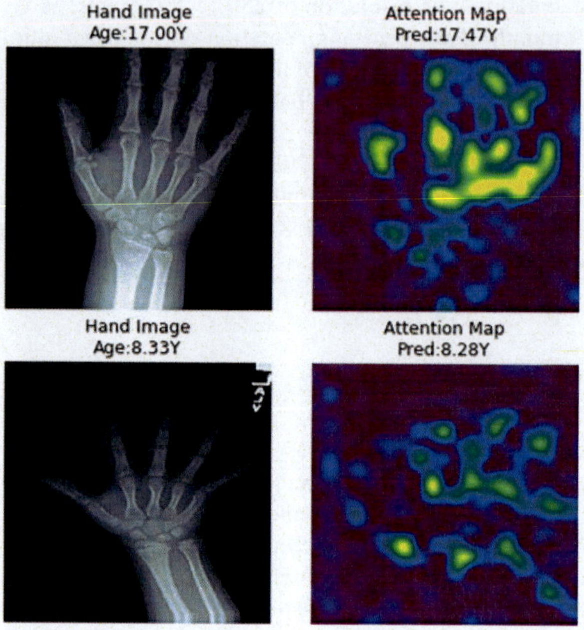

Fig. 6 Results of attention map

4.1 Comparative Study

In order to compare the proposed method with other recent reported literatures, Table 2 is recorded. MAE value and response time are recorded in each case. It is observed that the proposed method is on par with RCNN approach with response time being <2 s.

Table 2 Comparative analysis

Method	Proposed model	MAE in months	Response time in sec
Lee et al.	U-net architecture	8.08	<3
Iglovikov et al.	Ensemble learning	8.40	<1.5
Ren et al.	RCNN	7.35	3.5
Experienced radiologist	G&P OR TW	10.2	~480
Proposed architecture	Inceptionv3	7.61	<2

5 Conclusions and Future Scope

Overall, a model is developed which can successfully learn detailed hand features to predict skeletal bone age. Considering the limited size of our dataset, the performance of proposed model (MAE test error of 7.6 months) is very competitive with the other reported literatures. Over the last decade, developments in computer vision have increased our ability to extract information from images tremendously. This technological development has not only been a step forward in the computer science industry but also has changed the healthcare sector. Due to algorithmic and computing improvements, we are able to solve interdisciplinary problems throughout society. It is thus essential to understand what data and architectures are best for the task at hand and to apply them to problems that can have the most significant impact. As a future cope, one shall attempt for X-ray images with higher resolution with 16 bits to improve rate of processing. The model should be demographic generic that is it should work for all the data from all the regions. Deploy the model on mobile devices.

References

1. A. Kinnikar, M. Husain, S.M. Meena, Face recognition using gabor filter and convolutional neural network. Paper presented at the ACM International Conference Proceeding Series, 25–26 Aug 2016. https://doi.org/10.1145/2980258.2982104
2. V.J. Rathod, N.C. Iyer, S.M. Meena, A survey on fingerprint biometric recognition system. Paper presented at the Proceedings of the 2015 International Conference on Green Computing and Internet of Things, ICGCIoT 2015, pp. 323–326 (2016). https://doi.org/10.1109/ICGCIoT.2015.7380482
3. K.U. Sreekanth, S. Shetty, S.M. Meena, Head pose estimation using transfer learning. Paper presented at the Proceedings of the 2018 International Conference on Recent Trends in Advanced Computing, ICRTAC-CPS 2018, pp. 73–79 (2019). https://doi.org/10.1109/ICRTAC.2018.8679209
4. G.P. Mallya, A. Hiremani, G. Gajakosh, D. Badiger, M.P.S. Hiremath, Framework for automatic diagnosis of psychological disorders, in *2018 International Conference on Advances in Computing, Communication Control and Networking (ICACCCN)* (IEEE, 2018), pp. 929–934
5. S.G. Prateek, J. Jagadeesh, R. Siddarth, Y. Smitha, P.S. Hiremath, N.T. Pendari, Dynamic tool for American sign language finger spelling interpreter, in *2018 International Conference on Advances in Computing, Communication Control and Networking (ICACCCN)* (IEEE, 2018), pp. 596–600
6. E. Pietka, M.F. McNitt-Gray, M.L. Kuo, H.K. Huang, Computer-assisted phalangeal analysis in skeletal age assessment. IEEE Trans. Med. Imaging **10**(4), 616–620 (1991)
7. X. Pan, et al., Fully automated bone age assessment on large-scale hand X-ray dataset. Int. J. Biomed. Imaging (2020)
8. V.I. Iglovikov, A. Rakhlin, A.A. Kalinin, A.A. Shvets, Paediatric bone age assessment using deep convolutional neural networks, in *Deep Learning in Medical Image Analysis and Multimodal Learning for Clinical Decision Support*, ed. by D. Stoyanov (Springer, 2018), pp. 300–308
9. D.B. Larson, et al., Performance of a deep-learning neural network model in assessing skeletal maturity on pediatric hand radiographs. Radiology **287**(1) (2017)

10. C. Spampinato, et al., Deep learning for automated skeletal bone age assessment in X-ray images. Med. Image Anal. **36**, 41–51 (2017)
11. H. Lee, et al., Fully automated deep learning system for bone age assessment. J. Digital Imag. **30**(4) 427–441 (2017)
12. V. De Sanctis, S. Di Maio, A.T. Soliman, G. Raiola, R. Elalaily, G. Millimaggi, Hand X-ray in pediatric endocrinology: skeletal age assessment and beyond. Indian J. Endocrinol. Metab. **18**(Suppl 1), S63 (2014)
13. K. Simonyan, A. Zisserman, Very deep convolutional networks for large-scale image recognition. arXiv preprint arXiv:1409.1556 (2014).
14. E. Wu, B. Kong, X. Wang et al., Residual attention based network for hand bone age assessment, in *2019 IEEE 16th International Symposium on Biomedical Imaging (ISBI 2019)*, pp. 1158–1161, Venice, Italy, Italy (2019)
15. J. Han, Y. Jia, C. Zhao, F. Gou, Automatic bone age assessment combined with transfer learning and support vector regression, in *2018 9th International Conference on Information Technology in Medicine and Education (ITME)*, pp. 61–66, Hangzhou, China (2018)
16. S.H. Tajmir, H. Lee, R. Shailam, et al., Artificial intelligence assisted interpretation of bone age radiographs improves accuracy and decreases variability. Skeletal Radiol. **48**(2), 275–283 (2019)

A Novel Model for Disease Identification in Mango Plant Leaves Using Multimodal Conventional and Technological Approach

Lavanya B. Koppal, T. M. Rajesh, and K. B. Vedamurthy

Abstract Mango is considered as the king of fruits. India has the richest collection of mango cultivation and is an important fruit crop having socioeconomical significance. The fruit is admired because of the wide range of compliance, high nutritive value, medicinal values, excellent flavor and richness in variety. This has created high demand for mango in market. But, on the other hand, supply of mango to market is not sufficient, and the reasons could be many more, but plant disease problem stands first among all the problems. If there is no adequate yield of mango for export, there is increase of the price in market, which affects the common man to utilize the benefits of the same. Mango plants suffer from several infectious diseases and disorders including fungal, bacterial and other parasites of the tree as well as fruits. This drastically decreases yield and its quality. The identification of the diseases using conventional methods is time consuming, and there can be over usage of chemicals to overcome the diseases. The technological methods along with conventional methods can be used to identify the diseases efficiently and treat the disease time and cost effectively. This paper gives thorough knowledge to the readers/researchers on different types of mango plant diseases and the procedure followed in conventional and technological domains to identify the plant diseases.

Keywords Mango plant diseases · Conventional methods · Technological methods

L. B. Koppal (✉) · T. M. Rajesh
Department of CSE, Dayananda Sagar University, Bengaluru 560102, India
e-mail: lavanya-cse@dsu.edu.in

T. M. Rajesh
e-mail: rajesh-cse@dsu.edu.in

K. B. Vedamurthy
Karnataka Veterinary, Animal and Fisheries Sciences University, Bengaluru 560024, India

V. Bhateja et al. (eds.), *Data Engineering and Intelligent Computing*,
Advances in Intelligent Systems and Computing 1407,
https://doi.org/10.1007/978-981-16-0171-2_13

1 Introduction

Mango is one of the prominent crops in India. It accounts 57% of the total world production which ranks India, first among world's mango growing countries. Mango is popular because of splendid tropical flavor and richness in variety. Apart from the flavor, it provides high nutritional and medicinal value. There are numerous problems faced to get the maximum yield. Lack of nutrients and water in the soil makes the plant vulnerable to many diseases. No adequate water supply to the plants hinders the photosynthesis process of the plants, and nutrients required cannot be properly absorbed by them. Proper usage of fertilizers is essential so as to enrich the soil with adequate amount of nutrients. There are number of diseases that cause problem to the mango plants. This includes bacterial, fungal and parasite diseases. The yield of crops decreases to a greater extent. The quality of the yield gets compromised. The commonly identified diseases are mango powdery mildew, phoma blight, mango anthracnose, mango malformation, die-back, scab and sooty molds. The identification of diseases becomes a major task to overcome the effects so as to increase the yield and maintaining the quality of crops. The farmers used traditional methods to identify the diseases which include visual assessment of different parts of plant for disease symptoms. These traditional methods were not providing accurate identification results, and they were time consuming. So, conventional methods were used which gives accurate results compared to traditional methods. But the conventional methods are time consuming and cost ineffective. The technological methods can be used to identify the diseases with accurate results in very less duration of time compared to conventional methods [1]. This will help detecting the diseases in early stage of infection. The identification of diseases in plants through technological methods includes digital image processing, pattern recognition, soft computing and artificial neural networks.

2 A Review on Root Cause for Downfall of Profit in Agriculture Domain

The vulnerability of the plant attracts many disease causing agents such as bacteria, fungus or nematodes. This requires careful application of medicines so as to overcome the diseases. The downfall of profit pertaining to agriculture may be due to number of reasons. Some of the commonly identified reasons are discussed as below.

- **Climatic Conditions**: Sudden changes in the climate like heavy rain result in soil erosion which reflects on the plants growth when fertilized soil becomes loose soil, extension of sunny days which results in increase of too high temperature which reflects on converting the wet soil to dry soil, and heavy wind in agriculture field affects on the crops to fall down abruptly from the plants soil. If the plants do not get the proper amount of nutrients, the plants become vulnerable to diseases.

- **Infertile Soil**: One of the major reasons of plant diseases is the lack of nutrients in the soil. The parasites are likely to attack such plants. Fertilizers play an important role to supply essential nutrients to the plants. Careful application of the fertilizers either in natural or synthetic form is required. This will enhance nutritive value of the soil which is the major source of nutrients soil. If the plants do not get the proper amount of nutrients, the plants become vulnerable to diseases.
- **Intruders**: Avoiding intruders requires protecting the crops from wild animals by installing fence or sensors which is very expensive. Insects are the other types of intruders where they damage the crops to procure food where insecticides have to be used, which is expensive, may indirectly lessen the nutrients in the soil and can be chemically hazardous to the vegetation. Apart from these intruders, there is a requirement to safeguard the farm and crops from the thieves.
- **Diseases in plant Leaves**: Apart from all these concerns, specific medicines play an important role to overcome the problem of diseases and to increase the yield. The medicines have to be used in an efficient manner if over used may be hazardous for the consumer who consumes the produce of plants as food, the essential nutrients of the crop will be deteriorated, the nutritive value of the soil gets affected, and some of the beneficial agents like beetles, birds, earthworm, etc., which help in pollination and retaining the nutritive value of the soil get adversely affected. The photosynthetic area of the plant gets deteriorated; this will eventually defoliate the leaves. One of most important issues because of the attack of pathogens is in changes that occur for the plants where the flowering and fruit seasons that lead to biennale or alternate bearing of yield. The disease factor affects as chain link from growth of the plants to financial status and GDP of the country. Year by year % of GDP of India is as in Fig. 1 and Table 1.

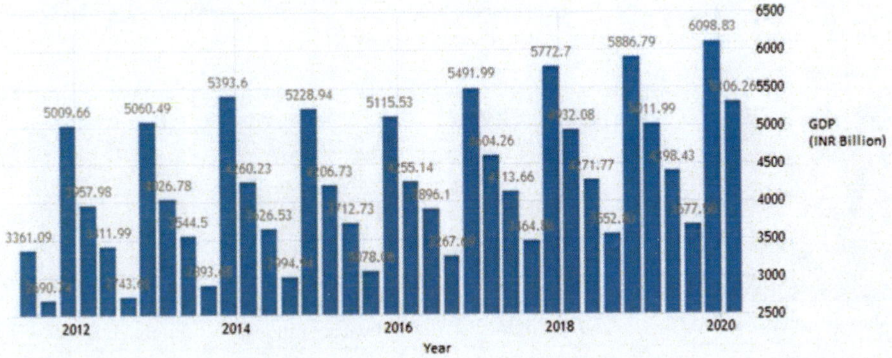

Fig. 1 Impact of agriculture on GDP (year wise). *Source* Tradingeconomics | Ministry of Statistics and Programme Implementation (MOSPI)

Table 1 Impact of agriculture on GDP (year wise)

S. No.	Year	GDP (Billion)
1	2011	3361.09
2	2012	5009.66
3	2013	5060.49
4	2014	5393.6
5	2015	5228.94
6	2016	5115.53
7	2017	5491.99
8	2018	5772.70
9	2019	5886.79
10	2020	6098.83

3 Background Study on the Diseases Which Affects the Mango Plants

Mango is one of the major growing fruits in India. It is also considered as the king of all fruits. So, we have considered mango plant pathology domain as a reference of this work. This section provides the researchers, information about most commonly identified diseases and their symptoms on the mango leaves. Table 2 provides the diseased leaf images.

- **Mango Anthracnose**

Table 2 Most commonly identified mango diseases in leaves

Examples of diseased leaves			
 Mango anthracnose	 Gray blight	 Powdery mildew	 Phoma blight
 Scab	 Red rust	 Leaf blight	 Bacterial canker

Mango anthracnose is a prominent disease caused by the *Colletotrichum gleosporioides* of Fungi Kingdom. Two types of symptoms can be observed on leaves. The first type of symptom is small dark to brown spots lesions that dry up and fall out which gives a form of perforated leaves. The second type of symptom observed is when lesions join together to form large necrotic regions along leaf margins. Leaves are severely affected and regularly curled. Anthracnosc appears as oval or irregular which are brown to deep brown spots of different size spread over the leaf surface. It forms extended brown necrotic regions which measures about twenty five millimeters in diameter [2].

- **Gray Blight or Pestalotia Leaf Spot**
  Gray blight is affected by the fungus, *Pestalotiopsis mangiferae* hence the name Pestalotia Leaf Spot. Gray blight is characterized by the occurrence of brown spots on the leaf lamina. These spots may expand from the margins or tip. Initially, spots are brown and small which slowly increase in size and become dark brown. Some of the spots form large lesions, with grayish white center and tan color margins [3].
- **Mango Powdery Mildew**
  Mango powdery mildew is originated by a Fungus *Oidium magniferae* is one of the commonly seen diseases of mango. It causes different symptoms on leaves and fruits. On some cultivars, new flushes of growth and younger leaves are highly susceptible and may curl up and become distorted. Grayish, necrotic lesions or large, irregularly shaped spots may form on leaves. A white superficial powdery growth of the fungus can be observed which later turns into gray. The youngest leaves may become completely covered with fungal spores and mycelium which will turn from brown to light green [4].
- **Phoma Blight**
  The phoma blight is occurred by the Fungus *Phoma glomarata*. The symptoms of phoma blight are visible on old leaves. Initially, the lesions are angular, irregular which are yellow to light brown and spread over the leaf lamina. As the lesions enlarge, their color changes from brown to cinnamon. These lesions become irregular. The fully developed spots are characterized by dark margin and gray necrotic centers. When the infection becomes severe, the spots combine to take a form with patches of about three and half to thirteen centimeters [5].
- **Scab**
  Mango scab is the affected by the Fungus *Elsinoe mangiferae*. Spots are circular, slightly angular, extended and take about two to four millimeters in diameter, brown. During rainy season, lesions differ in size, color and shape. Lesions are of different shape and color. Spots are smaller than in anthracnose, and the infection surface is covered with velvety down. Severe infection leads to crinkling and distortion of leaves [3].
- **Red Rust**
  The Red Rust is caused by the alga. The algal attack reduces the photosynthetic activity and defoliation of leaves which lowers life of the mango plant. The disease is readily identified by the presence of rusty red fructifications of the alga on the surface of leaves. Initially, the spots are greenish gray in color and velvety

in texture, and finally, they turn reddish brown. Spots are circular to irregular in shape measuring about two mm in diameter but when joined together. After spores shed off, the algal medium remains attached to the leaf surface, leaving a creamy white mark at the original rust spot [5].

- **Leaf Blight**
  The Leaf Blight is caused by the bacteria Pseudomonas syringae or in short syringae. The disease is characterized by the presence of yellowish pinhead spots on leaves. Soon after, the spots enlarge and become light to dark brown in color. Later, these spots become oval to irregular in shape, raised and dark purplish. Due to complete drying of leaf, the infection travels downward toward the petiole [6].
- **Bacterial Canker or Bacterial Black Spot**
  Bacterial Canker is caused by the bacteria Xanthomonas Campestris. The disease is featured by the water soaked irregular and angular lesions which is usually packed at the apex. On young leaves, halos are larger and are distinct. The older leaves are affected as thin halos. In severe conditions, the leaf turns yellow [7].

4 Conventional Versus Technological Way of Diagnosing Plant Diseases

The methodologies for diagnosing plant diseases can be broadly classified as conventional and technological.

- **Conventional methodology**: Conventional methods include the laboratory-based techniques of disease detection which helps in minimizing the disease provoked damage to the yield and ensures the maximum productivity and guarantee agricultural sustainability. Some of the conventional techniques such as polymerase chain reaction, immunofluorescence, fluorescence in-situ hybridization, enzyme-linked immunosorbent assay, flow cytometry, gas chromatography-mass spectrometry, thermography and fluorescence imaging are used [8]. Polymerase chain reaction is the process where specific DNA fragment from the genome of an individual plant helps the genesists to analyze the DNA. Immunofluorescene is a technique where fluorescence microscope is used to visualize the allocation of the target molecule for the collected samples. Fluorescence in-situ hybridization is a molecular technique. It is used to identify and localize the presence/absence of chromosomes and DNA sequences. Enzyme-linked immunosorbent assay is a laboratory method where detection of plant disease is done on the basis of antibody-antigen interaction by assessing enzyme activity. Flow cytometry is used to detect and measure physical and chemical characteristics of the cells where the sample containing cells are suspended in a fluid and introduced into the flow using cytometer. Gas chromatography–mass spectrometry combines the both gas and mass spectrometry features to identify the samples. Thermal imaging is used for early detection of plant disease when the diseases directly have an effect on transpiration rate. The conventional techniques also include the usage of biosensors based on highly selective bio-recognition elements such as enzyme, antibody, DNA or RNA and bacteriophage as a tool to identify crop diseases [9]. If the samples are collected from the field which are later tested in the laboratory, it is time consuming and expensive. Figure 2 depicts the steps involved in conventional methods for the identification and verification of diseases.

- **Technological methodology**: The technological methods of identifying plant diseases include image processing, pattern recognition, soft computing and artificial neural network. There are two phases training phase and testing phase. Figure 3 depicts the overall stages, whereas in the training phase, dataset is gathered either by the images from the Internet or by creating the own dataset. The raw acquired dataset, usually images, will not be clean and efficient to extract the features which are done in the preprocessing phase. Images may contain different type of noise, and the image quality may be low because of the external factors like weather changes which include fog, dew or light source. To overcome these problems, preprocessing stage plays an important role. Preprocessing techniques include removal of noise from the image using filters like low pass filter, median

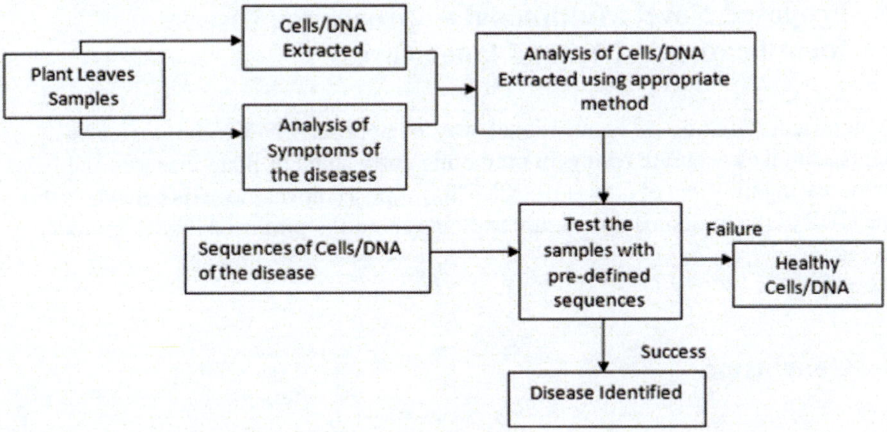

Fig. 2 Conventional method of identifying plant diseases

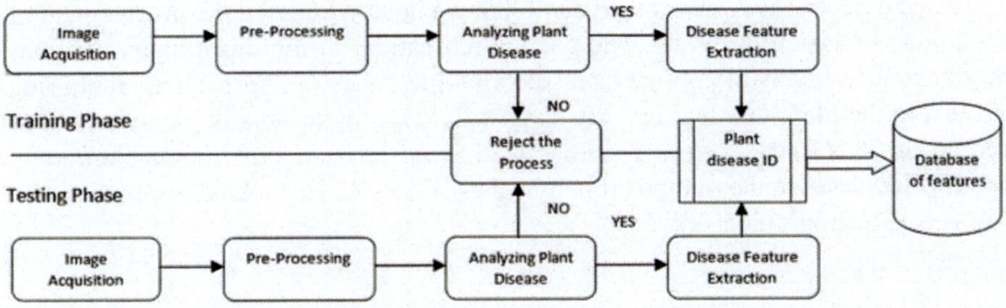

Fig. 3 Commonly used methodology for identification and verification of plant diseases

filter, Gaussian filter, Laplacian filter, Wiener filter, resizing the image, thinning and thickening of the images, etc. [10–13]. After the preprocessing process, plant diseases are analyzed to identify the plant disease in the leaf using various techniques like genetic algorithm, principal component analysis, and independent component analysis, etc. If the disease is not identified, it gets rejected at the identification and verification phase [14–17]. For the extracted features from the images, a distinct ID is provided and stored in the database. In the second stage, testing is carried out using the same procedure as in training stage with different distinct ID which is stored in the database. The IDs are matched up to identify and verify the detected plant disease so as to check whether they are referring to same disease for the particular plant species. This process requires technological methods like K-means clustering, random forest classifier, support vector machine, confusion matrix, different artificial neural networks [8, 10, 18]. Based on the results, identification, verification, classification and grading of diseases for different plant species and different types of diseases can be done.

5 Proposed Novel Multimodal Approach for Disease Identification in Mango Plant Leaves

A good combination of conventional way of approach and technological way of approach gives accurate results in predicting early stage of plant diseases and helps in increasing the yield of crops [19, 20]. This proposed novel model is believed to hike the GDP of the nation. Below figure explains about the proposed model approach as in Fig. 4.

6 Conclusion

Disease problem in agriculture domain is a major hit-back in any country. India is a country, which is majorly dependent on agriculture. Till late year 2000, the good old conventional methods were followed in agriculture domain to identify, verify and cure diseases. After the year 2000, we have witnessed the involvement of technology in agriculture field. The good combination of interdisciplinary domains like conventional way of approach and technological way of approach in identifying and curing the plant diseases at early stage will help the farmers in gaining yield of their crops. This analysis work gives knowledge for the researchers about commonly identified diseases in the mango crop, how these diseases are tackled in conventional way versus technological way.

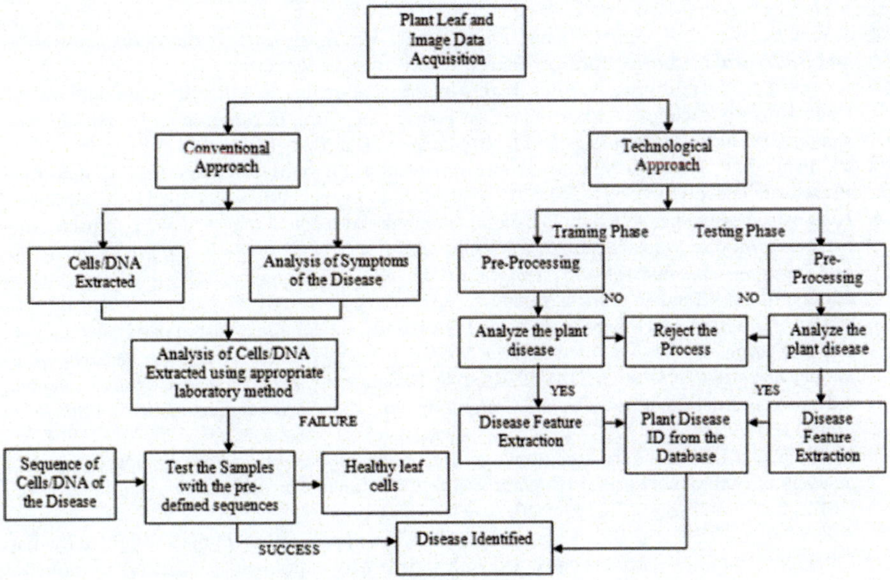

Fig. 4 Novel model for identification of plant disease

References

1. G. Li, Z. Ma, H. Wang, Image recognition of grape downy mildew and grape powdery mildew based on support vector machine, in *IFIP International Federation for Information Processing* (Springer, 2012), pp. 151–162
2. Haggag, W.M., Mango diseases in Egypt. Agric. Biol. J. N. Am.
3. A.K. Misra, B.K. Pandey, *Diseases of Mango*. Researchgate Publication 31188685 (2012)
4. Misra, A.K., *Powdery Mildew—A Serious Disease of Mango*. Researchgate Publication 281590887 (2001)
5. Om Prakash, A.K. Misra, *Diseases and Their Manangement*. Researchgate Publication 311715840 (2001)
6. S.I.A. Islam, M.R. Islam, K.M.G. Dastogeer, I. Hossain, Characterization of leaf blight pathogen, *Pseudomonas syringae* pv. syringae of mango in Bangladesh. Int. Res. J. Biol. Sci. **2**(6), 39–45 (2013)
7. P.R. Rothe, R.V. Kshirsagar, Cotton leaf disease identification using pattern recognition techniques, in *International Conference on Pervasive Computing (ICPC)* (IEEE, 2015)
8. V. Narayanan, B. Parsi, Center symmetric local descriptors for image classification. Int. J. Nat. Comput. Res. (IJNCR) **7**(4), 56–70 (2018)
9. M. Sarwar, Practices for integrated control of mango (*Mangifera indica* L.) diseases to protect in preharvest as well as postharvest phases. Biosci. Bioeng. **1**(3), 57–62 (2015)
10. V. Bhateja, A. Verma, K. Rastogi, C. Malhotra, S.C. Satapathy, Performance improvement of decision median filter for suppression of salt and pepper noise, in *Advances in Signal Processing and Intelligent Recognition Systems*, ed. by S. Thampi, A. Gelbukh, J. Mukhopadhyay. Advances in Intelligent Systems and Computing, vol. 264 (Springer, Cham, 2014). https://doi.org/10.1007/978-3-319-04960-1_26
11. D. Cui, Q. Zhang, M. Li, G.L. Hartman, Y. Zhao, Image processing methods for quantitatively detecting soybean rust from multispectral images. Biosyst. Eng. **107**, 186–193 (2010)

12. J. Zhu, A. Wu, X. Wang, H. Zhang, *Identification of Grape Diseases Using Image Analysis and BP Neural Networks.* Springer Nature 2019 (Springer, 2018)
13. A. Asfarian, Y. Herdiyeni, A. Rauf, K.H. Mutaqin, Paddy diseases identification with texture analysis using fractal descriptors based on Fourier spectrum, in *International Conference on Computer, Control, Informatics and Its Applications* (IEEE, 2013)
14. Yi. Fang, R.P. Ramasamy, Current and prospective methods for plant disease detection. Biosensors (Basel) **5**(3), 537–561 (2015). https://doi.org/10.3390/bios5030537
15. Himanshi, V. Bhateja, A. Krishn, A. Sahu, An improved medical image fusion approach using PCA and complex wavelets, in *2014 International Conference on Medical Imaging, m-Health and Emerging Communication Systems (MedCom)*, Greater Noida, pp. 442–447 (2014). https://doi.org/10.1109/MedCom.2014.7006049
16. A. Krishn, V. Bhateja, Himanshi, A. Sahu, PCA based medical image fusion in ridgelet domain, in *Proceedings of the 3rd International Conference on Frontiers of Intelligent Computing: Theory and Applications (FICTA) 2014*, ed. by S. Satapathy, B. Biswal, S. Udgata, J. Mandal. Advances in Intelligent Systems and Computing, vol. 328 (Springer, Cham, 2015). https://doi.org/10.1007/978-3-319-12012-6_52
17. H. Wang, G. Li, Z. Ma, X. Li, Image recognition of plant diseases based on principal component analysis and neural networks, in *8th International Conference on Natural Computation (ICNC)* (IEEE, 2012). 978-1-4577-2133-5/10
18. A.K. Shukla, V. Bhateja, R.L. Verma, M.S. Alam, An improved Directional Weighted Median Filter for restoration of images corrupted with high density impulse noise, in *2014 International Conference on Reliability Optimization and Information Technology (ICROIT)*, Faridabad, pp. 506–511 (2014). https://doi.org/10.1109/ICROIT.2014.6798376
19. T.M. Rajesh, K. Dalawai, N. Pradeep, *Automatic Data Acquisition and Spot Disease Identification System in Plants Pathology Domain: Agricultural Intelligence System in Plant Pathology Domain* (IGI Global, 2020). https://doi.org/10.4018/978-1-5225-9632-5.ch006
20. T. Jena, T.M. Rajesh, M. Patil, *Elitist TLBO for Identification and Verification of Plant Diseases* (Springer, 2019). https://doi.org/10.1007/978-981-13-6569-0_3

Combating COVID-19 Through Emerging Digital Technologies

Supriya Dubey and Pritee Parwekar

Abstract The COVID-19 pandemic is a threat not only to the health of millions of population, but also to the stability of global infrastructure and economies. The disease introduced a unavoidable burden on healthcare systems that cannot be efficiently dealt with the existing facilities or responses based on traditional approaches. Mostly, it is believed that a thorough clinical and common response can only be mounted by using intelligence derived from a variety of data sources to better utilize limited healthcare resources, provide personalized patient management plans, inform policy and clinical trials. This paper introduces some of the most important challenges in responding to COVID-19 and shows how each of them can be addressed by recent developments in digital technologies. It concludes that the integration of these techniques into local, national, and international healthcare systems will save lives and propose specific methods by which implementation can happen swiftly and efficiently. The paper also proposes a methodology to detect COVID-19 patients through their radiographs at early stage.

Keywords Artificial intelligence · Machine learning · Contact tracing · Surveillance · Radiographs

1 Introduction

Coronavirus disease 2019 (COVID-19) outbreak is an extraordinary health hazard around the world. One of the most basic purposes of this challenge is the management of COVID-19 patients requiring intense or potentially basic respiratory consideration. COVID-19 has emerged as one of the greatest challenges to confront the twenty-first

S. Dubey (✉) · P. Parwekar
Computer Science and Engineering, Faculty of Engineering and Technology, SRM Institute of Science and Technology, Delhi-NCR Campus, Delhi-Meerut Road, Modinagar, Ghaziabad, UP, India
e-mail: supriyad@srmist.edu.in

P. Parwekar
e-mail: priteep@srmist.edu.in

V. Bhateja et al. (eds.), *Data Engineering and Intelligent Computing*,
Advances in Intelligent Systems and Computing 1407,
https://doi.org/10.1007/978-981-16-0171-2_14

century world. Further complicating the response to this issue is simply the hazy areas encompassing the infection, as far as its spread and how to treat it.

Healthcare systems require the help of new advancements like artificial intelligence, Internet of things, big data and machine learning to combat with and making strategies against the new health hazard. This paper aims to highlight the role of latest digital technologies to combating with COVID-19.

2 New Threats, Old Lessons

SARS-CoV was distinguished in 2003 as the reason for an episode of extreme intense respiratory condition (SARS) that started in China close to the furthest limit of 2002. MERS-CoV was recognized in 2012 as the reason for Middle East respiratory condition (MERS).These Coronaviruses that cause serious respiratory diseases are pathogens naturally transmitted between animals and humans, which start in tainted creatures and are transmitted from creatures to individuals.

The SARS and MERS plagues have placed us in a superior situation to react to COVID-19. The straightforwardness exhibited in the quick transmission of the SARS-CoV-2 hereditary data has been basic in directing a worldwide way to deal with limit the spread of infection. This has empowered the fast advancement in diagnostic tests and their worldwide treatment. Thus, these indicative tests have encouraged the administration of cases in those regions generally influenced and the distinguishing proof of cases in different nations, all of which has supported in restricting the worldwide spread of SARS-CoV2. Nonetheless, current proof shows that these general health measures and strategies alone might be inadequate to combat with COVID-19. New medicines are urgently required, and it is the high time. One may consider these technologies, along-side practices from two past novel Coronavirus illness flare-ups—serious intense respiratory condition (SARS), in 2003 and Middle East respiratory disorder (MERS), in 2012—to add to quick control of the most recent one, COVID-19.

3 Digital Technologies Versus COVID-19

Advanced plans of action and new innovations assume a significant job in the nation's prosperity. In particular, these advances can be conveyed in four different ways.

A. *Limiting the spread of virus*
Technologies, for example, location-based administrations, vast information analysis, and robotics have been applied on an amazing scale to follow and distinguish high-hazard cases, limit development, and limit human-to-human contact.

B. *Informing the people*
Advanced stages and innovations that give straightforwardness and access to constant data are basic to facilitating open frenzy. In China, stages have been set up to total and disperse indispensable data. Area-based cautions have kept residents sheltered and educated.

C. *Ensuring quality of life*
Organizations are utilizing on the Web to disconnected administrations and stages, just as virtual and expanded reality arrangements, to not just fulfill fundamental necessities of self-isolated individuals, yet in addition assist them with beating the weariness and impediments related with remaining at home for broadened timeframes.

D. *Re-igniting the productivity*
Organizations across China have utilized advanced innovations to continue business tasks. Endeavors have included utilizing live-spilling occasions to keep customers connected with, utilizing IoT and apply autonomy for robotized producing, and applying flexible computerized workspace answers for empowering representatives to continue work from their homes.

4 Challenges to Be Addressed

As nearby and local governments around the globe work eagerly to successfully address the COVID-19 emergency, it is critical to recognize that, regardless of whether in or past occasions of emergency, digital tools must go about as an empowering influence to arrive at the objective of the protecting citizens' well-being, advancing social union, and the security of human rights, including at the digital level. There are seven primary classifications of issues they endeavor to address:

- Early warning and surveillance;
- Contact tracing;
- Quarantine and social control; and
- Managing limited healthcare resources
- Testing and responder capacity
- Research and cure
- Developing personalized patient management and treatment plans.

A. *Early warning and surveillance*
Early warning and surveillance to all the more likely comprehend the pathogen and screen the flare-up are basic segments of reacting to any pestilence. Ailment observation centers around following the occurrences of the infection and its way, which regularly corresponds with briefly following the individuals who contract or connect with a sickness.

B. *Contact tracing*
Contact tracing is the way toward following a pandemic's spread—and by a wide margin the need to follow the pathogen is the most well-known explanation

given to build the measure of information shared during a general health crisis. Generally, the manner in which contact tracing works is that a positive tested person for COVID-19 shares however much about their ongoing location and contacts during the irresistible duration as could be expected.

C. *Quarantine and social control*

As COVID-19 reaction increases far and wide, an increasing number of nations are imposing social controls that include a scope of authorization systems—from civic recommendations all the way to armed enforcement. Maybe the most significant, and regularly disregarded, effect of innovation centered intercessions in observation is to adequately incapacitate any keeps an eye on the administration utilizing them—they can employ these apparatuses without risk of punishment, which is by structure.

D. *Managing limited healthcare resources*

Medicinal services frameworks on the planet will before long face an extreme shortage of assets—specifically testing packs, emergency clinic beds, ICU beds, ventilators, and work force. Since assets are restricted and hazard and sickness movement so heterogeneous, it is essential to recognize—as ahead of schedule as could be expected under the circumstances—which people are destined to have been contaminated by the infection, which tainted people may encounter unfriendly occasions, which kinds of clinical assets those people will require, and when these assets will be required. There is accordingly an unmistakable and earnest need to send frameworks that can give early alerts to customized hazard and illness movement of people and that can illuminate clinical staff and social insurance frameworks about which patients would profit by what assets and when.

E. *Testing and responder capacity*

Another application of digital technology in health crisis is to improve, adjust, or put resources into clinical gadgets, tests, and defensive rigging. Endeavors to improve existing limit, be that as it may, are probably the best approaches to mediate, in light of the fact that they have nearly explicit issues to explain (say, improving the nature of defensive apparatus), existing pathways to dissemination (that is, trying, producing and strategic conveyance foundation) and clients with at any rate an elevated level comprehension of the hidden instrument.

F. *Patient treatment planning*

Immunizations and therapeutics for COVID-19 will set aside some effort to be created and presented. To spare lives and adequately utilize rare assets, it is accordingly basic to create the board and treatment designs that are customized for every patient, contingent upon their one of a kind attributes and needs.

G. *Research and cure*

While technologies may hold critical potential for misuse, contributing assets to manage inquire about is one of the most secure and most important ways for organizations and specialists to utilize technology to add to the COVID-19 response. The essential purpose behind the security is that there are vigorous market, legitimate and institutional checks activated at each phase between the development of an vaccine in a restricted biomedical research condition and the all around regulated distribution of a pharmaceutical item.

5 Tackling COVID-19 with Digital Technology

Mentioned below are ten different ways which are helping people in the battle against COVID-19: Some of these models are curated from government, industry, and media sources, accessible in the open space.

- Researchers can conceivably utilize AI [6] to skim through logical papers and insightful articles to help battle data over-burden. The exploration group has built up an AI stage that makes it simpler for colleagues to rapidly discover applicable investigations that can possibly prompt new bits of knowledge or ways to deal with address the COVID-19 flare-up.
- Blockchain [9] innovation can conceivably be utilized to adequately oversee clinical information, track the gracefully of infection counteraction materials (disinfectants, sanitizers), and counsel the general population.
- Chatbot [10] instruments can give an online triage in disease diagnostics. The device is principally utilized by medicinal services experts to collaborate with patients through remote visits.
- The COVID-19 episode has affected a significant number of the business' most significant gatherings and expos this year. Innovations like virtual reality/augmented reality [11] can be considered as computerized choices to address the issue.
- 3D printing [12] can be utilized to help produce imperative medical clinic supplies in the midst of the COVID-19 pandemic. Several 3D printing structure architects and masters have just made an open record mirroring their enthusiasm to aid the creation of respirators, valves, covers, and so on.
- Drones [13] can be conceivably used to convey test units and clinical supplies, cutting conveyance time and improving reaction time for clinical specialists. Automations outfitted with warm imaging innovation are being utilized to distinguish individuals with fever/influenza side effects.
- Robotics [14] can be put to extraordinary use in checking temperatures and distinguishing and sanitizing individuals out in the open spaces, in this manner disposing of human contact. The utilization of robots in clinics can give genuinely necessary alleviation to clinical staff and other social insurance laborers as they keep on giving consideration to influenced patients.
- Video conferencing [15] arrangements have been the backbone of organizations and instructive establishments. With workplaces shut and travel being limited, cloud gatherings have shielded organizations from self-destructing and have helped them limit the business sway as indicated by media reports.
- Social media [16] stages as of late have been a key vehicle of correspondence during an emergency. When the compass of customary media has been constrained because of social removing measures, online life clients have kept the world

refreshed and have assumed a significant job in dispersing dependable data. Mediums, for example, Twitter, and Facebook have found a way to guarantee that solitary accommodating and sound substance is dispersed on their foundation by checking the spread of phony and deluding content.
- 5G [17] can assume a roundabout job in the regulation of this infection episode. Colossal transmission capacity, rapid, and low inactivity take into account consistent correspondence in medical clinics and crisis war rooms. It might fill in as an establishment that empowers the usage of the previously mentioned advancements

6 Mobile Application for COVID-19

Mobile apps are indeed bridging the distance between citizens and organisations, but also have a potentially crucial role in tackling the next phase of Coronavirus response. There are five clusters: informational apps; self-assessment/medical reporting apps; contact tracing apps; multi-purpose apps, and other COVID-19-related apps. There are more than 108 mobile applications launched by 73 countries to deal with the present scenario. Indian government also launched many mobile applications to combat with COVID-19 which are listed below:

- **Aarogya Setu**: The Aarogya Setu is a following application that has been as of late propelled by the center to get some answers concerning Coronavirus disease cases utilizing a GPS framework and Bluetooth highlights present in all cell phones. The administration will utilize these cell phone highlights to decide whether an individual is living near a patient of COVID-19.
- **Chatbot**: The government had reported the dispatch of a WhatsApp chatbot with the goal that the residents can find moment and credible solutions to every one of their inquiries identified with the Coronavirus pandemic
- **Corona Kavach**: Another COVID-19 tracker application, this was made by the Union Ministry of Electronics and Information Technology in a joint effort with the Ministry of Health and Family Welfare. This application gives clients ongoing area of tainted clients who have initiated the 'Kavach' feature..
- **COVID-19 Feedback**: This application was created by the center to get immediate input from individuals who have experienced Coronavirus treatment in the nation

7 Discussion

We can aggregate the overall challenges to be addressed into two conventional health measures for tackling COVID-19:

- **Strategy 1** monitoring, surveillance, detection, and prevention of COVID-19; and
- **Strategy 2** mitigation of the impact to health care indirectly related to COVID-19.

Table 1 Impact of digital technology in public health strategies

	Strategy 1	Strategy 2
Internet of Things	HIGH	HIGH
Big data	HIGH	MODERATE
Artificial Intelligence	MODERATE	HIGH

Table 1 depicts the impact of various digital technologies on conventional health measures against COVID-19.

8 Proposed Methodology

Without explicit remedial medications or immunizations for COVID-19, it is fundamental to distinguish the illness at a beginning time and promptly segregate an affected patient from the sound populace. Early determination of COVID-19 is essential for infection treatment and control. Contrasted with RT-PCR, chest CT imaging might be a progressively reliable, commonsense, and quick technique to analyze and evaluate COVID-19, particularly in the pandemic zone.

Thinking about the reality, the subjects tested positive with COVID-19 may have pneumonia and can without much of a stretch be demonstrated by utilizing automated framework for recognition of COVID-19, and the worry division ought to think about prompt seclusion and treatment for the subjects. A subject with basic stage ought to endure with a changeless lungs harm, if not kicked the bucket. Agreeing to WHO, COVID-19 makes holes in the lungs like SARS which is not recoverable.

This paper proposes a methodology to detect COVID-19 using the chest X-ray of the patient. The road map for the proposed method is demonstrated in Fig. 1 (Fig. 2).

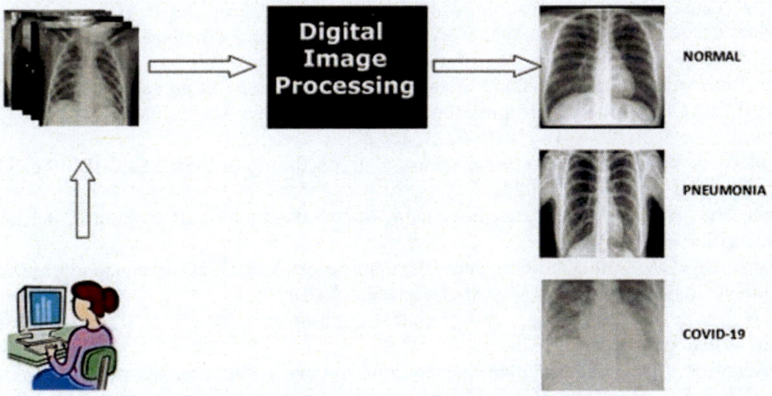

Fig. 1 Road map for the proposed methodology

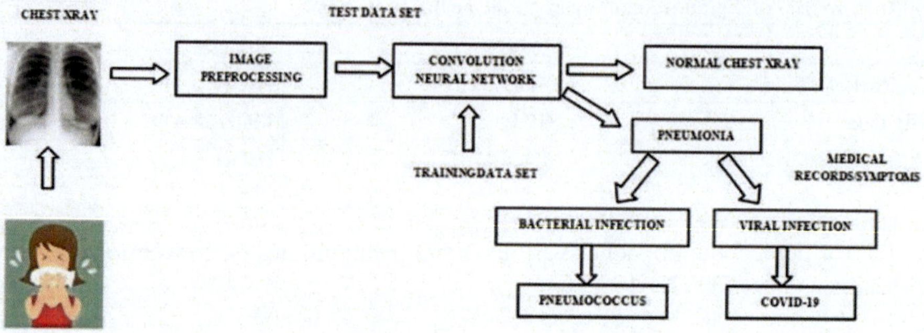

Fig. 2 Steps involved in proposed methodology

9 Conclusion

This is just a hint of something larger. All things considered, AI will assume a huge job in completion the COVID-19 pandemic. Advanced technology is essential in the battle against Coronavirus and future pandemics. Notwithstanding having the option to help displaying endeavors and anticipating the progression of a pandemic, artificial intelligence, AI, and other innovation can rapidly and viably break down information to help people on the bleeding edges make sense of the best. This paper highlights the role of modern digital technologies in combating with COVID-19 pandemic. It also proposes a novel approach of detecting COVID-19 patients through their X-ray images automatically using image processing and machine learning approaches.

References

1. D.S.W. Ting et al., Digital technology and COVID-19. Nat. Med. **26**(4), 459–461 (2020)
2. S. Keesara, A. Jonas, K. Schulman, Covid-19 and health care's digital revolution. N. Engl. J. Med. (2020)
3. X.B. Pan, Application of personal-oriented digital technology in preventing transmission of COVID-19, China [published online ahead of print, 2020 Mar 27]. Ir. J. Med. Sci., 1–2 (2020). https://doi.org/10.1007/s11845-020-02215-5
4. M. Ienca, E. Vayena, On the responsible use of digital data to tackle the COVID-19 pandemic. Nat. Med. **26**(4), 463–464 (2020)
5. Z. Hu, et al., Artificial intelligence forecasting of covid-19 inchina. arXiv preprint arXiv:2002.07112 (2020)
6. Vaishya, R., et al., Artificial Intelligence (AI) applications for COVID-19 pandemic, in *Diabetes Metabolic Syndrome: Clinical Research Reviews* (2020)
7. T.T. Nguyen, Artificial intelligence in the battle against coronavirus (COVID-19): a survey and future research directions (2020)
8. https://www.id-hub.com/2020/04/08/responding-to-covid-19-with-ai-and-machine-learning/
9. M. Torky, A.E. Hassanien, COVID-19 blockchain framework: innovative approach. arXiv preprint arXiv:2004.06081 (2020)
10. J. Espinoza, K. Crown, O. Kulkarni, A guide to chatbots for COVID-19 screening at pediatric health care facilities. JMIR Public Health and Surveillance **6**(2), e18808 (2020)

11. S. Carlaw, Impact on biometrics of Covid-19. Biometric Technol. Today **2020**(4), 8–9 (2020)
12. S.T. Flanagan, D.H. Ballard, 3D Printed Face Shields: A Community Response to the COVID-19 Global Pandemic. Acad. Radiol. **27**(6), 905 (2020)
13. R. Estrada, M. Arturo, The uses of drones in case of massive epidemics contagious diseases relief humanitarian aid: Wuhan-COVID-19 crisis. Available at SSRN 3546547 (2020)
14. G.-Z. Yang, et al., Combating COVID-19—The role of robotics in managing public health and infectious diseases (2020)
15. T. Greenhalgh, et al., Video consultations for covid-19 (2020)
16. M. Cinelli, et al., The covid-19 social media infodemic. arXiv preprint arXiv:2003.05004 (2020)
17. D. Soldani, Fighting COVID-19 with 5G enabled Technologies
18. https://en.wikipedia.org/wiki/COVID-19apps
19. https://www.healthimaging.com/topics/diagnostic-imaging/chest-x-ray-coronavirus-1st-us-patient

Android Malware Detection Using Convolutional Neural Networks

A. Lakshmanarao and M. Shashi

Abstract The usage of mobile devices has been increased in recent years. Android is the most widely used operating system on mobile devices. As millions of applications with varied veracity are made available for download in the play store, it is essential to automatically identify malicious mobile applications. Most of the existing works rely on detecting Android malware based on either static (permissions, intents, application program interface calls, etc.) or dynamic features (API calls and system call traces, etc.) of the APIs. In this paper, Android malware and benign apks are converted into images based on their sizes, and on top of it, convolutional neural network architecture for deep learning is proposed for detecting Android malware. Malware apks are collected from virusshare.com, and benign apks are collected from play store, apkpure.com. Extensive experimentation conducted in two different settings has shown that only classes.dex files of apks are sufficient for Android malware detection. The proposed deep learning framework with convolutional neural networks could achieve 97.76% accuracy in detecting Android malware with minimal information requirement.

Keywords Android · Malware · Convolutional neural networks

1 Introduction

Android is the most widely used mobile operating system, which dominates the mobile OS market with 74% of the global market share (December-2019) [1]. With the advent of mobile technology and connectivity at affordable prices, usage of mobile devices has become part and parcel of day to day life of most of the users who may be naive toward cyber security issues and threats. Specifically, people are

A. Lakshmanarao (✉)
Department of IT, Aditya Engineering College, Surampalem, A.P, India

M. Shashi
Department of CS&SE, AU College of Engineering, Andhra University, Visakhapatnam, A.P, India

V. Bhateja et al. (eds.), *Data Engineering and Intelligent Computing*,
Advances in Intelligent Systems and Computing 1407,
https://doi.org/10.1007/978-981-16-0171-2_15

prone to download third-party apps made available to the public free of cost which may contain malicious software referred to as malware. Hence, Android malware detection has to be addressed more carefully.

Similar to PC malware attackers, Android malware attackers are also trying to bypass the security walls created by defenders. Generally, Android applications are downloaded from Google play store. Before installing the application, the app installer will prompt for user permissions often going beyond the genuine limits. As it is difficult for the normal mobile users to understand the functionality of the application and the essential inputs required by it, a normal user cannot differentiate the malicious intent from a genuine one at that time.

Machine learning is one of the artificial intelligence approaches that enable a system to learn from its previous experience or often from a given collection of known instances. Machine learning is widely applied for Android malware recognition issues. Machine learning techniques when applied for malware detection require labeled instances describing the past experience to learn malware signatures through supervisory learning, which will be used later to match with the given data to detect if it has malware. With the advent of deep learning architectures as an advancement of ML, extraction of even the most complex features has become feasible provided that there is a sufficiently large dataset. Thus, one of the key challenges to developing successful machine learning solutions to cyber security problems is the availability of appropriate datasets.

This paper focuses on constructing an Android malware detection system using a convolutional neural network model for deep learning. Malware apks are collected from virusshare.com, and benign apks are collected from play store, apkpure.com. The input is given in two different settings namely.dex files and.apk files after converting them into image form. The performance of the proposed model is evaluated in terms of the detection accuracy of the model. The remaining sections of this paper are organized as follows: Sect. 2 presents related research on Android malware detection, Sect. 3 discusses research methodology, Sect. 4 presents experimentation and results, and Sect. 5 has concluding remarks and future scope.

2 Literature Survey

Several researchers applied machine learning techniques for Android malware detection. Wu et al. [2] applied *k*-means, *K*-NN machine learning algorithms on static features extracted from permissions, intents, API calls. Fatima et al. [3] proposed Android malware detection using genetic algorithm. A combination of genetic algorithms for the feature selection followed by support vector machines was used to achieve high detection accuracy. Li et al. [4] proposed a malware detection technique that is based on Android app permissions only. In [4], three levels of pruning have been done for identifying relevant permissions to distinguish between malware and benign apps. Wang et al. [5] extracted string features and structural features to build a more accurate detection model for Android malware detection and applied machine learning algorithms support vector machine, *k*-nearest neighbor, random forest, and achieved good accuracy. Feng et al. [6] proposed an Android malware detection system based on dynamic behavior features. In [6], the ensemble learning (stacking) technique is used and achieved the best classification performance. Arp et al. [7] proposed a lightweighted method that allows identifying malicious apps on a smartphone based on static analysis. The features which are enclosed in a joint vector space allow the ordinary patterns that indicate the malware to be identified automatically and are also used for explaining the decisions. DREBIN could identify 94% of the malicious cases with a few false alarms outperforming the other approaches of malware detection.

Kalash et al. [8] proposed a malware classification system based on deep learning. Twenty-five families of windows malware types are converted into gray scale pictures, and afterward, a convolutional neural networks (CNN) applied to malware pictures. CNN has applied on two distinctive datasets namely Malimg and Microsoft malware and achieved more than 95% accuracy on two datasets, respectively. Yerima et al. [9] created a model by preparing base classifiers at a lower level and afterward applied ranking-based algorithms on their predicted accuracies at the higher level to determine a final classifier. The proposed model is applied to four different datasets and shown adequate accuracy in all cases. Li et al. [10] proposed an Android malware detection approach using two types of features from the Android apps, namely permissions and API function calls. The authors analyzed Android apps to get the corresponding manifest files (*.xml) and the source files (*.java), and then, they extracted permissions, API function calls to constitute the features set of an Android app. Deep learning is applied along with nine grouping rules for differentiating malicious apps (malware) from the benign ones. Dasgupta et al. [11] proposed malware detection using Android using data mining techniques. Various classification algorithms are applied for malware detection and achieved good accuracy with random forest which takes only 1 s to identify 100,000 samples. Vinod and Viswalakshmi [12] applied various feature selection methods to detect Android malware. System calls are captured in two different settings via monkey runner, human interaction and achieved best results with ensemble learning.

Darus et al. [13] converted Android malware samples into gray scale images and extracted image features using the GIST descriptor. Later, linear classification algorithms namely decision tree, *K*-NN, random forest are applied and achieved an accuracy of 84%. This paper also used the same techniques as in [13] for converting Android apps into images, but the authors applied a convolution neural networks feature extraction from the input images followed by the densely connected neural network with Softmax for malware detection.

3 Research Methodology

Malware apps often involve repeated occurrences of certain types of methods and classes that lead to malicious activity upon activation in a system. The proposed methodology aims to recognize the repeated occurrences of such methods and classes in the form of code patterns in the.dex files. The contents of a code file are converted into image form by representing the sequence of code bytes as gray scale images for efficient pattern extraction from it. Since deep learning with CNN architecture is accepted as the best option for complex feature extraction from images for the purpose of classification, the authors developed an efficient framework for Android malware detection.

The Android apk file (mobile app) is a zip file. Extraction of an apk file produces a collection of files that includes assets, meta-inf, res, AndroidManifest.xml, classes.dex, etc. In these files, 'classes.dex' file contains classes or methods used within that app, and hence, 'classes.dex' is an important file for differentiating malware and non-malware apps. Experimentation was carried out to check whether the whole apk file has better discrimination capability than 'classes.dex' file to detect malware by creating input images in two different ways. In the first experimental setting, whole apk files are converted to images, whereas in the second experimental setting, only 'classes.dex' parts of apks are converted to images. After creating images, CNN model is applied on these two datasets separately. The block diagram of the malware detection framework for this experimentation is depicted in Fig. 1.

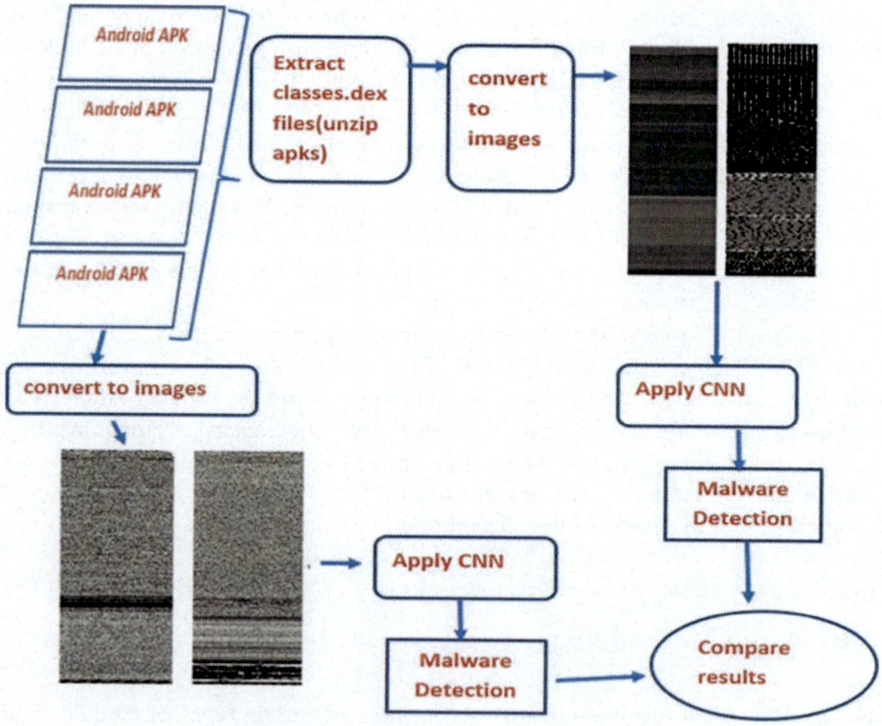

Fig. 1 Proposed framework for malware detection

3.1 Convolutional Neural Network

Convolutional neural network or Convnets introduced by LeCun [14] are very popular neural network methods for solving image classification problems. There are several layers in the CNNs which make them suitable for complex problems. The input images are processed in different layers of CNNs for feature extraction and classification. The proposed architecture involves one or more convolutional and pooling layers depending on the dataset for feature extraction followed by flattening of 2-D features to unidimensional form to provide input to a fully connected neural network with RELU for generating a linear combination of input features and sigmoid activation at the output neuron.

3.2 Convolutional Operation

In convolution operation, input images are applied with feature detectors (or filters) to get feature maps. The size of the feature detector in our experiments is selected as 3 by 3. The resulted feature maps summarize the presence of features detected in the input images. The feature map preserves the important features of input images, and they also reduce the size. After this step, the inputs are applied to the activation function ((RELU).

3.3 Pooling

CNNs are very powerful, and they can detect the features of original images with spatial invariance property. Pooling is a mechanism which makes CNNs to detect the actual features in the input images even though they are tilted or squashed. There are several types of pooling techniques available like max-pooling, min-pooling, and average-pooling. Max-pooling returns the maximum value from the portion of the image covered by the kernel. Max-pooling also performs noise suppression. So, a max-pooling strategy is selected for pooling. Pooling also reduces the size of convolved features.

3.4 Flattening

In this step, all features are converted into a single dimensional array. The output of flattening is used as input for the neural network.

3.5 *Full Connection*

After flattening, a vector is generated (which represents features of the images). This vector is passed as input to the neural network model.

4 Experimentation and Results

4.1 *Dataset*

Malware mobile apps are collected from virusshare.com. Benign mobile apps are collected from apkpure.com, CICAndMal2017 [15], and Google play store. The dataset contains 2474 malware mobile apps and 2499 benign mobile apps. All Android mobile apps have an extension of.apk. All these apks are converted into two different types of image datasets. In dataset-1, whole apks are converted into images, whereas in dataset-2, only classes.dex files of apks are converted into images. For each dataset, the images are divided into training and testing sets with 80% and 20% split, respectively. Tables 1 and 2 show the number of samples used in the experiments.

Table 1 Dataset-1 details

	Number of benign samples	Number of malware samples	Total
Number of images generated	2484	2474	4958
Number of images used for training	1955	1973	3928
Number of images used for testing	529	501	1030

Table 2 Dataset-2 details

	Number of benign samples	Number of malware samples	Total
Number of images generated	2499	2309	4808
Number of images used for training	1970	1810	3780
Number of images used for testing	529	499	1028

4.2 Conversion of apks/dex Files into Gray Scale Images

Android apps are by default zipped files with.apk extension. Apk contains several files in a zipped format. For creating dataset-1, we converted direct apks into images. Dataset-2 is created by applying to unzip operation on the apk files, and later, classes.dex files are selected from the unzipped folder of each file. Gray scale image datasets are created in these two different settings. The height of the gray scale image changes depending on the file size of the whole apk/only dex file. The width of the gray scale images is selected based on [16]. Image width varies from 32 to 1024 KB based on the apk/dex file size.

Algorithm for converting apk files/dex files into gray scale images:

```
  Step 1:open directory with apk/dex files
  Step 2: for all files in directory
  Step2.1:calculate file size
  Step 2.2:if filesize<10 KB then, width=32
    if filesize>=10 KB and filesize<=30 KB then, width=64
    if filesize>30 KB and filesize<=60KB then, width=128
    if filesize>60 KB and filesize<=100KB then, width=256
    if filesize>100KB and filesize<=200KB then, width=384
    if filesize>200KB and filesize<=500KB then, width=512
    if filesize>500KB & filesize<=1000KB then, width=768
    else, width=1024
  Step 2.3: Convert file into image using python numpy's
"fromfile" method
  Step 2.4 :Save the image file as .png file using py-
thon's imsave method
  Step 3: close the directory
```

After this step, images with different sizes are generated. Then, all the images are resized using bulkresize.com. After extracting images from Android apks in two different ways, convolutional neural networks are applied on these datasets.

4.3 Data Augmentation

Data augmentation plays a vital role in improving the accuracy of the CNN model. Before inputting images to CNN architecture, it is necessary to apply data augmentation techniques to the training image dataset [17]. In the image data augmentation, some operations like zooming, rotating, flipping are added artificially. This step creates a modified version of images with different transformations. The advantage of data augmentation is that the classifier can easily differentiate the images in the testing phase even though the positions of patterns in the testing set images are bit different from training set images. A feature scaling technique is applied by dividing each pixel value by 255, so that all the values fit between 0 and 1. In addition to feature scaling, 'shear range = 0.2' (geometrical transformation), 'horizontal_flip = True', 'zoom_range = 0.2' data augmentation techniques are also applied. The data augmentation step creates different variations of the training images that can improve the ability of the CNN model to generalize the learning process to new images. For the testing set, only the feature scaling technique is applied. The reason for not applying other transformations to testing data is that the CNN model tests the performance on the original testing set. 'ImageDataGenerator' class [18] of Keras is used for applying all these transformations.

4.4 Experiments with Dataset-1

Dataset-1 contains images of direct apks with a training set size of 3928 (1955 samples of benign, 1973 samples of malware) and a testing set size of 1030 (529 samples of benign, 501 samples of malware). 80% of the total samples are used for the training set and 20% for the testing set. All experiments are conducted on the Google Colab-Keras GPU environment.

4.5 CNN Architecture for Dataset-1

Figure 2 shows the CNN architecture for dataset-1. For dataset-1 (whole apk as an image), the input size of the image is taken as 128 × 128. The feature detector size is 32 with a kernel size as 3. In max-pooling step, we used a pool size of 2 with stride

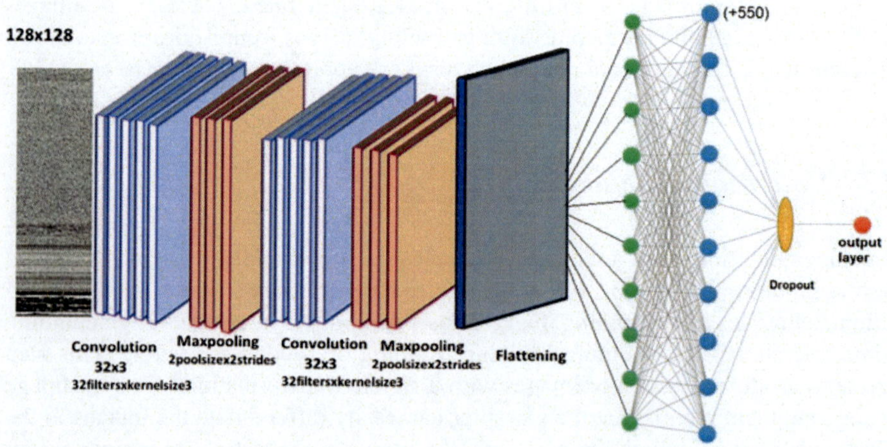

Fig. 2 CNN architecture for dataset-1

Table 3 Experimentation results

	Training accuracy (%)	Training loss	Testing accuracy (%)	Testing loss
CNN with whole apks	99.8	0.0010	96.50	0.6475
CNN with dex files	99.9	0.0062	97.76	0.2599

size 2. Two layers of convolutional and max pool operations followed by a flattening step are used in CNN for feature extraction. Five hundred and sixty neurons with RELU activation function are used in the fully connected dense layer followed by the output neuron with sigmoid activation function for detecting malware. 'Adam' optimizer is used for minimizing binary cross entropy which is used for assessing the loss. The network is trained for 150 epochs for learning the weights of edges.

4.6 Experiments with Dataset-2

Dataset-2 contains images of classes.dex files of apks with a training set size of 3780 (1970 samples of benign, 1810 samples of malware) and a testing set size of 1028 (529 samples of benign, 499 samples of malware). For processing the dex file in the image form of input size 128 × 128, the feature detector size is 32 with kernel size as 3 is used. In the max-pooling step, a pool size of 2 with stride size as 2 was used. One layer of convolutional and max-pooling operations is followed by a flattening step. In the fully connected layer, five hundred neurons are used with RELU for activation.

This paper uses the accuracy percentage as a measure for evaluating model performance. Accuracy is defined as the 'total number of correctly labeled samples by the total number of samples'.

4.7 Results of Experiments

Table 3 shows the results of two experiments. Training accuracy for both the datasets is above 99%. The training loss for both datasets is also nearly the same. For dataset-1, an accuracy of 96.5% achieved, and for dataset-2, an accuracy of 97.7% achieved.

4.8 Comparison of Results with Previous Work

Table 4 shows the accuracy comparison of the proposed work with previous work. In [13], three machine learning algorithms were applied and achieved an accuracy

Table 4 Accuracy comparison with previous work

Method	Accuracy (%)
K-NN [11]	80.69
Decision tree [11]	78.62
Random forest [11]	84.14
Proposed CNN with whole apks	96.50
Proposed CNN with dex files	97.76

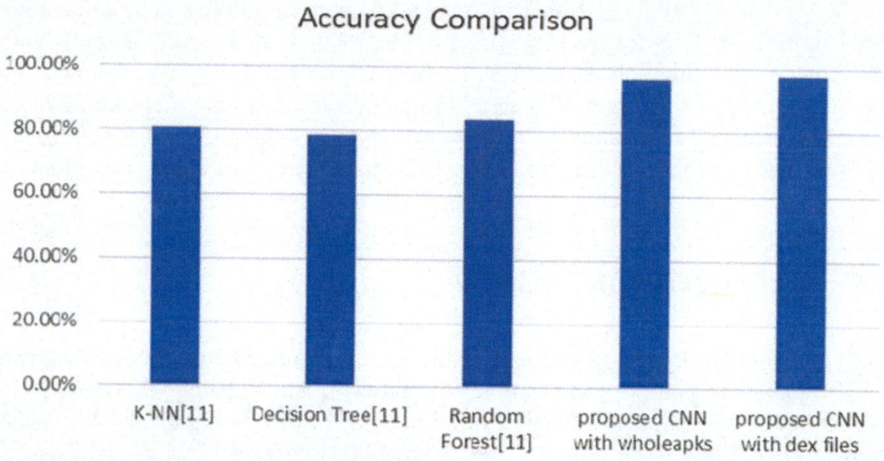

Fig. 3 Accuracy comparison with previous work

of 80.69% with *K*-NN, 78.62% with decision tree, and 84.14% with random forest classifier. In this paper, the authors applied CNN on two different input datasets and achieved an accuracy of 96.5%, 97.76% for direct apk images and dex images files, respectively.

From the results, it can be concluded that only classes.dex files of apks are sufficient for Android malware detection. Dex file image classification could efficiently classify Android apps as either malware or benign with fewer data only, whereas whole apk image classification has less accuracy with architecture (Fig. 3).

5 Conclusion and Future Scope

In this paper, the authors proposed a convolutional neural network approach for Android malware detection. Android apk samples are converted into gray scale images and then trained a CNN model for malware detection. Experimental results on different combinations of malware and benign apk images shown that convolutional neural networks outperform the traditional machine learning models used in previous works and could achieve an accuracy of over 97.76% on.dex files. Based on further experimentation, it was also concluded that the 'classes.dex' files rather than the whole Android apk files provide sufficient inputs for accurate detection of Android malware in this methodology. In future, we intend to conduct experiments on the larger dataset. It is also planned to apply this procedure for the classification of malware families.

References

1. Statista, The statistics portal. [Online], Available: https://www.statista.com (https://www.statista.com/statistics/272698/global-market-share-held-by-mobile-operating-systems-since-2009/)
2. D.J. Wu, C.-H. Mao, T.-E. Wei, H.-M. Lee, K.-P. Wu, DroidMat: Android mal-ware detection through manifest and API calls tracing, in *Proceedings of 7th Asia Joint Conference on Information Security (Asia JCIS)*, 2012, p. 62–69.
3. A. Fatima, R. Maurya, M.K. Dutta, R. Burget, J. Masek, Android malware detection using genetic algorithm based optimized feature selection and machine learning, 978-1-7281-1864-2/19 ©2019 IEEE.
4. J. Li, L. Sun, Q. Yan, Z. Li, W. Srisaan, H. Ye, Significant permission identification for machine-learning-based android malware detection. IEEE Trans. Ind. Inf. **14**(7) (2018)
5. W. Wang, Z. Gao, M. Zhao, Y. Li, J. Liu, X. Zhang, DroidEnsemble: detecting android malicious applications with ensemble of string and structural static features, 2169-3536, 2018, IEEE Access. https://doi.org/10.1109/ACCESS.2018.2835654
6. P. Feng, J. Ma1, C. Sun, X. Xu, Y. Ma, A novel dynamic android malware detection system with ensemble learning. 2169-3536 2018 IEEE-Access. https://doi.org/10.1109/ACCESS.2018.2844349
7. D. Arp, M. Hubner, M. Spreitzenbarth, K. Rieck, CERT Siemens, H. Gascon, Drebin: effective and explainable detection of android malware in your pocket. NDSS, 2014, 23–26 Feb 2014, San Diego, CA, USA
8. M. Kalash, M. Rochan, N. Mohammed, N.D.B. Bruce, Y. Wang, F. Iqbal, Malware classification with deep convolutional neural net-work. 978-1-5386-3662-6/18 ©2018 IEEE
9. S.Y. Yerima, S. Sezer, DroidFusion: a novel multilevel classifier fusion approach for Android malware detection. IEEE Trans. Cybern., **49**(2), (2019)
10. W. Li, Z. Wang, J. Cai, S. Cheng, An Android malware detection approach using weight-adjusted deep learning, in *2018 International Conference on Computing, Networking and Communications (ICNC): Communications and Information Security Symposium*, 978-15386-3652-7/18 ©2018 IEEE.
11. S. Dasgupta, S. Saha, S.K. Da, Malware detection in Android using data mining. Int. J. Nat. Comput. Res. **6**(2) (2017).
12. P. Vinod, P. Viswalakshmi, Empirical evaluation of a system call-based android malware detector. Arab. J. Sci. Eng. **43**(12), 6751–6770 (2018)
13. F.M. Darus, N.A. Ahmad, A.F.M. Ariffin, Android malware detection using machine learning on image patterns. 978-1-5386-7541-0/18, ©2018 IEEE.
14. Y. Lecun, P. Haffner, L. Bottou, Y. Bengio, Object recognition with gradient based learning, AT&T Shannon, 100 Schulz Drive, Red Bank, NJ 07701, USA
15. online: unb.ca/cic/datasets/andmal2017.html
16. L. Nataraj, S. Karthikeyan, G. Jacob, B.S. Manjunath, Malware images: visualization and automatic classification, in *Proceedings of the 8th International Symposium on Visualization for Cyber Security*, Pittsburgh, Pennsylvania, USA, 2011.

17. C. Shorten, T.M. Khoshgoftaar, A survey on image data augmentation for deep learning. J. BigData. https://journalofbigdata.springeropen.com/articles/10.1186/s40537-019-0197-0, Springer-open, 2019.
18. online:https://keras.io/api/preprocessing/image/#imagedatagenerator-class

Predicting the Accuracy of Accessibility of LTE Network Using ANN

Abhishek Tyagi, Asmita Singh, Sindhu Hak Gupta, and Mukesh Mishra

Abstract The excessive enhance in the number of subscribers in the wireless cell network, due to which the Mobile Network Operators (MNOs) facing the immense issues in giving the high-level executives of the broadband utility.cyc Long-Term Evolution (LTE) has capable of encountering the user's demands by giving a high level of data rate. The key performance indicators (KPIs) have been recorded and reform the network execution in request to give the high-grade utility and also to attain better resource usage. Accessibility is utilized to figure out the success ratio of the user equipment (UE) in penetrating the cell network. The accessibility is indicated as the possibility that user equipment (UE) will be apt to penetrate the cell network utility for a fixed time duration. In this paper, the ANN tool is used to find out the prediction accuracy level of the accessibility of the wireless cellular network. For this purpose, key performance indicators (KPIs) data are acquired from the real field measurement of approximately 50,000 BTS locations by the Nokia Network Pvt. Ltd., India.

Keywords Long term evolution (LTE) · Artificial neural network (ANN) · Key performance indicators (KPIs) · Accessibility

---

A. Tyagi (✉) · A. Singh · S. H. Gupta
Department of Electronics and Communication Engineering, Amity University, Noida, Uttar Pradesh 201313, India

S. H. Gupta
e-mail: shak@amity.com

M. Mishra
Nokia Network Pvt. Ltd., Gurugram, Haryana, India
e-mail: mukesh.2.mishra@nokia.com

V. Bhateja et al. (eds.), *Data Engineering and Intelligent Computing*,
Advances in Intelligent Systems and Computing 1407,
https://doi.org/10.1007/978-981-16-0171-2_16

1 Introduction

There is excessive enhancement in wireless technologies in the upcoming years. The mobile cell networks are moderately altering into the Internet of things (IoT) environment that combines with wireless user equipment (UEs) [1]. This unparalleled modification will not only operate an excessive development in the network congestion in the upcoming years, yet it will further show the way of the arrival of the latest and unproved wireless transmission resource utilize instances that considerably diverge from the standard grids or utility depend on voices [2]. For example, apart from the demand for large data rate, which has been the prime operator of the cell network advancement in the past several years, the upcoming technology networks will also dispatch ultra-reliable, low-latency transmission and that is flexible and in actual time to the spirited of the user's and the respective ecosystem [2, 3]. For now, in the upcoming technology, a high amount of data will be composed, repeatedly and in actual time, over an immense number of the user equipment. Such immense transmitting will lead to significant congestion across the network [4]. This eventually builds an extremely dissimilar networking ecosystem whose unusual implementation and inherent various QoS and security necessity fiat a rudimentary alter in the strategy in which cell grids are graved, perused, created, and upgraded [5].

LTE confronts the prime challenges such as the execution of the system, reliability, successive transmission, and the dropping of information transmitting while going at a very high speed. The execution level of the transmitting terminal depends on the enhancement in the transmitting terminal model. Due to the constructive transmission in the LTE network, high speed, high level of reliability, and large sequences transmitting terminal methods that change the numerous upcoming devices.

The demand to get through this continuing and speedy development of transmission utility has to guide to an appreciable device of exploration that evaluates what is the best wireless cell network structure will be inside the environment of the upcoming technology cell network [6]. The necessity of smart operations is apt to alterably utilize the transmission network utility and the created data, in place to upgrade the cell network performance and warranty, in the actual period, the quality of services (QoS) demands of appearing transmission and Internet of thing (IoT) utilities. Such a smart center can be perceived by the concepts of machine learning (ML), in specific artificial neural network method [7]. Artificial neural network is a mathematical nonlinear machine learning method that can be utilized for supervised learning, unsupervised learning, semi-supervised learning, and reinforcement learning, in numerous wireless networking situations [8].

Artificial neural networks have appeared to achieve an important role in the field of wireless communication and also in the wireless cell network, due to the excessive growth in the intelligent grids and the mobile subscribers of the wireless cellular network. Furthermore, due to the growth of mobile edge computing and the storing technologies build the BTS to keep and examine the efforts of the subscribers of the wireless cellular network. As well as, the exposure of the Internet of things (IoT) influences the utilization of artificial neural networks to enhance the method in which the cell network information is handled, gathered, and utilized for numerous observing and another occasion [5].

Artificial neural networks can be utilized to examine and predict the cell network subscriber's mobility and the other parameters, consequently permitting the BTS to reform the utilize of their utilities, that will be stored all over the cell network.

To achieve the desired aim of this paper, the following studies have been done. In [9], the author determines and examines the accessibility and retainability of 24 cell groups of the LTE network executed in Telecom of Kosovo (TK). In [10], the author predicts the execution level of the radio resource control setup rate by using the discrete-time Markov chain algorithm. In [11], the author examines the quality of service (QoS) based on the key performance indicators such as accessibility, retainability, PBS downlink used for the LTE cell network to calculate the issue using the bandwidth development.

In this paper, artificial neural networks is used for predicting the accuracy of accessibility KPI of the LTE wireless cellular network for getting a better transmission environment. This prediction is done on the data set obtained from the real field scenario of nearly 50,000 BTS locations by the Nokia Network Pvt. Ltd, India.

The remaining part of the paper is arranged as follows: Sect. 2 expresses the accessibility and its respective KPIs, Sect. 3 gives the short idea of artificial neural network (ANN), Sect. 4 gives the optimizing outcome, and Sect. 5 gives the conclusion.

2 Accessibility

Accessibility is used to put down the objective of the success rate of the subscribers in acquiring the cell network and recognized the issues for compensating them. The key performance indicators (KPIs) are attained by the help of the drive test. The accessibility KPI allows the cell network engineer to recognize if the resource needed by the UE can be acquired. If the assessments for accessibility are not contemplated well, then the cell network engineer should explore the several methods for enhancing the accessibility [10]. This KPI has undergone three different regions to lay out the resources, i.e., radio resource control (RRC) success ratio, E-UTRAN radio access bearer (E-RAB) success ratio, and S1 setup success ratio. RRC setup success ratio is generally predominant KPI which helps in measuring the accessibility. Accessibility is the possibility that the UEs of a resource after an appeal to a network obtain the begin to choose the signal within the set-out situations. For example, user appeals to acquire the cell network, penetrates the voice call, and data call.

Radio resource control (RRC) connection setup success (RRC Con. SSR) is used to estimate the connection success ratio in a cell network. It is estimated as the ratio of the RRC connection setup completion to the RRC connection requests [9]. The notion of RRC is utilized in Universal Mobile Telecommunication Service (UMTS) and Long-Term Evolution (LTE) on the air interface. It is a level that manages between the UE and e-node B (eNB) and also manages at the IP level. The main responsibility of the RRC concord involves link formation and deliver trails, telecast

Table 1 Probable levels of radio resource control setup success rate

Level	Situation	Accessibility
1	RRC_Con._SR > 99.999%	Acceptable
2	98% ≤ RRC_Con._SR < 99.999%	Adequate
3	RRC_Con._SR < 98%	Unacceptable

Table 2 Portable level of ERAB_SSR

Level	Situation	Accessibility
1	ERAB_SSR > 99.99%	Acceptable
2	98% ≤ ERAB_SSR < 99.99%	Adequate
3	ERAB_SSR < 98%	Unacceptable

of network information, radio bearer formation, rearrange and deliver, RRC link portability approach, paging informing and deliver, and external loop potential curb. By the signaling trails, the RRC composes the user and control planes as per the cell network stature and grants the radio resource management approach to be executed (Table 1).

$$\text{RRC Con. SSR} = \frac{\text{RRC connection setup completion}}{\text{RRC connection request}} \times 100 \tag{1}$$

E-UTRAN radio access bearer (E-RAB) setup success ratio is depending on the setup success ratio for entire utility in the cell network or clusters of cells. ERAB is a predominant specification in Long-Term Evolution (LTE) key performance indicators (KPIs) perusal. This specification is established by merging the S1 bearer and the commensurate radio bearer. E-RAB can be notable that the trail is cognate to the ERAB ID and the quality class identifier (QCI). The ERAB ID specification is utilized to pivotal the ERAB on the S1 interface. In a mobile cell network, some utilities alter the quality of service (QoS), like bit error rate (BER) and point-to-point slow down [12] (Table 2).

$$\text{ERAB SSR} = \frac{\text{ERAB Setup Successful}}{\text{ERAB Setup Attempts}} \times 100 \tag{2}$$

3 Artificial Neural Networks

As in the brain, the essential procedure components of an artificial neural network are the neurons (nodes). In the neural network, nodes are set as layers and are interconnected so that the data move unicellular, from the input nodes along the nodes placed in the hidden layers, to the nodes in the output layer. Input nodes give out the signal to the hidden nodes of the next layer. A neuron primarily executes linear regression heed by nonlinear tasks, $f(°)$. Neurons of dissimilar layers are interrelated with the commensurate weights. The level of different layer ANN system, whose node a' in layer '$k + 1$' capitulates

$$X_a^{k+1} = f\left(\sum_i w_{ia}^k x_i^k + w_{ba}^k\right) \quad (3)$$

where w_{ia}^k are the mass linking node 'i' in the layer 'k' to node 'a' in layer '$k + 1$', w_{ba}^k is the mass denominate of node 'a' in the layer 'k' and 'f' is a logistic awaking trait. The prognostication of the system for the specimen $x_i x_i$ is standing for $f(x_i)$. The objective of the method is to figure out a group of weight which confirms that, for each inserted vector, the outcome vector from the network is similar or pretty close, to the appropriated outcome vector. If there is a decided and limited group of input–output specimens, the comprehensive error in the working of the network with the specific group of weights can be estimated by contrasting the actual and preferred outcome vectors for each specimen, for example, by the technique of least squares. Drilling of an ANN demands an adopting organize (number of invisible layers and nodes per layers), the actual determining of the weights, learning rate, and the regulating variables to avert overfitting [13].

Figure 1 reveals the systematic structure of the artificial neural network. In this, the entire structure is divided into three sub-layers: The first layer is the input layer, the succeeding layer is known as the hidden layer, and the last layer which provides the results is the output layer. The entire computation is done in the hidden layer.

Artificial neural networks can be concurrently engaged for both the prediction and the smart functions, for such a framework in which two operations are highly interconnected. For such occurrence, the information is helpful in making the conclusion, and the creating conclusion helps in generating more information. For example, when assuming the real field functions across the cell network, so the operator can utilize artificial neural networks to predict the efforts of the subscribers. These predictions can aid an artificial neural network to depend on the reinforcement learning algorithm to assign the calculation and spectral utilities to the subscribers so enhancing the quality of services.

4 Simulation Results

The simulation is attained with the help of the ANN tool. The details were collected from the nearly 50,000 macro BTS positions. This detail is going on with the ANN tool. The design approach prophesied the regression level of the radio resource control (RRC) setup success rate and E-UTRAN access bearer (E-RAB) setup success rate.

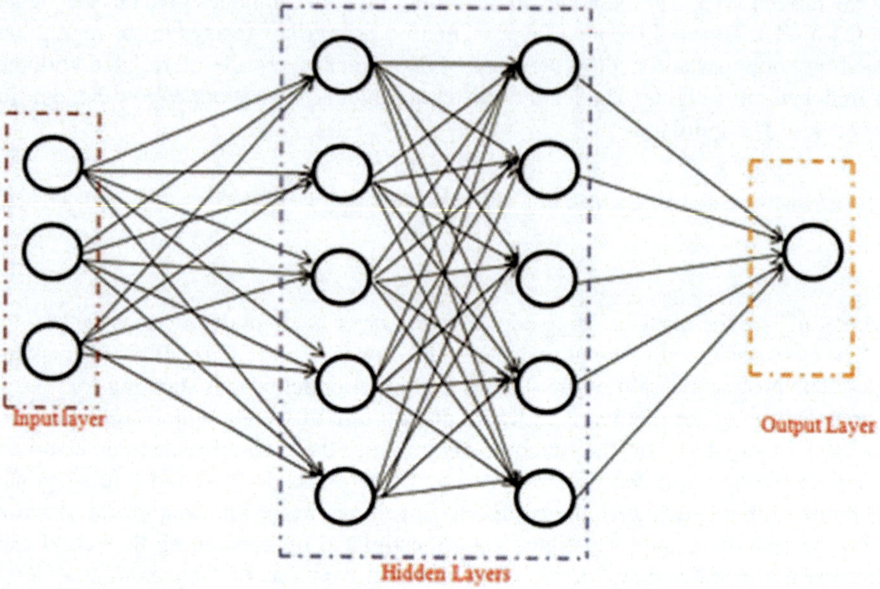

Fig. 1 Artificial neural network architecture

The different parameters set in the ANN tool, which help in attaining the desired objective, are given in Tables 3 and 4.

Figure 2 shows the different steps of the optimizing process. Figure 2a shows the error intensity during the training and testing of radio resource control. Similarly, Fig. 2b gives the performance graph of radio resource control during training and testing. The best validation of the performance is 0.036276 at the epoch 85. Now, the next is the regression plot of RRC_SSR which provides the accuracy level at the time of training and testing. Figure 2d shows the training plot. The mean square error of the RRC is 0.782.

Table 3 Parameters utilized

Parameters	Values
Number of hidden layer	20
Algorithm utilized	Scaled conjugate

Table 4 Parameters used

Input parameters		Output/target parameter
RRC connection setup completion	RRC connection request	RRC_SSR
ERAB setup successful	ERAB attempts	ERAB_SSR

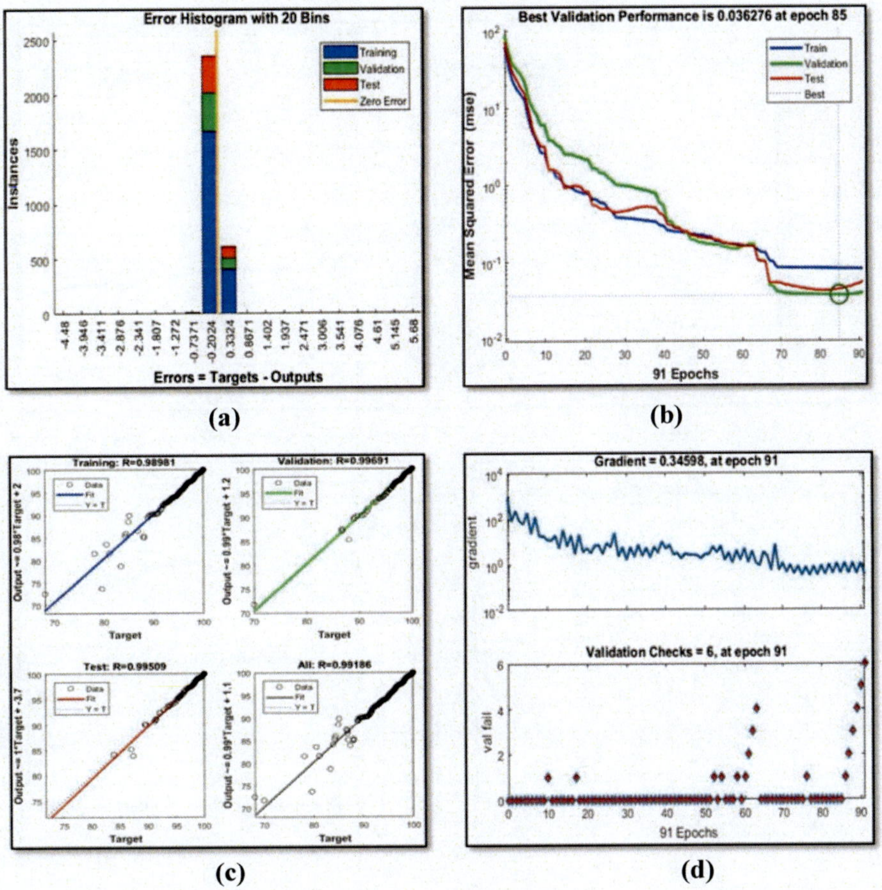

Fig. 2 Optimization of RRC_SSR **a** Error, **b** performance level, **c** regression, and **d** training level

Figure 3 shows the different steps of the optimizing process. Figure 3a shows the error intensity during the training and testing of E-UTRAN access bearer. Similarly, Fig. 3b gives the performance graph of ERAB during training and testing. The best validation of the performance is 0.29437 at the epoch 25. Now, the next is the regression plot of ERAB_SSR which provides the accuracy level at the time of training and testing. Figure 3d shows the training plot. The mean square error of ERAB is 2.13.

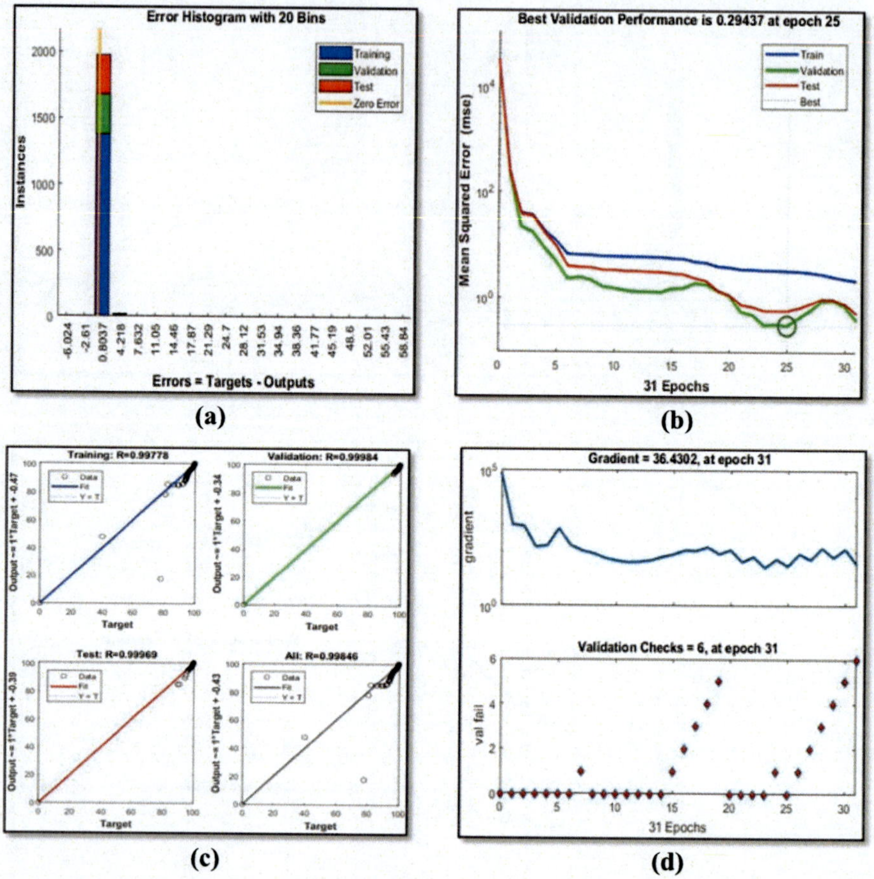

Fig. 3 Optimization of ERAB_SSR, **a** Error, **b** performance level, **c** regression, and **d** training intensity

5 Conclusion

In this paper, the artificial neural network (ANN) tool is used to prophesy the network implementation level. For achieving the desired objective, the actual time data records are used, which is the key performance indicators (KPIs) are obtained by the Nokia Networks Pvt. Ltd., India. The prime goal of this paper is to predict the accuracy of the accessibility KPI. As the data congestion level is enhancing in the upcoming technologies, there will be an immense demand for access to the cell network. So, this paper is to fulfill the desired objective. In the future, if the KPIs data set is obtained hourly than this scheme aid the network engineers to prophesy their network accessibility in advance and also makes the transmitting terminal smart, which is most helpful in the 5G.

References

1. N.C. Luong, D.T. Hoang, P. Wang, D. Niyato, D.I. Kim, Z. Han, Data collection and wireless communication in Internet of Things (IoT) using economic analysis and pricing models: a survey. IEEE Commun. Surv. Tutor. **18**(4), 2546–2590 (2016)
2. Z. Dawy, W. Saad, A. Ghosh, J.G. Andrews, E. Yaacoub, Toward massive machine type cellular communication. IEEE Wirel. Commun. **24**(1), 120–128 (2016)
3. T. Park, N. Abuzainab, W. Saad, Learning how to communicate in the Internet of Things: finite resources and heterogeneity. IEEE Access **4**, 7063–7073 (2016)
4. G. Durisi, T. Koch, P. Popovski, Toward massive, ultra-reliable and low-latency wireless communication with short packets. Proc. IEEE **104**(9), 1711–1726 (2016)
5. M. Chen, U. Challita, W. Saad, C. Yin, M. Debbah, Artificial neural networks-based machine learning for wireless networks: a tutorial. IEEE Commun. Surv. Tutor. **21**(4), 3039–3071 (2019)
6. J.G. Andrew, S. Buzzi, W. Choi, S.V. Hanly, A. Lozano, A.C. Soong, J.C. Zhang, What will 5G be? IEEE J. Sel. Areas Commun. **32**(6), 1065–1082 (2014)
7. T. Segaran, *Programming Collective Intelligence: building Smart Web 2.0 Applications* ("O'Reilly Media, Inc.", 2007).
8. B. Yegnanarayana, *Artificial Neural Network* (PHI Learning Pvt. Ltd., 2009)
9. F. Krasniqi, A. Maraj, E. Blaka, Performance analysis of mobile 4G/LTE networks, in *2018 South-Eastern European Design Automation, Computer Engineering, Computer Networks and Society Media Conference (SEEDA_CECNSM)*. IEEE, 2018, p. 1–5.
10. H. Hendrawan, Accessibility degradation prediction on LTE/SAE network using discrete time Markov chain (DTMC) model. J. ICT Res. Appl. **13**(1), 1–18 (2019)
11. F. Sirait, A.W. Dani, Y. Yuliza, U. Albab, Optimization in quality of service for lte network using bandwidth expansion. Sinergi **23**(1), 47–54 (2019)
12. A. Genc, *An Effective Solution of ERAB Problems in LTE*
13. V. Rodriguerz-Galiano, M. Sanchez-Castillo, M. Chica-Olmo, M.J.O.G.R. Chica-Rivas, Machine learning predictive models for mineral prospectivity: an evaluation of neural networks, random forest, regression trees and support vector machines. Ore Geol. Rev. **71**, 804–818 (2015)

Histopathological Image Classification Using Deep Neural Networks with Fine-Tuning

A. S. Vidyun, B. Srinivasa Rao, and J. Harikiran

Abstract The role of deep learning is predominant in classifying medical images. Histopathological image classification involves classifying tumor and cancer tissue subtypes, but it has many challenges like complexity of image, labeled data availability. Overcoming these challenges and accurately classifying the images require better approach than traditional one. In this work, we use fine-tuning-based deep neural network in classifying the histopathological image into eight different classes. The obtained results are compared with different other CNN and machine learning models. In addition, we trained a model using early stopping function which monitors the validation accuracy during training and prevents overfit, which makes the model work trained efficiently. This approach is tested on colorectal cancer histology image dataset which contains 5000 labeled images belonging to eight different classes. In our result, fine-tuning with early stopping performed better than other methods with accuracy of 91.2%.

Keywords Deep learning · Fine tuning · Early stopping · Histopathological image classification · Overfitting

1 Introduction

Colorectal cancer is the third most common type of cancer [1] with 1.3 million new cases every year. According to the American Cancer Society, about one in 23 men and one in 25 women will develop colorectal cancer [2] during their lifetime in the USA, and further, a new study shows that there are 440,000 deaths in a year in USA caused by medical errors [3]. Hence, accurate diagnosis in early stage is crucial for treatment. Diagnosis of cancer is quite challenging since there is a shortage of pathologists, China has one pathologist per 74,000 people [4]. The amount of labeled data availability is less so using lesser data and providing better accuracy are quite

A. S. Vidyun (✉) · B. Srinivasa Rao · J. Harikiran
School of Computer Science and Engineering, Vellore Institute of Technology, VIT-AP University, Amaravati, India
e-mail: vidyun.as@vitap.ac.in

V. Bhateja et al. (eds.), *Data Engineering and Intelligent Computing*, Advances in Intelligent Systems and Computing 1407,
https://doi.org/10.1007/978-981-16-0171-2_17

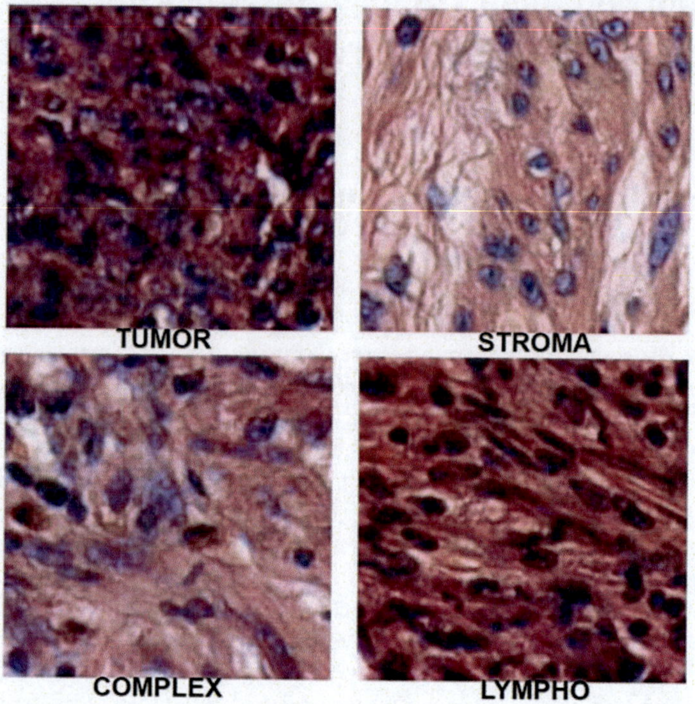

Fig. 1 Images of colorectal tissues of four types

challenging. As per the Harvard study by Professor Ashish Jha, over 5.2 million medical error cases have been accounted [5] in India annually, so accurate diagnosis with help of available data is key step for diagnosis. The images are complex [6] and hard to diagnose so it is obscure for pathologist as shown in Fig. 1. With the help of present technology and deep learning techniques, we can overcome these challenges and diagnose correctly with the amount of data available.

Artificial intelligence bridged the gap between machines and humans in all fields, particularly in the medical field to uncover the hidden opportunities and patterns in the clinical data, helping doctors diagnose the disease. In the literature, deep learning has provided better classification accuracy compared to other machine learning algorithms since they have layers of neurons which are connected to each other present in other layers and updates the weight automatically on every epoch, and the data is scaled which is much better than machine learning algorithms. In biomedical image classification, many scholars provided different approaches and ideas on handling data, and Pang et al. [7] proposed novel fused convolutional neural network for biomedical image classification. Zhang et al. [8] classified the biopsy images using support vector machine (SVM), and on validating the model with 361 images, he achieved better accuracy. Ver, Asri et al. [9] performed machine learning algorithms on Wisconsin breast cancer images where he used SVM, *K*-nearest neighbors (KNN), decision tree, Naive Bayes methods and achieved better accuracy in SVM method compared to other machine learning methods, so to compare our approach, we considered SVM classifier which outperformed other machine learning methods for Zhang et al. [8] and Ver, Asri et al. [9] as one of the models.

Further, Venubabu R et al. [10] proposed a lightweighted CNN model with dropout, max-pooling layers and validated against histopathological dataset which we used here. The lightweighted CNN model outperformed other machine learning with a minimum error rate of 22.7% so we considered traditional CNN as one of the methods to compare our proposed method. Kermany et al. [11] used a new approach in deep learning called transfer learning where models are pre-trained with 1.2 million images and 1000 separate object categories. Those model weights are used in classifying the images of retinal tomography and achieved better accuracy, so transfer learning is considered as one of the approaches for comparison. Proper feature selection [12] is the success factor in classification of images. Deep learning helps us in automatically extracting those features while training and performs better than others. Overfitting of model with more epoch is not an efficient way of training, therefore, in this work, we used early stopping function [13, 14] along with dropout which monitors the validation loss, and if the loss is not improved for certain number of epochs, the training is stopped, and best model is selected from the model checkpoint function. We use fine-tuning approach [15] where pre-trained weights (transfer learning) are taken, and some of the layers are freezed, and other layers are allowed to get trained, so the model works better than that of transfer learning model. The proposed method is performed on publicly available colorectal cancer histology dataset by kather [16] which contains eight different categories of H&E stained histopathological cancer images. The proposed model is compared with SVM model which performed well among the machine learning models and other CNN models. Here, we mainly focused on overfitting and accuracy of the model as our main priority.

2 Preliminaries

Zero Padding: Zero padding is defined as addition of a layer of zero values around the image, which helps to obtain data present at edges of images and helps in better performance of model without loss of data.

Convolution Layer: In this layer, convolution operation is performed on input image matrix. Convolution is a linear operation that involves multiplication of a set of weights with input matrix. These set of weights are called kernel or filter through which best features of the images are extracted automatically and feature map is created.

Activation Layer: Here, the neurons in the layers are activated based on the activation function which was used. This layer helps in production of nonlinear output to neurons. A neural network without an activation function works as a linear regression model, and therefore, the activation function transforms the data which is capable of learning and performing complex tasks.

Pooling Layer: It involves sliding a filter over the feature map to get summarized values of matrix. It is also used to reduce the feature map dimension without losing information, so the number of parameters to learn and the amount of computation performed are reduced in the network. Pooling methods namely average pooling and maximum pooling are used. In maximum pooling, maximum or average value of feature map on which filter is slided is considered as a new feature representing the region over which filter is placed. In the proposed approach, max-pooling with 2 × 2 filters and stride 2 is used.

Dropout Layer: This layer helps in prevention of overfitting problems where random neurons in the layer are deactivated so that all the neurons are weighed and trained equally, and therefore, all neurons in the layer have a major role in the training process. This makes the network more resistant [17, 18].

Softmax Function: The Softmax function is mainly used in classification problems where it provides a probability distribution for given input to each class. The number of neurons in the layer must be equal to the number of class labels. Every neuron represents a class label, and every neuron value represents probability of image belonging to the class.

Fully Connected: The neurons present in this layer are connected to every neuron in previous layer and apply weight to predict the label or pass the value to next layer.

Transfer Learning: In this technique, a model which is already pre-trained with the ImageNet dataset [19] is trained by 1.2 million images and 1000 separate object categories. It is reused, and the last layer is replaced with a layer of neurons equal to the number of classes present in our working dataset, so the model classifies images only among the provided classes [20]. Here, the VGG19 pre-trained model architecture along with filter size is shown in Fig. 2.

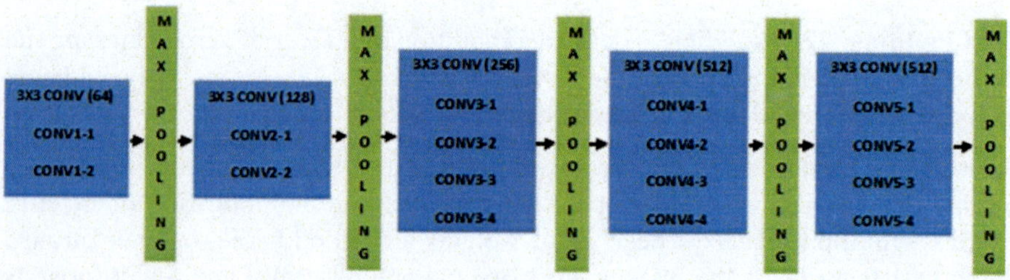

Fig. 2 Architecture of VGG19

3 Proposed Approach—Fine-Tuning

In fine-tuning [15], the pre-trained model from transfer learning is taken which is already trained with ImageNet dataset containing 1.2 million images and 1000 separate object categories. Many pre-trained models are available that are trained with different architectures. Here, we used VGG19 architecture as shown in Fig. 2. It consists of 16 layers made up of convolutional layer and fully connected layer.

In this paper, we propose fine-tuning approach using VGG19 architecture. Among the 16 layers of VGG19, last five layers of the network are fine-tuned or allowed to get trained, whereas other 11 layers and their weights are fixed. This helps the model to adapt its weights in the last five layers and weights of other layers fixed. This approach allows the model to adapt and perform the new task with the help of half trained (11 layers with fixed weights) rather than training the whole model from scratch. The convolutional layers in the model help to create feature map, followed by max-pooling layer which summarizes the feature map by considering only maximum value. We added fully dense connected layer with 1024 neurons with an activation function RELU which has reduced likelihood of vanishing gradient, dropout layer with a probability of 50% where probability of neuron deactivation is 0.5. This is connected to fully dense connected layer with 1024 neurons with the activation function RELU and dropout layer with the probability of 50% followed by dense layer of 1024 neurons, and finally, the output layer is added with activation function 'Softmax' is used to classify images. The model is fit with early stopping function [13] which monitors the validation loss and stops the training, if the loss did not improve for three epochs. The best model is saved using model checkpoint function. Figure 3 shows the fine-tuning model and its architecture which we used.

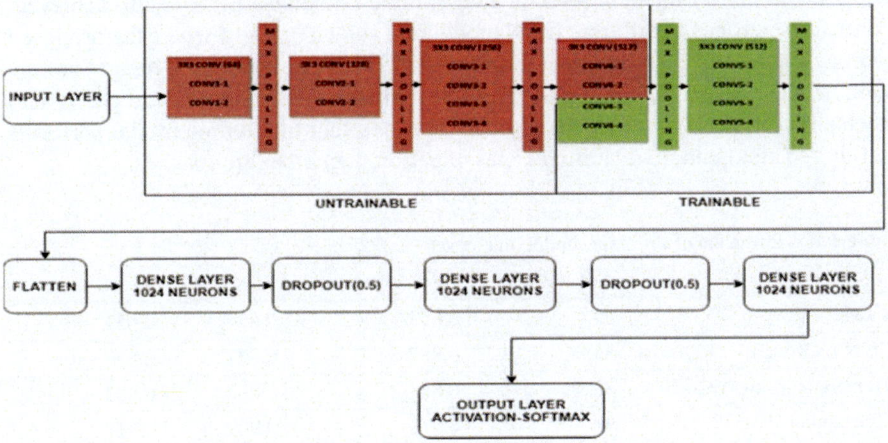

Fig. 3 VGG19 fine-tuning (proposed approach)

4 Experimental Results

4.1 Considered Dataset

The colorectal cancer histology image dataset [16] used in this paper consists of cancer patient images with eight different categories of tissues. The tissues are (a) stroma, (b) debris, (c) adipose, (d) mucose, (e) tumor, (f) lympho, (g) complex, and (h) empty. All the images are RGB images of size $150 \times 150 \times 3$.

4.2 Results

To validate the proposed approach, we divided the dataset of size 5000 images into 3900 for training, 100 for validation, and 1000 for testing. While testing, the outputs obtained are in the form of list containing probability distribution to each label. The maximum probability from the list is considered as the prediction, and the respective class label is assigned. Fine-tuned model performed better than other models and achieved an accuracy of 91.2%, precision score 0.913, F1 score 0.909, and recall score 0.91. The traditional CNN method classified the images with an accuracy of 85.9%, transfer learning VGG19 model with added layers classified the images with an accuracy of 88.5%, and support vector machine classifier classified images with accuracy of 65.4%. Early stopping function stops the model training at 12th epoch as validation loss does not improve for three consecutive epochs. Model trained during 9th epoch is the best model which is saved by model checkpoint function which is then used in classifying the colorectal image. Table 1 tabulates all the results achieved. Confusion matrices of different CNN models are shown in Fig. 4 where the proposed approach achieved least miscalculation of 0.088. The classification of images among STROMA, COMPLEX, and LYMPHO was performed less compared to other class labels. Therefore, from the results, it can be stated that fine-tuning model performs well in the histopathological image classification than other approaches.

Table 1 Performance of different model on dataset

	Accuracy %	Precision score	F1-score	Recall score
SVM	65.4	0.657	0.65	0.651
CNN-traditional	85.9	0.857	0.857	0.859
Transfer learning-VGG19	88.5	0.884	0.882	0.883
Fine-tuning VGG19	91.2	0.913	0.909	0.91

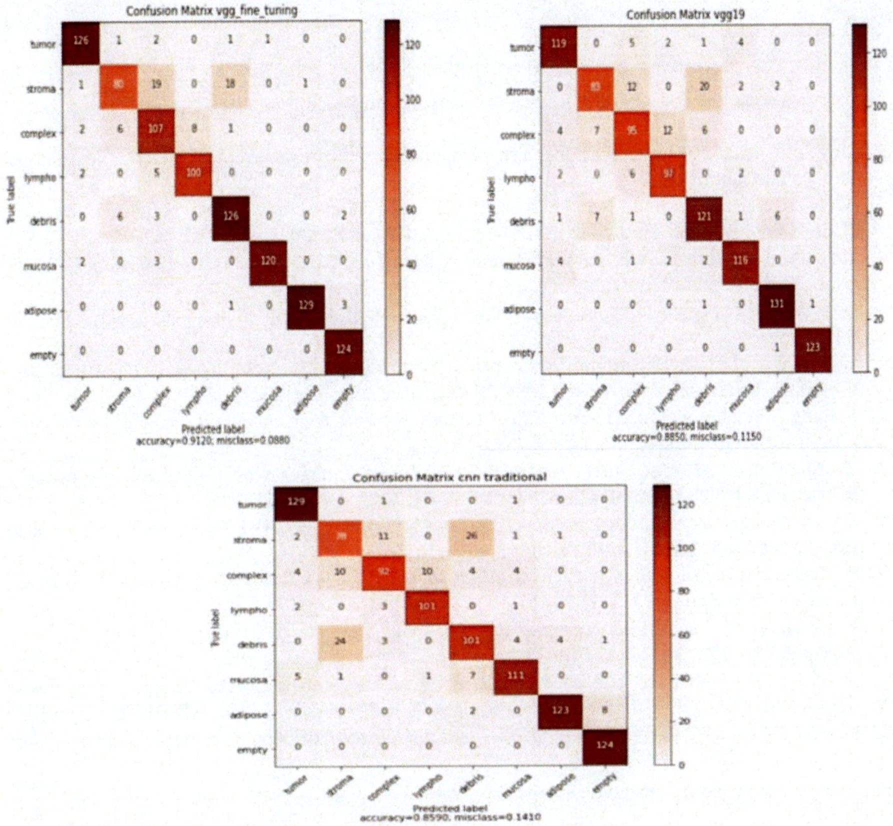

Fig. 4 Confusion matrix of different CNN models

5 Conclusion

This paper proposes better approach among deep neural network for histopathological images, and we used VGG19 pre-trained model for fine-tuning approach along with early stopping function to prevent overfitting. The proposed approach is tested on colorectal cancer histology image dataset corresponding to eight different cancer tissues. Four different approaches were used to compare performance in terms of accuracy, recall, precision, F1 score, and confusion matrix. From the result, fine-tuning model has outperformed other approaches without overfitting with nine epoch and achieved an accuracy of 91.2%, precision score 0.913, F1 score 0.909, and recall score 0.91. In future, the image augmentation can be used to increase the dataset size, and if COVID-19 dataset is made publicly available, this approach can be experimented to provide better and faster results as current test results arrive only after 24 h, but deep learning model provides faster result compared to the present testing technique.

References

1. https://www.cnbc.com/2018/02/22/medical-errors-third-leading-cause-of-death-in-america.html
2. Key Statistics for Colorectal Cancer, https://www.cancer.org/cancer/colon-rectal-cancer/about/key-statistics.html
3. M. Ganz et al., Automatic segmentation of polyps in colonoscopic narrow-band imaging data. IEEE Trans. Biomed. Eng. 59(8), 2144–2151(2012). (440,000 death)
4. How Telemedicine Answers Global Pathology Demands, 2018, https://proscia.com/blog/2015/07/14/global-crisis-digitalsolution
5. https://www.expresshealthcare.in/lab-diagnostics/illegal-path-labs-indias-open-secret/413141/
6. H. Mittal et al., *Classification of Histopathological Images Through Bag-of-Visual-Words and Gravitational Search Algorithm*, LNSCPS (Springer, 2019), p. 231–241.
7. S. Pang et al., A novel fused convolutional neural network for biomedical image classification. Med. Biol. Eng. Comput. **57**(1), 107–121 (2018)
8. Y. Zhang et al., Breast cancer diagnosis from biopsy images with highly reliable random subspace classifier ensembles. Springer MVA **24**, 1405–1420 (2013)
9. H. Asri et al., Using machine learning algorithms for breast cancer risk prediction and diagnosis. Elsevier PCS **83**, 1064–1069 (2016)
10. V. Rachapudi et al., *Improved Convolutional Neural Network Based Histopathological Image Classification* (Springer E.I. 2020).
11. Kermany et al., Identifying medical diagnoses and treatable diseases by image-based deep learning. Cell **172**, 1122–1131 (2018).
12. R. Pal et al., Enhanced bag of features using alexnet and improved biogeography-based optimization for histopathological image analysis, in *Proceedings of IC3* (IEEE, 2018), p 1–6.
13. Neural Network Early Stopping, https://chrisalbon.com/deep_learning/keras/neural_network_early_stopping/
14. Yao et al., On early stopping in gradient descent learning. Constr. Approx. **26**(2), 289–315 (2007).
15. Fine-tuning convolutional neural network on own data using keras tensorflow https://cv-tricks.com/keras/fine-tuning-tensorflow/
16. J.N. Kather et al., Multi-class texture analysis in colorectal cancer histology. Sci. Rep. **6**, 27988 (2016)
17. Dropout Neural Network Layer In Keras Explained https://towardsdatascience.com/machine-learning-part-20-dropout-keras-layers-explained-8c9f6dc4c9ab
18. Further going on CNN, https://classroom.udacity.com/courses/ud187/lessons/1771027d-8685-496f-8891-d7786efb71e1/concepts/db0b93a6-402d-4f13-8869-cf5fc1fe89ad
19. J.D.J. Deng et al., ImageNet: a large-scale hierarchical image database, in *Proceedings of IEEE ICCVPR, 2009*, p. 2–9.
20. Transfer learning, https://classroom.udacity.com/courses/ud187/lessons/a915f824-ce4a-4f5e-9897-a78ccbff313d/concepts/7fd8f8d5-979a-44ac-8ead-6c19a359f767

Prophetic Probe of Accidents in Indian Smart Cities Using Machine Learning

Ganesh Khekare and Pushpneel Verma

Abstract With the advancement in transportation technology, the number of vehicles is increasing rapidly, but concerning that, the number of accidents is also increasing in the same proportion. Accidents are very common nowadays. Infrastructure is redefining and on-road speed of vehicles is increasing. Accident-prone areas are increasing. Millions of people losing their life due to accidents. This article analyses the data of India's 11 states which are accident prone considering different time zone and proposes a new system that predicts the accident-prone areas and provides the most dangerous time for traveling along that route. A machine learning concept is used to train your system to make the best decisions based on experience. DBSCAN data algorithm with negative sampling is used to apply the machine learning concept in a better way. Through simulation results, it is cleared that number of accidents will reduce after applying the proposed concept.

Keywords Traffic analysis · Accidents · Machine learning · Intelligent transportation system · DBSCAN algorithm · Negative sampling

1 Introduction

The number of accidents is increased vastly in last decade. Developed countries with innovative infrastructure are unable to control the death rate caused due to accidents [1]. Lower and middle economical countries have more accidents [2]. In countries like Pakistan, almost 20 people are losing their lives per day due to road accidents. As per the data provided by Pakistan Bureau in the public domain, around 10,000 accidents were confirmed and in that 5000 deaths occurred [3]. It becomes important to focus on road safety for developing countries. Advancement of the Internet of

G. Khekare (✉)
G H Raisoni College of Engineering, Nagpur, India
e-mail: khekare.123@gmail.com

P. Verma
Bhagwant Universiry, Ajmer, India

V. Bhateja et al. (eds.), *Data Engineering and Intelligent Computing*,
Advances in Intelligent Systems and Computing 1407,
https://doi.org/10.1007/978-981-16-0171-2_18

Everything (IoE) boosts up the concept of an intelligent transportation system and promoted the concept of smart cities [4]. Through the global positioning system, it becomes possible to identify the position of vehicles and traffic scenarios. Nowadays, most of the vehicles come with the in-built global positioning system setup [5]. Fog computing [6] helps to save the large amount of data generated in real-time [7]. The extension of IoT and significant use of sensors made things much easier [8].

Developing countries are focusing on to create an intelligent transportation system that will save travel time and reduce the driver's headache as well as pollution [9]. Decision making in real-time scenarios plays a major role in the intelligent transportation system [10]. Government organizations are focusing on the infrastructure systems which will identify accidents red zone and after accidents intimation system is required to provide quick relief. Sometimes sensors are also damaged, so they are unable to identify the situation [11]. The main goal of this article is to provide the predictor system which provides the best route avoiding red zone or accident-prone areas during rush hours. Our proposed system uses a machine learning algorithm to train the system using DBSCAN algorithm with negative sampling and will make future predictions. The rest of the article is organized as follows: Sect. 2 describes the literature review. In Sect. 3, proposed methodology is explained. In Sect. 4, experimental setup and discussion of the results are done, and in Sect. 5, conclusion is made with future scope.

2 Literature Review

The need for smart city infrastructure:

- Garbage collection: Garbage collection and maintenance is a big issue in metro cities. Smart city infrastructure provides better management of this to reduce the garbage and maintains cleanliness in the smart city.
- Enriched life—Better communication and getting information on right time will ease the overall travel fatigue which results to improve quality of life [12].
- Minimized pollution—It provides the shortest congestion-free paths to vehicles resulting in reduced pollution. Through an intelligent transportation system, it becomes easier to share personal vehicles as well as public transports.
- Green city: Smart city infrastructure also provides the facility to maximize the use of natural resources like solar energy plants, biogas plants, etc., to operate remotely.
- Individual traveling safety and security—In case of any on-time emergency, it would be helpful to maintain smooth traffic flow.
- Remote places—Market in the smart city gives scope to Android applications for business purposes. People in the city can easily know what things are trending in the market.

Smart cities improving their performance by using a smart transportation system. It consists of:

- Data processing systems that gather information regarding the current traffic scenario. By using this system, traveling will become easier and more efficient.
- Innovative transportation method which modified the current methods of transportation and come up with an innovative one. The way Android applications have come up like Uber, Ola, etc., which makes transportation easier for the traveler as well as the driver. Vehicle sharing, rental, outstation, and fuel consumption becomes efficient by this. Public transport becomes more organized.

Various countries provide their accidental data to researchers, so that analysis can be done, and the conclusion can be made.

3 Methodology

The predictor is implemented by using the machine learning concept which will predict future accidents. This predictor is capable of doing the following things:

- To provide the optimal congestion-free path from source to destination.
- To allow the user to select the path considering the time factor and which is not an accidental red zone on that particular timing.

Information Gathering and Data Processing.

The inverse function is used to find out the desired output. It consists of three parts: The inverse problem in mathematical model has three factors: inspectional data (dins), an underlying natural law (OF), and model variables of interest (V). It consists of method variables as $v \in V$, which creates the inspectional data (dins):

$$\text{dins} = \text{OF}(v)$$

Our data consists of 14 years of accidental data available on the official Web site of the Ministry of Road Transport and Highways of the Government of India in the public domain. This data consists of various columns like the type of accident, year, state, number of accidents as per 3 h span in a day, etc. Let us consider traveler has to do round trip, i.e., traveler starts from point X and go to Y, then after some time return journey is done via the same route, i.e., now starting point is Y and ending point is X. Around 57% of data is of round trip, and around 92% data of round trips satisfy our consideration during the simulation. Considering this ridership information, a new dataset is created.

A new structure of the database is created which is referred to as "vehicle movement," which satisfies our assumption regarding the round trip which consists of two tuples: start time and end time, both at origin o.

$$T_v^o(d) = (T_s|o)(T_e|o)$$

where v, o, d, T_i; and T_e indicate vehicle, station ID (origin), date, start time, and end time, respectively.

Then, the ridership data for a particular date and origin o is denoted by

$$R^o(d) = \sum_v T_v^o(d).$$

Now let us suppose, to calculate ridership data for the years 2018 and 2019, then the equation becomes

$$R(2018, 2019) \sum_d \sum_o \sum_v T_v^o(d)$$

where 2018 and 2019 are the years for which you wish to calculate ridership. Round trip must be considered rather than one way trip which is explained in the next section. A tuple of a vehicle transit called data structure for the round travel of a vehicle is defined as

$$(T_v^{O_0}(d)|(T_v^{O_1}(d) = (T_s|o_0)(T_e|o_0)|(T_s|o_1)(T_e|o_1)$$

However, the starting point of the tuple is the transit entry of travel from point O_0 to point o_1, and the next element is the exact opposite of that which is the return trip. Our target is to rebuild $T_v(d)$ for every traveler record, R is the metro subway inspectional data which requires identification of each travel. The average distribution of outdoor duration time (ODT) of traveler is denoted by $A\left(T_v^o\right)$, which represents the median of transit data in terms of vehicles. It depends on the origin point and the endpoint of travel and known as the measure of vehicles and denoted as $T_v^O(d)$.

Assume that a vehicle starts its journey from point X at time h_X^s and reach to destination Y at time h_Y^e. After waiting for some time on station Y, again it becomes ready for the next travel, so it becomes ready at station Y at time h_Y^s and travel to destination X and reach at time h_X^e. Then, in this case, measure of ODT is denoted as

$$\text{ODT} = h_X^e - h_Y^s$$

The major problem here is to accurately identify which passenger enters the station at what time. While return journey, the same vehicle data must be taken which is considered on the first trip. In short in heavy traffic it is important to uniquely identify the vehicles on both ways. This is done by providing unique ID to each vehicle which will remain permanent in the entire journey. If this is not done, the authenticity of data is not guaranteed.

Normalized matrix Ω is represented by the equation,s

$$\Omega = \begin{matrix} w_{0,0} & \cdots & w_{0,n} \\ \dots & \dots & \dots \\ w_{n,0} & \cdots & w_{n,n} \end{matrix}$$

where n is the total number of elements and $w_{i,j}$ is weighted function represented by

$$w_{i,j} = \frac{\alpha_{i,j}}{\sum_n \sum_m \alpha_{m,n}}$$

So, $\alpha_{i,j}$ from Ω is denoted as

$$\alpha_{ij} = F * w_{ij}$$

where F indicates the underline natural value.

DBSCAN Data Clustering.

After analyzing the dataset available for Indian accidents, it is observed that accidents happen almost in every city perhaps on every road. Some roads are accidents prone. They become the red zone. The road accidental data taken for analysis is of the last 14 years 2006 to 2019 of India. After analyzing the data, the area in which more accidents occur is converted into clusters. The cluster has been defined with an area size of 50 m each. DBSCAN algorithms are taken into consideration because it is quickly responsive, it forms the cluster very easily and it is robust to outliners as compared to other machine learning algorithms available. The accidents happen outside the cluster area is considered as outliners. So the main focus is on the area which is more accident-prone perhaps on the states in which more accidents occur. So the data of positive sampling is there, but negative sampling needs to generate data of non-accidents. It generates four negative entries after every one positive entry for the cluster. With the preprocessing step, now ready to move toward the experimental setup, results, and discussion.

4 Experimental Setup and Results

The complete data is divided in the ratio 60:40 as a training data set and testing data set. The supervised learning method is applied to the data set. Python's scikit-learn library is used as a platform for data testing training and result generation. Generalized classification models like support vector machine algorithms, random forest, and logistic regression are used to process the data. The random forest model performs well on our dataset as per the parameters set. The prediction model performance is analyzed, and the area under curve and receiver operating characteristic curve are generated. Then, this scikit-learn model is provided as a Web site using Flask which is a Python-based Web platform creator. On the Flask application, we can provide the source and destination points of travel. It provides the desired result for a model trained. Google Maps API also plays a major role in this. For creating a cluster in simulation, a circle is drawn having radius 50 m. Darksky function is used to create the same. A binary prediction is done to provide final results to activate the red zone or not.

For positive sampling, the data provided by the Indian ministry of road transport and highways in the public domain is taken. This data gives the state-wise accidents which occur in India every year. 14 years of data are available. It is segregated in the form of 3 h span throughout the day. Through experimental status, this positive sampling data is analyzed.

Figure 1 shows a road accident analysis of 11 states in India which are more accident prone. The *x*-axis represents the state name, the *y*-axis represents the number of tragedies occurred, and the z-axis represents time duration. Each time slots have 3 h span, and in this way, whole day is divided into eight parts time slots. As clear from the graph, two-time slots have a maximum number of accidents, i.e., 15–18 h (Day) and 18–21 h (Night). This time zone will be considered as a red zone while applying negative sampling.

Average of all 14 years data is taken as shown in Fig. 2 graph that is plotted. In Fig. 2, the *x*-axis represents the state, whereas the *y*-axis represents the number of accidents. Look at the linear average, it is gradually increasing. Tamilnadu has the highest number of accidents per year. After generating the negative samples and applying DBSCAN clustering as discussed in the methodology, random negative

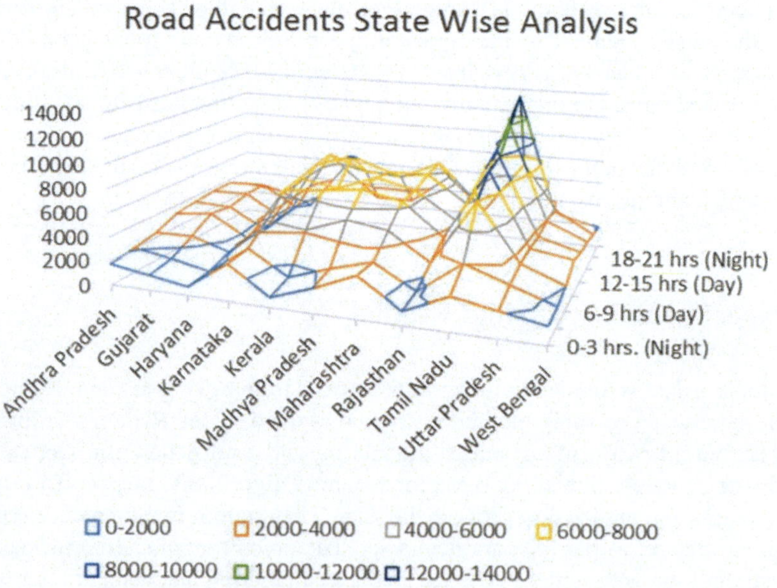

Fig. 1 Road accidents state-wise analysis India

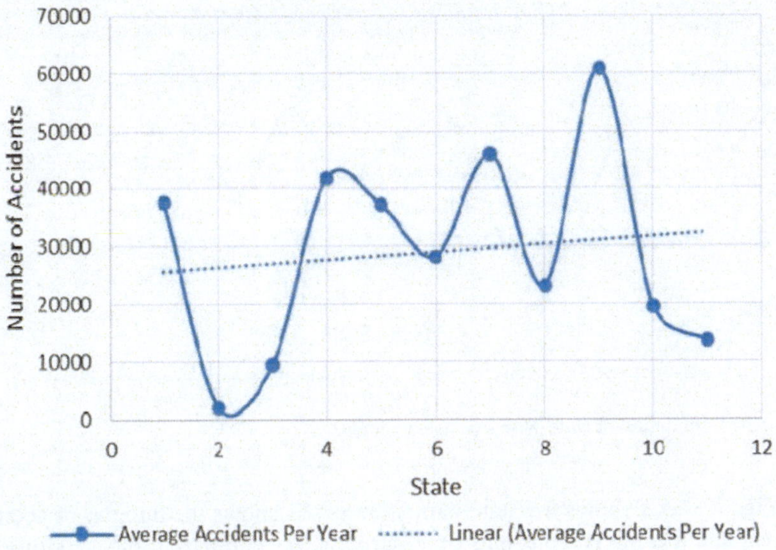

Fig. 2 State-wise average accidents per year

samples are generated by using the predictor system. The red zone has been identified based on accidents and timings through simulation based on machine learning. While providing a path to the vehicles, these red zones are avoided by the predictor system. Random samples generated and various iterations are performed. Based on the trace file generated, following results have been generated as shown in Figs. 3 and 4.

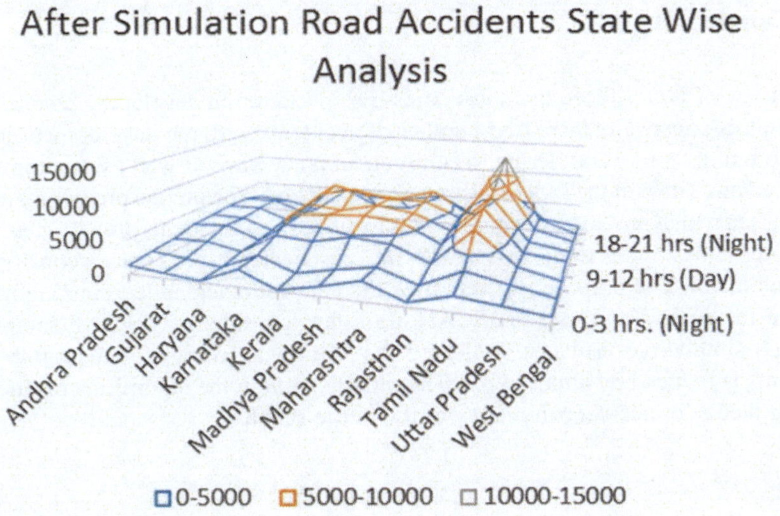

Fig. 3 After simulation of state-wise analysis of road accidents

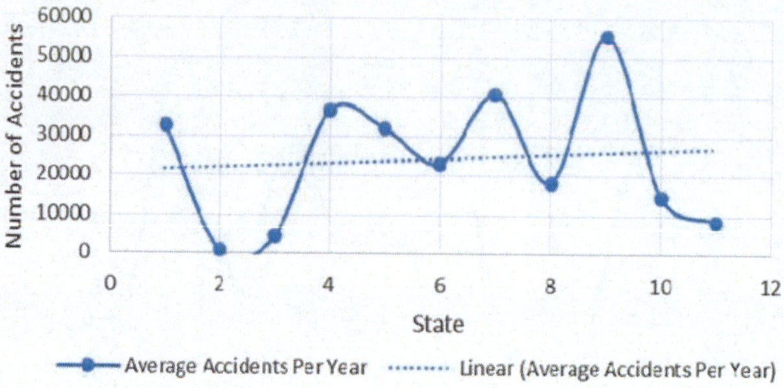

Fig. 4 After simulation of state-wise average accidents

In Fig. 3, *x*-axis shows the state name, the *y*-axis shows the number of accidents occurred, and the *z*-axis represents time duration. By comparing Fig. 3 with Fig. 1, the state-wise analysis is done, and it is observed by the proposed system, and the number of accidents has been highly reduced. The red zone has been avoided during peak hours.

In Fig. 4, the *x*-axis shows the state, whereas the *y*-axis displays the number of accidents. After comparing our results with Fig. 2 which is data of only positive sampling, almost a 17% reduction in the accidents is observed by applying the proposed method.

5 Conclusion

The survey of issues faced by smart cities transportation in developing countries is done in this paper. The increased number of accidents with the development in the transportation sector is bothering to common human beings as well as governments. The accidental data of the Indian 11 states are analyzed. The unique predictor system is proposed which provides the route avoiding the red and accidents prone zones. This system provides routes to the driver with the least accident-prone areas considering the time factor. The proposed system uses the concept machine learning algorithm to train the system and uses DBSCAN algorithm considering negative sampling. Through simulation results, it is cleared that the linear average of the number of accidents is reduced by almost 17%. The modification in the algorithm considering various factors of accidents remains for the future scope.

References

1. A. Khan, F. Bibi, M. Dilshad, S. Ahmed, Z. Ullah, H. Ali, Accident detection and smart rescue system using Android Smartphone with real-time location tracking. Int. J. Adv. Comput. Sci. Appl. **9**, 341–355 (2018)
2. L. Jackson, R. Cracknell, *Road Accident Casualties in Britain and the World* (House of Commons Library, London, UK, 2018).
3. IEEE. Pbs.gov.pk., Traffic Accidents (Annual) Pakistan Bureau of Statistics. Available online: https://www.pbs.gov.pk/content/traffic-accidents-annual (accessed on 30 January 2019).
4. A. Kirimtat, O. Krejcar, A. Kertesz, M.F. Tasgetiren, Future trends and current state of smart city concepts: a survey. IEEE Access **8**, 86448–86467 (2020). https://doi.org/10.1109/ACCESS.2020.2992441
5. J. Cuena, J. Hernández, M. Molina, Knowledge oriented design of an application for real time traffic management: the TRYS system. Eur. Conf. Artif. Intell. **96**, 308–312 (1996)
6. G.S. Khekare, Design of emergency system for intelligent traffic system using VANET, in *International Conference on Information Communication and Embedded Systems (ICICES2014)*, Chennai, 2014, p. 1–7, https://doi.org/10.1109/ICICES.2014.7033910.
7. F. Wang, M. Zhang, X. Wang, X. Ma, J. Liu, Deep learning for edge computing applications: a state-of-the-art survey. IEEE Access **8**, 58322–58336 (2020). https://doi.org/10.1109/ACCESS.2020.2982411
8. F. Bhatti, M.A. Shah, C. Maple, S. Ul Islam, A novel Internet of Things-enabled accident detection and reporting system for smart city environments. Sensors **19**, 2071 (2019). https://doi.org/10.3390/s19092071
9. G.S. Khekare, U.T. Dhanre, G.T. Dhanre, S.S. Yede, Design of optimized and innovative remotely operated machine for water surface garbage assortment. Int. J. Comput. Sci. Eng. **7**(1), 113–117 (2019)
10. M. Gohar, M. Muzammal, A. Ur Rahman, SMART TSS: Defining transportation system behavior using big data analytics in smart cities. Sustain Cities Soc. **41**, 114–119 (2018).
11. S.P. Gohane, G.S. Khekare, Reconfiguration of industrial embedded system in WSN, in *2015 IEEE 9th International Conference on Intelligent Systems and Control (ISCO)*, Coimbatore, 2015, p. 1–5, https://doi.org/10.1109/ISCO.2015.7282284.
12. R.A. Gonzalez, R.E. Ferro, D. Liberona, Government and governance in intelligent cities, smart transportation study case in Bogotá Colombia. Ain Shams Eng. J. **11**, 25–34 (2020).

Identifying Handwriting Difficulties in Children in Devanagari Script Using Machine Learning

Sailaja Mulakaluri and G. S. Girisha

Abstract Dysgraphia is a medical disorder in children, due to which children face difficulties in writing. Children suffering from dysgraphia find it difficult in holding a pencil, alignment, differentiating the strokes, curves, and sizing of the alphabets, etc. It is more challenging in scripts where there are more alphabets, curves, and strokes. In India, about 75% of school-going kids write in Devanagari script (includes Hindi, Marathi, Gujarati) as a medium of study or second language. Diagnosis of writing difficulties in the Devanagari script can help the child to change the language or get assistance. Many researchers have come up with diagnosing dysgraphia in languages like English, Hebrew, etc. The current study aims at developing a model to identify children with dysgraphia and without dysgraphia in Devanagari script based on their performance characteristics as per the BHK algorithm. 52 students handwriting samples were collected, and feature analysis is done based on BHK and machine learning model using the KNN algorithm for the predictions.

Keywords Dysgraphia · Writing difficulties · Devanagari script · Hindi script · BHK · Machine learning · KNN

1 Introduction

1.1 Dysgraphia

Learning difficulties are a neurological disorder, which can be categorized as language processing (Dyslexia), writing difficulties (dysgraphia), and math disabilities (dyscalculia) [1]. Dysgraphia is specific to handwriting, it is a medical disorder where children face difficulties concerning poor letter formation, spacing of the

S. Mulakaluri (✉)
Department of Computer Science, St. Francis De Sales College, Bengaluru, India

G. S. Girisha
Department of CSE, Dayananda Sagar University, Bengaluru, India
e-mail: girisha-cse@dsu.edu.in

V. Bhateja et al. (eds.), *Data Engineering and Intelligent Computing*,
Advances in Intelligent Systems and Computing 1407,
https://doi.org/10.1007/978-981-16-0171-2_19

letters, ungrammatical structuring of sentences, disorganized writing which leads to incomplete school work and frustration in writing [2]. Such students are generally labelled as negligent or lazy in writing as these students may be good in oral answering [3, 4]. It should be noted that dysgraphia need not be Dyslexia [5, 6]. In short, dyslexia is a reading disorder, whereas dysgraphia is writing difficulty [7]. Early detection and assistance can avoid children from emotional stress and going to low confidence level and inferiority complexes, there are various assisting tools to help children with dysgraphia such as different mechanisms of writing and textured paper which can help kids for a better pen/pencil grip [8]. With the advancement in technology, writing should not be a hindrance for the growth of a child with dysgraphia, so early diagnosis can help the child in taking any of these assistances based on their level of the difficulty [9]. Technology not only offers tools to help individuals to cope with these difficulties but can also help in diagnosing, as the first step of any treatment is a proper diagnosis.

1.2 Devanagari Script

In India, most commonly used script is the Devanagari script, around 300 million people in India use Devanagari script for documentation, and third mostly used script in the world. Several Indian languages such as Sanskrit Hindi, Sindhi, Marathi, Konkani, etc., use Devanagari script for writing and documentation [10]. In India, around 51% of schools use the Devanagari script as their medium of instruction. There are total 52 alphabets in all, 11 nowels, 33 consonants, 4 combined consonants, anusvar or chandra bindu—1, aphas—1, and binary consonant—2. Around 70% of students either have Hindi/Devanagari script as a second language or medium education. Schoolchildren (especially with non-Hindi as mother tongue) face a lot of issues not only with reading and understanding the language but writing it would be challenging. In this scenario, it would be more stress full for the children to cope with multiple languages. [11] At the primary school level, early diagnosis of such difficulties especially with writing would help them to opt for other alternatives.

2 Method

BHK Beknopte Beoordelingsmethode Voor Kinderhand-Schriften (BHK) is one of the most used screening tools which was designed to measure handwriting quality [8]. In this study, we extend the concepts of BHK to test handwriting quality in the Devanagari script [7]. We consider 12 significant features for analysing handwriting for diagnosis dysgraphia [9, 11, 12]. Based on the 12 features, data is collected and processed, and then, machine learning model is applied where 70% of the data is training data and 30% is test data [13]. *K*-nearest neighbour algorithm is used for

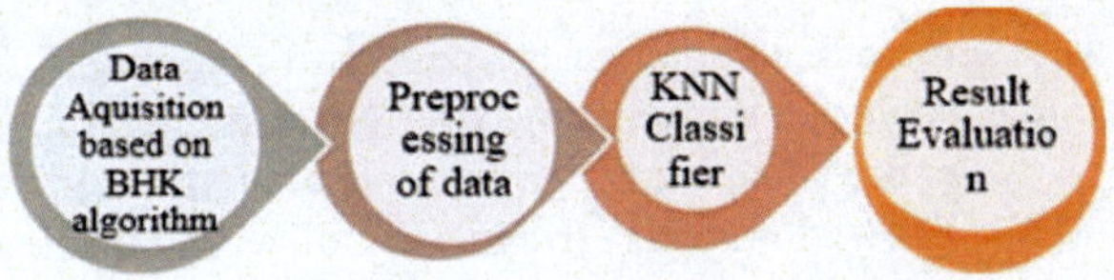

Fig. 1 Process flow [14]

prediction, since data points are separated into several classes to predict the classification of a new sample point and feature similarity. The overall process followed is depicted in Fig. 1 [14].

2.1 Data Acquiring

52 students of age group 6–14 years were instructed to copy five lines of given Hindi (Devanagari script) text based on their age [15]. Based on 12 features like (1) Size of the text, (2) Top alignment, (3) Nonalignment of left margin, (4) Skewed writing, (5) Insufficient spacing between the words, (6) Curves and strokes, (7) Collisions between two letters and Matras, (8) Irregular size of the letters, (9) Atypical letters, (10) Ambiguous letters, (11) Striking of words/traced letters, (12) Unstable track data is collected [7, 13, 16].

(1) **Size of the text**: Devanagari script is aligned basically with reference to the top line. Size of the text is considered with respect to the top line. Based on child's age and the size of the text from the top line, the scores are tabulated [7] (Fig. 2).

(2) **Top alignment**: Top stripe in Devanagari script is one of the important parameter in writing analysis, as based on that the readability of the text. If the top stripe or line is not straight or is at an angle of 45° or above deviation with the top line in a ruled page, then it is 1, or else it is 0 (Fig. 3).

(3) **Alignment of left margin**: The Devanagari script text also need to be aligned to the left margin. Sometimes students start from left margin and end up to

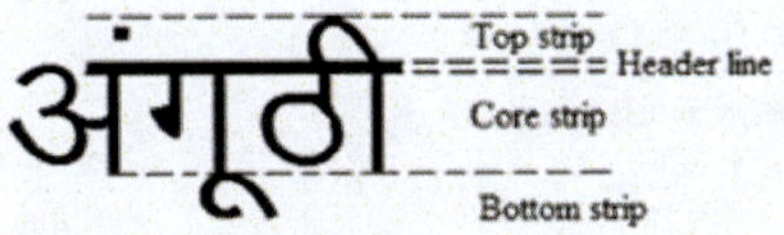

Fig. 2 Sample image showing size of the text from the top line in Devanagari script [17]

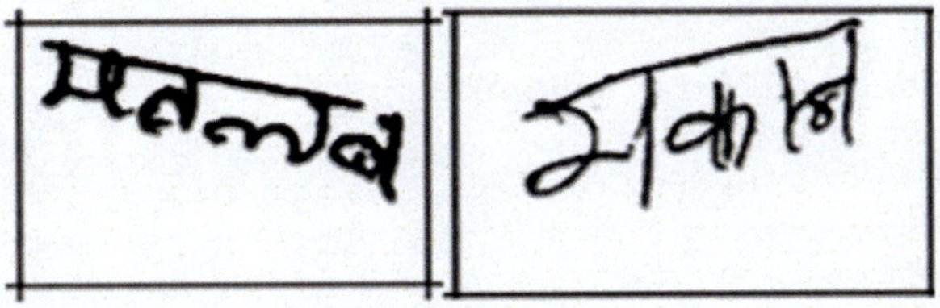

Fig. 3 Top non-aligned Hindi text [18]

the other corner of the page and completely tilted to the right. The feature score is 0 if it is inclined to right or irregular margin. The score is based on the entire written text. An example of an inclined margin is shown below (Fig. 4).

(4) **Skewed writing**: Irregularity of words in the sentences, where words goes up and down. Undulated writing, as shown in Fig. 5. In this example, a word goes up or down, if some characters are too far. If irregularity is detected, then the score is 1 else 0.

(5) **Insufficient space between words**: In some cases, children does not have the space estimation between the words and letters. In such cases, spacing between the letters is more than the spacing between the words. If the gap between two words is not adequate, or space between letter is more, then the score is 1 else 0 (Fig. 6).

(6) **Curves and strokes**: In Devanagari script, matras play an important role, for example, when students write ◌ॆ and ◌े both with same curves in the top,

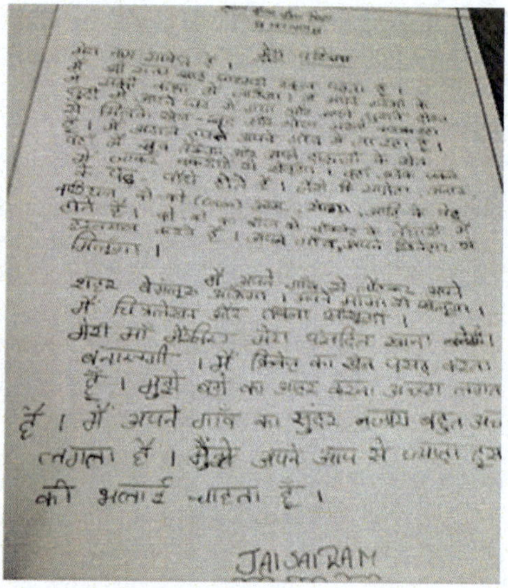

Fig. 4 Non-aligned Hindi text

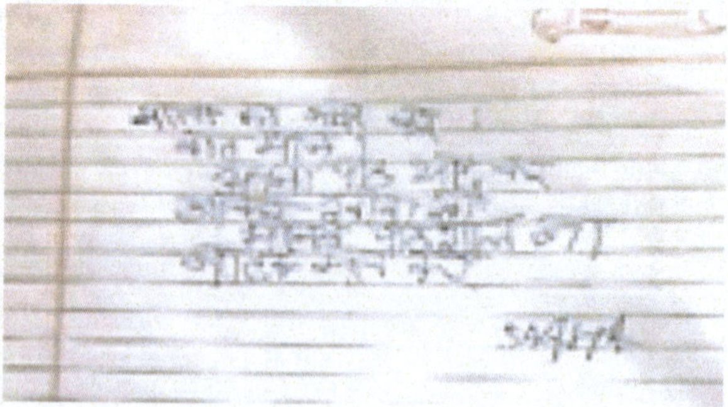

Fig. 5 Skewed Hindi writing

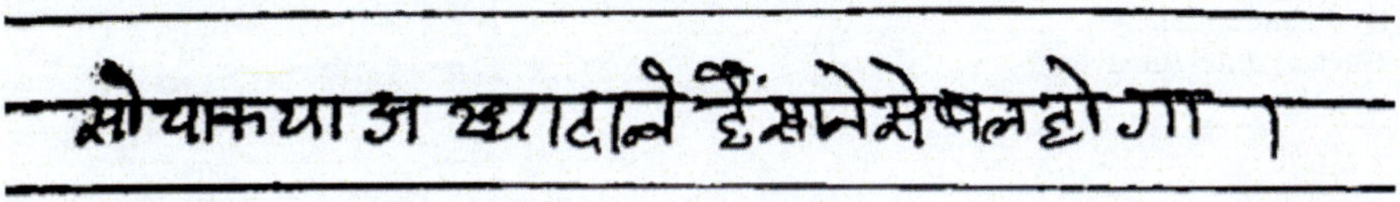

Fig. 6 Writing, without proper spacing

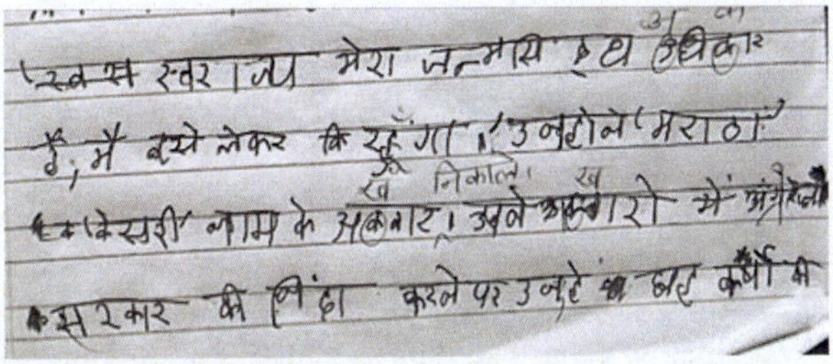

Fig. 7 Collision of letters and matras

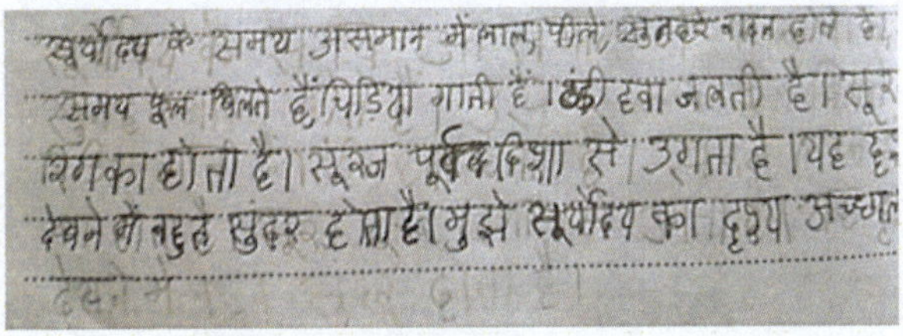

Fig. 8 Inconsistent size

and not clear with ि ी ु ू ृ ॄ. Elongated in the alphabets curves which may lead to interchange of afphabets, for example, घ छ ध and रा with श. If the curves and strokes are not clear, then the feature score is 1 else 0 [19].

(7) <u>**Collisions between two letters and Matras**</u>: In Devanagari script, collision of two characters or collision of *Matras*[1] and *Halants*[2] (half akshara) to create an overlaying area then the feature score is 1 else 0 (Fig. 7).

(8) <u>**Irregular size of the letters**</u>: Inconsistency in the alphabet height, same alphabet is written in different sizes in the text. Irregular text, with respect to size of the alphabets, placed unevenly. The score is 1 if irregular, otherwise 0 (Fig. 8).

(9) <u>**Atypical letters**</u>: When the characters or letters are written differently compared to common style, such letters are atypical letters. If such letters are more, then we assign a score of 1 else 0. In the below fig, you can find different ways of झakshar which is written (Fig. 9).

(10) <u>**Ambiguous letters**</u>: Many alphabets in Devanagari script are similar with very slight variation. So interpreting the letters without ambiguity is scored

[1] Matra: A unit of metrical quantity equal to a short vowel in Devanagari script.

[2] Halant :A diacritic used to signify the lack of an inherent vowel in Devanagari script.

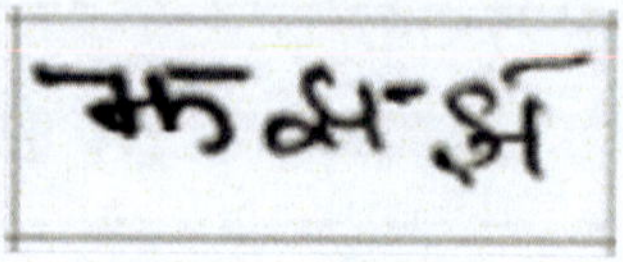

Fig. 9 Same alphabet written in different style [18]

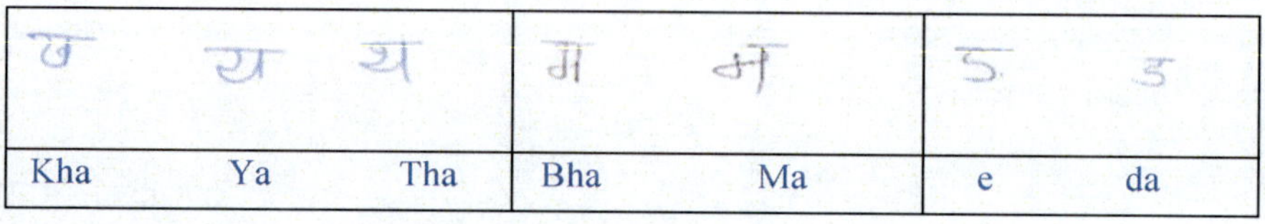

Fig. 10 Ambiguous letters

as 0, if the letters are identified or interpreted differently, then the score is 1, that is, value 1 will be assigned if ambiguous letters exist; otherwise, a value of 0 will be assigned (Fig. 10).

(11) **Striking of words/traced letters**: Feature score is 1 for the following cases, else score is 0 (Fig. 11):

1. Frequent striking of words
2. Rewriting or tracing the letters partly or completely
3. Redrafted to fit its form

(12) **Unstable track. Wavery/wobbly writing**: With an unstable track or non-continuous writing, feature score of 1 is assigned, else 0 (Fig. 12).

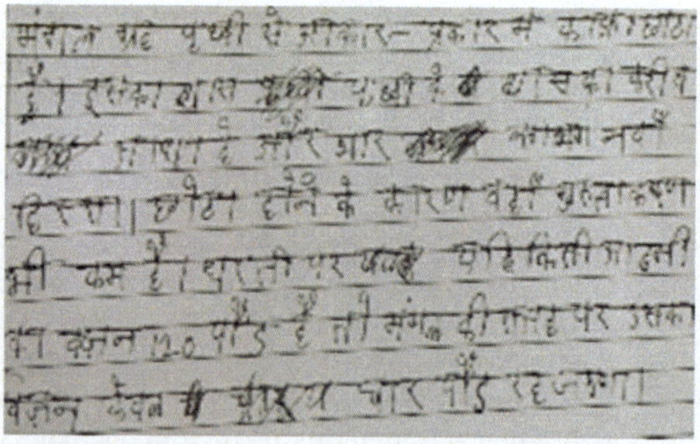

Fig. 11 Striking of words

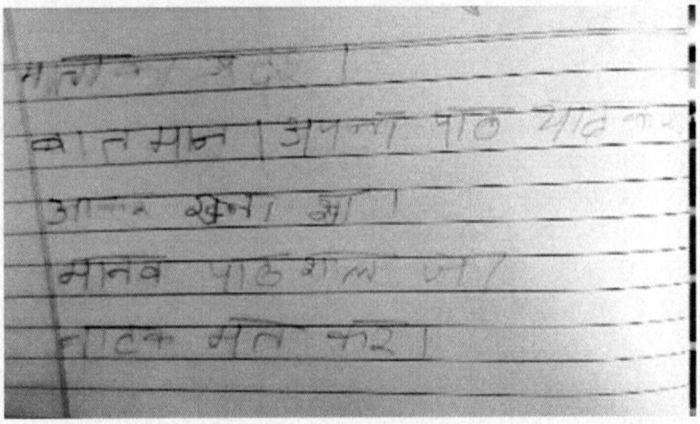

Fig. 12 Unstable track

	Age	p1	p2	p3	p4	p5	p6	p7	p8	p9	p10	p11	p12	OutCome
0	12 to 14	0	0	0	1	0	0	1	0	1	1	1	1	0
1	10 to 12	0	0	1	1	0	1	1	0	1	0	0	0	0
2	08 to 10	0	0	0	0	0	0	0	0	1	0	0	0	0
3	12 to 14	0	0	0	0	0	0	0	1	1	0	0	0	0
4	12 to 14	0	1	1	1	1	1	1	1	1	1	1	0	1
5	06 to 08	0	0	1	0	1	0	0	0	1	1	0	0	0
6	08 to 10	0	0	0	1	1	1	0	1	0	1	0	0	0
7	08 to 10	0	0	1	1	0	1	0	0	1	0	0	0	0
8	06 to 08	1	1	0	1	0	1	0	1	1	1	1	0	0
9	08 to 10	0	0	0	0	1	1	0	0	0	1	1	0	0

Fig. 13 Labelled features with binary values

Table 1 Writing size reference

Age years	<3 mm	4–6 mm	6–8 mm	9–11 mm
6–8	0	0	0	1
8–10	1	0	0	1
10–12	1	0	0	1
12–14	1	0	0	1

2.2 *Data Pre-processing*

Parameters p1 to p12 are assigned to the data collected as given in Table 1. The sample binary values for p1 to p12 are shown in Fig. 13. Age is a categorical value and is converted to binary values in Fig. 14, Table 2.

2.2.1 Data Analysis

The data is explored in various aspects based on responses received as per the age, parameters, and statistical analysis. The data is explored in various aspects based on responses received as per their age, various parameters, and statistical analysis. Out of 52 students, seven students are of the age group 12–14 years, 23 are 10–12 years, 13 are 8–10 years and 9 are 6–8years which is shown in Fig. 15.

The positive responses given for each of the parameter are tabulated as follows (Table 3).

	Age_06 to 08	Age_08 to 10	Age_10 to 12	Age_12 to 14	p1	p2	p3	p4	p5	p6	p7	p8	p9	p10	p11	p12	OutCome
0	0	0	0	1	0	0	0	1	0	0	1	0	1	1	1	1	0
1	0	0	1	0	0	0	1	1	0	1	1	0	1	0	0	0	0
2	0	1	0	0	0	0	0	0	0	0	0	0	1	0	0	0	0
3	0	0	0	1	0	0	0	0	0	0	0	1	1	0	0	0	0
4	0	0	0	1	0	1	1	1	1	1	1	1	1	1	1	0	1

Fig. 14 Age converted to binary format

Table 2 List of attributes

S. No.	Feature	Parameter
1.	Size of the text	P1
2.	Top alignment	P2
3.	Nonalignment of left margin	P3
4.	Skewed writing	P4
5.	Insufficient spacing between the words	P5
6.	Curves and strokes	P6
7.	Collisions between two letters and matras	P7
8.	Irregular size of the letters	P8
9.	Atypical letters	P9
10.	Ambiguous letters	P10
11.	Striking of words/traced letters	P11
12.	Unstable track	P12

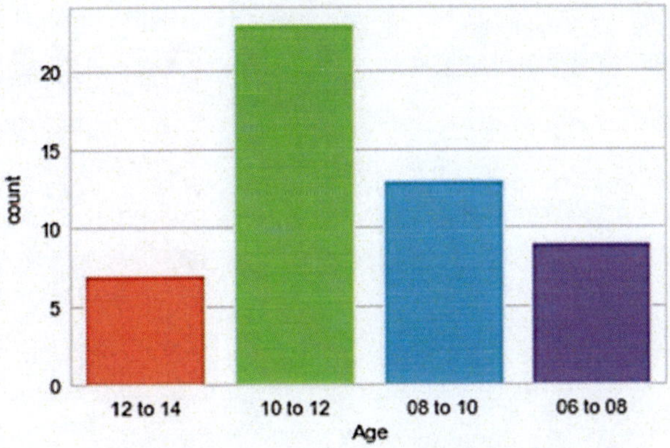

Fig. 15 Age-wise responses received

Table 3 Number of students with positive response for each feature, p1 to p12

Age	p1	p2	p3	p4	p5	p6	p7	p8	p9	p10	p11	p12
06–08	3	6	5	5	5	5	3	6	7	7	5	2
08–10	3	2	4	6	3	10	4	5	6	7	5	4
10–12	4	7	11	7	6	8	10	9	10	10	9	8
12–14	0	0	0	1	0	0	1	0	1	1	1	1

2.2.2 Statistical Analysis

The statistical analysis of the data is shown in Fig. 16.

Based on the available data, it is observed that there is an imbalance between dysgraphic and non-dysgraphic students as depicted in the below graph. ~20% of the data is dysgraphic, and 80% is not dysgraphic. The outcome graph is given in Fig. 17.

	p1	p2	p3	p4	p5	p6	p7	p8	p9	p10	p11	p12	OutCome
count	52.000000	52.000000	52.000000	52.000000	52.000000	52.000000	52.000000	52.000000	52.000000	52.000000	52.000000	52.000000	52.000000
mean	0.192308	0.384615	0.423077	0.384615	0.365385	0.538462	0.403846	0.500000	0.538462	0.538462	0.442308	0.307692	0.192308
std	0.397959	0.491251	0.498867	0.491251	0.486236	0.503382	0.495454	0.504878	0.503382	0.503382	0.501506	0.466041	0.397959
min	0.000000	0.000000	0.000000	0.000000	0.000000	0.000000	0.000000	0.000000	0.000000	0.000000	0.000000	0.000000	0.000000
25%	0.000000	0.000000	0.000000	0.000000	0.000000	0.000000	0.000000	0.000000	0.000000	0.000000	0.000000	0.000000	0.000000
50%	0.000000	0.000000	0.000000	0.000000	0.000000	1.000000	0.000000	0.500000	1.000000	1.000000	0.000000	0.000000	0.000000
75%	0.000000	1.000000	1.000000	1.000000	1.000000	1.000000	1.000000	1.000000	1.000000	1.000000	1.000000	1.000000	0.000000
max	1.000000	1.000000	1.000000	1.000000	1.000000	1.000000	1.000000	1.000000	1.000000	1.000000	1.000000	1.000000	1.000000

Fig. 16 Statistical analysis

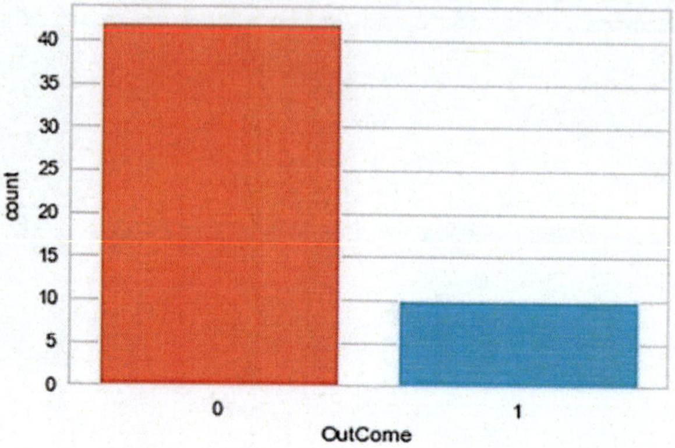

Fig. 17 Graphical representation of outcome

2.3 Implementation

Machine learning (ML) is a sub-field of artificial intelligence, which is used for predictions in various fields including medical. The main idea of ML is collecting the data in which 70% is used for training the machine learning model KNN. Test data of 30% is tested using the model [6] (Fig. 18).

The K-nearest neighbour algorithm is a classification method that does not make assumptions about the form of relationship between the class membership (Y), i.e., outcome and the predictors p1, p2, p12. In this model, we consider training data: 70% and test data: 30%. There is no need to eliminate any feature as we have technically taken all those predicator parameters that are only required to predict the target label. Value k shows strong effect on KNN performance [20]. Using trial and error method, k value is determined as 5 (Fig. 19).

$K = 5$ and the distance formula used is Euclidian distance

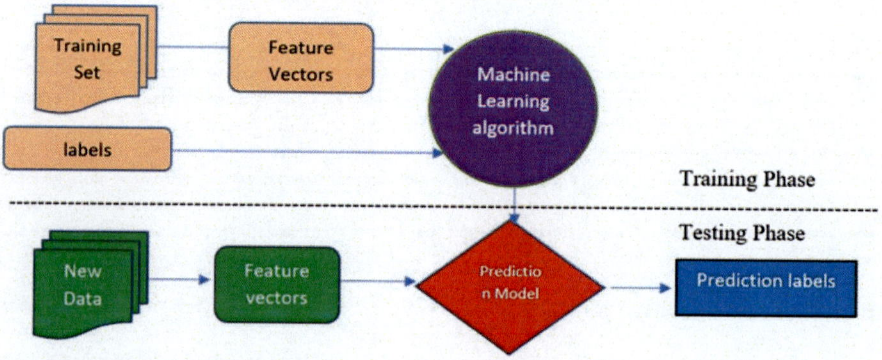

Fig. 18 Machine learning model

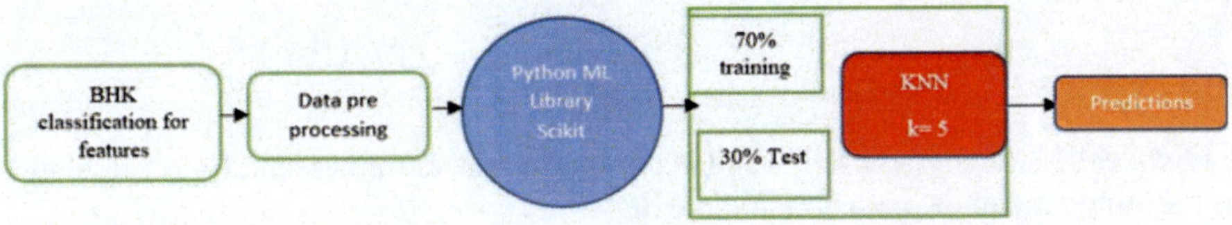

Fig. 19 Implementation model

```
[1 0 0 0 1 0 0 0 0 0 0 0 0 0 0 1]
[0 0 0 0 1 0 0 0 0 0 0 0 0 0 0 1]
```

Fig. 20 Prediction matrix

$$d = \sqrt{\sum_{i=1}^{k} (x_i - y_i)^2}$$

2.4 Prediction Results

The accuracy of KNN on the given data is 94%, and the prediction matrix is given as (Fig. 20).

2.4.1 Error Analysis

KNN model for predicting students with and without writing difficulties is 94% accurate. Considering test data set of 16 students, 13 students were predicted non-dysgraphic and who were not having dysgraphia, that is 13 are true negative, One student is predicted as dysgraphic were as the child is not dysgraphic that is False Positive. No students were predicted non-dysgraphic who were dysgraphic that is false negative = 0. And two students who were predicted positive and were ttrue positive. The confusion matrix is as follows Table 4.

Precision Score: Precision is the accuracy of positive predictions that is [21]

Table 4 Confusion matrix

$N = 16$ (test data)	Predicted: No	Predicted: Yes
Actual: No	13 (TN)	1 (FP)
Actual: Yes	0 (FN)	2 (TP)

$$\text{Precision} = \text{TP}/(\text{TP} + \text{FP})\text{and recall} = \text{TP}/(\text{TP} + \text{FN})$$

F1 Score: F1 score is generally used in comparing the classifier, and F1 is calculated as harmonic mean of precision and recall.

$$\text{F1} = ((\text{recall} \times \text{precision})/(\text{recall} + \text{precision})) \times 2$$

The precision recall and F1-score are given in Table 5.

The receiver operating characteristic graph, ROC curve [22], which plots TP true positive rate against FP false positive, for KNN classifier to predict students with and without writing difficulties in Devanagari script is depicted in Fig. 21.

Table 5 Accuracy matrix

	precision	recall	f1-score	support
0	1.00	0.93	0.96	14
1	0.67	1.00	0.80	2
accuracy			0.94	16
macro avg	0.83	0.96	0.88	16
weighted avg	0.96	0.94	0.94	16

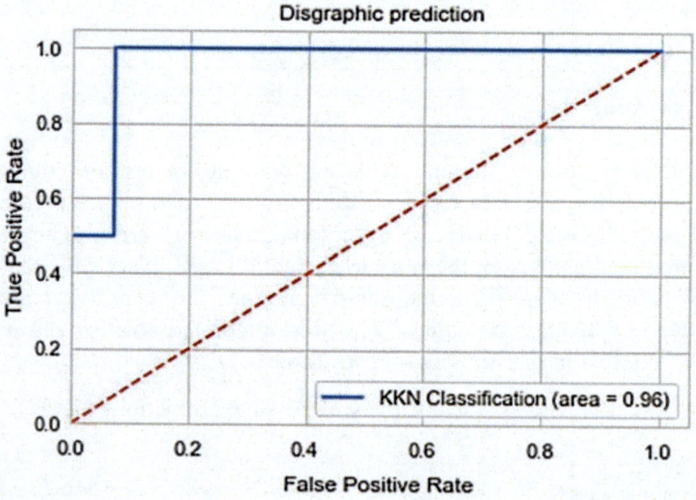

Fig. 21 ROC curve

3 Conclusion and Future Scope

Identifying handwriting difficulties in children in Devanagari script, using machine learning algorithm KNN, has given an accuracy of 94%. Identifying such difficulties at an early age or at the beginning of introduction to the language can help children to opt out the language or seek assistance.

Future scope: This concept can be extended to scan the handwriting, analyse and process the data as per the required features to identify the handwriting difficulties with respect to Devanagari script.

References

1. "Specific Learning Disabilities Auditory Processing Disorder (APD)," p. 1–5 (2013)
2. R.K. Deuel, Developmental dysgraphia and motor skills disorders. J. Child Neurol. **10**, 6–8 (1995). https://doi.org/10.1177/08830738950100S103
3. S. Graham, K.R. Harris, B. Fink, Is handwriting causally related to learning to write? Treatment of handwriting problems in beginning writers. J. Educ. Psychol. **92**(4), 620–633 (2000). https://doi.org/10.1037/0022-0663.92.4.620
4. S. Rosenblum, P.L. Weiss, Handwriting evaluation for developmental dysgraphia: process versus product. May 2014, 2004, https://doi.org/10.1023/B:READ.0000044596.91833.55
5. E.P. Assessment, Diagnosis of dysgraphia. vol. 2016, 2016, [Online]. Available: http://www.educational-psychologist.co.uk/sen-resources-blog/2016/1/22/cannot-pass-the-knowledge-of-life-and-language-in-the-uk-test-knoll-1
6. P. Tamboer, H.C.M. Vorst, S. Ghebreab, H.S. Scholte, Machine learning and dyslexia: classification of individual structural neuro-imaging scans of students with and without dyslexia. NeuroImage Clin. **11**, 508–514 (2016). https://doi.org/10.1016/j.nicl.2016.03.014
7. G. Dimauro, V. Bevilacqua, L. Colizzi, D. Di Pierro, TestGraphia, a software system for the early diagnosis of dysgraphia. IEEE Access **8**, 19564–19575 (2020). https://doi.org/10.1109/ACCESS.2020.2968367
8. H. Van Waelvelde, T. Hellinckx, W. Peersman, B.C.M. Smits-Engelsman, SOS: A screening instrument to identify children with handwriting impairments. Phys. Occup. Ther. Pediatr. **32**(3), 306–319 (2012). https://doi.org/10.3109/01942638.2012.678971
9. K. Zolna, The Dynamics of Handwriting Improves the Automated Diagnosis of Dysgraphia arXiv : 1906. 07576v1 [cs . CY] 12 Jun 2019, no. June, 2019
10. M.O.F. Education, A study of learning difficulties faced by non Hindi speaking students of standard VIII studying in English medium schools in Vadodara City guide Prof. Thomas Vadaya Mrs. Suvarna Jagtap school of science and education (2015)
11. I.M. Bosga-Stork, J. Bosga, R.G.J. Meulenbroek, Dysgraphic handwriting development and inclusive education: the role of interdisciplinary counseling. Open J. Soc. Sci. **03**(08), 35–47 (2015). https://doi.org/10.4236/jss.2015.38003
12. M. McCloskey, B. Rapp, Developmental dysgraphia: an overview and framework for research. Cogn. Neuropsychol. **34**(3–4), 65–82 (2017). https://doi.org/10.1080/02643294.2017.1369016
13. R. Kariyawasam, M. Nadeeshani, T. Hamid, I. Subasinghe, P. Samarasinghe, P. Ratnayake, Pubudu: deep learning based screening and intervention of Dyslexia, Dysgraphia and Dyscalculia, in *2019 IEEE 14th International Conference on Industrial and Information Systems Engineering Innovations for Industry 4.0, ICIIS 2019—Proceedings*, p. 476–481, 2019, https://doi.org/10.1109/iciis47346.2019.9063301
14. S.W. Sihwi, K. Fikri, A. Aziz, Dysgraphia identification from handwriting with support vector machine method Dysgraphia identification from handwriting with support vector machine method (2019). https://doi.org/10.1088/1742-6596/1201/1/012050

15. S. Rosenblum, Identifying Developmental Dsygraphia characteristics utilizing handwrtiing classification methods 00(00), 1–4 (2016)
16. I. Adubasim, Brainfeed intervention programme: an alternative approach for supporting people living with Dyslexia. Online Submiss **5**(2), 124–143 (2018)
17. K. Dutta, P. Krishnan, M. Mathew, C.V. Jawahar, Offline handwriting recognition on devanagari using a new benchmark dataset, in *Proceedings—13th IAPR International Workshop on Document Analysis Systems DAS 2018*, p. 25–30, 2018. https://doi.org/10.1109/das.2018.69
18. G. Hegde, M. Seetha, N. Hegde, Gabor filter matching score level fusion. Springer Int. Publ. Switz. **5**(December), 47–57 (2015). https://doi.org/10.1007/978-3-319-26832-3
19. M.S. Nehra, N. Nain, M. Ahmed, Proceedings of the international conference on recent cognizance in wireless communication & image processing. Proc. Int. Conf. Recent Cogniz. Wirel. Commun. Image Process. (January) (2016). https://doi.org/10.1007/978-81-322-2638-3
20. M.M.T,M. Hanumanthappa, A. Sangamithra, Intelligent predicting learning disabilities in school going children using fuzzy logic K mean clustering in machine learning. Int. J. Recent Technol. Eng. **8**(4), 1694–1698 (2019). https://doi.org/10.35940/ijrte.c5620.118419
21. C. Goutte, E. Gaussier, Ch10_Witnesses[8463].Pdf. no. April, 2005. https://doi.org/10.1007/978-3-540-31865-1
22. T. Fawcett, An introduction to ROC analysis. Pattern Recognit. Lett. **27**(8), 861–874 (2006). https://doi.org/10.1016/j.patrec.2005.10.010

Recognizing Fresh and Rotten Fruits Using Deep Learning Techniques

Archana Chougule, Apurva Pawar, Rahul Kamble, Juned Mujawar, and Akshay Bhide

Abstract Agriculture 4.0 aims at automation in all parts of supply chain of agriculture. In precision agriculture, precise execution of any task is done by using machine learning techniques. These techniques work on historical information. In horticulture, efficient technique for automated identification of fresh and rotten fruits is required. Automated fruit recognition will help farmers, traders, and customers with quality grading. It will help to reduce efforts and improve accuracy in fruit sorting. The time required for sorting fruits will be reduced which in turn will reduce total supply chain time. Customers will get more fresh fruits, and all stakeholders will be benefited. Automating fruit classification is the need of time for exporting good quality fruits. In this line, this chapter explains about classification of fruits using deep learning techniques. Deep learning is a kind of machine learning, based on artificial neural networks. Artificial neural network technique is inspired from working of human brain neurons. Convolution neural networks have inevitable capability of classifying images in required number of categories. It works on number of features and gives good results for classification task. This chapter shows how deep learning does fruit recognition for two categories as fresh fruit and rotten fruit.

Keywords Automation · Fruit recognition · Machine learning · Image processing · Classification · Convolution neural network

1 Introduction

As artificial intelligence is bringing revolution in every field including agriculture, automation in all processes of agriculture is must. Automatic identification of fresh and rotten fruits helps fruit markets to automate identification fruits. When fruits are produced and sold in large quantity, counting of fruits is also required in supply chain. Manual counting requires time and efforts. Object detection techniques can be used for such tasks [1–3]. Automated identification of fruits can be extended to automate

A. Chougule (✉) · A. Pawar · R. Kamble · J. Mujawar · A. Bhide
Sanjay Ghodawat University, Kolhapur, India

V. Bhateja et al. (eds.), *Data Engineering and Intelligent Computing*,
Advances in Intelligent Systems and Computing 1407,
https://doi.org/10.1007/978-981-16-0171-2_20

counting of fruits also. Total supply chain time will be reduced. Classification and counting of fruits manually may incur errors in the process.

This task of fruit recognition can be automated using deep learning techniques [4, 5]. Deep learning architecture consist of multi-layer feed-forward networks. Deep learning is the most popular technique for image processing tasks. Images of fruits can be processed by applying basic image processing techniques and then deep learning. Traditionally, classification techniques [6–9] from classical machine learning algorithms were used for this task. This paper shows how deep learning can give better results for fruit recognition.

2 Design and Implementation

This section gives overview of system design and implementation details. Section 2.1 discusses fresh fruit recognition implementation for nine types of fruits. Section 2.2 describes model building for five types of rotten fruits.

The user must give an input image to the system by actually specifying the path of the image. After that, the user will get the output as the name of fruit along with the input image. Fruit prediction is done using the algorithm convolution neural network in Keras.

2.1 Fresh Fruit Recognition

There are total nine classes of fruits used for training and validation purposes. These nine classes of fruits include apple, blueberry, lemon, mango, orange, pear, pineapple, pomegranate, and walnut. The required dataset of these fruits were taken from the Fruit-360 dataset of Kaggle.

The first step is to provide an image dataset to the system. The dataset comprises of nine classes. The second step comprises of storing the dataset into its memory. This image dataset is used for pre-processing. Pre-processing of the image is done to remove the noise and get clear image. The third step is feature extraction from it. The features such as edge detection, texture, and color are included. Edge detection can be done by detecting the edges of the image. Texture is used to detect the skin disease and defects present on the fruit. Color features are used to identify the color of the fruit including mean and variance of the RGB color model. Based on the features extracted, convolution neural network (CNN) in Keras [10] is trained. The model uses layers of CNN such as Conv2D and MaxPooling2D. Conv2D is also called as 2D convolution layer. This layer produces tensor of outputs by creating convolution kernel that is convolved with the input layer. MaxPooling2D is used for max pooling operation for spatial data. Selecting the maximum element from the region of the feature map covered by the filter is the operation performed by max pooling layer. To reduce the dimensions of the feature maps, pooling layers are used so. Thus, it can be concluded that use of pooling layers reduces the number of parameters to learn and the amount of computation performed in the network.

In compiling the model, a loss function, an optimizer, and a list of metrics are used. The model includes categorical_crossentropy as a loss function and adamax as an optimizer. These two were used because after trying with several other loss functions and optimizers, among them, they gave the best results. A summary of the model is generated using the function model.summary() after carrying out the first three steps. Figure 1 shows the output.

The fourth step is predicting the image based on trained model. The test image is used in this step. Trained model then checks for all the possibilities from the given image. The possibilities include getting the model and weights.

For the purpose of graphical user interface, Flask was used to build a Web application. This Web application is designed and implemented with sample input and output as depicted in Fig. 2.

2.2 Rotten Fruit Recognition

There are total three classes of fruits used for training and validation purposes. These three classes of fruits include apple, orange, and banana. The required dataset of these fruits were taken from the Fruit-360 dataset of Kaggle. The procedure used is the same as used for fresh fruit recognition. In compiling the model, a loss function, an optimizer, and a list of metrics are used. The model includes categorical_crossentropy as a loss function and adamax as an optimizer. These two were used because after trying with several other loss functions and optimizers, among them, they gave the best results. A summary of the model is generated using the function model.summary() after carrying out the first three steps. The sample inputs and outputs of the working of system for rotten fruit recognition are shown in Fig. 3.

3 Performance Evaluation

The implemented algorithm is checked for performance and accuracy with the help of different optimizers. This is done by representing the values for training accuracy, training loss, validation accuracy, and validation loss, respectively. These variations are calculated for fresh fruit recognition with use of different optimizers as follows (Figs. 4, 5 and 6).

After comparing with optimizers, the best results were obtained from adamax optimizer than others. Hence, use of adamax optimizer for fresh fruits recognition was done. Also, false predictions for some fruits were observed. They were as follows.

The image of pomegranate was inputted as the test image as shown in Fig. 7, and the output was displayed as apple with 83.88% of accuracy as shown in same figure. Thus, prediction was false because the model considered the color feature of

Layer (type)	Output Shape	Param #
conv2d_1 (Conv2D)	(None, 128, 128, 16)	208
activation_1 (Activation)	(None, 128, 128, 16)	0
max_pooling2d_1 (MaxPooling2	(None, 64, 64, 16)	0
conv2d_2 (Conv2D)	(None, 64, 64, 32)	2080
max_pooling2d_2 (MaxPooling2	(None, 32, 32, 32)	0
conv2d_3 (Conv2D)	(None, 32, 32, 64)	8256
max_pooling2d_3 (MaxPooling2	(None, 16, 16, 64)	0
conv2d_4 (Conv2D)	(None, 16, 16, 128)	32896
max_pooling2d_4 (MaxPooling2	(None, 8, 8, 128)	0
dropout_1 (Dropout)	(None, 8, 8, 128)	0
flatten_1 (Flatten)	(None, 8192)	0
dense_1 (Dense)	(None, 150)	1228950
activation_2 (Activation)	(None, 150)	0
dropout_2 (Dropout)	(None, 150)	0
dense_2 (Dense)	(None, 9)	1359

Total params: 1,273,749
Trainable params: 1,273,749
Non-trainable params: 0

Fig. 1 Summary of training fresh fruit recognition model

the input image and the prediction for the color of apple was more in percentage as compared to pomegranate.

Also, the image of green mango was inputted as the test image as shown in Fig. 8, and the output was displayed as pear with 60.76% of accuracy as shown in same figure. Thus, prediction was false because the model considered the color feature of the input image and the prediction for the color of pear was more in percentage

Fig. 2 Sample input and output for fresh fruit recognition

Fig. 3 Sample inputs and outputs for rotten fruits recognition

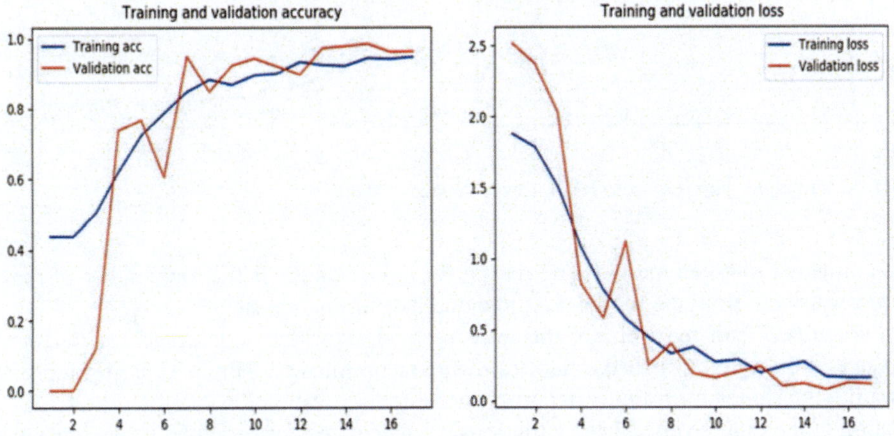

Fig. 4 Variations with respect to SGD optimizer

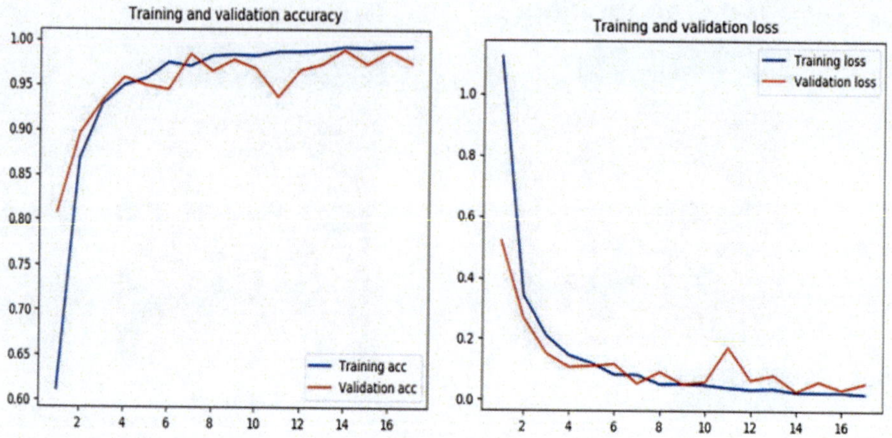

Fig. 5 Variations with respect to adamax optimizer

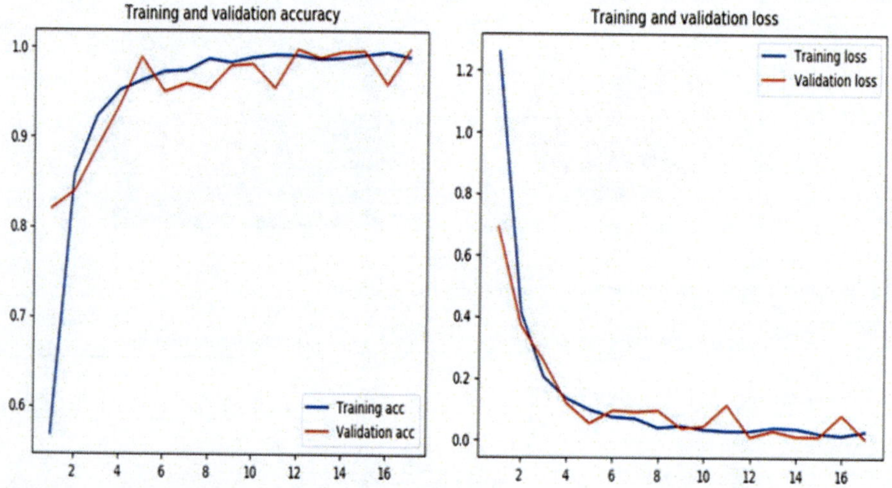

Fig. 6 Variations with respect to RMSprop optimizer

as compared to green mango. The reason for low accuracy is the low quality of the images used to train the model, i.e., low quality training dataset.

For rotten fruit recognition, the implemented algorithm is checked for performance and accuracy with the help of different optimizers. This is done by representing the values for training accuracy, training loss, validation accuracy, and validation loss, respectively. These variations are calculated for rotten fruit recognition with use of different optimizers and similar kind of results as for fresh fruit recognition were obtained. After comparing with optimizers, the best results were obtained

Fig. 7 Example of false prediction for pomegranate

Fig. 8 Example of false prediction for green mango

from adamax optimizer than others. Hence, use of adamax optimizer for rotten fruits recognition was done.

4 Conclusions

The paper emphasizes on the use of deep learning technique for fruit recognition. Experiment on different types of fruits for fresh fruit recognition and rotten fruit recognition shows that CNN gives good results, and Keras library is a good option for implementation. Model was experimented with different types of optimization techniques, and results are shown in performance evaluation section. During this work, knowledge was gained on deep learning and optimization techniques.

References

1. C. Li, Y. Zhang, Y. Qu, Object detection based on deep learning of small samples, in *2018 10th ICACI*, March 2018, p. 449–454
2. M.T. Islam, S. Rahman, B.M. Nafiz Karim Siddique, T. Jabid, Image recognition with deep learning, in *ICIIBMS*, pp. 106–110, 2018
3. C. Tang, Y. Feng, X. Yang, C. Zheing, Y. Zhou, The object detection based on deep learning, in *International Conference on Information Science and Control Engineering*, pp. 723–728, 2017
4. C. Liu, Y. Cao, Y. Luo, G. Chen, V. Vokkarane, Y. Ma, P. Hou, A new deep learning based food recognition system for dietary assessment on an edge computing service infrastructure, IEEE, pp. 1–13 (2016)
5. S. Mao, Y. Li, Y. Ma, B. Zhang, J. Zhou, K. Wang, Automatic cucumber recognition algorithm for harvesting robots in the natural environment using deep learning and multi-feature fusion. Comput. Electron. Agric., pp. 1–12, 07 February 2020
6. S. Jana, S. Basak, R Parekh, Automatic fruit recognition from natural images using color and texture features, in *IEEE, Devices for Integrated Circuit (DevIC)*, p. 620–624 (2017)
7. A. Bhargava, A. Bansal, Fruits and vegetables quality evaluation using computer vision: a review. J. King Saud Univ.—Comput. Inf. Sci., pp. 1–15, 8, May 2018
8. E.K. Nyarko, I. Vidovic, K. Radocaj, R. Cupec, A nearest neighbor approach for fruit recognition in RGB-D images based on detection of convex surfaces. Expert Syst. Appl., pp. 454–466, 24 July 2018
9. Y., Tao, J. Zhou, Automatic apple recognition based on the fusion of color and 3D feature for robotic fruit picking. Comput. Electron. Agric., pp. 388–396, 26 September 2017
10. https://keras.io/

Deep Learning-Based Software Tools for Tuberculosis Detection in Chest X-Ray Images

Muralikrishna Puttagunta and S. Ravi

Abstract Deep learning is a rising phenomenon in data analysis and one of the ten innovative technologies. Deep learning allows for multi-layered computational models to learn data representations with multiple levels of abstraction. Deep learning provides the best solution to visual object recognition. Deep learning offers the best solution to tuberculosis (TB) detection in X-ray images. In a shortage of qualified radiologists, these new technologies increase the capacity to improve overall TB diagnosis and treatment. This paper aims to review deep learning-based software techniques, such as CAD4TB, qXR, and Lunit INSIGHT used for detecting chest X-ray (CXR) abnormalities. Deep learning-based CAD4TB software highlights the abnormal region in the form of a heat map image. qXR software identifies 15 abnormalities from abnormal chest X-ray images. Lunit INSIGHT CXR software identifies ten abnormalities from chest radiography. The human observer must validate deep learning-based computer-aided detection (CAD) systems used in real diagnosis service. The accuracy of CAD software is analysed by using the receiver operating characteristic (ROC) curve. Deep learning-based systems in medicine must meet ethical values and validate the results.

Keywords Deep learning · CAD4TB · Chest radiographs · TimBre app

1 Introduction

Early recognition is essential for the efficient prevention and regulation of TB. The high TB rate is noticed because of poverty, overcrowding population at slums, poor nutrition, mental health, drug usage, and the human immunodeficiency virus (HIV) [1]. In several countries, poor health care conditions and lack of trained persons to read CXRs may lead to an increase in the TB detection gap. Many screening

M. Puttagunta · S. Ravi (✉)
Department of Computer Science, School of Engineering and Technology, Pondicherry University, Pondicherry, India

V. Bhateja et al. (eds.), *Data Engineering and Intelligent Computing*,
Advances in Intelligent Systems and Computing 1407,
https://doi.org/10.1007/978-981-16-0171-2_21

methods used for the detection of TB include tuberculin skin test (TST), chest X-rays (CXRs), symptom questionnaires, sputum culture, acid-fast bacillus (AFB) test, interferon-gamma release assays (IGRAs), and blood test.

The World Health Organization (WHO) suggests that a chest X-ray is an initial examination tool for screening of TB. Chest radiographs are important tools used to detect chest diseases [2]. Limited X-ray equipment, limited trained persons for equipment maintenance, and poor specificity of observers limited the role of X-ray images for the detection of TB. The emergence of digital radiography revolutionized the automated screening of TB detection using the computer-aided detection system [3]. The diagnostic accuracy of detection TB from CXR depends on the radiologist.

2 Role of Deep Learning

Deep learning (DL) provides a powerful way of automating medical imaging research and diagnosis. It might reduce the burden on radiologists in radiology practice [4]. The applications of deep learning in radiology comprise medical image registration, segmentation, neurological disease diagnosis, computer-aided detection systems for X-ray, computerized tomography (CT), mammography and magnetic resonance imaging (MRI) images, and natural language processing (NLP) for text analysis of reports [5]. Conventional machine learning systems required domain knowledge to extract features from raw data. These features form as feature vectors to train the classifier. It could classify patterns in the input [6].

Neural networks and deep learning become popular because of the large training data set and computational resources [7]. DL methods could automatically extract features from a raw dataset. Deep learning-based convolutional neural networks are used in image classification and recognition also applied in CAD systems for the detection of diseases from different medical images [8, 9]. Convolutional neural networks (CNNs) have proven to be a reliable tool for computer vision tasks. Medical image analysis groups are entering the field of deep learning to apply CNNs and other deep learning methodologies to a variety of medical applications [10]. CNN can be divided into two parts; namely (i) feature extraction and (ii) classifier. In a CNN, convolution layers are used for feature extraction. Fully connected layers act as a classifier [11].

The rest of this paper is organized as follows. Section 3 describes the CAD4TB software for TB screening and the comparison of different versions of CAD4TB. Sections 4 and 5 give a brief view of qXR and Lunit INSIGHT softwares used for TB detection from CXR images. Machine learning-based TimBre app for TB detection is described in Sect. 6. Section 7 deals with discussion, and Sect. 8 concludes the paper.

3 CAD4TB

CAD4TB software is a screening tool developed by Diagnostic Image Analysis Group, Radboud University Medical Center, and Nijmegen, Netherlands. Delft imaging systems released CE-certified the CAD4TB [12]. CAD4TB software is trained using machine learning algorithms with labelled images to differentiate the 'normal' and 'abnormal' X-ray images. CAD4TB software takes DICOM (digital imaging and communications in medicine) chest X-ray images as input. The output of CAD4TB includes an abnormality score between 0 and 100, and a heat map exhibits abnormality area in the image. Using a threshold value CXR image is classified as normal or abnormal. Generally, this threshold value is greater than 50. Figure 1 shows the example output of CAD4TBV6.

In [13], the accuracy of screening digital CXR using CAD4TBv1.08 is compared with clinical officers for the detection of TB. The CAD4TB is designed as a combination of different detection systems that are textural abnormality detection (TAD), shape abnormality detection (SAD), lung field detection, and clavicle detection (CD). A TAD system operating at the pixel level is combined with the CD system to reduce the false-positive response, fused to an image decision. The SAD system operating at the image level produces an abnormality score based on the lung shape.

The SAD system abnormality score was compared with the TAD system abnormality score to produce the final abnormality score. Texture and shape abnormality scores were used as characteristic image features to train a k-NN classifier which was used to determine the abnormality score of a new CXR image [14]. Figure 2 shows the abnormality score detection system of CAD4TBv1.

The following are the steps to get an abnormality score for a new CXR image for CAD4TBV 1.08:

- SAD and TAD detection systems automatically segment the un-obscured lung field for the new CXR image;

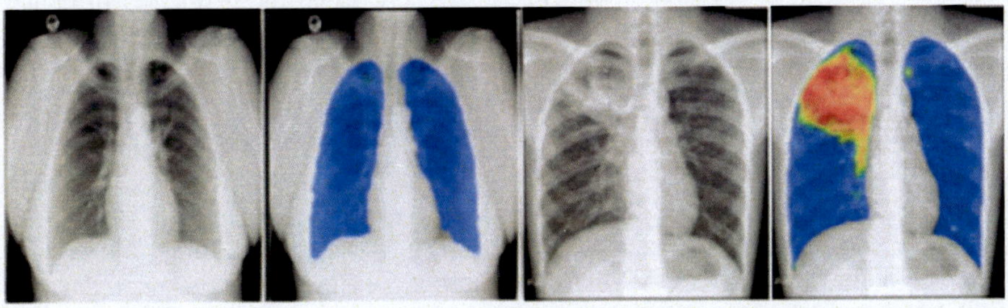

(a) (b) (c) (d)

Fig. 1 Two example CXRs with corresponding output heat maps using CAD4TBv6. **a)** CXR image with negative TB, an abnormality score of 17. **b)** Corresponding abnormality heat map. **c)** CXR image with positive TB, an abnormality score of 82. **d)** Corresponding abnormality heat map

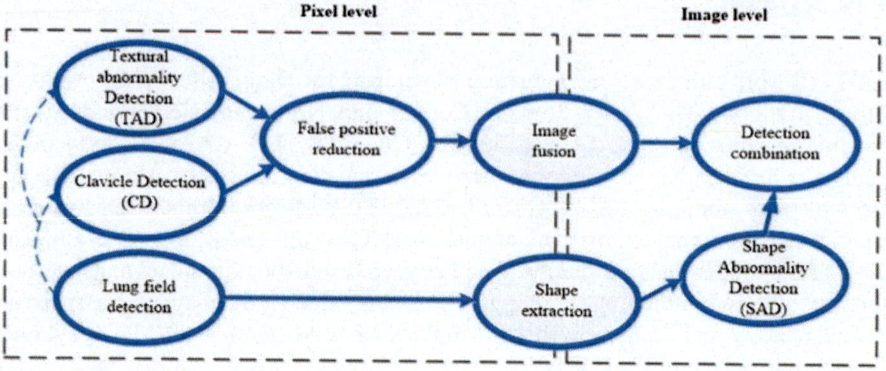

Fig. 2 Combination of detection systems [14]

- TAD system extracts the descriptive features from circular patches to differentiate the normal and abnormal images;
- The SAD system computes a shape abnormality score based on lung shapes;
- For a new CXR image, the abnormality score is estimated based on the training data of texture and shape abnormality scores.

Muyoyeta et al. [15, 16] had successfully tested the CAD4TBv1.02 program for TB detection in a real-world setting. The authors had shown that CAD4TBv1.02 had high sensitivity and low specificity. In Steiner et al. [17], the authors compared the performance of CAD4TB v3.07 software in chest X-ray images screening with human readers (radiologists). In Philipsen et al. [18] and Melendez et al. [19], authors evaluated cheap pre-screen test for TB detection using automated digital chest radiography (ACR) prior to Xpert MTB test.

Melendez et al. [20] compared the performance of automatic software CAD4TB version 5 in chest X-ray (CXR) screening with that of radiologists' readers. CAD4TBv5 used support vector machines as a classifier for better performance as compared to previous versions of CAD4TB. Murphy et al. [21] presented the four versions of CAD4TB v3.0, CAD4TBv4.0, CAD4TBv5.0, and CAD4TBv6.0 software that were compared to the Xpert test as a reference standard. CAD4TB v6 had a sensitivity of 90% and specificity of 76%. CAD4TB v6 had shown a considerable increase in sensitivities without loss of specificity related to its previous CAD versions.

CAD4TBv6 is software based on the concept of deep learning released in 2018.The main features of CAD4TBv6 are processing time per image, reduction of 60 s processing time to 15 s, and the ability to process images of subjects four years and older, whereas the lower age limit for previous versions was 16 years [22]. In CAD4TB software, input images had the following requirements (i) DICOM image format, (ii) the ability to process images of subjects should be at least four years and older, (iii) posterior anterior view, (iv) digital images either from a CR or DX source (scanned films are not acceptable).

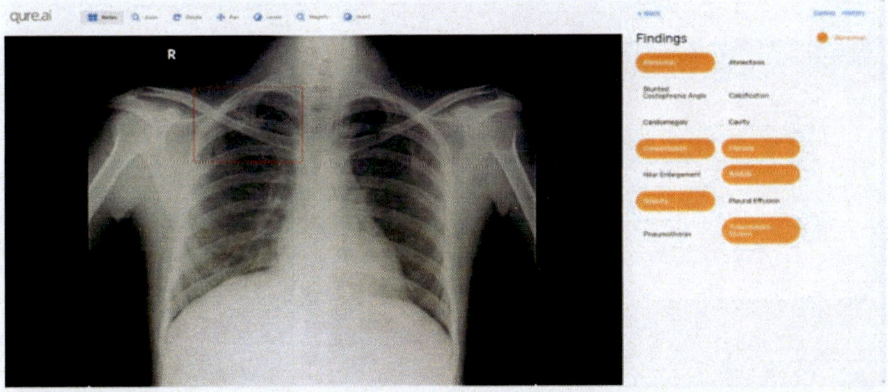

Fig. 3 qXr software output image with abnormalities

4 qXR

Qure.ai is an artificial intelligence-based start-up company in India, which applies machine learning and deep learning technology for rapid and accurate disease diagnosis to radiology imaging. qXR is a chest X-ray screening tool that detects signs of consolidation, cardiomegaly, pulmonary, cavity, hilar enlargement, and pleural tuberculosis [23]. In [24], the authors studied the diagnostic accuracy of a CAD software qXR (Qure.ai, Mumbai, India) using a reference standard of microbiologically confirmed TB. Figure 3 shows the output of qXR software to the input of the DICOM chest X-ray image. qXR software works with different X-ray images of varying quality and exposure from all X-ray machine models. A commercial CAD software qXR received CE-certification in 2018 [25]. In the public health screening program in India, qXR software is used for TB detection. The area under a receiver operating characteristic (ROC) curve (AUC) of qXR for the detection of microbiologically confirmed that PTB was 0.81. For the other abnormalities, the AUC ranged from 0.75 to 0.94. The qXR software achieved an AUC of 0.87 using a radiologist as a reference standard. For the latest version of CAD4TB, an AUC was greater than the qXR.

5 Lunit INSIGHT CXR

The aim of Lunit INSIGHT CXR 3 is to assist in the identification of ten chest abnormalities that are consolidation, nodule, calcification, fibrosis, pneumothorax, pneumoperitoneum, atelectasis, cardiomegaly, pleural effusion, mediastinal widening, and tuberculosis by screening on the chest X-ray image. Lunit INSIGHT is CE certified [26]. Figure 4 shows output JPG format images (abnormality score of different

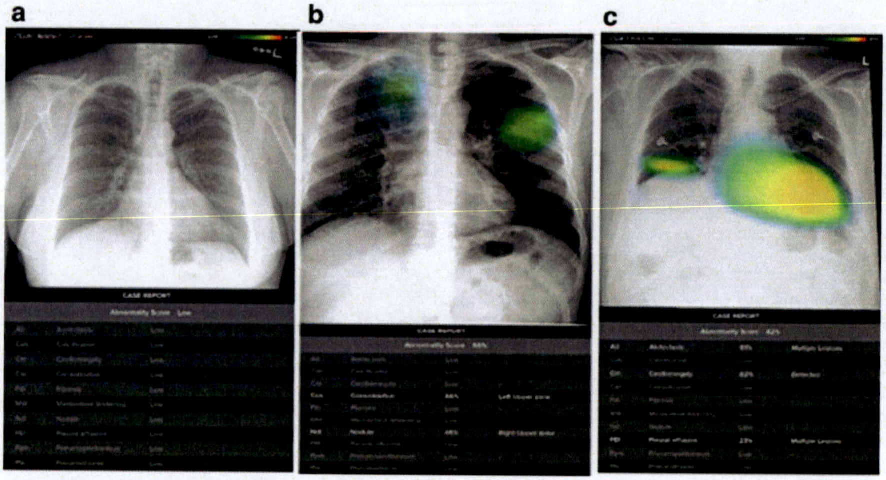

Fig. 4 Lunit INSIGHT CXR 3 software output JPG format images for DICOM chest X-ray images. Three images are shown in the above of figure **a)** normal image, **b)** the consolidation image, indicating nodule with an abnormality score of 66%, and **c)** image with an abnormality score of 82%

chest abnormalities) of Lunit INSIGHT CXR 3 software by uploading DICOM Chest X-ray Images.

In Qin et al. [27], the authors tested three deep learning systems CAD4TB, qXR, and Lunit INSIGHT for identifying TB-associated abnormalities in chest X-ray images. The authors comparatively evaluated the three DL systems. These were CAD4TB (V.6), qXR (v.2), and Lunit INSIGHT (V 4.7.2) for the chest X-ray image developed by Lunit (South Korea). Lunit and CAD4TB read the DICOM format of CXR images. CAD4TB output was in the abnormality score from 0 to 100 for TB detection. From the CXR image, cavitation, nodule, pneumothorax, etc., like pulmonary abnormalities were identified by qXR and Lunit. CAD4TB, qXR, and Lunit, three deep learning-based softwares, outperformed experienced human readers in differentiating people having TB and those without TB.

6 TimBre

Docturnal is a non-invasive point of care diagnostics and screening provided for tuberculosis, diabetic retinopathy under the umbrella of telemedicine leveraging mobility and machine/deep learning. Docturnal is a private limited, which developed a screening app TimBre for TB detection. This app makes the detection of TB easy and hassle-free [28]. It required a medical professional to record the cough of a patient using a microphone array and along with some clinical data like demographic details, living with TB, pre-existing health conditions, family history of TB, sleep, cough patterns, and HIV which were given to the TimBre app. Thereafter, applying machine learning, the recording was processed in real-time to predict whether it was TB positive or not. If any person is found positive on screening, doctors often recommended for an X-ray of the chest for diagnosis. When TimBre app was cross-validated with existing TB detection methods, it had an accuracy of 85%. TimBre was deployed as an experimental project in a few chest hospitals at Hyderabad and trials are going on at Narayana Hrudayala in Bangalore.

7 Discussion

Deep learning-based software tools are used as pre-screening systems in low-income country regions and work as an accurate system in detecting TB from CXR images. Different software, dataset, and performance metrics are used for TB detection which are given in Table 1. Performance metrics used for TB detection by various authors in their publications are AUC abbreviated as "The Area Under a receiver operating characteristic (ROC) Curve", sensitivity (SN), and specificity (SP).

8 Conclusion

Deep learning-based CAD systems used in the automatic screening of CXRs can expand the tuberculosis diagnosis. Deep learning-based CAD systems can classify many normal CXRs in high population areas for TB screening with high sensitivity [34]. These CAD systems could make radiographic TB screening is less cost. Future work in deep learning-based screening systems will concentrate on further increasing the precision and specificity of CAD systems for TB detection [35].

Table 1 Different software's used for TB detection

Author's Name	Database used	CAD software	Reference standard	Performance metrics
Maduskar et al. [13]	Prospective, study-specific	CAD4TB (v 1.08)	Liquid culture, AFB smear	AUC = 0.73
Muyoyeta et al. [15]	Prospective, study-specific	CAD4TB (v 1.08)	NAAT	AUC = 0.714
Muyoyeta et al. [16]	Prospective, study-specific	CAD4TB (v 1.08)	NAAT for CXR+, AFB smear for CXR	–
Steiner et al. [17]	TB REACH project	CAD4TB (v 3.07)	Human reader CXR	AUC = 75.13%
Philipsen et al. [18]	TB-NEAT collaborative study	CAD4TB (v 3.07)	NAAT, liquid culture	Sensitivity = 78.9% Specificity = 98.1%
Melendez et al. [19]	TB-NEAT collaborative study	CAD4TB (v 3.07)	Liquid culture	AUC = 0.84
Melendez et al. [20]	Zambia national TB prevalence survey	CAD4TB (v 5)	Liquid culture/NAAT for CXR+	Sensitivity = 48.1% Specificity = 98.5% (at threshold = 70)
Murphy et al. [21]	Healthy life Karachi Pakistan	CAD4TB (v 6)	Human reader CXR, NAAT	Sensitivity = 90% Specificity = 98%
Nash et al. [24]	Kasturba Hospital, Manipal	qXR	Microbiologically-confirmed PTB	AUC = 0.81 (for TB)
Qin et al. [27]	Nepal Cameroon	CAD4TB Lunit, qXR	Human reader NAAT	Lunit AUC = 0.94 qXR AUC = 0.94 CAD4TB AUC = 0.92
Breuninger et al. [29]	TB Cohort and TB CHILD study	CAD4TB (v 3.07)	Liquid culture, AFB smear	AUC = 0.84

(continued)

Table 1 (continued)

Author's Name	Database used	CAD software	Reference standard	Performance metrics
Rahman et al. [30]	TB-NEAT collaborative study	CAD4TB (v 3.07)	NAAT	AUC = 0.74
Mohammad et al. [31]	Healthy life Karachi Pakistan	CAD4TB (v3.07)	NAAT	AUC = 0.83
Melendez et al. [32]	–	CAD4TB (v5)	Human reader CXR, TB treatment	AUC = 0.90
Koesoemadinata et al. [33]	Prospective study-specific	CAD4TB (v 5)	Liquid culture, NAAT	AUC = 0.89

References

1. K. Paquette, M.P. Cheng, M.J. Kadatz, V.J. Cook, W. Chen, J.C. Johnston, Chest radiography for active tuberculosis case finding in the homeless: a systematic review and meta-analysis. Int. J. Tuberc. Lung Dis. **18**(10), 1231–1236 (2014). https://doi.org/10.5588/ijtld.14.0105
2. J.Y. Wu et al., The role of chest radiography in the suspicion for and diagnosis of pulmonary tuberculosis in intensive care units. Int. J. Tuberc. Lung Dis. **13**(11), 1380–1386 (2009)
3. C. Schaefer-Prokop, U. Neitzel, H.W. Venema, M. Uffmann, M. Prokop, Digital chest radiography: an update on modern technology, dose containment and control of image quality. Eur. Radiol. **18**(9), 1818–1830 (2008). https://doi.org/10.1007/s00330-008-0948-3
4. M. Bigelow et al., Integrating AI into radiology workflow: levels of research, production, and feedback maturity **7**(1) (2020). https://doi.org/10.1117/1.JMI.7.1.016502
5. M.A. Khan et al., Lungs cancer classification from CT images: an integrated design of contrast based classical features fusion and selection. Pattern Recogn. Lett. **129**, 77–85 (2020). https://doi.org/10.1016/j.patrec.2019.11.014
6. S. Wang, R.M. Summers, Machine learning and radiology. Med. Image Anal. **16**(5), 933–951 (2012). https://doi.org/10.1016/j.media.2012.02.005
7. Gautham A, Bhateja V, Tiwari A, Satapathy SC (2018) An improved mammogram classification approach using back propagation neural network. Data Eng. Intell. Comput. **542**, 283–291 (2018). https://doi.org/10.1007/978-981-10-3223-3
8. V. Bhateja, M. Misra, S. Urooj, Non-linear polynomial filters for edge enhancement of mammogram lesions. Comput. Methods Programs Biomed. **129**, 125–134 (2016). https://doi.org/10.1016/j.cmpb.2016.01.007
9. V.P. Vianna, Study and development of a computer-aided diagnosis system for classification of chest X-ray images using convolutional neural networks pre-trained for ImageNet and data augmentation 1–7 (2018)
10. H. Greenspan, B. Van Ginneken, R.M. Summers, Guest editorial deep learning in medical imaging: overview and future promise of an exciting new technique. IEEE Trans. Med. Imaging **35**(5), 1153–1159 (2016). https://doi.org/10.1109/TMI.2016.2553401
11. L. Pigou, S. Dieleman, P.J. Kindermans, B. Schrauwen, Sign language recognition using convolutional neural networks, in Lecture Notes in Computer Science. Lecture notes in artificial intelligence. Lecture notes in bioinformatics, vol. 8925 (2015), pp. 572–578. https://doi.org/10.1007/978-3-319-16178-5_40
12. WHO/UNITAID, *Tuberculosis Diagnostics Technology and Market Landscape* (UNITAID Secr. World Health Organization, 2014)
13. P. Maduskar, M. Muyoyeta, H. Ayles, L. Hogeweg, L. Peters-Bax, B. Van Ginneken, Detection of tuberculosis using digital chest radiography: automated reading vs. interpretation by clinical officers. Int. J. Tuberc. Lung Dis. **17**(12), 1613–1620 (2013). https://doi.org/10.5588/ijtld.13.0325
14. L. Hogeweg, C. Mol, P.A. De Jong, R. Dawson, H. Ayles, B. Van Ginneken, Fusion of local and global detection systems to detect tuberculosis in chest radiographs, in Lecture Notes in Computer Science. Lecture notes in artificial intelligence. Lecture notes in bioinformatics, vol. 6363, LNCS, No. Part 3 (2010), pp. 650–657. https://doi.org/10.1007/978-3-642-15711-0_81
15. M. Muyoyeta et al., The sensitivity and specificity of using a computer aided diagnosis program for automatically scoring chest X-rays of presumptive TB patients compared with Xpert MTB/RIF in Lusaka Zambia. PLoS ONE **9**(4), 16–18 (2014). https://doi.org/10.1371/journal.pone.0093757
16. M. Muyoyeta et al., Digital CXR with computer aided diagnosis versus symptom screen to define presumptive tuberculosis among household contacts and impact on tuberculosis diagnosis. BMC Infect. Dis. **17**(1), 1–8 (2017). https://doi.org/10.1186/s12879-017-2388-7
17. A. Steiner et al., Screening for pulmonary tuberculosis in a Tanzanian prison and computer-aided interpretation of chest X-rays. Public Health Action **5**(4), 249–254 (2015). https://doi.org/10.5588/pha.15.0037

18. R.H.H.M. Philipsen et al., Automated chest-radiography as a triage for Xpert testing in resource-constrained settings: a prospective study of diagnostic accuracy and costs. Sci. Rep. **5**(March), 1–8 (2015). https://doi.org/10.1038/srep12215
19. J. Melendez et al., An automated tuberculosis screening strategy combining X-ray-based computer-aided detection and clinical information. Sci. Rep. **6**(October 2015), 1–8 (2016). https://doi.org/10.1038/srep25265
20. J. Melendez, R.H.H.M. Philipsen, V. Sunkutu, N. Kapata, Automatic versus human reading of chest X-rays in the Zambia National Tuberculosis Prevalence Survey **21**(April), 880–886 (2017)
21. K. Murphy et al., Computer aided detection of tuberculosis on chest radiographs: an evaluation of the CAD4TB v6 system (2019), pp. 1–11 [Online]. Available: https://arxiv.org/abs/1903.03349
22. Our Improved CAD4TB Software is Now Reliable for Children | Delft Imaging. https://www.delft.care/cad4tb6/
23. Qure.ai Chest X Ray Study. https://qure.ai/qxr/
24. M. Nash et al., Deep learning, computer-aided radiography reading for tuberculosis: a diagnostic accuracy study from a tertiary hospital in India. Sci. Rep. **10**(1), 1–10 (2020). https://doi.org/10.1038/s41598-019-56589-3
25. Qure.ai | qXR Becomes First AI-Based Chest X-ray Interpretation Tool to Receive CE Certification. https://qure.ai/news/2018/05/31/qXR-CE.html
26. Lunit INSIGHT for Chest Radiography. https://insight.lunit.io/cxr/
27. Z.Z. Qin et al., Using artificial intelligence to read chest radiographs for tuberculosis detection: a multi-site evaluation of the diagnostic accuracy of three deep learning systems. Sci. Rep. 1–10 (2019). https://doi.org/10.1038/s41598-019-51503-3
28. TimBre | Docturnal. https://www.docturnal.com/products/timbre/
29. M. Breuninger et al., Diagnostic accuracy of computer-aided detection of pulmonary tuberculosis in chest radiographs: a validation study from sub-Saharan Africa. PLoS One **9**(9) (2014). https://doi.org/10.1371/journal.pone.0106381
30. M.T. Rahman et al., An evaluation of automated chest radiography reading software for tuberculosis screening among public- and private-sector patients. Eur. Respir. J. **49**(5) (2017). https://doi.org/10.1183/13993003.02159-2016
31. S. Mohammad, A. Zaidi, S.S. Habib, B. Van Ginneken, Evaluation of the diagnostic accuracy of computer-aided detection of tuberculosis on chest radiography among private sector patients in Pakistan. Sci. Rep. (July), 1–9 (2018). https://doi.org/10.1038/s41598-018-30810-1
32. J. Melendez et al., Accuracy of an automated system for tuberculosis detection on chest radiographs in high-risk screening. Int. J. Tuberc. Lung Dis. **22**(5), 567–571 (2018). https://doi.org/10.5588/ijtld.17.0492
33. R.C. Koesoemadinata et al., Computer-assisted chest radiography reading for tuberculosis screening in people living with diabetes mellitus. Int. J. Tuberc. Lung Dis. **22**(9), 1088–1094 (2018). https://doi.org/10.5588/ijtld.17.0827
34. M.H. Id et al., A systematic review of the diagnostic accuracy of artificial intelligence-based computer programs to analyze chest X-rays for pulmonary tuberculosis 1–19 (2019)
35. P. Putha et al., Can artificial intelligence reliably report chest X-rays? Radiologist validation of an algorithm trained on 2.3 million X-rays 1–13 (2018). [Online]. Available: https://arxiv.org/abs/1807.07455

Cancer Prediction Using Machine Learning

Shona Afonso, Anusha Pai, and Razia de Loyola Furtado Sardinha

Abstract Breast cancer is a common form of cancer experienced in women across different age groups and accounts for 14% of cancers in Indian women. The high mortality rate is due to the lack of awareness of the disease and delay in screening and diagnosis. Early detection is the solution which implies that breast cancer can be treated, and patients have the chance to live a healthy life post-recovery. The proposed system aims at providing a diagnostic analysis of the cytological features derived from a biopsy-based image. The designed model attempts to predict cases of malignancy or benign on the Wisconsin Diagnostic Breast Cancer (WDBC) dataset. The dataset used for analysis and implementation consists of attributes obtained from images of fine needle aspiration (FNA) tests on an affected breast tissue. The feature selection algorithm implemented was recursive feature elimination based on random forest approach. A comparative analysis was carried out against several other classification methods in an attempt to compare the accuracy of random forest approach against each of the other models. The neural network constructed using the features selected exhibited an accuracy of 96.51% while predicting malignant and benign cases in the test set. The main objective of this study was to devise an efficient model that a pathologist can effectively use in determining cases of malignancy based on the most critical cytological features. This ensures high accuracy, thereby enabling a patient to avail of the required medical attention at the earliest.

Keywords Data analysis · Visualization · Recursive feature elimination · Random forest · Neural network · Regularization

S. Afonso (✉) · A. Pai
Department of Computer Engineering, Padre Conceicao College of Engineering, Verna, Goa, India

R. de Loyola Furtado Sardinha
Department of Information Technology, Padre Conceicao College of Engineering, Verna, Goa, India

V. Bhateja et al. (eds.), *Data Engineering and Intelligent Computing*,
Advances in Intelligent Systems and Computing 1407,
https://doi.org/10.1007/978-981-16-0171-2_22

1 Introduction

The rural and urban sectors of India have seen a rise in cases of breast cancer. Statistics suggest a prevalence rate of 25.8 women per 100,000 women in India. If there is increasing awareness and detection amongst women, the number of cases identified would increase by a large factor. Recent times have seen the onset of better techniques of diagnosis and effective methods of treating the disease since early diagnosis can save lives [1]. Reports suggest that experienced medical practitioners can diagnose cancer with a higher accuracy using machine learning techniques. Hence, there has been continuous effort to devise effective data mining and machine learning techniques for breast cancer detection and classification.

Research indicates that the accuracy of prognosis is only 60% for pathologists. Hence, it is necessary to overcome this hurdle and create a high accuracy diagnostic system. Machine learning provides a solution by employing a variety of statistical, probabilistic and optimization techniques enabling a system to be trained on known examples for prediction and classification.

Data mining approaches are a very effective way of classifying the data which is imperative in decision making. Research tools can enable identifying patterns and relationships among large number of variables based on historical data, thereby predicting the outcome of a disease. With predictive medicine taking a lead and the growing reliance on machine learning to make predictions, the study included a detailed review of cases that used machine learning based prediction methods to design the model. The objective of the proposed system aims at providing a diagnostic analysis of the cytological features derived from a biopsy-based image to predict cases of malignancy or benign on the Wisconsin Diagnostic Breast Cancer (WDBC) dataset.

2 Literature Survey

A study in [2] provided a diagnosis for breast cancer using application of machine learning techniques to the classification of cancer-related data. Various classification methods were explored such as *K*-nearest neighbors, probabilistic neural networks and support vector machines classifiers and applied to selected feature subsets. The best overall accuracy for breast cancer diagnosis was achieved using support vector machines classifier models.

Machine learning classification techniques to categorize benign and malignant tumor were implemented in [3], in which the machine learns from stored historical data and predicts the new input category. Models like logistic regression and random forest were applied on the UCI dataset achieving the best accuracy of 96% with the random forest model. This study provided a basis for a complete understanding of the random forest model and was used to implement the proposed system.

In [4], the research work undertaken provided the prediction of recurrence of breast cancer using KNN, SVM, Naive Bayes and logistic regression techniques. Each algorithm was evaluated and an understanding was provided that depending on the dataset and the parameter selection more accuracy can be achieved with different models.

A comprehensive view of the implementation of a diagnostic system for breast cancer was provided in [5]. Performances of multilayer perceptron neural network (MLPNN), combined neural network (CNN), probabilistic neural network (PNN), recurrent neural network (RNN) and support vector machine (SVM) were evaluated.

An analysis conducted in [6] explored and compared the accuracy of algorithms such as J48, Naive Bayes, RBF, to build cancer prediction models. Analysis of results obtained showed that Naive Bayes prediction was the best with an accuracy of 97.36%.

A comparative study was conducted in [7] evaluating the accuracy of Naive Bayes classifier, support vector machine (SVM), AdaBoost tree, artificial neural networks (ANN), to find a powerful model for breast cancer prediction. Dimensionality reduction was implemented using principal component analysis.

The authors in [8] explored the advantages and short comings of using machine learning methods like SVM, Naive Bayes, neural network and decision trees for breast cancer prediction. The study revealed that ANN was the most preferred technique as compare to methods like decision tree, regression tree, etc. It was also ascertained that ANN technique provides a robust solution to real-time predictions.

A survey [9] conducted on various neural network technologies for classification of cancer provided valuable insights into selecting an appropriate structure for the proposed model. Different types of neural network configurations were explored and the pros and cons discussed. The survey also stated that multilayer perceptron gives 97.1% accuracy and probabilistic neural network provides 96% accuracy for a given dataset.

The research undertaken in [10] presented a study of different machine learning algorithms: GRU-SVM, linear regression, multilayer perceptron, nearest neighbor search, softmax regression and support vector machine on the Wisconsin Diagnostic Breast Cancer dataset. It also included their classification test accuracy, sensitivity and specificity values. Results showed that all the presented machine language algorithms performed well on the classification task with a high-test accuracy. However, prediction accuracy could be improved by better feature selection algorithms which was attempted in the proposed model.

3 Software Requirements

Python has been the language chosen for machine learning and artificial intelligence for this study. Google TensorFlow and Keras libraries have been used to implement the machine learning algorithms. Other scientific computing libraries used were matplotlib, pandas, graphviz and scikit-learn. Anaconda environment along with Jupyter notebook was used on a Windows operating system.

The breast cancer dataset was obtained from the UCI Machine Learning Repository. Data from five hundred sixty-nine patients has been used as input into the diagnostic model. Each observation represents FNA test measurements on affected breast masses. Areas of the FNA with minimal overlap between nuclei were selected for scanning, and using a software called Xcyt [11], approximate boundaries were created. Finally, calculations were made resulting in 30 features of the dataset.

The first two attributes of the dataset correspond to the "identifier number" of the patient and the diagnosis status. Attributes of nuclear size are described by the radius and area features. Nuclear shape is defined in terms of concave points, smoothness, concavity, symmetry, compactness and fractal dimension features. Size and shape of the cell nuclei are expressed by the perimeter feature. Nuclear texture is measured by computing the variance of the gray-scale intensities of the image pixels. The mean value, worst and standard error of each feature were computed for each image sample.

4 Research Methodology

The proposed prediction model in Fig. 1 for cancer classification includes the following stages:

- Exploratory analysis and data preprocessing
- Feature selection
- Neural network design and implementation
- Performance analysis and improvements of the model.

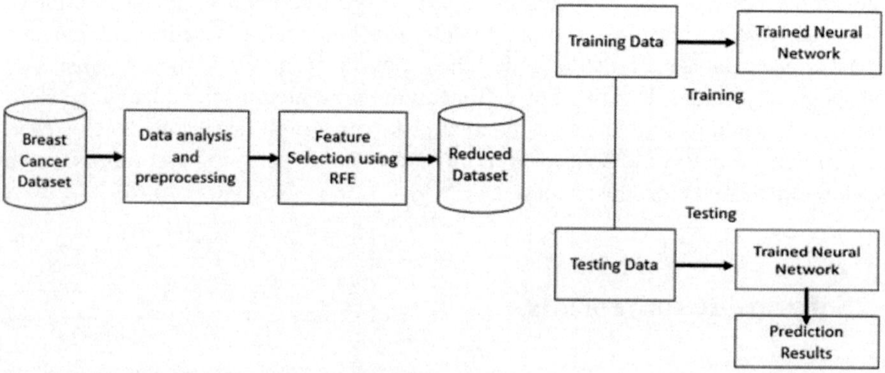

Fig. 1 General framework for cancer prediction model

4.1 Exploratory Analysis and Data Preprocessing

The dataset was explored to analyze the various attributes and data values for the purpose of data cleaning and other preprocessing activities. Data visualization was undertaken that enabled analyzing data by placing it in a visual context so that patterns, trends and correlations could be observed.

(a) **Dataset analysis**

The dataset under consideration has dimensions (32 × 569) where 32 features describe each of the 569 rows. Thirty features that provide information of the cell nuclei are of the same data type (float). "Diagnosis" is the class label of type object which can be converted from labeled form to numeric form for the purpose of processing. "Id" is of no consequence as it serves only as an index for the dataset rows and hence can be dropped off during processing.

Class imbalance is the problem in machine learning where the total number of examples in one class of data is far less compared to the other. Optimized results cannot be obtained as the samples taken during validation and testing may not provide an effective representation of the dataset. The breast cancer dataset is a well-balanced case with 357 benign cases and 212 malignant cases. Diagnosis is the class label of type object which can be converted from labeled from to numeric form for the purpose of processing. Malignant cases corresponding to "M" are set to numeric value 1 and benign cases corresponding to "B" are set to numeric value 0.

(b) **Visual analysis**

A heat map was used for exploring the correlation of features in a dataset. If there is a correlation value which is close to 1, it indicates the features are directly proportional and a value which is equal to 0 indicates that there is no relationship between two features. A partial heat map generated in Fig. 2a shows that a number of features in the dataset are highly correlated such as radius_mean has a high correlation to perimeter_mean and area_mean. The correlation among all 30 features was explored to identify the critical features that need to be selected to construct the model.

A swarm plot is another visual representation that can be drawn to show all observations along with some representation of the underlying distribution. Swarm plots attempt to avoid plotting points on each other by calculating non-overlapping positions instead of adding random jitter. As indicated in Fig. 2b, the data points appeared as a swarm of bees or honeycomb.

Data points were plotted on a horizontal and vertical axis describing relationships between features, observable patterns, exhibiting outliers and various other aspects through a scatter plot. A linear relationship, for example, can be seen between radius_worst and perimeter_worst as shown in Fig. 2c. However, radius_worst against concavity_mean showed the presence of some outliers along with a certain overlap in the two classes.

The visual representation of the statistical five-number summary in a boxplot indicates that smoothness_mean, compactness_se, etc. have outliers. Analysis of the

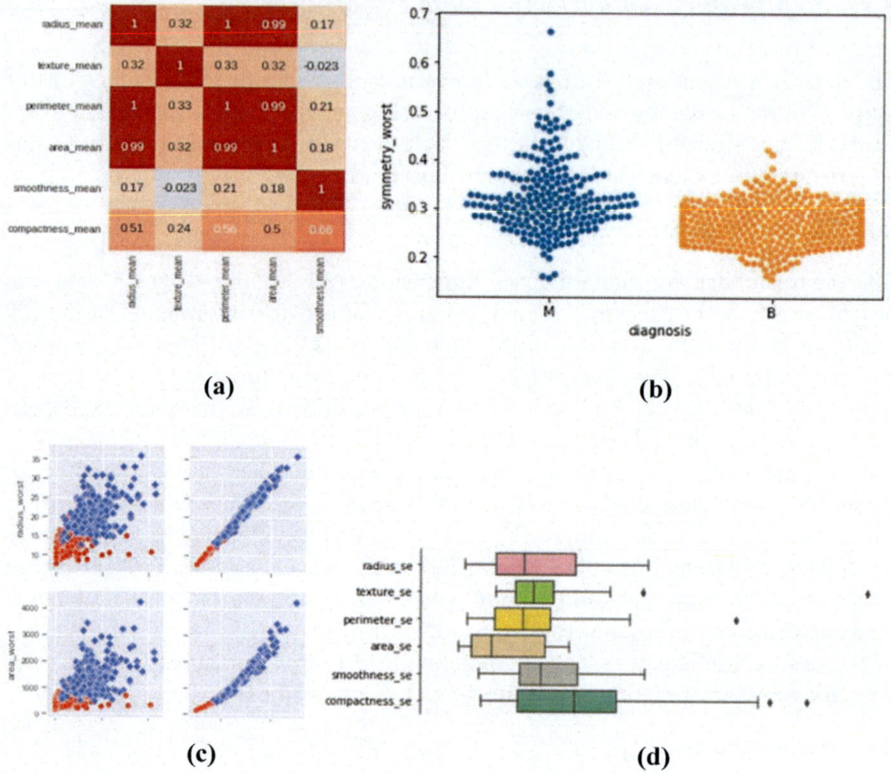

Fig. 2 **a** Correlation matrix, **b** swarm plots, **c** scatter plots, **d** boxplots

box plot in Fig. 2d revealed the distribution of data in different quartiles for each of the features under consideration. Features with large set of outliers are undesirable.

4.2 Feature Selection

(a) **Feature selection algorithm—recursive feature elimination (RFE)**

The RFE algorithm is a greedy optimization technique that attempts to find the best feature subset. At each iteration, the algorithm repeatedly creates models and considers the best set of features. It then constructs the next model with the reduced set of features until no more features are left to be considered for evaluation.

(i) Algorithm for recursive feature elimination (RFE)

1. Initially consider the entire set of features for selection.
2. Define a model and apply the set of features to train the model.
3. At every iteration, unimportant features will be dropped off.

4. A final subset of the most relevant features or best features is obtained.
5. Features can be evaluated and assigned an importance score based on the order of elimination of features.

(ii) Comparison of various classification models
In an attempt to identify the best classification model to implement RFE, seven various classification methods were implemented on the dataset [3–5, 7]. The accuracy was computed and confusion matrix was generated that showed a contention between random forest, SVM with linear function and logistic regression as indicated in Table 1.

(b) Visual representation of the decision tree and optimal number of features

Random forest approach uses Gini importance that gives an indication of how well the model fits and accuracy decreases when you drop a variable. The greater decrease in accuracy, the more significant is the variable. As indicated in Fig. 3a, the most importance features are concave points_worst, area_worst,area_se, etc. These features are selected as the splitting attributes at each node.

The method of principal component analysis [6, 7] was used to determine the number of features to be selected by the RFE algorithm. The variance curve in

Table 1 Confusion matrix values for different classifiers

No.	Classifier	TB	TM	FM	FB	Accuracy
1	Logistic regression	106	59	2	4	96.491
2	Support vector machine (SVM)	106	59	2	4	96.491
3	Nearest neighbor	107	57	1	6	95.906
4	Kernel SVM	108	0	0	63	63.157
5	Naive Bayes	104	57	4	6	94.152
6	Decision tree algorithm	106	58	2	5	95.906
7	Random forest classifier	107	58	1	5	96.491

TB True benign; *TM* true malignant; *FM* false malignant, *FB* false benign

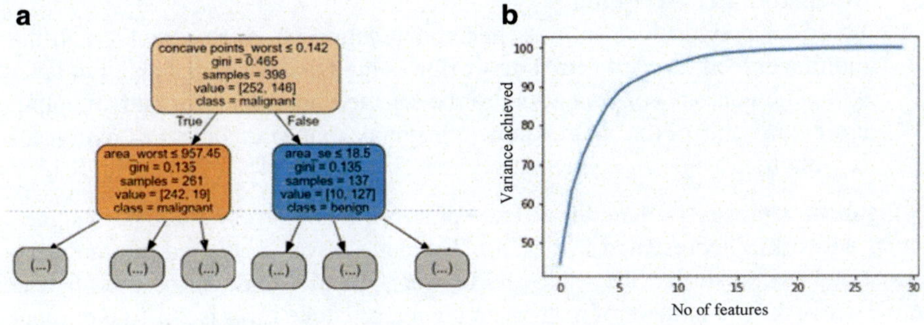

Fig. 3 **a** Tree representation of random forest model. **b** Plotting variance

Fig. 3b indicated that 100% variance was attained with 16 features and above. This indicated that the ideal number of features that RFE should select were 16 out of 30. Using the 16 features selected by each model, the random forest classifier performed better on the dataset as compared to logistic regression model [12] and SVM with a higher accuracy of 97.076, true benign = 107, true malignant = 59, false malignant = 1, false benign = 4 making it the final choice for feature selection using RFE. Further investigation revealed high correlation amongst some of 16 selected features which required further reduction of the set of features based on the feature importance scores provided by the RFE method. The final nine selected features set was generated "concave_points_mean," "texture_worst," "concave_points_worst," "perimeter_worst," "texture_mean," "area_se," "area_worst," "concavity_mean," "concavity_worst."

4.3 Neural Network Construction

Artificial neural network (ANN) is an intelligent paradigm that can process information and can be configured for a specific applications [8, 13, 14]. A choice of feed forward networks was made wherein the first layer receives n inputs and the last layer produces m outputs and learning takes place through weight adjustments and other techniques.

(a) **Algorithm of an artificial neural network**

1. Define a neural network with an input layer of n input nodes.
2. Determine the number of hidden layers and the neurons in the hidden layer.
3. Set the learning rate, bias value and the activation function.
4. Define the number of epochs to pass the entire training dataset.
5. Train the network for given set of training data.
6. Apply the test data to the trained network.
7. Train the network until the number of epochs is completed and calculate accuracy.

(b) **Model design description**
Based on previously mentioned recommendations from experts [7, 15], the multilayer feed forward neural network was built comprising of an input layer with nine neurons corresponding to the selected input features and an output layer consisting of one neuron that corresponds to the outcome, i.e., malignant or benign.

Figure 4a represents a general structure of a feed forward network. The designed neural network is represented in Fig. 4b. The dataset was converted into an array format and scaled so that the values lie between 0 and 1 and can be used by the neural network. The number of neurons in the intermediate layer is 2/3 * no of input neurons = 2/3 * 9 = 6. A bias value of 1 and learning rate of 0.15 was set by default. These values can be adjusted in subsequent iterations to improve accuracy

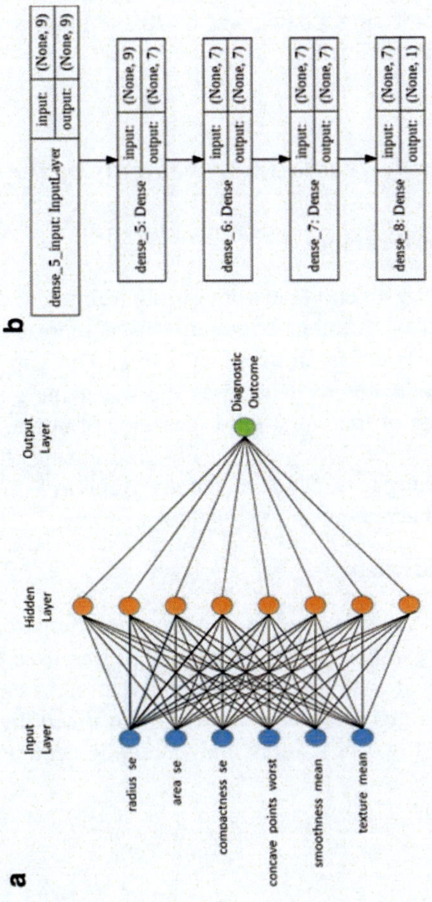

Fig. 4 **a** General neural network. **b** Layers of the feed forward network

of the model. The activation function chosen was rectified linear unit (ReLU). It has the advantage of providing computational simplicity and representational sparsity. Binary_crossentropy was the loss function used. The termination condition of the network is defined by the number of epochs. As a prerequisite to training and testing, the model data preprocessing was carried out on the dataset to convert it into an array of values that the neural network can process. The dataset has been split into training (70%), testing (15%) and validation (15%) sets. Hyperparameters like optimizers, number of epochs and number of hidden layers were adjusted to achieve best performance and accurate results. Using a batch size of 20 and number of epochs as 150, a satisfactory accuracy was obtained.

4.4 *Performance Analysis and Improvements of the Model*

(a) **Visualizations of performance**

A loss curve is used to debug a neural network during training and helps to monitor the network training process. Another commonly used representation to observe the progress of neural networks is an accuracy curve. The gap between training and validation accuracy indicates overfitting as the wider the gap, the higher the overfitting. Below are some of the worst-case scenarios of neural network training with the designed model. Model loss graph in Fig. 5a indicates that the validation loss is much ahead of training loss. Model accuracy graph in Fig. 5b shows a clear divergence between model accuracy and validation accuracy.

(b) **Performance enhancement**

L2 regularization [16] was implemented to minimize cost function and control the increase in weights that are already high. Dropout [16] was used to set a threshold value and drop off neurons at each layer. Post-regularization as observed in Fig. 6a training and validation loss gradually decreased without much divergence between them. Also as indicated in Fig. 6b, training and validation accuracy increased with each epoch.

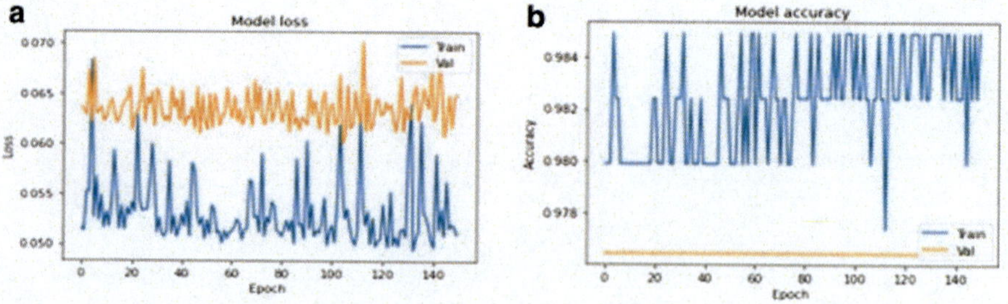

Fig. 5 **a** Model loss graph. **b** Model accuracy graph

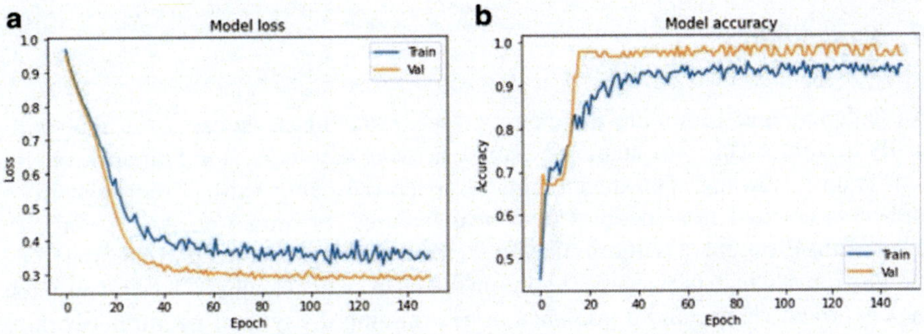

Fig. 6 **a** Model loss graph. **b** Model accuracy graph

(c) Final results and observations

As indicated in Fig. 7a, an improved accuracy of 96.51 is achieved. For interpretation of results, the X-axis indicates predicted values and Y-axis indicates actual values. The results show true benign = 51, true malignant = 32, false benign = 3, false malignant = 0. The model performed fairly well with the selected architectural parameters for a test set of 86 samples. An interesting observation was that there were no cases falsely classified as malignant. This was not achieved with any of the previously explored standard classifiers. Moreover, the number of cases falsely classified as benign was reduced to 3. Predictions were made for all test examples in the dataset, and the input rows and predicted class value for the first ten examples was compared to the expected class value. We can see in Fig. 7b that most of the rows are correctly predicted.

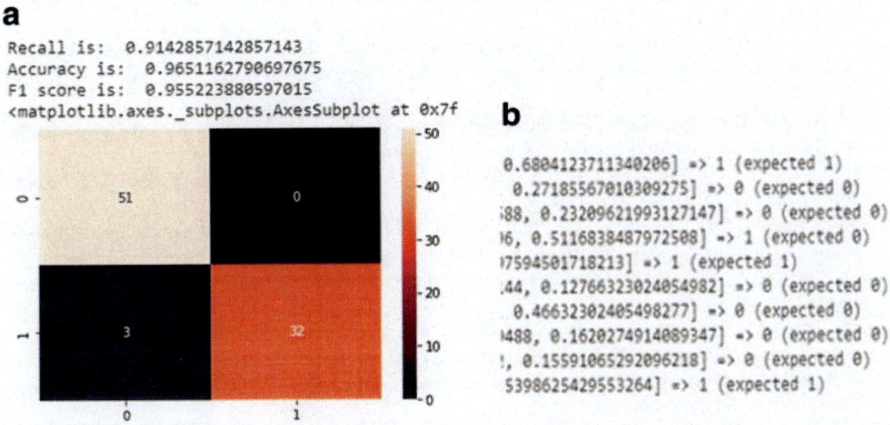

Fig. 7 **a** Improved neural network performance. **b** Predicted values by the model

5 Conclusion

A prediction model for a life-threatening disease like breast cancer needs to be optimally accurate. This was achieved using extensive dataset exploration and conclusive feature learning. The data analysis carried out using various representations and plots assisted in selecting the critical features for model design. A comparison among different classifiers enabled selection of the most effective classifier, i.e., random forest which derived the minimum subset of critical features. Several research papers provided substantial guidance in determining the type of neural network to be employed for classifying cases of malignancy and benign. The neural network model implemented has acceptably good performance in comparison to standard classification models explored. Also, the number of benign cases falsely classified as malignant was zero. Repeated executions of the network with random test sets showed accuracy in the range of 95.34 to 96.51. This shows that the model is stable and does not display erratic performance.

The designed prediction model serves as an intelligent healthcare assistant and can provide valuable assistance to physicians to diagnose patients with greater accuracy. Accurate classification can further assist clinicians to prescribe the most appropriate treatment regime and provide more meaningful benchmarks in assessing the medical situation of a cluster of patients.

References

1. Y.-S. Sun, Z. Zhao, H.-P. Zhu, Risk factors and preventions of breast cancer. Int. J. Biol. Sci.
2. A. Osarech, B. Shadgar, A computer aided diagnosis system for breast cancer. Int. J. Comput. Sci. **8**(2) (2011)
3. S. Dhivya, R. Arulprabha, M. Kowsalya, G.H. Kumar, M.B.P. Gandhi, Survival outcome prediction for breast cancer patients. Int. J. Recent Technol. Eng. (IJRTE) **9**(1) (2020). ISSN: 2277-3878
4. M. Rana, P. Chandorkar, A. Dsouza, Breast cancer diagnosis and recurrence prediction using machine learning techniques. Int. J. Res. Eng. Technol. **04**(04) (2015)
5. E.D. Übeyli, Implementing automated diagnostic systems for breast cancer detection. Expert Syst. Appl. **33**(4), 1054–1062 (2007)
6. V. Chaurasia, B.B. Tiwari, S. Pal, Prediction of benign and malignant breast cancer using data mining techniques. J. Algorithms Comput. Technol.
7. H. Wang, S.W. Yoon, Breast cancer prediction using data mining method, in *IEEE Conference Paper*
8. D. Dubey, S. Kharya, S. Soni, Predictive machine learning techniques for breast cancer detection. Int. J. Comput. Sci. Inf. Technol. **4**(6), 1023–1028 (2013)
9. S. Agrawal, J. Agrawal, Neural network techniques for cancer prediction: a survey. Procedia Comput. Sci. **60**, 769–774 (2015)
10. A.F. Agarap, On breast cancer detection: an application of machine learning algorithms on the Wisconsin diagnostic dataset
11. N. Street, Xcyt: a system for remote cytological diagnosis and prognosis of breast cancer (2000). https://doi.org/10.1142/9789812792488_0008
12. C.-Y. Joanne Peng, K.L. Lee, G.M. Ingersoll, An introduction to logistic regression analysis and reporting **96**(1) (September/October 2002)

13. A. Gautam et al.,An improved mammogram classification approach using back propagation neural network, in *Data Engineering and Intelligent Computing* (Springer, Singapore, 2018), pp. 369–376
14. S.A. Mojarad, S.S. Dlay, W.L. Woo, G.V. Sherbet, Breast cancer prediction and cross validation using multilayer perceptron neural networks
15. W. Yue, Z. Wang, H. Chen, A. Payne, X. Liu, Machine learning with applications in breast cancer diagnosis and prognosis. Designs **2**, 13 (2018). https://doi.org/10.3390/designs2020013
16. E. Phaisangittisagul, An analysis of the regularization between L2 and dropout in single hidden layer neural network, in *2016 7th International Conference on Intelligent Systems, Modelling and Simulation (ISMS)*, Bangkok (2016)

Prediction of Admission in Master's Program in Foreign Countries Using Machine Learning Algorithms

Devesh Kumar Srivastava, Akhilesh Kumar Sharma, and Venkatesh Gauri Shankar

Abstract In the present competitive era, admissions into master's programmed in prestigious universities has become an arduous task because of the ever-increasing competition leading to higher standards while shortlisting aspirants and challenging for grabbing an admit. Considering the cumbersome nature of the admission process, students often get muddled because of its various factors under consideration. Keeping this view in context, a sound model is proposed that could predict the chances of getting an admit thereby also reducing workload for the admissions committee by automating this tedious task, along with students leveraging the fact of predicting their chances of an admit beforehand applying. The proposed predictive model makes the review process quite efficient by reducing the time spent on a particular application by reviewers. The focus can be fixated to the parts essential for determining the applicant's candidature. There are several parameters which are to be taken into consideration while applying for Master Program. For the development of model, we used a public open-sourced dataset (www.kaggle.com) [1] which consists the data of Indian students applying abroad for Master of Science program. Our approach is to develop several regression and classification models to predict the chances of a student getting an admit based on various parameters. We computed the performance of the models and compared them to determine the best model.

Keywords GRE · TOEFL score · CGPA · Classification · Regression · LOR · SoP

1 Introduction

A recent trend of increase in the number of applicants is observed for Master of Science (MS) program among the engineering graduates in India. This trend is burgeoning every year. A lot of factors contribute to such a development whose reasons can be attributed to scope and ample of job opportunities along with a higher paycheck and less competition. Also, the standard of living along with the urge for

D. K. Srivastava (✉) · A. K. Sharma · V. G. Shankar
SCIT, Manipal University, Jaipur, Rajasthan, India

V. Bhateja et al. (eds.), *Data Engineering and Intelligent Computing*,
Advances in Intelligent Systems and Computing 1407,
https://doi.org/10.1007/978-981-16-0171-2_23

better education are possibly some reasons for such a surge in demand of applicants. Whatever the reasons possibly are, most of the students aim to study abroad right after graduation. At such a tender age, the cumbersome admission process turns out to be really tedious for students [2]. In this paper, our aim is to propose sounds models that could predict the students' chance of getting an admit given various parameters on which it depends, such as "GRE score," "TOEFL score", "CGPA." This model also targets the admission committee for them to evaluate student's application automatically that are exceptional in nature, maybe outstanding, or pretty weak. By predicting those first using our models, they could easily move onto the average ones and the process becomes less time consuming and efficient. Our objective was to compare various regression and classification models. The status of admission in terms of "admit" and "reject" can also be determined by the proposed model, hence acting as an aid for students and admission committee by turning the process highly decisive in nature. Although, a pretty good number of applicants usually get an admit who have 1–2 years of work experience after graduation. We do not consider such applicants as our audience as their chances of admission are pretty high. We target our audience who are fresh graduates to help them cross check their capabilities for getting an admit in their "dream college".

2 Literature Survey

Moore, J. S. discussed about the performance prediction-related capability and uses personal demographic variables with optimal rule structure, and the results are also suitably validated [3]. Austin and Miikkulainen used the historical admission data and neural networks and machine learning methods for predictions and performance calculations [4]. Judith et al. used the bootstrapping and variable aggregation embedded in a double-layered cross-validation loop for better accurate and accuracy. The results were very satisfying in this research [5]. Chithra Apoorva et al. developed a model by using the KNN, linear regression and ridge regression, and random forest methods for better predictability and enhanced performance calculation to get graduate admission in U I [6]. Avwerosuoghene Enughwure and Ebitiminipre Ogbise focused on student's performance by using machine learning techniques. They concluded to predict the performance of any students the whole academic program which is required instead of academic data and educator comprehensive data [7].

3 Methodology

Dataset description: The "Graduate Admissions" dataset is taken from Kaggle, and it contains 5000 rows and 9 columns which contains several parameters which are considered important during the application process for admission into master's programs, such as "GRE Score," "TOEFL Score," "University Rating," "SOP", "LOR", "CGPA", and "Research Papers published". The dataset does not contain any null values.

4 Proposed Method

Initially, we preprocess the dataset available to us, by checking consistency in data types and imputing any null or missing values in the dataset, and we normalize the dataset to remove any bias present in it. While examining data types of variables, we see that the dependent or target variable, i.e., "chance of admit", is a numerical or continuous variable. While this is needed condition for a regression model, for any classification model, the target variable is supposed to discrete or categorical to partition it into classes, either binary (having 2 classes, 0 and 1) or multi-class (having more than two classes). To convert "chance of admit" to a categorical variable, we add a condition that draws a decision boundary in the numerical values. The condition being, if chance of admit ≥ 0.75, then the instance would belong to the positive class or 1, which means the applicant will get an admit, else the instance would belong to the negative class or 0, inferring that the applicant would not get an admit. Next step in preprocessing is splitting the dataset into training and test sets. We do this in the ratio of 67:33. The model would be trained on the training set and evaluated or validated on the test set where the model would give predicted values based on the independent variables in the test set. We will be applying both regression and classification models.

Regression models: The regression models used are linear regression, decision tree regression, and random forest regression. Classification models: The classification models used are logistic regression, decision tree classifier, and K-nearest neighbors. In case of regression, our model is fitted on the training set, and a best-fit regression line is generated. The model is applied on the test set, and predicted values are computed which are evaluated by calculating the error of the model. The performance metric taken into consideration for regression is R^2 error, which is a statistical measure of how close the predicted data points are to the best-fit line. The lesser the value of R^2, better is the model. In case of classification, we use the discretized "chance of admit" variable. In case of random forests, we used the number of tress as 80 for better results. Our classification model is fitted on the training set and evaluated using the test set, by using the predicted instances. A confusion matrix is plotted that depicts the count of the number of correctly and incorrectly classified instances, and performance metrics (accuracy, precision, recall, $F1$-score) are applied along with it. Finally, we print the classification report for each model, and we compare the performance of each model with a bar graph for better interpretation.

Algorithm

1. Preprocess the data set, by checking data consistency and imputing missing or null values, and make it fit for training the models.

2. Perform exploratory data analysis (EDA) to gain valuable insights by studying the relationship between dependent variables and independent variable, thereby getting an idea of how these variables affect the chance of grabbing an admit.
3. This dataset containing past records of applicants is split into training and test sets. The training set is trained with different algorithms and evaluated on the test set. The results obtained evaluate the proficiency of the applicant in terms of probability, which is predicted by the classifier based on which admission for each new applicant will be processed. The information generated by the model is sent to the admission committee.
4. The information can be used to decide on which files to be focused for verification contributing to downsizing of time to a very significant level.
5. The resulting predicted instances obtained from regression model can be used to explain the data points by plotting the best-fit line. To achieve the best-fit graphically, a scatter plot is used. The formula for generating the best-fit line is $y = a + b * x$, which is used to define the relationship between the independent and dependent data points.
6. To actuate the regression, there is a three-step process followed that includes analysis of correlation of data, estimation of the model (best-fit line), and evaluating the model.
7. A recursive case is carried out to generate a regression tree model, wherein the independent variable is rived at various split points. The R^2 error is calculated by comparing the model to a constant baseline, which is chosen by computing the mean of the data and making a line at the mean. The lower the R^2 error, the better is the model.
8. In classification models, the modification of the dataset target value to a discrete/categorical value is carried out, and then, the dataset is fitted to the training set for each model (Table 1).

5 Experimental Result

In this section, our penultimate focus is on gathering the connection among various parameters and how they alter and affect an individual's admission chance. This identification does not make sense out of a dataset to how an overlook on each parameters (Figs. 1, 2 and 3).

1. More than 50% of students that apply for post-graduation programs have completed some type of research previously.
2. The most frequently occurring range of CGPA is over 8.

Table 1 Partial dataset (as a sample from www.kaggale.com)

GRE score	TOEFL score	University rating	SOP	LOR	CGPA	Research	Chance of admit
337	118	4	4.5	4.5	9.65	1	0.92
324	107	4	4	4.5	8.87	1	0.76
316	104	3	3	3.5	8	1	0.72
322	110	3	3.5	2.5	8.67	1	0.8
314	103	2	2	3	8.21	0	0.65
330	115	5	4.5	3	9.34	1	0.9
321	109	3	3	4	8.2	1	0.75
308	101	2	3	4	7.9	0	0.68
302	102	1	2	1.5	8	0	0.5
323	108	3	3.5	3	8.6	0	0.45
325	106	3	3.5	4	8.4	1	0.52
327	111	4	4	4.5	9	1	0.84
328	112	4	4	4.5	9.1	1	0.78
307	109	3	4	3	8	1	0.62
311	104	3	3.5	2	8.2	1	0.61
314	105	3	3.5	2.5	8.3	0	0.54
317	107	3	4	3	8.7	0	0.66
319	106	3	4	3	8	1	0.65
318	110	3	4	3	8.8	0	0.63
303	102	3	3.5	3	8.5	0	0.62
312	107	3	3	2	7.9	1	0.64
325	114	4	3	2	8.4	0	0.7
328	116	5	5	5	9.5	1	0.94
334	119	5	5	4.5	9.7	1	0.95
336	119	5	4	3.5	9.8	1	0.97
340	120	5	4.5	4.5	9.6	1	0.94
322	109	5	4.5	3.5	8.8	0	0.76
298	98	2	1.5	2.5	7.5	1	0.44
295	93	1	2	2	7.2	0	0.46
310	99	2	1.5	2	7.3	0	0.54

3. The students who have undertaken some research work get opportunities from universities having higher ratings which in turn are followed by higher CGPA values.
4. If a student has previously completed a research work, his chance of getting selected are significantly higher.

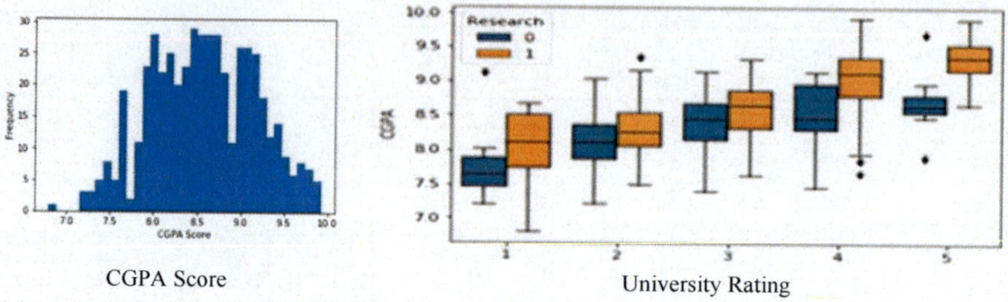

Fig. 1 Data visualization through graph of CGPA score and university rating

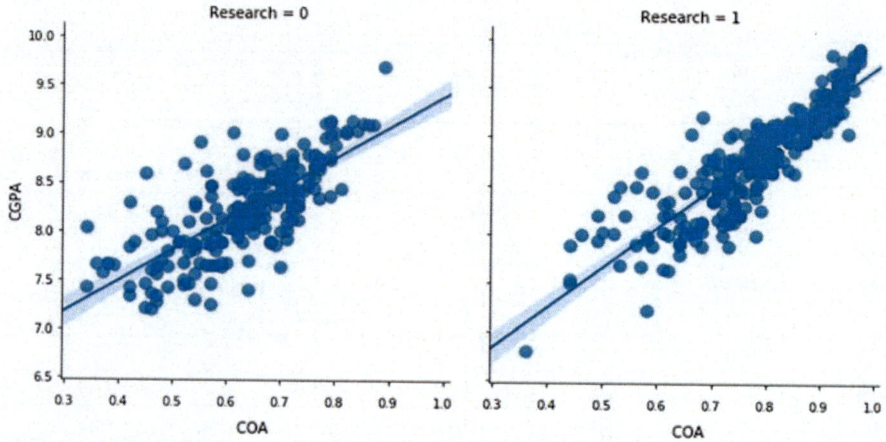

Fig. 2 Graph shows that visualization of progress of research work if not accepted (research = 0) and if accepted (research = 1)

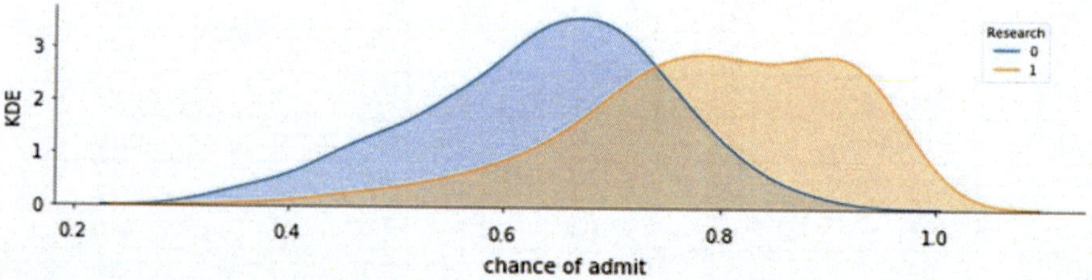

Fig. 3 Graph visualization shows that chances of admission for MS program

5. Candidates that have an average CGPA of 7.5 and above and have published a research paper have a higher chance of getting admitted.

Predictive Analysis

(a) **Regression Model**

(b) **Classification Models**

Logistic Regression

Precision score: 0.79710144927536

Recall score: 0.9016393442622951

$F1$_score: 0.8461538461538461

Decision	Predicted yes	Predicted no
Actual yes	90	14
Actual no	6	55

Decision Tree Classifier

Precision score: 0.774647887

Recall score: 0.901639344262

$F1$_score: 0.83333333333333

Decision	Predicted yes	Predicted no
Actual yes	90	14
Actual no	5	56

Random Forest Classifier

Precision score: 0.88333333333

Recall score: 0.86885245901639

$F1$_score: 0.8760330578512396

Decision	Predicted yes	Predicted no
Actual yes	88	16
Actual no	6	55

Support Vector Machine

Precision score: 0.797101449275

Recall score: 0.90163934426229

$F1$_score: 0.8461538461538461

Decision	Predicted yes	Predicted no
Actual yes	97	7
Actual no	8	53

6 Conclusions

1. More than 50% of applicants that apply for the master's program have completed or published some research work. This maybe because of the reason that most of the students know that the value of research is immense which may push them through to their field of interest.
2. Most frequently occurring CGPA is in the range of 8 and 9.
3. The students who have undertaken some research work get opportunities from universities having higher ratings which in turn are followed by higher CGPA values.
4. As foreign universities give precedence to practical knowledge over theoretical, the chances of students getting selected on the basis of research work is very high.
5. Even if a student has as low CGPA as 7.5, his chances of admission is very good if he/she has undertaken a research. This shows the impact that a research has.
6. The regression models depict that the r squared value is more in linear regression which is then followed by random forest and last decision tree as per Fig. 4. This recommends that the linear regression model is a better fit to the given data and is much more accurate in predicting the chance of admission given different parameters.
7. In the classification models, random forest yielded the best performance in predicting the outcome value followed by logistic regression and support vector machine on the same level, and decision tree had the least accurate results in comparison as per Fig. 5.
8. The performance of decision tree was below par mainly due to the problem of over fitting which is usually observed while working with this algorithm as per Fig. 5.

Model name	R square score
Linear Regression	0.8242314445551622
Decision tree Regression	0.57799511002445
Random Forest Regression	0.7903684367359411

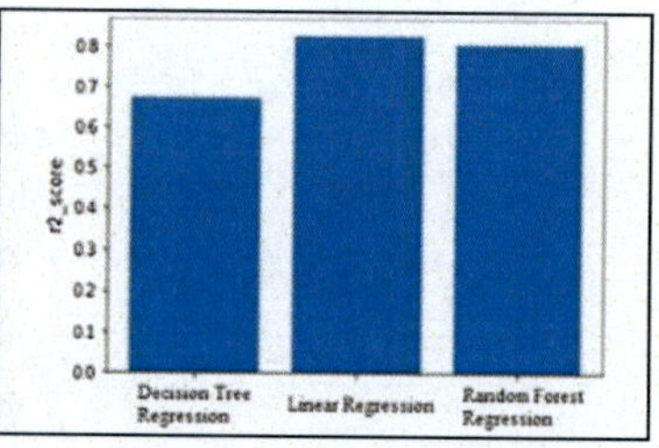

Fig. 4 Graph shows r^2 values against classifiers

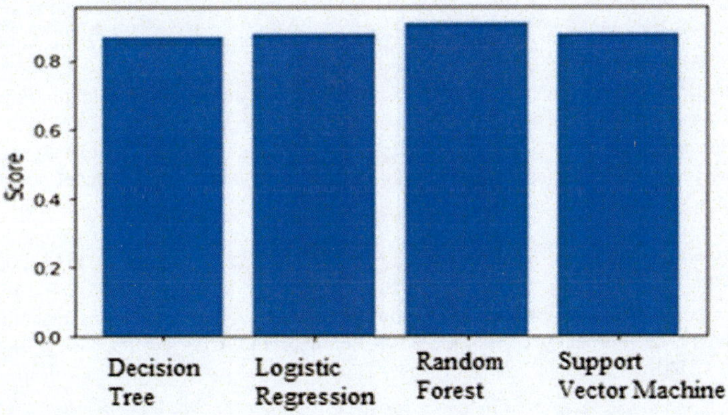

Fig. 5 Graph shows variation of score values against classifiers

7 Future Work

- The field of machine learning has seen a major growth in the past decade, with models getting constantly updated to increase the accuracy to meet the ever-growing demands. In this paper, we have worked toward regression, classification models, and unsupervised learning models through which we can compare the performance results.
- Each and every university has its own criteria and priorities for admission in a master's program. The limitations of this dataset are the only reason which we are able to provide the overall view for master's program admissions. In order for better prediction of graduate admissions for master's program, we need to be further specific
- This dataset is firmly applicable to the Indian students who apply for an international master's program. These models could be applied to the international students as well if the appropriate data is provided.
- Indian students usually prefer to gain a couple years of experience through a job before applying for MS, the study would have been much more efficient if it were to be included in the study as to calibrate its effects on the chances of a student's admission, but due to the available limited dataset, this study only consists of students that directly apply after graduation.

References

1. https://www.kaggle.com/naishalthakkar/gre-dataset-analysis
2. T.H. Bruggink, V. Gambhir, Statistical models for college admission and enrollment: a case study for a selective liberal arts college. Res. High. Educ. **37**(2), 221–240 (1996)

3. J.S. Moore, An expert system approach to graduate school admission decisions and academic performance prediction. Omega **26**(5), 659–670 (1998)
4. A. Waters, R. Miikkulainen, GRADE: machine learning support for graduate admissions. AI Mag. **35**, 64–75 (2014)
5. J. Zimmermann, K.H. Brodersen, H.R. Heinimann, J.M. Buhmann, A model-based approach to predicting graduate-level performance using indicators of undergraduate-level performance. J. Educ. Data Min. **7**(3) (2015)
6. D.A. Chithra Apoorva, M. ChanduNath, P. Rohith, S. Bindu Shree, S. Swaroop, Prediction for university admission using machine learning. Int. J. Recent Technol. Eng. **8**(6) (2020). ISSN: 2277-3878
7. A. Avwerosuoghene Enughwure, M. Ebitiminipre Ogbise, Application of machine learning methods to predict student performance: a systematic literature review. Int. Res. J. Eng. Technol. V **07**(05) (2020)

A VQA System for Medical Image Classification Using Transfer Learning

C. Dhanush, D. Pradeep Kumar, and Anita Kanavalli

Abstract The Visual Question Answering (VQA) is based on Computer Vision and Natural Language Processing (NLP). The goal of VQA system is to predict the textual answer to a question based on the image. The VQA system takes images and questions as input and combines the information of the input to generate readable answers as output. The medical VQA system has the following advantages. The radiologists can use the VQA system for their inference about the medical image. It can also help the patients to get basic information about the clinical image prior to doctor consultation. In this paper, we discuss about the VQA on medical images using the ImageCLEF 2019 VQA dataset. We build a medical VQA system using transfer learning on radiology images using MobileNet for input images on the plane class and predict the answer. The proposed VQA model is evaluated on the test dataset and the accuracy obtained is 80.8% on the plane class.

Keywords Convolutional neural network · Medical images · MobileNet · Transfer learning · Visual question answering

1 Introduction

Computer Vision is a field of artificial intelligence where the machines are trained to interpret, understand and analyze the images. The semantic information from the image is collected by knowledge acquisition. Computer Vision finds application in areas like image identification, classification, segmentation, motion analysis, scene reconstruction etc. [1]. Natural Language Processing is the automatic processing of

C. Dhanush (✉) · D. P. Kumar · A. Kanavalli
Ramaiah Institute of Technology, Bangalore, Karnataka, India

D. P. Kumar
e-mail: pradeepkumard@msrit.edu

A. Kanavalli
e-mail: anithak@msrit.edu

V. Bhateja et al. (eds.), *Data Engineering and Intelligent Computing*,
Advances in Intelligent Systems and Computing 1407,
https://doi.org/10.1007/978-981-16-0171-2_24

natural language by the machines. NLP includes text summarization, language translation, optical character recognition, speech recognition, text to speech conversion, parts of speech tagging, lemmatization etc. NLP helps in the effective communication between machine and human being [2].

The VQA system in the field of medical domain is the most challenging task. The reasons are as follows: There are large number of images available in general domains but limited to few in case of medical domain. Hence training on smaller dataset does not yield good results. The VQA system finds applications in the field of medical domain as it could help patients to get better understanding of their health conditions from the medical images [3].

In recent years, to promote research and development in the field of medical domain, medical image VQA is conducted by ImageCLEF 2019. The MedVQA System using transfer learning on the MobileNet model obtained accuracy of 80.8% on the plane class dataset. In Sect. 2, we briefly discuss about literature review, Sects. 3 and 4 deals with problem definition and model architecture respectively. The model results and evaluation metrics are discussed in Sects. 5 and 6 respectively. Conclusion and future work are discussed in the final section.

2 Literature Review

Recent advancement and research have been carried in the field of medical domain. ImageCLEF has organized different challenges in medical domain. In our research, we have considered the visual question and answering task on the medical data.

Lubna et al. [4] in their paper discussed about the medical VQA system on ImageCLEF 2019 dataset. The dataset consists of 2000 training and 300 test images related to modality class. The dataset was categorized into 8 classes based on modality: X-ray, Ultrasound, CT (Computerized Tomography), PET (Position Emission Tomography), Mammography, Angiogram, MRI (Magnetic Resonance Imaging), GI (Gastro Intestinal). They have used a simple CNN network to classify the images based on modality. For text processing, decision tree is used. The output from both CNN network and decision tree is combined to generate the final output. The validation accuracy of 0.838 was obtained on the modality class.

3 Problem Definition

The ImageCLEF conducted a VQA challenge on medical images [5]. The distribution of the medical dataset is shown in Table 1. The medical dataset consists of questions which belong to four classes: Modal, Plane, Organ system and Abnormality. Example of the question answer pair is as follows: (what modality is shown?, cta—ct angiography), (what plane was used in this ct scan?, axial) etc. We have considered only plane-based class for the VQA task.

Table 1 Distribution of the medical dataset

	Training	Validation	Test
Images	3200	500	500
Questions	12,792	2000	500
Answers	12,792	2000	

Table 2 Plane based classification of dataset

Plane classes	Training	Validation
Axial	1562	213
Sagittal	481	81
Coronal	392	65
Ap	197	48
Lateral	151	30
Frontal	120	23
Pa	92	20
Transverse	76	7
Oblique	50	11
Longitudinal	43	3
3d reconstruction	22	4
Mammo—cc	11	1
Mammo—mlo	9	2
Mammo—mag cc	2	1
Total	3207	513

The training and validation images distributed among the fourteen plane classes are as shown in Table 2.

4 Model Architecture

Here we consider plane-based class for VQA task. The model architecture primarily consists of 3 main components: Pretrained Image Model, Question Classifier Model and VQA Classifier Model.

4.1 Pretrained Image Model

Transfer learning is used to train the model on the dataset for source task and use it on target task. Image models like VGG, ResNet, Inception, Xception models were pretrained on ImageNet for different tasks such as VQA, image captioning etc. In the transfer learning, the knowledge obtained by learning on one domain can be applied to the other domain as the model has learned the extraction of features from images. Transfer learning learns a new task through the knowledge that it has learned from previous related tasks. The advantage of using transfer learning compared to traditional ML is that the learning process is faster, lesser training time and more accurate results can be obtained on the test data [6].

We use the concept of transfer learning using MobileNet. The MobileNet model has convolution layers which are made of several neurons. These neurons are capable of learning features from images using weights and biases. The convolution layer consists of set of n independent filters which produce set of n feature maps. Another component is the batch normalization which normalizes the output of previous activation layer. The batch normalization allows each layer to learn by itself more independently of other layers.

MobileNet architecture [7] is based on depthwise and pointwise convolutions. The first layer in MobileNet is full convolution whereas other layers are built on depthwise separable convolutions. The depthwise convolution is performed independently over each channel and pointwise convolution uses 1 * 1 convolution to project the output produced by depthwise convolution to another space. A depthwise separable convolution uses two layers for filtering and combining. All the layers have ReLU activation followed by a batch normalization except for the last fully connected layer which uses SoftMax activation. The input to the MobileNet is an image resized to the dimension (224, 224). The output from the MobileNet is fed to a global max pooling layer followed by a dense layer with 512 units in the output layer and ReLU activation which is defined by:

$$R(z) = \text{Max}(0, z) \tag{1}$$

The proposed model uses a pretrained MobileNet network because of following reasons:

- The pre-trained network is simple and easy to incorporate to the image classification task.
- Instead of random weight initialization, the model is pretrained with ImageNet to reduce the training time and improve the efficiency.

4.2 *Question Classifier Model*

Each word in the question is converted into sequences using word embedding. Answers are assigned a unique id and are converted into one hot encoding. The question classifier model is shown in Fig. 1. The pre-processed question sequence is passed to an embedding layer, followed by 2 layers of Bidirectional LSTM and a dense layer with 1024 units. To reduce overfitting, a dropout was added to the hidden layer with 50% probability. Finally, we added a dense layer with 512 units to give the final output. This layer uses tanh as the activation function.

Layer (type)	Output Shape	Param #
input_2 (InputLayer)	[(None, 20)]	0
embedding (Embedding)	(None, 20, 50)	2350
bidirectional (Bidirectional	(None, 20, 256)	183296
bidirectional_1 (Bidirection	(None, 512)	1050624
dense_6 (Dense)	(None, 1024)	525312
dropout (Dropout)	(None, 1024)	0
dense_7 (Dense)	(None, 512)	524800

Total params: 2,286,382
Trainable params: 2,286,382
Non-trainable params: 0

Fig. 1 Question classifier model

4.3 *VQA Classifier Model*

The output from the image model and question classifier model is concatenated and is fed into a dense layer with 1000 units followed by a dropout layer with data drop probability of 20%. The dropout was followed by dense layer of 1000 units. A dense layer is followed by dropout layer with 50% of data drop probability. The final dense layer had 500 output units with SoftMax as the activation function. The proposed model is shown in Fig. 2.

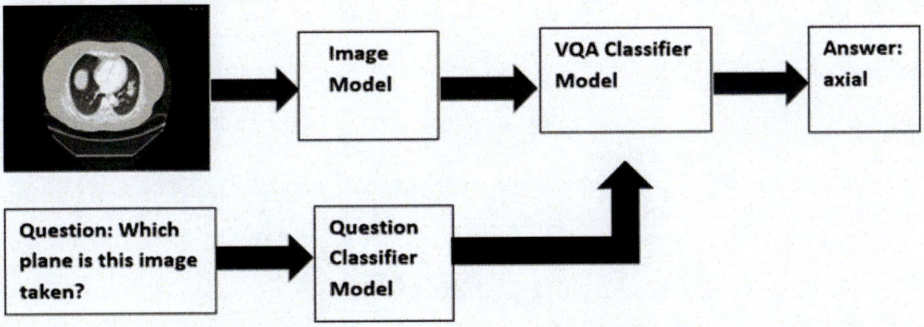

Fig. 2 Proposed MedVQA model

5 Model Results

The plane class dataset consists of 3207 training images, 513 validation images. We run the model for 90 epochs, batch size of 64, categorical cross entropy as the loss function, 0.0001 learning rate and Adam optimizer. The accuracy of 96.47% was obtained on the training dataset and accuracy of 80% on the validation dataset. The accuracy and loss for the training and validation dataset is shown in Figs. 3 and 4 respectively. The orange and blue curve shows accuracy and loss for training and validation data respectively. The model was evaluated on the testing dataset and the accuracy of 80.8% was obtained.

We have used different models like VGG16, VGG19 and MobileNet for the MedVQA task. Each of these pretrained models along with the question model were combined and the results were obtained. The result for plane-based classification

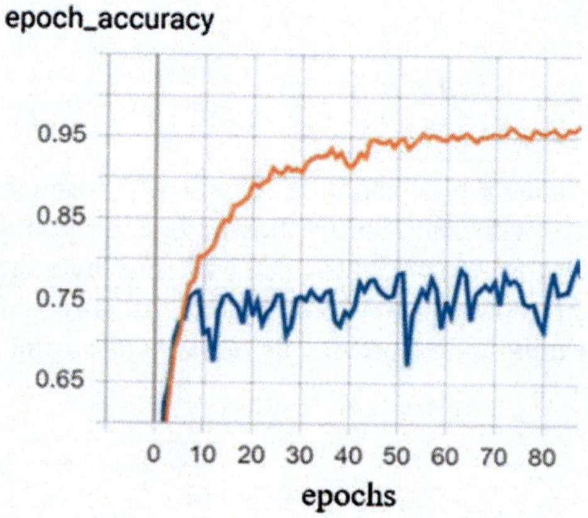

Fig. 3 Accuracy graph

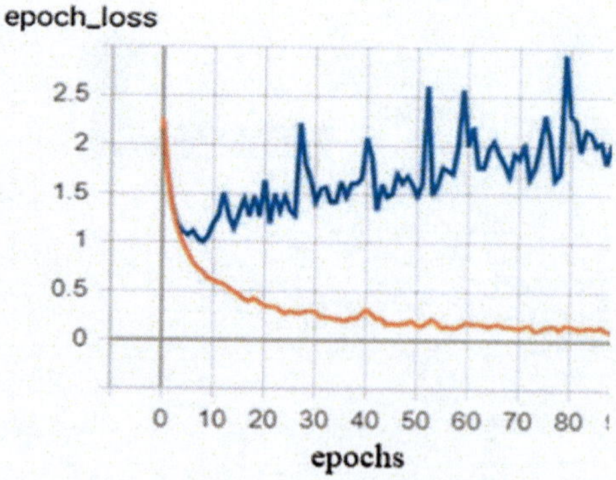

Fig. 4 Loss graph

Table 3 Comparison with other models

Models	Validation accuracy	Test accuracy
VGG-16	0.756	0.72
VGG-19	0.762	0.72
MobileNet	0.8	0.808

using different models is shown in Table 3. The validation and test data accuracy computed using MobileNet is better when compared to other models.

6 Evaluation Metrics

Accuracy is the ratio between number of observations that were predicted correctly to the total number of observations. We consider other parameters like Precision, Recall and $F1$-scores to evaluate the performance of the model. Precision is defined as the number of predicted positive class predictions that belong to the positive class. Recall is defined as the number of predicted positive samples to all the examples in the dataset. $F1$-score is a single score that balances both precision and recall [8]. The plane-based classification report for the test data is shown in Table 4. The plane class axial has highest values, which may be due to the large number of images in train and validation set. The plane classes at the bottom show lower score due to fewer images in training set. The screenshot of the program execution is shown in Figs. 5 and 6.

Table 4 Plane based classification report

Plane classes	Precision	Recall	$F1$-Score	Support
Axial	1.00	0.97	0.98	64
Sagittal	0.88	0.93	0.90	15
Coronal	0.89	0.80	0.84	10
Ap	0.33	0.67	0.44	6
Lateral	0.62	0.62	0.62	8
Frontal	0.17	0.50	0.25	2
Pa	0.67	0.33	0.44	6
Transverse	0.50	1.00	0.67	4
Oblique	0.00	0.00	0.00	3
Longitudinal	0.00	0.00	0.00	1
3d reconstruction	1.00	0.25	0.40	4
Mammo—cc	0.00	0.00	0.00	1
Mammo—mlo	0.00	0.00	0.00	1
Mammo—mag cc	0.00	0.00	0.00	1

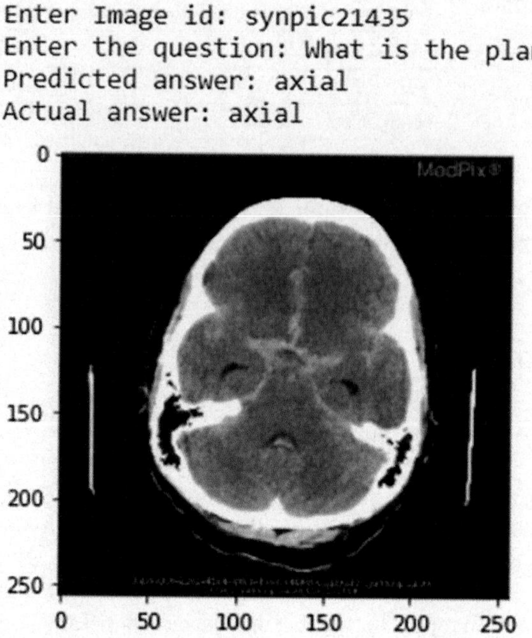

Fig. 5 Correct result predicted by the model

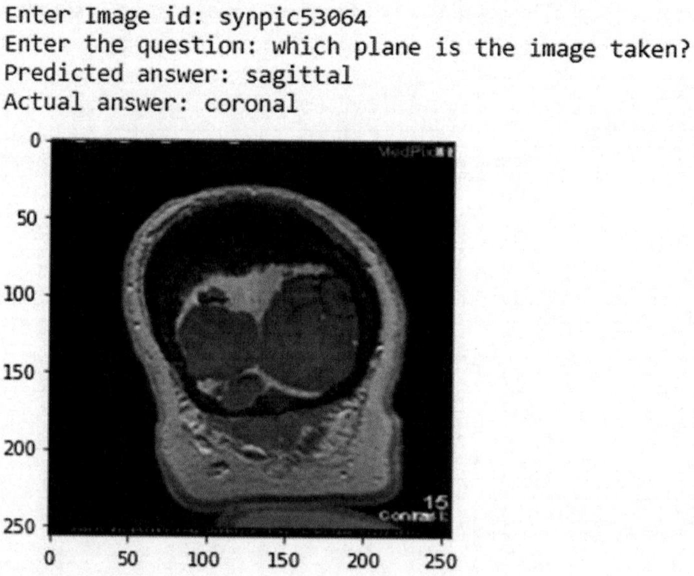

Fig. 6 Incorrect result predicted by the model

7 Conclusion

A MobileNet based medical VQA system was built using transfer learning pretrained on ImageNet. The pretrained image model is fused with question classifier model to predict the answer. The accuracy obtained on the test dataset was 80.8% for the plane class images. The model was also evaluated using different evaluation metrics like Precision, $F1$-score and Recall. The accuracy obtained was better compared to other pretrained network models like VGG16 and VGG19. As part of future work, we are planning to explore other image models and make comparative study of the same. We can upgrade our model for different categories like modality, abnormality and organ system.

References

1. X. Li, Y. Shi, Computer vision imaging based on artificial intelligence, in *2018 International Conference on Virtual Reality and Intelligent Systems (ICVRIS)*, Changsha (2018), pp. 22–25
2. M.C. Surabhi, Natural language processing future, in *2013 International Conference on Optical Imaging Sensor and Security (ICOSS)*, Coimbatore (2013), pp. 1–3. https://doi.org/10.1109/ICOISS.2013.6678407
3. K. Kafle, C. Kanan, Visual question answering: datasets, algorithms, and future challenges. Comput. Vis. Image Underst. 3–20 (2017)
4. A. Lubna, S. Kalady, A. Lijiya, MoBVQA: a modality based medical image visual question answering system, in *TENCON 2019—2019 IEEE Region 10 Conference (TENCON)*, Kochi, India (2019), pp. 727–732. https://doi.org/10.1109/TENCON.2019.8929456
5. A. Ben Abacha, S.A. Hasan, V.V. Datla, J. Liu, D. Demner-Fushman, H. Müller, VQA-Med: overview of the medical visual question answering task at ImageCLEF 2019, in *CLEF* (2019)
6. F. Liu, Y. Peng, M.P. Rosen, An effective deep transfer learning and information fusion framework for medical visual question answering, in *Experimental IR meets multilinguality, multimodality, and interaction, CLEF* ed. by F. Crestani et al. (2019)
7. B. Khasoggi, E. Ermatita, S. Sahmin, Efficient mobilenet architecture as image recognition on mobile and embedded devices. Indonesian J. Electr. Eng. Comput. Sci. **16**, 389–394 (2019). https://doi.org/10.11591/ijeecs.v16.i1
8. R. Joshi, Accuracy, precision, recall and F1 score: "interpretation of performance measures" (2016)

TagIT: A System for Image Auto-tagging and Clustering

Aditi Katiyar, V. Srividya, and B. K. Tripathy

Abstract In today's world, most of the things we buy, whether we talk about buying apparels, home decor or kitchen utensils, are through the online mode. E-commerce websites play a major role in providing a platform to the vendors for online selling of products. Generally, the process of selling the product on e-commerce websites passes through three steps: Firstly, vendors input the product information including pictures and tag them manually. Secondly, the e-commerce websites perform refinement of tagging using some efficient automated algorithm. Thirdly, the refining of tagging manually and classification of products is done on the basis of their use, popularity, sales, and many more. In this paper, we propose a two-phase process: First, vendor's input images are taken and passed though the deep neural network-based auto-tagging algorithm that tags the input images based on the different features. Next, with the use of dimensionality reduction and clustering algorithms, the tagged images are clustered into different domains. This process helps the vendors and the service provider to automatically tag their product images and provide recommendations on the basis of domain-specific searches without going through the hassle of manual tagging.

Keywords Automatic image tagging · Convolution neural network (CNN) · Pre-trained model · Fashion modified national institute of standards and technology database (MNIST) · Fashion product Images dataset · T-distributed stochastic neighbor embedding (t-SNE)

A. Katiyar · V. Srividya
School of Computer Science and Engineering (SCOPE), Vellore Institute of Technology, Vellore, Tamil Nadu 632014, India
e-mail: aditi.katiyar2018@vitalum.ac.in

V. Srividya
e-mail: srividya.v2018@vitalum.ac.in

B. K. Tripathy (✉)
School of Information Technology and Engineering (SITE), Vellore Institute of Technology, Vellore, Tamil Nadu 632014, India
e-mail: tripathybk@vit.ac.in

V. Bhateja et al. (eds.), *Data Engineering and Intelligent Computing*, Advances in Intelligent Systems and Computing 1407,
https://doi.org/10.1007/978-981-16-0171-2_25

1 Introduction

During the past one decade or so, images are used more frequently in the context of image sharing or image capturing. More recently, digitization has enhanced the frequency of such uses. The present generation is addicted to social networking sites such as Twitter, Facebook, Instagram, and other aspects such as online shopping, gaming and many more, where every second they scroll at least one picture or come across an image. The images that we see on online shopping sites or Instagram almost 80% of them are tagged manually by a person with the labels or keywords relevant to that particular picture or an image. The tags and the keywords associated with the particular image are important for the keyword-based searches on e-commerce websites, social networking sites, or organizational websites. The outdated technique that was earlier used for tagging the images with multiple keywords mainly suffers from two issues: Firstly, in every half an hour social networking sites scan approximately 105 terabytes of data, which include 300 million photos per day so manually tagging these images is time-consuming and error-prone task. Secondly, human resources required will be costly. Therefore, the concept of automatic tagging of an image comes into the picture; it takes care of tagging single as well as multiple keywords. Using the image tagging as the base, one can also get ease in classifying the images accurately in the specific domain taking the factors and keywords into consideration. As it is a bit difficult to predict in which particular domain the image with multiple tags belongs to, we use the prediction algorithms. Prediction can be used in situations where there are multiple possibilities to a single problem or we want to automatically recognize the value of a parameter. There are many types of prediction algorithms like Bayesian.

Here the labeled images are used to train the model, and then the test images are passed in the model, so that they can be categorized or classified into the particular domains. For example, if we are assigning the images either to 'Indian Wear' or 'Western Wear' domains, since there are only two choices, it is called two-class or binomial classification. In this paper, we are assigning the images in classes like 'Casual dresses', 'Formal dresses', 'Festive dresses', 'Party dresses', and many more; therefore, it is known as multi-class classification.

Generally, the image annotation or we can say predicting multiple tags for the given image is a bit of the challenging task. Nowadays, mostly used algorithms are purely based on the concept of large sets of handcrafted features. In many of the experiments and in published papers, it is shown that deep convolutional neural networks have outperformed humans in image classification, and these networks can be used to extract features highly predictive of an image's tags. In this study, we analyze semantic information in features derived from pre-trained deep network classifiers by evaluating their performance in nearest neighbor-based approaches to tag prediction. We are able to generally exceed performance of the model using the deep features in place of the manual features.

2 Literature Survey

In [1], the authors propose to use deep learning techniques that utilize the set of queries provided by the users and their clicks to determine better tagging of products on the e-commerce website. The proposed method used is a search-boosted tag recommender. The tags obtained from this recommender are highly effective in suggesting relevant tags when compared to manual tags entered by the sellers which contain lots of noise. This in turn helps the users with better search suggestions. In [2], the authors have performed a performance evaluation on the Fashion MNIST using a two-layered convolutional neural network method. The model used is a pre-trained VGG-16 model. The dataset consists of 70,000 images of size 28 × 28. The accuracy of the model is 88.24%. In [3], two research areas are connected, namely automatic tagging on the web and statistical key phrase extraction. A demonstration of the automatic document tagging using the statistical key phrase extraction algorithm is shown. The new algorithm used in this work is 'Maui', which utilizes the semantic information got from Wikipedia. In [4], the authors propose a content-based automatic music tagging algorithm using fully convolutional neural networks (FCNs).

Different neural network architectures are evaluated consisting of 2D convolutional layers and subsampling layers only. The datasets used are the MagnaTagATune dataset and the Million Song dataset. The evaluation metrics are area under the curve (AUC) receiver operating characteristics (ROC) curve scores. The input tested is the Mel-spectrogram. In [5], an approach to automatically tag recommendations for blogs is proposed. Collective intelligence from Web 2.0 collaborative tagging along with word semantics is extracted. The insights from the collected information help to predict the best set of tags using a hybrid artificial neural network (ANN). The tags have been provided on everything on the web such as photos, videos, music, blogs and so on. In [6], the authors attempt to address the issues involved in summarizing reviews and readable tags by proposing an unsupervised method to generate tags and rank them based on relevance. The generated tags are refined by making use of the Natural Language Processing (NLP) syntactic rules so that they become more informative. In [7], the authors experiment by using different feature extraction methods like VGG-16 and VGG-19 to predict image descriptions for the e-commerce industry. The performances of these models are compared using the metrics bilingual evaluation understudy (BLEU)-1, BLEU-2, BLEU-3 and BLEU-4. The experiments involve varying hyperparameters such as number of layers, learning rate, dropout and so on. In [8], a machine-learning-driven visual content management system for extremely large e-commerce catalogues is proposed. A superior image set is created for each product by collecting images from various suppliers. Some other applications of deep neural networks can be found in [7–11].

3 Experiment

The process followed in the creation of an image auto-tagging and clustering system is listed below in the form of the algorithm which is divided into two phases, that is phase-1 and phase-2. As shown in Figs. 1, 2, 3 and 4.

Algorithm
Phase-1

Step 1: **Input**: Images provided by the vendor's.

Step 2: **Processing**: Using the VGG-16, the pre-trained model for extracting the features and then adding an extra 2-layer neural network model over the VGG-16 model for tagging the images with the keywords.

Step 3: **Output**: Images with relevant multiple tags.

Phase-2

Step 1: **Input**: Images with relevant multiple tags.

Step 2: **Processing**: Applying the t-SNE algorithm on the images and tags associated with them to reduce the dimensionality of the images and clustering them into precise domain specific clusters.

Step 3: **Output**: Clusters are generated on the basis of different domains and each domain includes the images with similar tags.

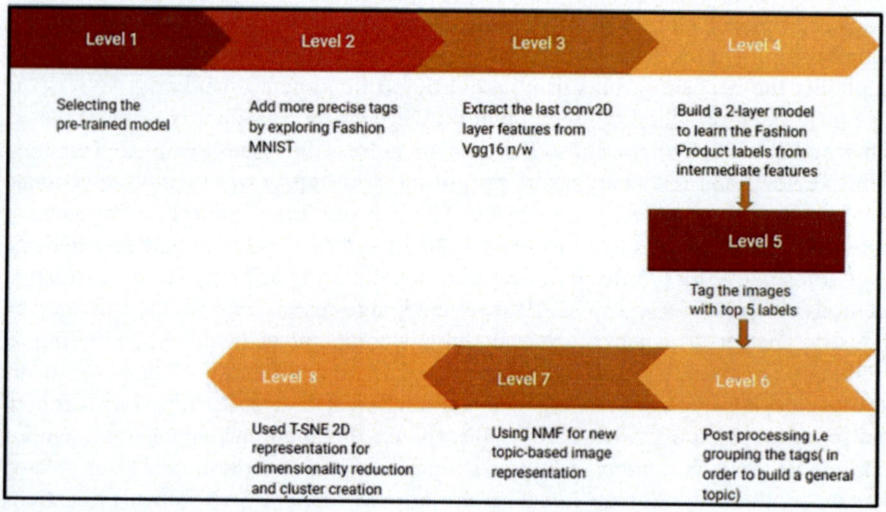

Fig. 1 Proposed system

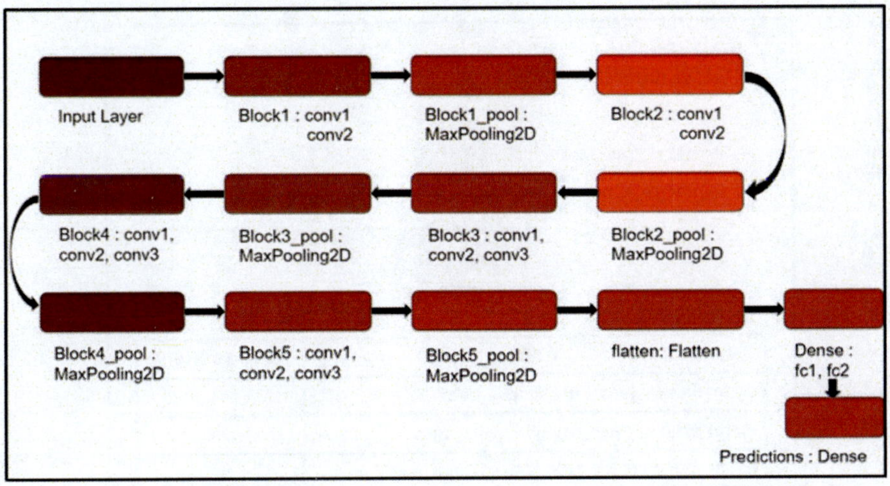

Fig. 2 Plot of layers in VGG model

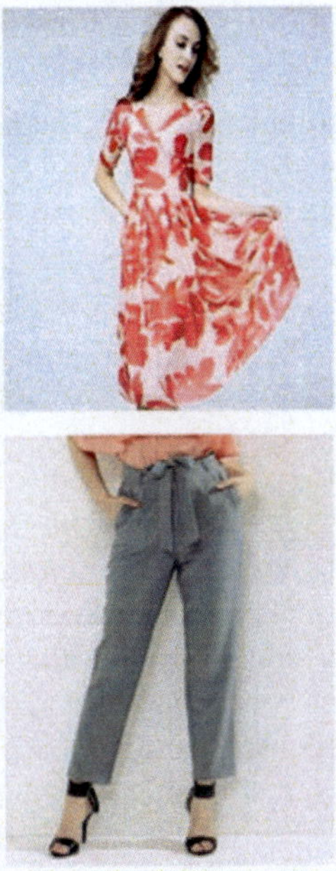

	Labels	Proba
0	Pink_dress	0.436545
1	Floral_print	0.567234
2	V-neck_dress	0.641923
3	Knee_length	0.653189

	Labels	Proba
0	grey_trousers	0.572545
1	bow_belt	0.257234
2	loose_fitting	0.639923
3	regular_trouser	0.486189

Fig. 3 Images tagged with the VGG-16 neural network model trained on the fashion product image dataset

TOPIC	TOPIC SCORE(%)
Long_dress - gown - full_length_frock	44
Floral_print - flower_design - roses_print	29
Black_dress - black_blue_dress - night_party_dress	27

Fig. 4 Scored images in the topics space

3.1 Building the Model for Automatically Tagging the Images and Classifying Them in Particular Domains

In this paper, we use a pre-trained model that is available at the Caffe Model Zoo. A pre-trained model is a model that has been trained on a very large benchmark dataset in order to solve a problem that is similar to the one that we are trying or want to solve. In order to reduce the high computational cost involved, usually published models from the literature are imported and used like the VGG-gate voltage. Transfer learning targets to enhance the procedure of learning by utilizing experience gained through the results of a similar problem. It is true that one cannot always find a model that corresponds exactly to one's data. Therefore, it is a better practice to map one's problem to an existing similar one.

The model that we have selected for this paper has been trained on the Fashion MNIST dataset that contains more than 2 million labeled images for clothes of men and women, with categories such as dresses, t-shirt, jackets, sarees and many more. This Fashion MNIST dataset fits perfectly with our Fashion Product Image dataset. The pre-trained model that we have used in our paper is the VGG-16 architecture.

VGG-16 is a convolution neural network (CNN) architecture, that instead of having a large number of hyper-parameters, focused on having the convolution layers of 3×3 filter with a stride 1 and maximum time it uses the same padding and maxpool layer of 2×2 filter of stride 2. It follows this arrangement of convolution and max pool layers consistently throughout the whole architecture. In the end of the model, it has 2 FC (fully connected layers) followed by a softmax for output. The 16 in VGG-16 refers to the 16 layers with which the weights are associated. This network is a pretty large network, and it has about 138 million (approx) parameters.

We have added more precise tags by exploring the Fashion Product Image dataset. It is composed of 70,000+ images for product recognition. Since we do not have sufficient numbers of images to train a powerful neural network, we will be using transfer learning—we have extracted the features from the last convolutional layer of the prior mentioned pre-trained model that is VGG-16 network and build a two-layer network model in order to learn the Fashion Product Image labels from these intermediate features [12, 13]. Since the dataset consists of 27 unique labels, we are restricting our model to work with only top 5 labels for each image in the dataset based on the probability associated with the label.

3.2 *Post-processing and Topics Extraction*

Once we have tagged our images, now there is a need for a post-processing step. As we can see, after tagging the images, we have some tags that can bring interdependent information such as dress—long—floral_print, while other tags bring superfluous information such as gown—long_frock—middy, t-shirt—vneck_shirt—casual_shirt and many more. By grouping the tags on the basis of what they represent will be more useful in order to create a more generalized topic. For an instance, summer_dresses = dress—long—floral_print. On the other side, one tag should not be bound to belonging to one exclusive or particular topic.

Non-negative matrix factorization (NMF or NNMF) is a collection of algorithms of linear algebra and multivariate analysis, where a matrix V is factorized usually into two matrices that is W and H, having elements with no negative values. This non-negativity makes the resulting matrices easier to inspect. An image is now represented by 30 relevant topics that is the output of the NMF is the new topic-based image representation.

3.3 *t-SNE 2D Representation*

In this paper, we set the number of topics at 30 for each image along with t-SNE transformation algorithm for 2-dimension visualization. This algorithm is very useful for visualization because of two reasons: Firstly, it reduces the dimension of your data, and secondly, it makes precise clusters. We have done the post-processing on the output of the pre-trained model before applying the t-SNE transformation. The post-processing can be done by three different methods listed below:

1. We directly apply the t-SNE transformation on intermediate outputs of the neural network that is the features of the last conv2D layer of VGG-16 model).
2. We perform an NMF on the top 5 label matrix.
3. We perform an NMF on the intermediate outputs of the neural network (the last conv2D layer features).

Fig. 5 Test image for auto tagging and classification

In this paper, we have done the post-processing using the third method listed above that is the last layer features along with the NMF.

4 Results

Figure 5 shows the test image consisting of 18 layers in total (16 layers of VGG-16 and 2 add on layers of neural networks). Figure 6 shows the test image with the top five labels along with their probability. Figure 7 shows the output after the post processing and topic extraction, which is the image with the topic and their corresponding topic score in percentage. Figure 8 shows the precise clusters of different clothes.

5 Conclusions

We have come up with TagIT, a system for image auto-tagging and classification, that can be used by e-commerce websites for automatic tagging the product images with multiple keywords and classifying them into clusters on the basis of the tags or grouped-tags associated with each and every image. In comparison to traditional tagging of the images, TagIT takes advantage of massive-scale, publicly available and time saving. In addition, TagIT leverages state-of-the-art neural networks and machine learning techniques. Usability analysis shows that TagIT has helped in improving tagging systems with accuracy of 76–82% with time of processing reduced to 3–4 h.

	Labels	Proba
0	white_top	0.676545
1	long_sleeves	0.467234
2	strap_top	0.341923
3	floral_print	0.553189

Fig. 6 Output of the vgg16 model along with the extra 2-layer neural network

Fig. 7 Test image's topic score

	TOPIC	TOPIC SCORE(%)
0	White_top - tank_top - casual_top	0.676545
1	Long_sleeves - full_sleeves - loose_sleeves	0.467234

Fig. 8 Output of the t-SNE transformation—precise clusters for different domains like casual wear, party wear and formal wear

References

1. F.M. Belém, R.M. Silva, C.M. de Andrade, G. Person, F. Mingote, R. Ballet, M.A. Gonçalves, "Fixing the curse of the bad product descriptions"—search-boosted tag recommendation for E-commerce products. Inf. Process. Manage. **57**(5), 102289 (2020)
2. J. Rawat, D. Logofătu, S. Chiramel, Factors affecting accuracy of convolutional neural network using VGG-16, in *International Conference on Engineering Applications of Neural Networks* (Springer, Cham, 2020), pp. 251–260
3. O. Medelyan, E. Frank, I.H. Witten, Human-competitive tagging using automatic keyphrase extraction, in *Proceedings of the 2009 Conference on Empirical Methods in Natural Language Processing* (2009), pp. 1318–1327
4. K. Choi, G. Fazekas, M. Sandler, Automatic tagging using deep convolutional neural networks (2016). arXiv preprint arXiv:1606.00298
5. S.O.K. Lee, A.H.W. Chun, Automatic tag recommendation for the web 2.0 blogosphere using collaborative tagging and hybrid ANN semantic structures. ACOS **7**, 88–93 (2007)
6. A. Konjengbam, N. Kumar, M. Singh, Unsupervised tag recommendation for popular and cold products. J. Intell. Inf. Syst. 1–22 (2019)
7. A. Adate, B.K. Tripathy, Deep learning techniques for image processing, in *Machine Learning for Big Data Analysis*, ed. by S. Bhattacharyya, H. Bhaumik, A. Mukherjee, S. De (De Gruyter, Berlin, Boston, 2018), pp. 69–90
8. A. Bose, B.K. Tripathy, Deep learning for audio signal classification, in *Deep Learning Research and Applications*, ed. by S. Bhattacharyya, B.K. Tripathy, S. Saha, A.E. Hassanian (De Gruyter Publications, 2020), pp. 105–136
9. A. Adate, B.K. Tripathy, D. Arya, A. Shaha, Impact of deep neural learning on artificial intelligence research, in *Deep Learning Research and Applications*, ed. by S. Bhattacharyya, B.K. Tripathy, S. Saha, A.E. Hassanian (De Gruyter Publications, 2020), pp. 69–84
10. A. Adate, B.K. Tripathy, S-LSTM-GAN: shared recurrent neural networks with adversarial training, in *Proceedings of the 2nd International Conference on Data Engineering and Communication Technology. Advances in Intelligent Systems and Computing*, ed. by A. Kulkarni, S. Satapathy, T. Kang, A. Kashan, vol. 828 (Springer, Singapore, 2019), pp. 107–115
11. A. Adate, B.K. Tripathy, Understanding single image super-resolution techniques with generative adversarial networks, in *Soft Computing for Problem Solving. Advances in Intelligent Systems and Computing*, ed. by J. Bansal, K. Das, A. Nagar, K. Deep, A. Ojha, vol. 816 (Springer, Singapore, 2019), pp. 833–840
12. I. Mandal, A. Dwivedi, Deep learning algorithms for accurate prediction of image description for e-commerce industry, in *Data Management, Analytics and Innovation* (Springer, Singapore, 2020), pp. 401–418
13. A. Chaudhuri, P. Messina, S. Kokkula, A. Subramanian, A. Krishnan, S. Gandhi, et al., A smart system for selection of optimal product images in e-commerce, in *2018 IEEE International Conference on Big Data (Big Data)* (IEEE, 2018), pp. 1728–1736

A Novel Approach for Image Encryption Using Zaslavskii Map and Arnold's Cat Map

R. N. Ramakant Parida, Binod Kumar Singh, and Chittaranjan Pradhan

Abstract Recently, transformation of information is an emerging area of interest because of efficient Internet technology and faster communication medium. Nowadays, huge numbers of data transferred over many transferring mediums, out of which image is one of them. To maintain confidentiality and provide ownership of the data, we need to do any cryptographic technique like encryption. In this paper, we proposed a new cryptography technique for image encryption which is a combination of two different techniques for image encryption to increase the security of the data. The two different techniques are Zaslavskii map and Arnold's cat map. Both combinedly used for encryption as well as decryption. In this work, we first encrypt the image using Zaslavskii method, and then again, that encrypted image re-encrypted using Arnold's cat map to re-encrypt that, to get the required image using two times encryption. Then, in the reverse process, the final encrypted image undergoes first decryption using Arnold's cat map, and that decrypted image further undergoes Zaslavskii decryption to get back the final derived image which should be like the original image. This proposed technique gives users more key values for better security and unpredictability, which shows good entropy value and good correlation value. The efficiency of image encryption techniques is verified by number of pixels change rate (NPCR) and unified average chaning intensity (UACI) which is used for any small change in pixels, along with entropy and correlation coefficient values.

Keywords NPCR · UACI · Encryption · Decryption · Entropy · Correlation coefficient

R. N. Ramakant Parida (✉) · B. Kumar Singh
NIT Jamshedpur, Jamshedpur, India
e-mail: bksingh.cse@nitjsr.ac.in

C. Pradhan
KIIT(DU), Bhunaneswar, India
e-mail: chittaranjanfcs@kiit.ac.in

V. Bhateja et al. (eds.), *Data Engineering and Intelligent Computing*,
Advances in Intelligent Systems and Computing 1407,
https://doi.org/10.1007/978-981-16-0171-2_26

1 Introduction

Nowadays, it is impossible to live without the Internet, and a lot of things happen over a minute. Many things are transmitted over the network in various forms. Out of all, image files, one of them which contains data/information, have the major contribution in the Internet for data transformation. More than the millions of images are created and also being transmitted by various forms over various communication mediums such as television, smart phones, personal computers, tablets, and satellites. With the advancement of new technology, faster transmission medium and new handheld small devices like mobile or tablets, which allow people to manage and run their own Web resources, image copyright, security, play a major concern in these areas. So, providing the security image encryption acts like a safeguard to the original data from unauthorized access. Cryptography is the study of information security like data integrity, authentication based on certain mathematical tools or well-defined algorithms [3, 27]. Image encryption is a technique where location of pixels is transformed to a new location and changes its value so that the originality will be missed or unpredictable, whereas decryption is technique which reconstructs the original form of that image from encrypted image [7, 23–25]. So, out of many techniques, image encryption is a possible solution to this problem which is based on the pixels of the image [11, 12]. Many techniques exist to prove the ownership of the image, authentication, and security, but each of them has its own advantages and disadvantages too [1, 15, 19]. So, image encryption is a possible solution to this problem which purely depends on pixels of an image.

Many researchers have worked and proposed many varieties of chaotic systems based on many dimensions. Based on dimension, chaotic map techniques may be discussed in two different sub-groups. One is a high-dimensional map, and another is a one-dimensional map [6, 10]. The major difference between one-dimensional chaotic map and high-dimensional chaotic map is multiple number of dependant variables with good chaotic behavior which shows more unpredictability to small change in pixel leads to huge impact in output.

Wang et al. [22] define an improved 1D chaotic map which is combination of two sine maps to resist plaintext attack depends upon a small number of key space. Parvaz and Zarebnia [16] proposed a method which is based on logistic, sine and tent map techniques for generating random numbers. Because of these combined techniques here, key space is more which gives encryption is secure but its scheme of encryption is not satisfactory. Balaska et al. [4] proposed an image encryption technique by considering grain-128a algorithm and Zaslavsky chaotic map to prove the security of image of any size and any type. Pradhan et al. [17] proposed imperceptible watermarking technique using two different chaotic techniques like Arnold's transformation as well as cross chaos map in DCT domain to prove imperceptible watermarking. From the above work, we inspired to work for a novel approach for image encryption using a combination of two different chaotic maps with double encryption.

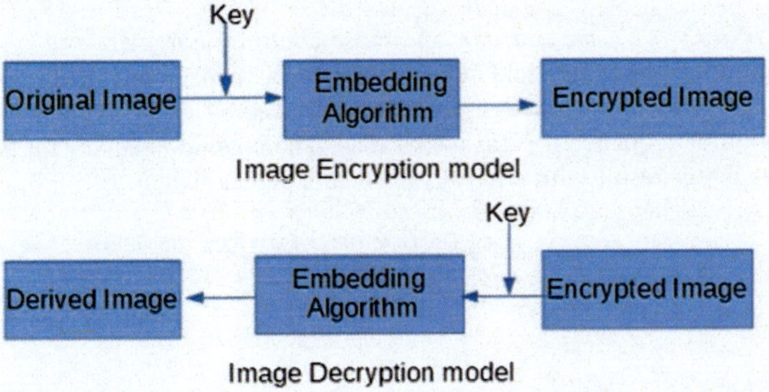

Fig. 1 Extraction algorithm

In this paper, our focus is on digital image encryption techniques to encrypt a digital image using chaotic maps and change the original location of pixels in that image. There are a number of keys associated with each image based on chaotic maps which are used to produce random numbers using mathematical functions, to encrypt a digital image. Then, the same keys are used for decryption purposes too. Here, without using keys, we cannot prove the originality of image, authentication or security. Here, we have used two different techniques to provide better security and robustness by using Zaslavskii [5, 28] and Arnold's cat map [10, 12] techniques together.

Sections 2 and 3 give brief information about Zaslavskii map and Arnold's cat map. Section 4 describes the encryption and decryption on gray scale images as well as color image. Section 5 contains result analysis, whereas conclusion and future scope mentioned in Sect. 6 part (Fig. 1).

2 Zaslavskii Map

There are many different types of chaotic map techniques available which could be used for producing random number generators(PRNG) [20] for encryption and decryption of digital images [5, 8, 28]. Out of many, we selected Zaslavskii map because it has three control parameters which is one of the important parameters for image encryption and decryption process. The Zaslavskii map (used for pseudo-random number generator) is a discrete-time nonlinear dynamical system introduced by George M. Zaslavskii which shows chaotic behavior by taking a point (x_i, y_i) in the plane and maps it to a new point by the following Eq. (1):

$$\begin{cases} x_{i+1} = [x_i + \nu(1 + \mu y_j) + \epsilon\nu\mu\cos(2\pi x_i)](\text{mod}1) \\ y_{i+1} = e^{-r}(y_i + \epsilon\cos(2\pi x_i)) \end{cases} \tag{1}$$

where $\mu = (1 - e^{-r})$/r, and ν, ϵ and r are control parameters and x_i, y_i, x_{i+1}, y_{i+1} are initial pixel position and new pixel position, respectively, and initial values of $i = 0, 1, 2 \ldots$the initial values are x_0, y_0 is replaced by x_1, y_1 in the next iteration for gereration of sequence for Zaslavskii map. The chaotic behavior for Zaslavskii map shows better result with control parameters values, where $r = 3.0$, $\nu = 400/3$ and $\epsilon = 0.3$.

$$H(x) = \begin{cases} 0, & 0 < m(i, j) \leq 0.5 \\ 1, & 0.5 < m(i, j) \leq 1 \end{cases} \quad (2)$$

3 Arnold's Cat Map

Arnold's 2D cat map was proposed by Arnold (1960) using an image of a cat. The main principle in this technique is to shift the pixel position in an image over iteration but it is observed that after a number of iterations we will get back the original state, called Arnold period, which depends on size of an image [10, 12]. This technique shows mixing behavior typical for chaotic systems. The equation for 2D Arnold's cat map is as follows (3):

$$\begin{cases} x_{i+1} = (x_i + ay_i) \mod N \\ y_{i+1} = (bx_i + (ab + 1)b_i) \mod N \end{cases} \quad (3)$$

where a and b are two system control parameters. Equation (3) is using on the old pixel coordinate (x_i, y_i) to shift the pixel value to a new location (x_{i+1}, y_{i+1}) in each of Arnold's 2D cat map iterations. Now, let us consider a $M \times M$ image having four pixels values represented as $M(x_i, y_i)$ which is a square matrix of size 2×2 having $M = [239; 10271]$. Using Eq. (3), change in pixel position of the image is calculated. Each pixel transferred to a new location as is shown in Table 1.

Table 1 Arnold transformation

Iteration	Matrix values
1	M = [71 23; 9 102]
2	M = [102 71; 23 9]
3	M = [9 102; 71 23]
4	M = [23 9; 102 71]

4 Proposed Method

As we have observed, in many papers, researchers are using more than one technique to increase the key space and total number of keys for encryption and decryption of images which assures high security [4, 8, 13, 14, 16–18, 21, 22, 27].

4.1 Proposed Work on Gray Scale Image

The image consists of pixels (having values ranging from 0–255) arranged in a matrix. In gray color image, the pixel value represents a single number that represents the brightness of the pixel, where 0 represents black and 255 represents white. Values in between represent shades of gray [9] (Fig. 2).

4.1.1 Encryption Process

First of all, create pseudo-random numbers by using Eq. (1) with control parameter values and then that is quantized by using quantization Eq. (2) that generates a binary matrix. Then, the gray image $M \times M$ and quantized matrix are XORed with each bit which generates the encrypted image.
For experimental result, we have been used "lena64.bmp" image with initial values x_i = 0.587201561347, $y_j = -0.28432144902$ along with control parameters $\nu = 400/3$, $\epsilon = 0.3$ and $r = 3.0$. Figure 3 shows the image encryption using 2D Zaslavskii map.

Equation (3) is applied on the encrypted image received from above step which shifted all its pixels present in that encrypted image to form a new encrypted image using Arnold's cat map key which depends on image size (Arnold period).

The experimental result used with control parameters $x = 2$ and $y = 3$ with Arnold period/key as 20 which generates the final encrypted image after a two step encryption process [10, 12].

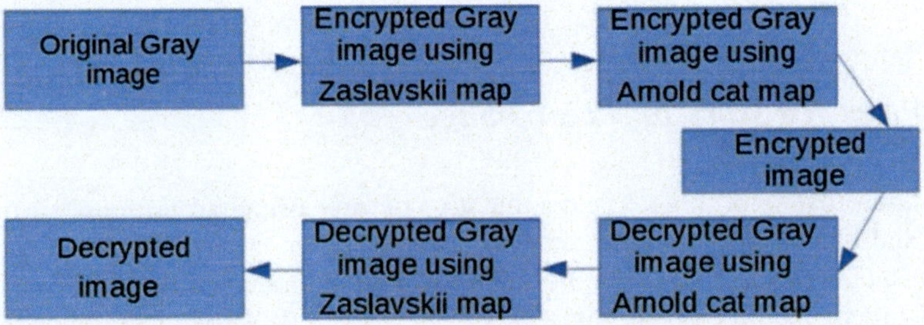

Fig. 2 Process diagram for proposed method using encryption–decryption technique for gray scale image

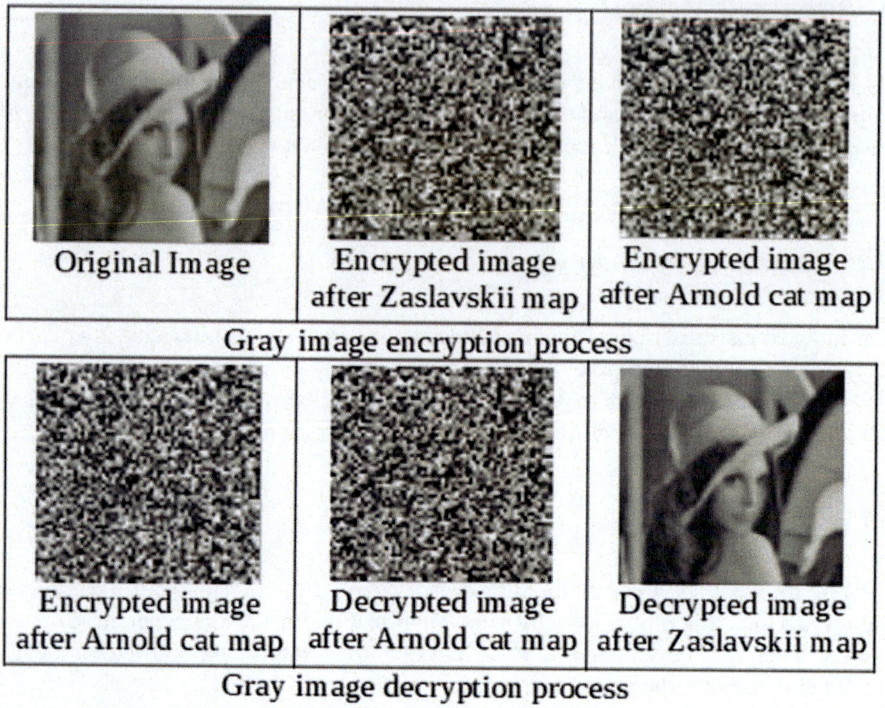

Fig. 3 Encryption and decryption on gray image

4.1.2 Decryption Process

Encrypted image is being decrypted using Arnold's cat map with secret key which is the first step in the decryption, where it will again shift each pixel to another new location in that image. In the very next step, that resultant decrypted image undergoes another decryption process to get back the final derived image after two successive steps of decryption.

4.2 Proposed Work on Color Image

As we know that color image is a combination of three principal components of red, green, and blue which indicates that it means every image must have these three components. It means that each color image is able to produce three matrices which are red, green, and blue pixels containing values [15, 19]. So like a gray image now each matrix we will apply successively encryption two times and decryption two times to get its derived component matrix. After getting all three derived metrics, again we will combine these three components to get back its derived color image.

Step 1: Find RGB components of a color image
Step 2: Apply encryption technique using Zaslavskii on each component
Step 3: Apply further encryption using Arnold's cat map on encrypted image with its each component
Step 4: Combine all the three components to get back colored encrypted image
Step 5: Find RGB components of encrypted image
Step 6: Applying decryption technique using Arnold's cat map on each component
Step 7: Applying further decryption using Zaslavskii map on decrypted image with its each components
Step 8: Combine all three components to get back the derived color image.

All these eight steps are explained in Fig. 4. We have been used "lena64.bmp" image with initial values $x_i = 0.587201561347$, $y_j = -0.28432144902$ along with control parameters $\nu = 400/3$, $\epsilon = 0.3$, $r = 3.0$ for Zaslavskii and control parameters $x = 2$ and $y = 3$ with Arnold period/key as 20 for Arnold's cat map used for both encryption and decryption (Fig. 5).

5 Analysis of Security

5.1 NPCR and UACI

We have implemented a proposed new technique using Zaslavskii and Arnold's cat map. As we know that chaotic maps are very sensitive to initial condition and control parameters, so a small change in source/original/initial image has a huge impact on destination/derived/final image. So, to do the pixel analysis, we used number of pixel changing rate (NPCR) and unified average Chang index (UACI) between encrypted image and original image [26].

The NPCR value and UACI value may be calculated using Eqs. (4) and (5) using following (where all matrix should have similar size):

$$\text{NPCR} = \frac{\sum_{i,j} K(i,j)}{a * b} * 100\% \tag{4}$$

$$\text{UACI} = \frac{\sum_{i,j} \left\{ \frac{|M(i,j) - M(i,j)|}{255} \right\}}{a * b} * 100\% \tag{5}$$

It has been observed that higher the value of NPCR and UACI, better is the algorithm. Also, we have calculated the time taken for encryption and decryption for different types of image. Tables 2 and 3 represent all above information. In the table, NPCR and UACI values are close to their ideal values which are 99.6 and 31.4%. These values indicated the proposed approach is good for encryption and decryption.

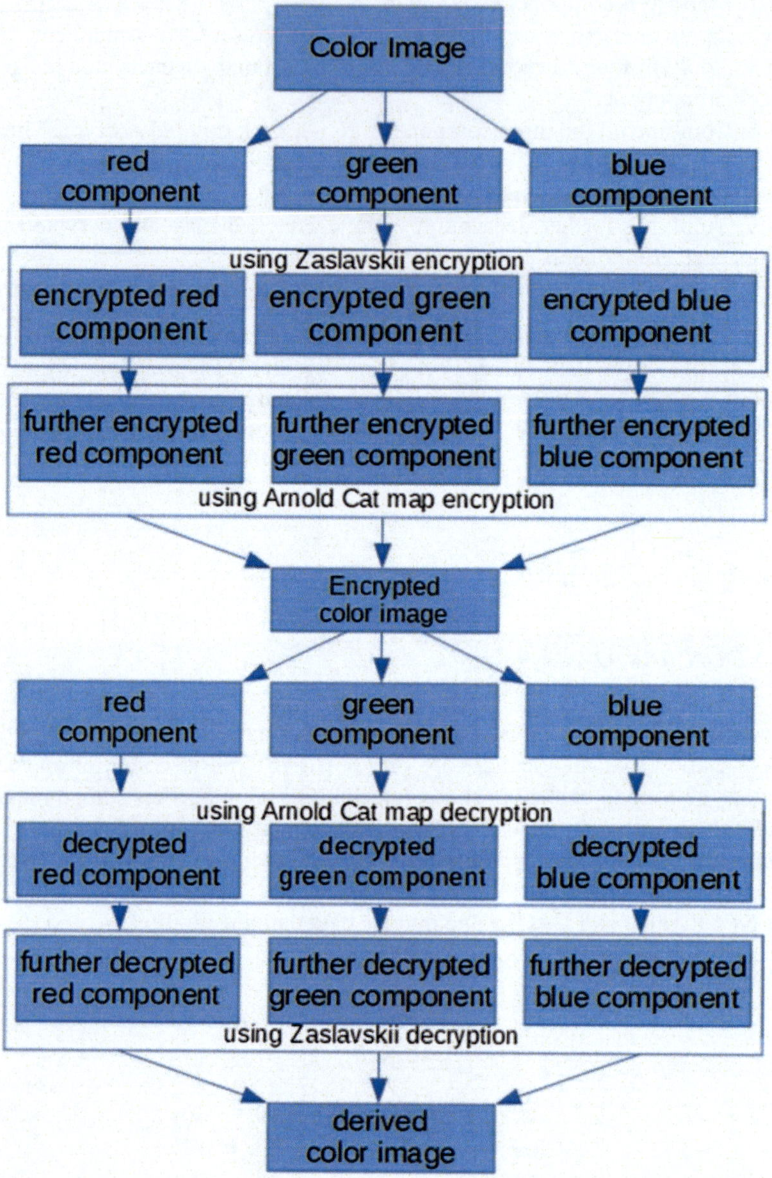

Fig. 4 Process diagram for proposed method using encryption–decryption technique for color image

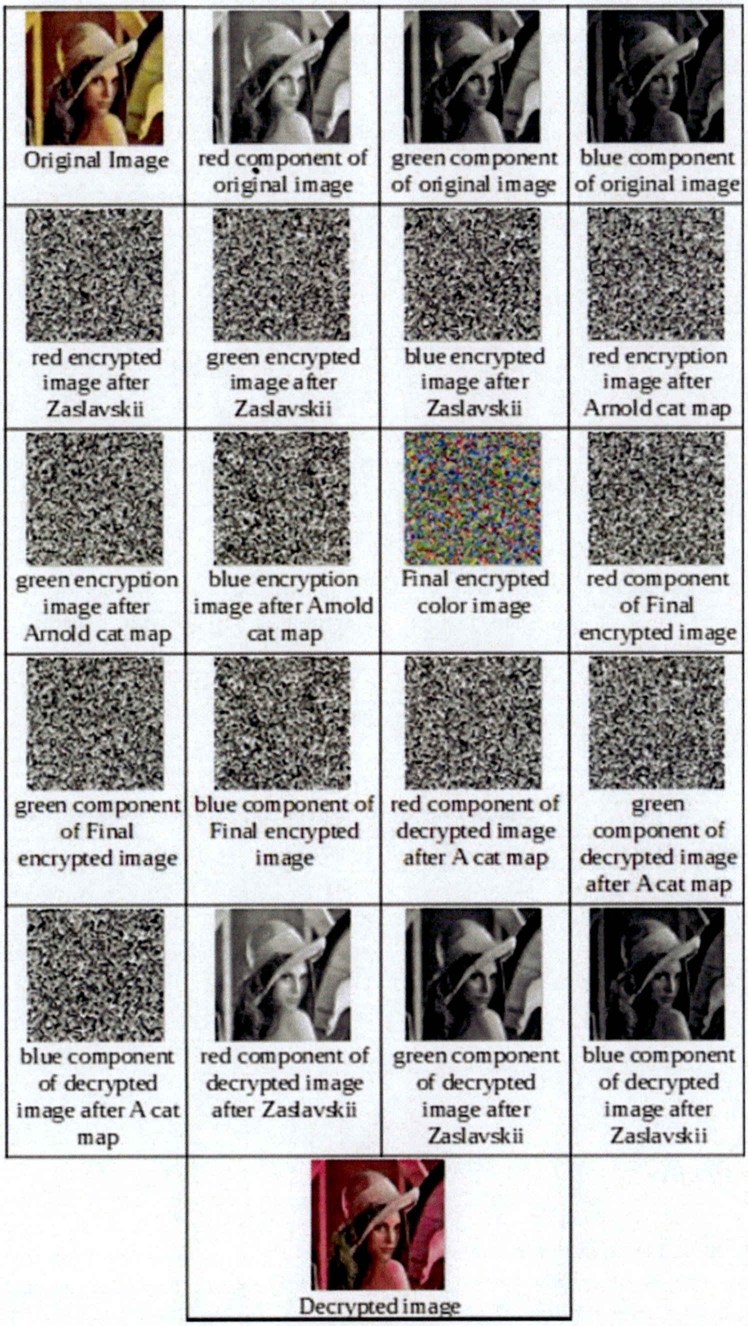

Fig. 5 Encryption and decryption on color image

Table 2 NPCR, UACI values along with encryption and decryption time for gray image

Image name	Image size	NPCR values	UACI values	Encryption time (s)	Decryption time (s)
Lena	[64 × 64]	99.56	32.04	0.5928	0.6552
Barbara	[64 × 64]	99.58	31.18	0.6084	0.6708
Peppers	[64 × 64]	99.63	32.37	0.5928	0.6553
Lena	[128 × 128]	99.68	31.43	2.34	2.964
Barbara	[128 × 128]	99.66	31.91	2.932	2.94
Peppers	[128 × 128]	99.6	32.25	2.3712	2.98
Lena	[256 × 256]	99.61	31.75	8.7637	12.38
Barbara	[256 × 256]	99.6	32.37	8.7049	13.18
Peppers	[256 × 256]	99.63	31.57	8.5801	12.3854
Lena	[512 × 512]	99.59	32.27	46.75	78.5933
Barbara	[512 × 512]	99.6	32.52	47.61	77.5169
Peppers	[512 × 512]	99.63	32.64	46.488	77.37

Table 3 NPCR, UACI values along with encryption and decryption time for color image

Image name	Image size	NPCR values	UACI values	Encryption time (s)	Decryption time (s)
Lena	[64 × 64]	99.5	33.7	2.012	1.875
Barbara	[64 × 64]	99.62	32.87	1.622	1.841
Peppers	[64 × 64]	99.59	32.84	1.747	1.841
Lena	[128 × 128]	99.60	31.18	5.843	7.675
Barbara	[128 × 128]	99.59	31.94	6.34	9.36
Peppers	[128 × 128]	99.61	32.29	6.286	9.0948
Lena	[256 × 256]	99.61	32.41	32.568	39.14
Barbara	[256 × 256]	99.61	32.49	30.248	43.82
Peppers	[256 × 256]	99.58	31.39	29.608	42.759
Lena	[512 × 512]	99.6	32.82	167.75	190.25
Barbara	[512 × 512]	99.6	32.83	182.2	331.89
Peppers	[512 × 512]	99.51	32.71	167.91	294.37

5.2 *Entropy*

Distribution of pixel values for a digital image is the entropy information. Digital images are represented by 8-bit intensity levels, and the ideal entropy should be 8, which indicates true random image [2]. Therefore, for an ideal image encryption technique, the entropy value should be close to 8 for encrypted image. Entropy is defined as

Table 4 Entropy of the original and encrypted images of different size (gray image)

Image name	Image size	Original gray image	Encrypted gray image
Lena	[64 × 64]	7.2670	7.2894
Barbara	[64 × 64]	6.9767	7.0033
Peppers	[64 × 64]	7.4934	7.5704
Lena	[128 × 128]	7.3914	7.4171
Barbara	[128 × 128]	7.0853	7.1114
Peppers	[128 × 128]	7.5490	7.6420
Lena	[256 × 256]	7.4801	7.5068
Barbara	[256 × 256]	7.1923	7.2282
Peppers	[256 × 256]	7.5689	7.6702
Lena	[512 × 512]	7.3478	7.3966
Barbara	[512 × 512]	7.3579	7.4051
Peppers	[512 × 512]	7.5945	7.6986

Table 5 Entropy of the original and encrypted images of different size (color image)

Image name	Image size	Original color image	Encrypted color image
Lena	[64 × 64]	7.0882	7.5047
Barbara	[64 × 64]	7.3992	7.5117
Peppers	[64 × 64]	7.3269	7.6599
Lena	[128 × 128]	7.1482	7.5715
Barbara	[128 × 128]	7.4838	7.5895
Peppers	[128 × 128]	7.3302	7.6828
Lena	[256 × 256]	7.1772	7.5884
Barbara	[256 × 256]	7.5624	7.6521
Peppers	[256 × 256]	7.3238	7.6768
Lena	[512 × 512]	5.1014	6.0576
Barbara	[512 × 512]	7.6441	7.7193
Peppers	[512 × 512]	7.2977	7.6510

$$H(M) = \sum_{i=0}^{n-1} P(M_i) \log_2 \frac{1}{P(M_i)} \tag{6}$$

Where M_i represents the pixel values (between 0 and 256), $P(M_i)$ is the probability of the symbol M_i and n is the number of symbols, i.e., is 256 (gray image). Tables 4 and 5 show the information of entropy indicates values are close to ideal value that means the proposed method is also close to a random source.

Table 6 Correlation coefficient value between original and encrypted images for gray and color images)

Image name	Image size	Gray image	Color image
Lena	[64 × 64]	0.0045	0.0014
Barbara	[64 × 64]	0.0101	0.0024
Peppers	[64 × 64]	0.0245	0.0081
Lena	[128 × 128]	0.01131	0.0032
Barbara	[128 × 128]	0.0010	0.0068
Peppers	[128 × 128]	0.0038	0.0022
Lena	[256 × 256]	0.0025	0.0037
Barbara	[256 × 256]	0.0044	0.0026
Peppers	[256 × 256]	0.0035	0.0081
Lena	[512 × 512]	0.0036	0.0003
Barbara	[512 × 512]	0.0046	0.0031
Peppers	[512 × 512]	0.0084	0.0034

5.3 *Correlation Coefficient*

In the digital image encryption process, an original image is converted to an encrypted image where comparison between original image and encrypted image occurs using correlation coefficient technique [2]. Quality of cryptography technique is measured by various techniques, out of which, correlation coefficient is one of them, where correlation between two variables occurs. The correlation coefficient value r varies in between −1 to 0 to 1, where r value 0 indicates both the images are completely unrelated with each other (ideal case), −1 indicates both are anti-correlated with each other, and 1 means both the images are completely identical. Table 6 indicates r value to zero that implies good cryptographic process. Correlation coefficient can be calculated by using follows Eq. (7)

$$r_{xy} = \frac{\text{cov}(x, y)}{\sigma_x \sigma_y} = \frac{\frac{1}{N_S}\sum_{i=1}^{N_S}(x_i - E(x))(y_i - E(y))}{\sqrt{\frac{1}{N_S}(x_i - E(x))^2}\sqrt{\frac{1}{N_S}(y_i - E(y))^2}} \tag{7}$$

where E(x) $=\frac{1}{N_S}\sum_{i=1}^{N_S} x_i$ and E(y) $=\frac{1}{N_S}\sum_{i=1}^{N_S} y_i$

6 Conclusion

A novel approach for digital image encryption technique based on Zaslavskii map along with Arnold's cat map is proposed. This proposed image encryption method consists of two stages encryption to get final encrypted image from a original image

using two different techniques like Zaslavskii followed by Arnold's cat map and again in the reverse way two stages decryption to get back retrieve image from final encrypted image using first Arnold's cat map followed by Zaslavskii map. Here, the proposed method supports a large number of keys which are difficult to predict the control parameters imply the system is safe from different types of attack. Because of the above-mentioned reason, we proposed two levels of encryption process to encrypt the image. The experimental results demonstrate that proposed method gives large key space, close to ideal NPCR and UACI values with good correlation value and entropy for better encryption standard algorithm.
In future work, we are thinking to extend 3D approaches by using both the techniques for image encryption as well as speech. Another future scope is to use the proposed encryption technique in video steganography.

References

1. A.A. Abd El-Latif, L. Li, T. Zhang, N. Wang, X. Song, X. Niu, Digital image encryption scheme based on multiple chaotic systems. Sens. Imag. Int. J. **13**(2), 67–88 (2012)
2. J. Ahmad, F. Ahmed, Efficiency analysis and security evaluation of image encryption schemes. Computing **23**, 25 (2010)
3. G. Alvarez, S. Li, Some basic cryptographic requirements for chaos-based cryptosystems. Int. J. Bifurc Chaos **16**(08), 2129–2151 (2006)
4. N. Balaska, Z. Ahmida, A. Belmeguenai, S. Boumerdassi, Image encryption using a combination of grain-128a algorithm and Zaslavsky chaotic map. IET Image Proc. **14**(6), 1120–1131 (2019)
5. F. Farsana, K. Gopakumar, A novel approach for speech encryption: Zaslavsky map as pseudo random number generator. Procedia Comput. Sci. **93**, 816–823 (2016)
6. M. François, T. Grosges, D. Barchiesi, R. Erra, A new image encryption scheme based on a chaotic function. Sig. Proc.: Image Commun. **27**(3), 249–259 (2012)
7. Z.H. Guan, F. Huang, W. Guan, Chaos-based image encryption algorithm. Phys. Lett. A **346**(1–3), 153–157 (2005)
8. G. Hanchinamani, L. Kulakarni, Image encryption based on 2-d Zaslavskii chaotic map and pseudo Hadmard transform. Int. J. Hybrid Inf. Technol. **7**(4), 185–200 (2014)
9. C.K. Huang, C.W. Liao, S. Hsu, Y. Jeng, Implementation of gray image encryption with pixel shuffling and gray-level encryption by single chaotic system. Telecommun. Syst. **52**(2), 563–571 (2013)
10. K.K. Kabi, C. Pradhan, B.J. Saha, A.K. Bisoi, Comparative study of image encryption using 2d chaotic map, in *2014 International Conference on Information Systems and Computer Networks (ISCON)*, pp. 105–108. IEEE (2014)
11. A. Kanso, M. Ghebleh, A novel image encryption algorithm based on a 3d chaotic map. Commun. Nonlinear Sci. Numer. Simul. **17**(7), 2943–2959 (2012)
12. S. Kumar, B. Sinha, C. Pradhan, Comparative analysis of color image encryption using 2d chaotic maps, in *Information Systems Design and Intelligent Applications* (pp. 379–387) (Springer, 2015)
13. Y. Luo, J. Yu, W. Lai, L. Liu, A novel chaotic image encryption algorithm based on improved baker map and logistic map. Multimedia Tools Appl. **78**(15), 22023–22043 (2019)
14. K. Muhammad, J. Ahmad, N.U. Rehman, Z. Jan, M. Sajjad, Cisska-lsb: color image steganography using Stego key-directed adaptive LSB substitution method. Multimedia Tools Appl. **76**(6), 8597–8626 (2017)

15. R.R. Parida, S. Singh, C. Pradhan, Analysis of color image encryption using multidimensional bogdanov map, in *Histopathological Image Analysis in Medical Decision Making*, pp. 202–225. IGI Global (2019)
16. R. Parvaz, M. Zarebnia, A combination chaotic system and application in color image encryption. Opt. Laser Technol. **101**, 30–41 (2018)
17. C. Pradhan, V. Saxena, A.K. Bisoi, Imperceptible watermarking technique using Arnold's transform and cross chaos map in dct domain. Int. J. Comput. Appl. **55**(15), (2012)
18. Y. Şekertekin, Ö Atan, An image encryption algorithm using Ikeda and Benon chaotic maps, in *2016 24th Telecommunications Forum (TELFOR)*, pp. 1–4. IEEE (2016)
19. S. Singh, R. Parida, C. Pradhan, Comparative analysis of image encryption using 2d and 3d variations of duffing map, in *2018 International Conference on Communication and Signal Processing (ICCSP)*, pp. 0751–0754. IEEE (2018)
20. B. Stoyanov, K. Kordov, Novel zaslavsky map based pseudorandom bit generation scheme. Appl. Math. Sci. **8**(178), 8883–8887 (2014)
21. X.J. Tong, Design of an image encryption scheme based on a multiple chaotic map. Commun. Nonlinear Sci. Numer. Simul. **18**(7), 1725–1733 (2013)
22. H. Wang, D. Xiao, X. Chen, H. Huang, Cryptanalysis and enhancements of image encryption using combination of the 1d chaotic map. Sig. Proc. **144**, 444–452 (2018)
23. L. Wang, H. Song, P. Liu, A novel hybrid color image encryption algorithm using two complex chaotic systems. Opt. Lasers Eng. **77**, 118–125 (2016)
24. X. Wang, L. Teng, X. Qin, A novel colour image encryption algorithm based on chaos. Sig. Proc. **92**(4), 1101–1108 (2012)
25. X.Y. Wang, Y.Q. Zhang, X.M. Bao, A novel chaotic image encryption scheme using dna sequence operations. Opt. Lasers Eng. **73**, 53–61 (2015)
26. Y. Wu, J.P. Noonan, S. Agaian, et al., Npcr and uaci randomness tests for image encryption. Cyber J. Multi. J. Sci. Technol. J. Selected Areas Telecommun. (JSAT) **1**(2), 31–38 (2011)
27. T. Yang, A survey of chaotic secure communication systems. Int. J. Comput. Cognit. **2**(2), 81–130 (2004)
28. S.F. Yousif, Speech encryption based on zaslavsky map. J. Eng. Appl. Sci. **14**(17), 6392–6399 (2019)

Text Extraction from Electronic Health Records for Predicting the Patient Diabetics Level by Machine Learning

R. Vineeth, R. Rithish, D. V. S. N. Sai Varma, N. Monish Gowda, and B. V. Ajay Prakash

Abstract In the modern world with the increase in the technical terms used in the medical field, the problem in understanding these medical terms arises and the need for the electronic health records (EHR) for easy understanding by common people is needed. EHR gives the complete and updated details of a particular patient. EHR is cost efficient as it decreases the required paperwork and reduces the redundancy of testing. The electronic health records (EHRs) mainly focus on making the task easier for the people in interpreting the medical reports which seem complex in understanding. The EHR contains valuable data for identifying health outcomes. EHRs are real-time entities where patients provide medical records that are interpreted and stored digitally so that they are readily available whenever required. The aforementioned feature would be helpful for patients to get an idea about the type of diabetes that includes Type-1, Type-2, pre-diabetes and no_diabetes which will be predicted using various machine learning algorithms such as random forest and LightGBM which showed a prominent result and best-obtained accuracy.

Keywords Electronic health record (EHR) · Artificial intelligence · Lightgbm · Random forest · Image processing · Pytesseract · Django

1 Introduction

Artificial intelligence is based on the principle that human intelligence can be simulated and programmed such that the system demonstrates characteristics of humans such as proficiency of reasoning, problem solving, planning and learning. Artificial intelligence plays a prominent role in medical fields such as healthcare chatbots, treatment design, managing health records and predicting health outcomes. The concepts of artificial intelligence inline with image processing help in collecting the relevant data from medical records and to build a machine learning model that can

R. Vineeth (✉) · R. Rithish · D. V. S. N. Sai Varma · N. Monish Gowda · B. V. Ajay Prakash
Department of Computer Science and Engineering, S J B Institute of Technology, Bengaluru, Karnataka, India

V. Bhateja et al. (eds.), *Data Engineering and Intelligent Computing*, Advances in Intelligent Systems and Computing 1407,
https://doi.org/10.1007/978-981-16-0171-2_27

predict the laboratory test results. Machine learning incorporates algorithms that evolve automatically through experience and comes out as a fragment of artificial intelligence which accomplishes the task without explicitly programmed to do so. Diabetes mellitus, habitually known as diabetes, is a metabolic disease that insinuates high blood sugar. Diabetes falls under four categories which include Type-1, Type-2, pre_diabetes and no_diabetes. Patients find it tedious to identify under which category they belong to. The paper is organized as follows: Sect. 2 describes the various authors who have performed similar research in diagnosing a disease and exhibiting an EHR platform. The structure of the EHR model being implemented is described in Sect. 3. The implementation and prominent results obtained for the proposed EHR model is shown in Sect. 4.

2 Literature Survey

In Jenna Wong et al. [1] reviews, four habitual schema that analysts may find useful for reasoning carefully about the tasks machine learning may be used to recognize medical conditions from health reports. The authors have presented instances of medical conditions in each schema to produce motivating results. In Riccardo Miotto et al. [2], the EHR uses an application of deep learning to procure resilient patient rendering. The article exemplifies the uses of EHR for prognosis of diseases encountered by the patients using deep learning techniques. The work in this paper is limited to liver cancer, diabetes and heart failure as it provided favorable outcomes for these medial domains. In this study of Ellinor hakansson [3], multi-layer perceptron prototypes are used for prognosis categorization of EHRs. The categorization accuracy of 48.1% is considered as a threshold for comparison of different architectures. Random forest holds 13.7% lower than that of the threshold value. Simon Suster [4] presents a new datafile with around 10,000 gap filling uncertainties for medical case reports. The analysis of the substantial gap in the performance between the best machine and the best human reader with the numerous thresholds and the condition of the artificial neural readers is observed to be 20% F1 score. In Mathai et al. [5], proposes to do a substantial assessment of previous literature on the challenges and suppositions concerning users and medical practitioners in the growth of EHR. Recognizing the EHR problems, concerns and difficulties allied to doctors and patients is especially important in the growth of EHR. In Stavros Pitoglou et al. [6], shows the use of machine learning techniques on information to portray the prognosis challenges in the Medical sector. The originality of this method contained in the health data is exemplified from the prognosis datafile. Furthermore, there exists a bounded necessity for segregation of the user group based on specific medical conditions.

3 Proposed Methodology

From the literature survey, it is seen there is a need for an enhanced framework depicting the EHR and an added feature which predicts the prognosis from the medical report uploaded. The proposed methodology is as shown in Fig. 1.

3.1 Users and Their EHR

The end users are the patients or the users who wish to store their medical record electronically. The users can also predict the outcomes from the EHR, i.e., the disease risk which is being the prime feature of this work. The users could also store all of their health records in one place so that they are very helpful to maintain the complete medical history oneselves with their own user credentials. The users can also be updated with the latest news and blogs in regard to the medical field as shown in Fig. 2. Users could also interpret them by updating and removing them in accordance with their interest.

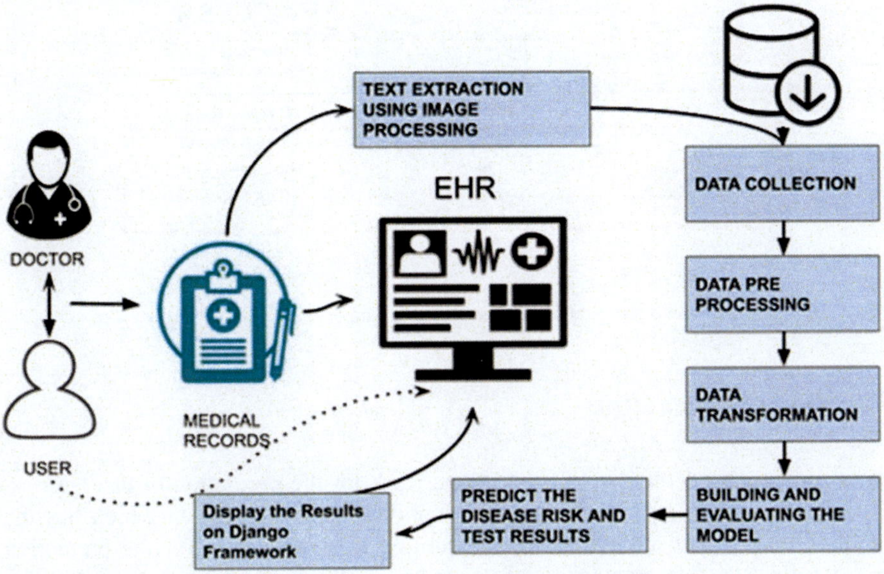

Fig. 1 Proposed methodology for electronic health record for prediction of prognosis

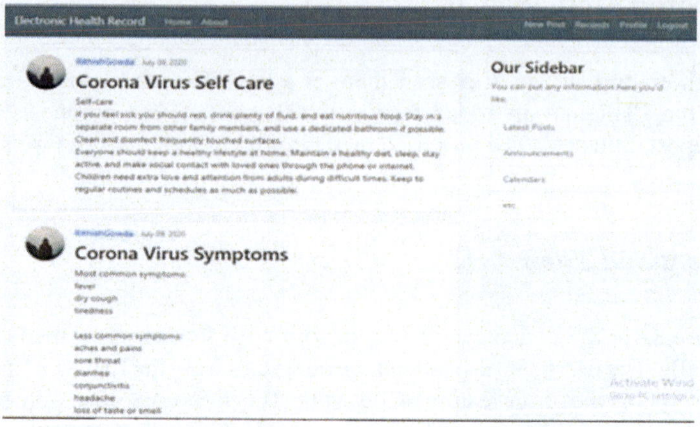

Fig. 2 Webpage showing the blogs concerning medical field in EHR Framework

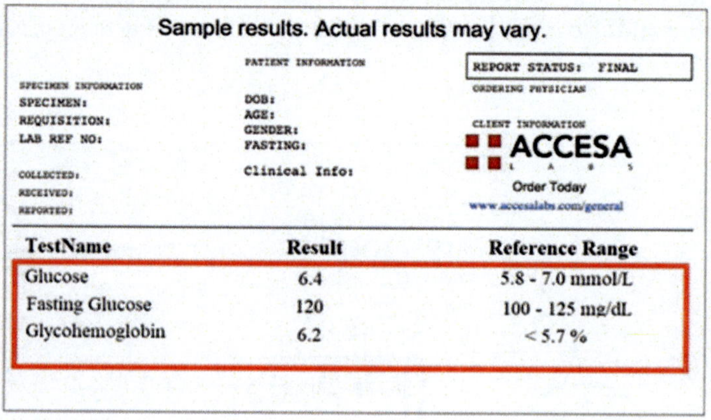
Sample results. Actual results may vary.

PATIENT INFORMATION

REPORT STATUS: FINAL

ORDERING PHYSICIAN

SPECIMEN INFORMATION
SPECIMEN:
REQUISITION:
LAB REF NO:

DOB:
AGE:
GENDER:
FASTING:

CLIENT INFORMATION

ACCESA LABS

COLLECTED:
RECEIVED:
REPORTED:

Clinical Info:

Order Today
www.accesalabs.com/general

TestName	Result	Reference Range
Glucose	6.4	5.8 - 7.0 mmol/L
Fasting Glucose	120	100 - 125 mg/dL
Glycohemoglobin	6.2	< 5.7 %

Fig. 3 Text extraction using region of interest and pytesseract

3.2 *Image Processing*

Image processing is the pivotal step in this work for the prediction of disease risk. Only the important part of the medical report as shown in Fig. 3 which has the result column of the laboratory test is required to be extracted for the prediction purpose. Region of Interest (ROI) of image processing method is used to fulfill this requirement. ROI is the specific region of the image that is required to be processed which is done by binary masking whereas for the interested region the pixels are set to 1 and the rest of the pixels set to 0.Please check whether the sentence "This way the values ..." conveys the intended meaning. This way the values extracted from the ROI are used as input as shown in Table 1 for the classification machine learning models built which is for predicting the disease risk.

Table 1 Values extracted from image processing stored in a .csv file

Glucose	Fasting glucose	Glycohemoglobin
6.4	120	6.2
6.1	110	5.9
6.0	108	5.8
6.6	145	6.4
5.8	100	5.6

3.3 Data Collection

Since the work is dealt with medical data, it is difficult to fetch a perfect dataset as they are confidential. The dataset for this work was obtained from Kaggle-National Health and Nutrition Examination Survey (NHANES) which comprised the health and nutrition status of children and adults in the USA. The dataset consisted of 424 columns and around 9800 rows of data, out of which 3 columns are chosen considering the diabetes data to predict the risk of diabetes which was then preprocessed to obtain accurate results.

3.4 Data Preprocessing

Three key attributes on which the diabetes disease test rely on are chosen for this work. The details of the patients from the NHANES dataset for these attributes are further preprocessed and filtered down to 6492 rows of patients which is done by removing unwanted and empty data. As the diabetes test results were not disclosed for these patients, the output column is computed using K-means clustering, classifying the data output to four types, namely no_diabetes, pre-diabetes, Type1 and Type2 which makes up the primary diabetes outcomes. The preprocessed dataset is as shown in Table 2 depicting first five entries of the preprocessed diabetes dataset, respectively. The output from the clustering is precise to further work with the machine learning models. The column headers being used in the dataset are LBDS GLSI-Glucose, LBX GLU-Fasting Glucose, LBXGH-Glycohemoglobin.

3.5 Data Transformation

The trained dataset sample is split into 80:20 of training and testing based on which model is being tested. The data being used is normalized in such a way that the values are in the same range and are computable for further evaluation. As the dataset is

Table 2 First Five entries of preprocessed diabetes dataset

Sl. No.	LBDSSGLSI	LBXGLU	LBXGH	Output
0	12.16	219	9.1	Type1
1	10.16	183	8.9	Type2
2	5.77	104	4.9	No_Diabetes
3	5.77	104	5.5	No_Diabetes
4	5.94	107	5.5	No_Diabetes

being preprocessed to a maximum extent, the dataset is seemed to be exceptionally reliable, which is also highly compatible for making most accurate predictions made by the built machine learning models.

3.6 Building the Machine Learning Models

In this work, random forest and LightGBM are proposed and built. **Random Forest** Algorithm also known as Random decision forest is a supervised classification model used in constructing decision trees for training and to get the mean prediction of all involved decision trees. Unlike decision trees which use information of gain index for calculating root nodes, here in the random forest process of finding root nodes and splitting featured nodes will happen randomly. In this work, we used random forest to predict the diabetic test report and the results were found out to be corresponding to the actual output with the accuracy of cent percent.

LightGBM is a gradient boosting framework which is a type of tree-based learning algorithms. Unlike other algorithms which follow level-wise tree growth, LightGBM grows the tree leaf-wise. LightGBM is designed to be well organized with many aspects and advantages such as more capability to handle large data with minimal memory usage, faster training speed with good productivity, and hold up parallel and Graphical Processing Unit (GPU) learning.The histogram-based splitting, gradient-based One sided sampling (GOSS) and exclusive feature bundling (EFB) attributes makes LightGBM accelerated than other algorithms. Operations such as binary and multi-class classification, regression, lambdrank and cross-entropy are supported by LightGBM. In this work, we employ multi-class classification where the output prediction is categorized to four disease risks of diabetes.

3.7 Evaluating the Models

The built machine learning models random forest and LightGBM developed in this work can be evaluated with most of the metrics wherein primarily in this work the algorithms are evaluated using accuracy, precision, recall, and F1 score. As accuracy is the chief metric that has to be pleased, the random forest model computed cent percent accuracy with LightGBM being 99.92% accuracy.

3.8 Prediction

Prediction is the key feature of this work. As said, the disease is to be predicted after processing the medical report which must be most accurate as we are dealing with medical work. The prediction made by the built models is of cent percent accurate and is evaluated by the metrics mentioned earlier.

3.9 Django Framework

Django is a Python-based non-proprietary web framework for development of composite, database-driven websites. Django is a lightweight and independent web server for building and testing. The electronic health record implemented in this paper uses Django framework as an web interface to enable users to perform certain essential operations such as authentication, store the medical records, review the records and prognosis of the laboratory test results of a diabetes report.

4 Implementation and Results

The work carried out in this research uses Python programming. Initially, we gathered the data from NHANES medical data [9]. The data cleaning was done by filling the missing values with the average taken from other data values. The data values thus obtained were normalized to keep up with the same scale. The normalized values were used to predict the output based on the K-means clustering. Based on this clustering method, the predicted output values were divided into four clusters—pre-diabetes, Type-1, Type-2, no_diabetes. The user uploads a medical record into the web interface, and image processing is done using region of interest concept on this medical record along with pytesseract module. The values obtained from image processing are used to build the machine learning model such as LightGBM, random forest. The predicted values obtained from these machine learning models correspond to the output in the dataset used. A confusion matrix has been used to outline the relation between actual and predicted prognosis of diabetes as shown in Fig. 4. Thus, EHR provided prominent results for varying hyperparameters for machine learning models such as LGBM and random forest as shown in Tables 3 and 4.

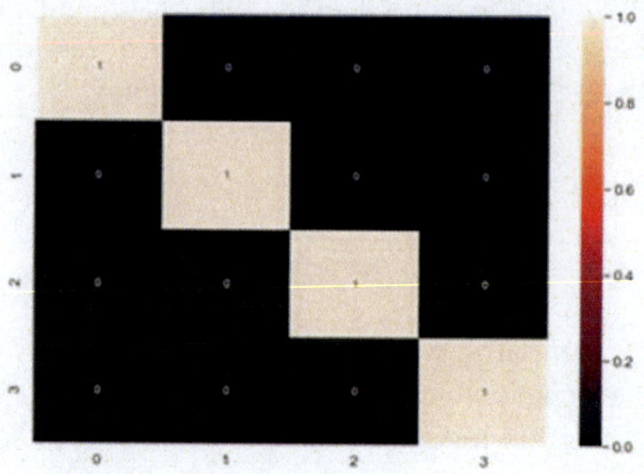

Fig. 4 Confusion matrix depicting the actual and predicted prognosis of diabetes

Table 3 Accuracy of random forest algorithm

N_estimators	max_leaf_node	max_depth	min_samples_split	min_samples_leaf	Accuracy (%)
10	2	2	5	2	95.84
100	2	3	5	2	98.92
10	3	3	3	1	99.23
100	2	4	3	1	99.76
10	3	4	2	1	99.92
10	None	None	2	1	100

Table 4 Accuracy of LightGBM algorithm

Learning_rate	manum_leaves	max_depth	min_data	bagging_fraction	Accuracy (%)
0.01	200	50	5	0.7	99.46
0.01	100	50	5	0.8	99.54
0.05	200	50	10	0.9	99.62
0.09	100	50	10	1.0	99.69
0.05	100	20	10	0.9	99.85
0.05	200	20	10	1.0	99.92

5 Conclusion

In the modern world, different medical conditions of patients are recognized using innovative EHR systems that consist of boundless amounts and categories of clinical data. In this work, the efficacy in diagnosis of laboratory test results to recognize the different types of diabetes has shown prominent results. The EHR system being developed can also be used for various applications such as blog, communicating with health professionals effectively. The users can store their medical records electronically which forbids them from visiting the doctor redundantly and to keep track of their health progress. Further enhancements can be done to EHR systems by adopting methods or machine learning models in predicting various diseases that the people come across in their medical reports. For instance, EHR can be used in the ongoing pandemic situation for accurate prognosis of whether the person is suffering from COVID-19 or not.

References

1. J. Wong, M.M. Horwitz, L. Zhou, S. Toh, *Using Machine Learning to Identify Health Outcomes from Electronic Health Record Data* (Springer Nature Switzerland AG, 2018)
2. R. Miotto(B), L. Li, J.T. Dudley, Deep learning to predict patient future diseases from the electronic health records, in *Springer International Publishing Switzerland 2016* ed. by N. Ferro, et al. *ECIR 2016*, LNCS 9626, pp. 768–774 (2016)
3. E. Hakansson, *A Deep Learning Approach to Predicting Diagnosis Code from Electronic Health Records* (STOCKHOLM SWEDEN, 2018)
4. S. Suster, W. Daelemans, CliCR: a dataset of clinical case reports for machine reading comprehension association for computational linguistics (2018)
5. N. Mathai, M.F. Shiratudin, S.F. Johel, Electronic health record management: expectations, issues, and challenges health med inform. An Open Access J. **8**(3) (2017) ISSN: 2157-7420
6. S. Pitoglou, Y. Koumpouros, A. Anastasiou, Using Electronic Health Records and Machine Learning to Make Medical-Related Predictions from Non-Medical Data International Conference on Machine Learning and Data Engineering (iCMLDE) (2018)
7. S. Zhou, *Extracting phenotypes of cancer patients from Electronic Health Records*. (IEEE, 2019)
8. S. Gehrmann, F. Dernoncourt, Y. Li, E.T. Carlson, J.T. Wu, J. Welt, et al., Comparing deep learning and concept extraction based methods for patient phenotyping from clinical narratives. PLoS One **13**(2) (2018)
9. National Health and Nutrition Examination Dataset. https://www.kaggle.com/cdc/national-health-and-nutrition-examination-survey

Hybrid Machine Learning Model Using Decision Tree and Support Vector Machine for Diabetes Identification

K. Dhasaradhan, R. Jaichandran, K. L. Shunmuganathan, S. Usha Kiruthika, and S. Rajaprakash

Abstract This paper presents a hybrid machine leaning model for diabetes identification. The hybrid machine learning model is compared with other machine learning models such as linear regression, logistic regression, support vector machine, naive Bayes, *K*-nearest neighbour, decision tree, random forest, and convolution neural network. It is trained and tested using prima Indian diabetes dataset using true positives, false negatives, true negatives and false positives as metrics. Performance of the hybrid machine learning model is evaluation using accuracy, precision, recall, and *F*-score. The hybrid machine learning model reported better results in diabetes identification compared to other machine learning models.

Keywords Diabetes · Machine learning · Linear regression · Logistic regression · Support vector machine · Naive bayes · K-nearest neighbour · Decision tree · Random forest · And convolution neural network

1 Introduction

Today, most of the hospitals are digitalized and patient's data are stored in database as electronic health records. Patient's electronic health records are asserts and can be used to train and test machine learning models in disease identification. Trained machine learning models can effectively identify and classify diseases [1].

K. Dhasaradhan · R. Jaichandran (✉) · K. L. Shunmuganathan · S. Rajaprakash
Department of Computer Science and Engineering, Aarupadai Veedu Institute of Technology, Vinayaka Mission's Research Foundation (Deemed to be University), Chengalpattu District, Chennai, Tamil Nadu 603104, India

K. L. Shunmuganathan
e-mail: klsnathan@avit.ac.in

S. Rajaprakash
e-mail: rajaprakash@avit.ac.in

S. Usha Kiruthika
SRM Institute of Science and Engineering, Kattankulathur, Chennai, Tamil Nadu, India

V. Bhateja et al. (eds.), *Data Engineering and Intelligent Computing*, Advances in Intelligent Systems and Computing 1407,
https://doi.org/10.1007/978-981-16-0171-2_28

Pima Indian diabetes dataset is used to train and test various machine learning models such as linear regression (LNR), logistic regression (LGR), support vector machine (SVM), naive bayes (NB),), *K*-nearest neighbour (KNN), decision tree (DT), random forest (RF), and convolution neural network (CNN). Above machine learning models are trained using Pima Indian diabetes dataset with 590 rows of data which includes 386 non-diabetic patient's data and 204 diabetic patient's data. The patient's data include seven fields such as blood glucose levels, insulin level, blood pressure, diabetes pedigree function, body mass index, age, and actual result.

Trained machine learning models are tested using 178 rows of data which includes 114 non-diabetic patient data and 64 diabetic patient's data. The patient's data include six fields such as blood glucose levels, body mass index, diabetes pedigree function, blood pressure, insulin level, and age. Machine learning models predicted results for the test data. Performances of machine learning models are evaluated by comparing actual results with predicted results. True positives, false negatives, false positives, true negatives, accuracy, recall, precision, and *F*-score are used as metrics in evaluating the performance of machine learning models. Hybrid machine learning using DT and SVM reported better results in diabetes identification compared to other machine learning models.

2 Literature Review

This section discusses various machine learning models used in diabetes identification.

Samanta et al. [2] proposed CNN model for early detection of diabetic retinopathy. Authors trained the machine learning model with 3050 images. The model identifies disease at early stage with accuracy of 88% by extracting the features such as hard exudates, blood vessels, and texture.

Chittaragi et al. [3] proposed single and ensemble classifiers for dialect identification. The classification algorithms SVM and DT are used. Spectral and prosodic features are cepstral coefficients, spectral flux, entrophy, pitch, and energy in dialect identification. Results show that ensemble method performed better over single independent SVM.

Mercaldoa et al. [4] developed machine learning models for diabetes identification using SVM, RF, NB, DT, and CNN. The models are evaluated using real-world diabetes dataset of prima Indians. Precision, recall, *F*-score, and RocAreas are used as metrics in evaluating the performance of the machine learning models. Authors found Hoeffding tree and SVM reported high precision and recall ratio compared to other machine learning models.

Garcia-Carretero et al. [5] evaluated the performance of KNN and RF in diabetes identification. A total of 1647 diabetes patients' datasets are used for training and testing the machine learning models. Authors found KNN accurately classified diabetes patient compared to RF.

Joseph et al. [6] proposed SVM for identifying brain tumor by reconstructed images. Authors found SVM able to reconstruct brain encephalic tissues quickly.

Lai et al. [7] built a machine learning model for diabetes identification using LGR and gradient boosting machine technique. Authors found the proposed model reported high sensitivity compared to RF and DT. The model is trained and tested using canadian diabetes patients' dataset.

Mujumdar et al. [8] evaluated the performance of different machine leaning methods in diabetes identification. LGR, DT, KNN, SVM, RF, and CNN are the different methods evaluated using prima diabetes dataset. Authors found LGR reported high accuracy.

Wei et al. [9] used deep neural network and SVM in diabetes identification. It is trained and tested the machine learning models with prima Indian diabetes dataset and evaluated their performance using accuracy as metric. Authors found deep neural network reported high accuracy with tenfold cross-validation.

Bhateja et al. [10] used SVM for breast cancer identification in mammograms using Haralick features. Authors used nonlinear sigmoidal transformation technique in mammograms. It shows better performance in identification of breast cancer compared to other methods.

Xu et al. [11] evaluated the performance of RF and NB in diabetes identification using datasets of University of Virginia. Authors found RF reduced the problem of overfitting and shown better prediction accuracy compared to NB.

Zheng et al. [12] proposed a machine learning framework for diabetes identification using KNN, NB, DT, RF, SVM, and LGR. Proposed framework is trained and tested using 300 patients' data of Chinese electronic health records. Accuracy, sensitivity, specificity, precision are the metrics used for evaluation. Compared to other state-of-the-art machine learning algorithms, the proposed machine learning frame is found effective in diabetes identification.

Barik et al. [2], Sarwar et al. [13], and Shahriare Satu et al. [14] proposed hybrid machine learning algorithms for disease identification.

3 Methodology

This section describes the methodology of machine learning models in diabetes identification.

3.1 Linear Regression

LNR method can be used for finding relationship between variables expressed as follows.

$$Y = b_0 + b_1 * x \tag{1}$$

Let x, Y are variables and b_0 and b_1 are coefficients, where Y is target variable and x is predictor variable. If b_1 is greater than zero; then, x has positive relationship.

3.2 Logistic Regression

LGR predicts binary outcome and probability of event occurrence. Equation for logistic regression is given as follows.

$$g(E(y)) = \alpha + \beta x1 + \Omega x2 \tag{2}$$

The link function g() links target variable and predictor variables.

3.3 Naive Bayes

NB is supervised learning algorithm predicts the probability of occurrence of a event based on the event prior knowledge. Bayes theorem is summarized as follows.

$$P(A|B) = \frac{P(A|B)P(A)}{P(B)} \tag{3}$$

Let A is a event, and its probability is $P(A)$. Similarly, B is a event, and its probability is $P(B)$. $P(A|B)$ is likelihood of A occur given B is true. $P(B|A)$ is likelihood of B occur given A is true.

3.4 Random Forest

RF constructs multiple decision trees using ensemble methods and predicts an individual tree as output [4, 8].

3.5 Support Vector Machine

SVM separates data points of diabetic class and non-diabetic class using a hyperplane [8, 15].

3.6 Decision Tree

DT classifies diabetic and non-diabetic dataset using rule set. It starts at root node and traverses down the tree based on the rule set [8, 15].

3.7 K-Nearest Neighbour

In KNN, diabetic and non-diabetic data points are converted into feature vectors, and the distance between feature vectors and test data is computed to classify diabetic and non-diabetic data points [4, 8].

3.8 Convolution Neural Network

CNN is a deep neural network consists of multiple hidden layers. It performs edge detection using filters and reduces dimensionality using pooling. Diabetic and non-diabetic data points are classified using softmax function [2].

4 Experiments and Results

LNR, LGR, NB, RF, SVM, DT, KNN, and CNN algorithms are developed using scikit-learn machine learning library in Jupyter notebook. The machine learning models are trained using Pima Indian diabetes dataset with 590 rows of data which includes 386 non-diabetic patients' data and 204 diabetic patient's data. The patient's data include seven fields such as blood glucose levels, diabetes pedigree function, blood pressure, insulin level, body mass index, age, and actual result. Trained machine learning models are tested using 178 rows of data which includes 114 non-diabetic patient data and 64 diabetic patient's data. The patient's data include six fields such as blood glucose levels, diabetes pedigree function, blood pressure, insulin level, body mass index, and age. Performances of machine learning models are evaluated by comparing actual results with predicted results. True positives, false negatives, true negatives, false positives, accuracy, recall, precision, and F-score are used as metrics in evaluating the performance of machine learning models.

Let N is the number of diabetes patient's data used to test machine learning models. Let TP is true positives, FN is false negatives, TN is true negatives, FP is false positives, and now recall, precision and F-score can be computed as follows.

$$\text{Accuracy} = \frac{\text{TP} + \text{TN}}{\text{TP} + \text{TN} + \text{FP} + \text{FN}} \tag{4}$$

$$\text{Recall} = \frac{\text{TP}}{\text{TP} + \text{FN}} \tag{5}$$

$$\text{Precision} = \frac{\text{TP}}{\text{TP} + \text{FP}} \tag{6}$$

$$\text{F score} = 2 * \frac{\text{Precision} * \text{Recall}}{\text{Precision} + \text{Recall}} \tag{7}$$

If a diabetic patient correctly identified as diabetic by a machine learning model, then it is TP else FN. If a non-diabetic patient correctly identified as non-diabetic by a machine learning model, then it is TN else FP. When the actual diabetes result is yes, how often the machine learning model predicts the diabetes result as yes is known as recall. When the machine learning model predicts the diabetes result as yes, how often is it correct is known as precision. Weighted harmonic mean of recall and precision is *F*-score.

Machine learning models are tested with $N = 178$ rows of data in which 64 are diabetes patient data and 114 are non-diabetes patient data. Tables 1 and 2 show results of the machine learning models. Figures 1, 2, 3 and 4 are based on Table 1.

Table 1 Results of machine learning models

	TP	TN	FP	FN
LNR	37	104	10	27
LGR	34	105	9	30
NB	39	96	18	25
RF	31	101	13	33
SVM	0	114	0	64
DT	40	89	25	24
KNN	37	90	24	27
CNN	32	84	30	32

Table 2 Results of machine learning models

	Accuracy (%)	Recall (%)	Precision (%)	*F*-score (%)
LNR	79.21	57.81	78.72	66.66
LGR	78.08	53.12	79.06	63.54
NB	75.84	60.93	68.42	64.46
RF	74.15	48.43	70.45	57.40
SVM	64.04	0	0	0
DT	72.47	62.5	61.53	62.01
KNN	71.34	57.81	60.65	59.20
CNN	65.16	50	48.38	49.18

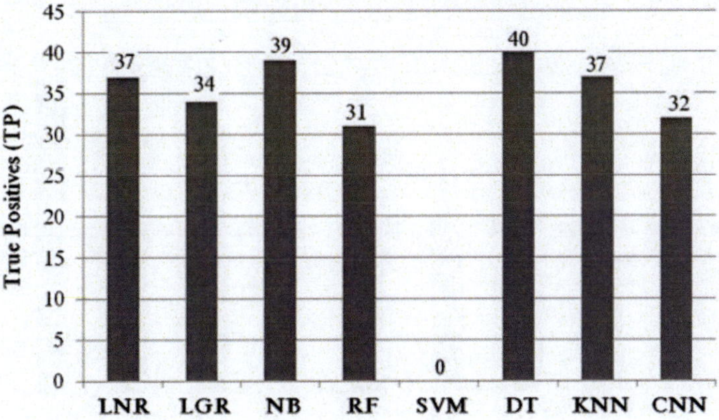

Fig. 1 True positives reported by machine learning models

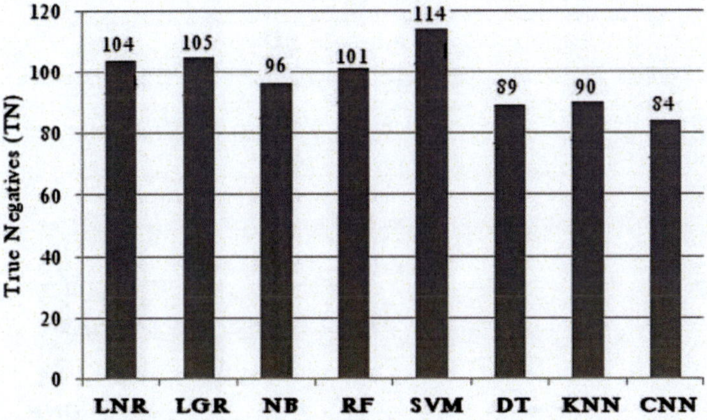

Fig. 2 True negatives reported by machine learning models

Figure 1 shows TP reported by machine learning models in diabetes identification. TP reported by machine learning models are 37 by LNR, 34 by LGR, 39 by NB, 31 by RF, 0 by SVM, 40 by DT, 37 by KNN, and 32 by CNN. DT reported highest TP and SVM reported least TP. Figure 2 shows TN reported by machine learning models in diabetes identification. TN reported by machine learning models are 104 by LNR, 105 by LGR, 96 by NB, 101 by RF, 114 by SVM, 89 by DT, 90 by KNN, and 84 by CNN. SVM reported highest TN and CNN reported least TN. Figure 3 shows FP reported by machine learning models in diabetes identification. FP reported by machine learning models are 10 by LNR, 9 by LGR, 18 by NB, 13 by RF, 0 by SVM, 25 by DT, 24 by KNN, and 30 by CNN. CNN reported highest FP and SVM reported least FP. Figure 4 shows FN reported by machine learning models in diabetes identification. FN reported by machine learning models are 27 by LNR, 30 by LGR,

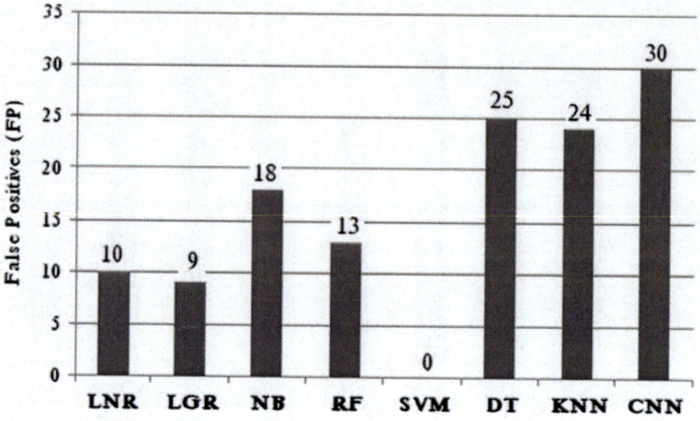

Fig. 3 False positives reported by machine learning models

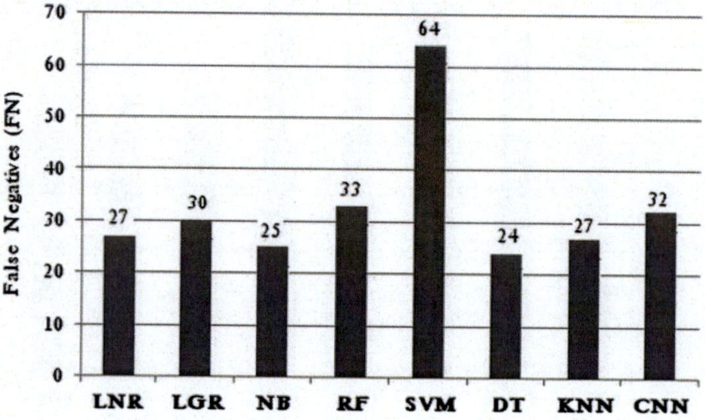

Fig. 4 False negatives reported by machine learning models

25 by NB, 33 by RF, 64 by SVM, 24 by DT, 27 by KNN, and 32 by CNN. SVM reported highest FN and DT reported least FP.

Figures 5, 6, 7 and 8 are based on Table 2. Figure 5 shows accuracy of machine learning models in diabetes identification. Percentages of accuracy reported by machine learning models are 79.21 by LNR, 78.08 by LGR, 75.84 by NB, 74.15 by RF, 64.04 by SVM, 72.47 by DT, 71.34 by KNN, and 65.16 by CNN. LNR reported highest accuracy and SVM reported least accuracy. Figure 6 shows recall of machine learning models in diabetes identification. Percentages of recall reported by machine learning models are 57.81 by LNR, 53.12 by LGR, 60.93 by NB, 48.43 by RF, 0 by SVM, 62.5 by DT, 57.81 by KNN, and 50 by CNN. DT reported highest recall and SVM reported least recall. Figure 7 shows precision of machine learning models in diabetes identification. Percentages of precision reported by machine learning models are 78.72 by LNR, 79.06 by LGR, 68.42 by NB, 70.45 by RF, 0 by SVM,

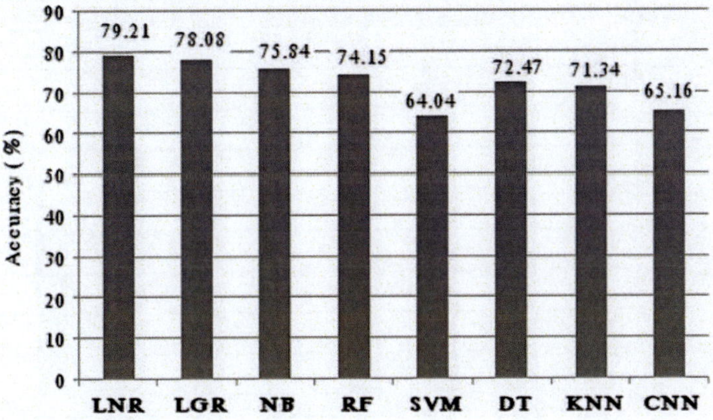

Fig. 5 Accuracy reported by machine learning models

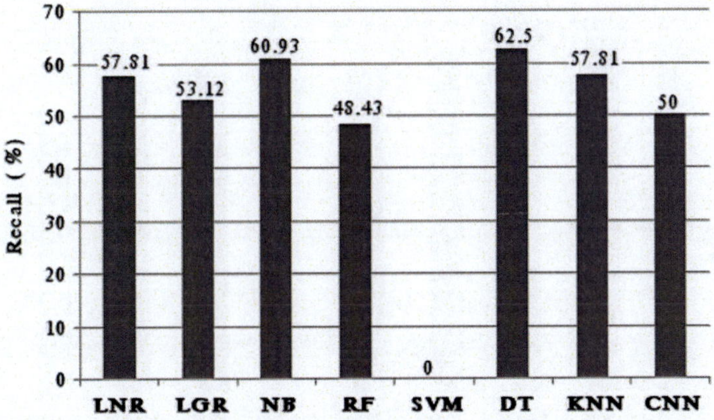

Fig. 6 Recall reported by machine learning models

61.53 by DT, 60.65 by KNN, and 43.38 by CNN. LGR reported highest precision and SVM reported least precision. Figure 8 shows F-score of machine learning models in diabetes identification. Percentages of F score reported by machine learning models are 66.66 by LNR, 63.54 by LGR, 64.45 by NB, 57.4 by RF, 0 by SVM, 62.01 by DT, 59.19 by KNN, and 49.17 by CNN. LGR reported highest F score, and SVM reported least F-score.

Results show no single machine learning model is better in diabetes identification. Hence, hybrid machine learning model using two or more effective machine learning methods can be used for better results. Effective machine learning methods can be identified using TP, TN, FP, and FN as metrics. Machine learning methods that reports

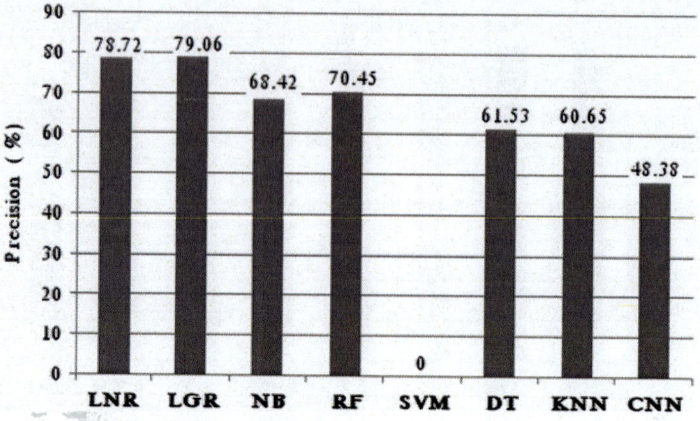

Fig. 7 Precision reported by machine learning models

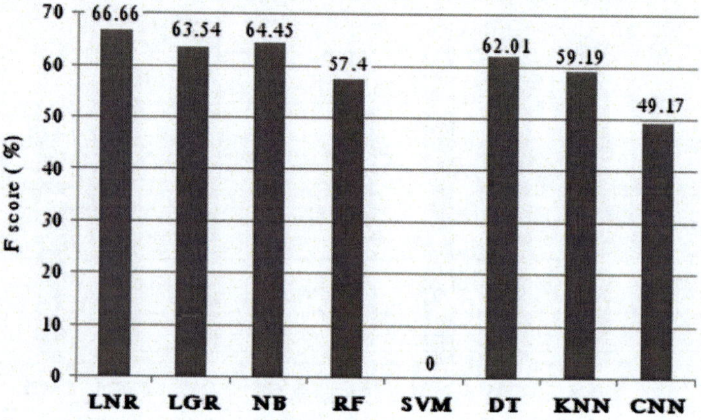

Fig. 8 *F*-score reported by machine learning models

high TP, TN and least FP and FN are considered as effective. Results in Figs. 1 and 4 shows DT reported high TP and least FN. Results in Figs. 2 and 3 show SVM reported high TN and least FP. Therefore, hybrid machine learning model using DT and SVM is proposed for diabetes identification, and results are shown in Tables 3 and 4. Hybrid machine learning model using DT and SVM reported 40 TP, 114 TN, 0 FP, 64 FN, 86.51% accuracy, 62.5% recall, 100% precision and 76.92% *F*-score. Figures 9, 10, 11 and 12 are based on Tables 2 and 4. Results in Figs. 9, 10, 11, and

Table 3 Results of hybrid machine learning model using DT and SVM

	TP	TN	FP	FN
Hybrid model	40	114	0	64

Table 4 Results of hybrid machine learning model using DT and SVM

	Accuracy (%)	Recall (%)	Precesion (%)	*F*-score (%)
Hybrid model	86.51	62.5	100	76.92

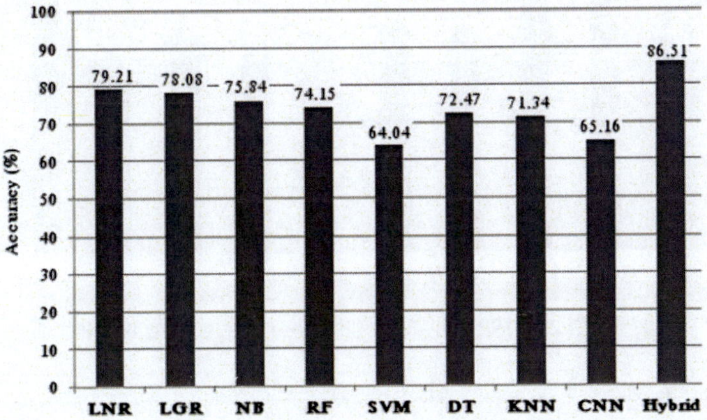

Fig. 9 Accuracy reported by hybrid model and other machine learning models

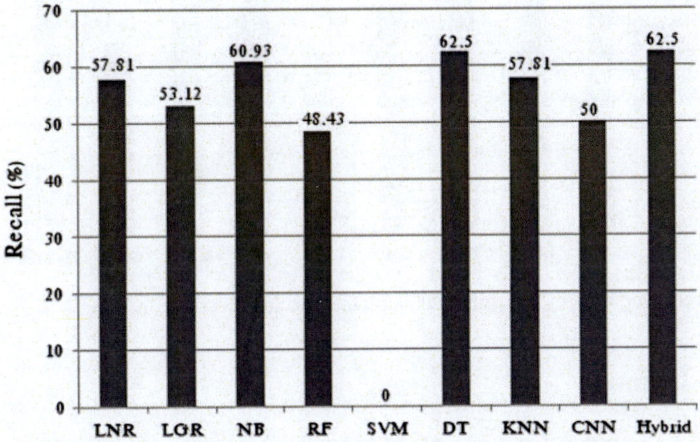

Fig. 10 Recall reported by hybrid model and other machine learning models

12 show hybrid machine learning model using DT and SVM reported high accuracy, recall, precision, and *F*-score in diabetes identification compared to other machine learning models.

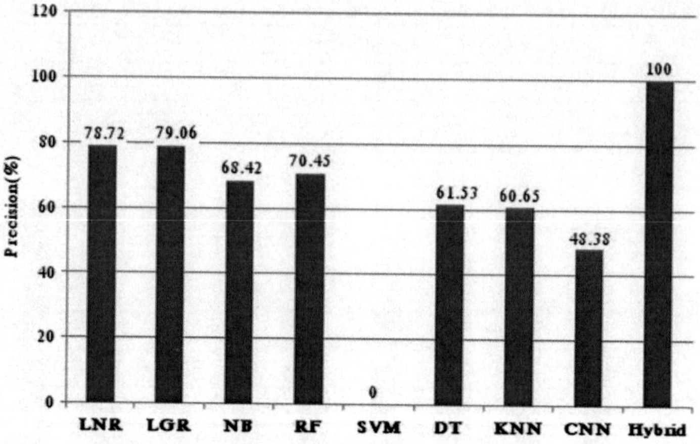

Fig. 11 Precision reported by hybrid model and other machine learning models

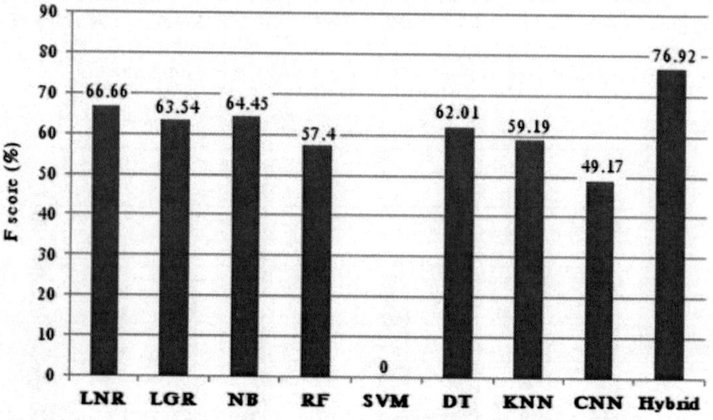

Fig. 12 *F*-score reported by hybrid model and other machine learning models

5 Conclusion

Machine learning models are developed for diabetes identification using various methods such as LNR, LGR, NB, RF, SVM, DT, KNN, and CNN. The machine learning models are trained and tested for diabetes identification using Pima Indians diabetes dataset. Performance of the machine learning models in diabetes identification is evaluated using different types of metrics such as TP, TN, FP, FN, accuracy, precision, recall, and *F*-score. Results show DT reported high TP, least FN and high recall. SVM reported high TN and least FP. LNR reported high accuracy, and LGR reported high precision. No single machine learning model is effective in identifying diabetes. Machine learning models that report high TP, TN and least FP and FN are considered as effective. Hence, hybrid machine learning model using DT and SVM is proposed for diabetes identification. Results show the proposed hybrid machine learning model reported high TP, TN, accuracy, recall and precision and least FP, FN compared to other machine learning models.

References

1. P. M. S. Sai, G. Anuradha, V.P. Kumar, Survey on Type 2 diabetes prediction using machine learning, in *2020 Fourth International Conference on Computing Methodologies and Communication (ICCMC)*, p. 771–775 (2020).
2. A. Samanta, A. Saha, S.C. Satapathy, S.L. Fernandes, Y.-D. Zhang, Automated detection of diabetic retinopathy using convolutional neural networks on a smalldataset. Pattern Recogn. Lett **135**, 293–298 (2020).
3. N.B. Chittaragi, A. Prakash, S.G. Koolagudi, Dialect identification using spectral and prosodic features on single and ensemble classifiers. Arab. J. Sci. Eng. **43**(8), 4289–4302 (2017)
4. F. Mercaldoa, V. Nardoneb, A. Santoneb, Diabetes Mellitus affected patients classification and diagnosis through machine learning techniques. Procedia Comput. Sci. **112**, 2519–2528 (2017).
5. R. Garcia-Carretero, L.Vigil-Medina, I. Mora-Jimenez, C. Soguero-Ruiz, O. Barquero-Perez, J. Ramos-Lopez, Use of a *K*-nearest neighbors model to predict the development of type 2 diabetes within 2 years in an obese, hypertensive population. Med. Biol. Eng. Comput. (2020).
6. S.S. Joseph, D. Aju, A comparative objective assessment on mesh-based and SVM-based 3D reconstruction of MRI brain. Int. J. Nat. Comput. Res. (IJNCR) **8**(3), 41–54 (2019)
7. H. Lai, H. Huang, K. Keshavjee, A. Guergachi, X. Gao, Predictive models for diabetes Mellitus using machine learning techniques. BMC Endocr. Disord. **19**(1) (2019)
8. A. Mujumdar, V. Vaidehi, Diabetes prediction using machine learning algorithm. Procedia Comput. Sci. **165**, 292–299 (2019)
9. S. Wei, X. Zhao, C. Miao, A comprehensive exploration to the machine learning techniques for diabetes identification, in *2018 IEEE 4th World Forum on Internet of Things (WF-IoT)*, Singapore, pp. 291–295 (2018).
10. V. Bhateja, B. Le Nguyen, N.G. Nguyen, S.C. Satapathy, D.-N. Le, Haralick features-based classification of mammograms using SVM. Inf. Syst. Des. Intell. Appl. Adv. Intell. Syst. Comput. **672**, 787–795 (2018).
11. W. Xu, J. Zhang, Q. Zhang, X. Wei, Risk prediction of type II diabetes based on random forest model, in *2017 Third International Conference on Advances in Electrical, Electronics, Information, Communication and Bio-informatics* (2017).
12. T. Zheng, W. Xie, L. Xu, X. He, Y. Zhang, M. You, Y. Chen, A machine learning-ased framework to identify type 2 diabetes through electronic health record. Int. J. Med. Inf., **97**, 120–127 (2017).
13. A. Sarwar, M. Ali, J. Manhas et al., Diagnosis of diabetes type-II using hybrid machine learning based ensemble model. Int. J. Inf. Technol. **12**, 419–428 (2020)
14. M. Shahriare Satu, S.T. Atik, M.A. Moni, A novel hybrid machine learning model to predict diabetes Mellitus in ed. by M. Uddin, J. Bansal, *Proceedings of International Joint Conference on Computational Intelligence. IJCCI 2019. Algorithms for Intelligent Systems* (Springer, Singapore, 2020).
15. S. Barik, S. Mohanty, S. Mohanty, D. Singh, Analysis of prediction accuracy of diabetes using classifier and hybrid machine learning techniques, in ed. by D. Mishra, R. Buyya, P. Mohapatra, S. Patnaik, *Intelligent and Cloud Computing. Smart Innovation, Systems and Technologies*, vol 153 (Springer, Singapore, 2021).

Feature Selection Using Equilibrium Optimizer

Gaurav Tikhe, Tanvi Joshi, Ashwin Lahorkar, Aamod Sane, and Jayaraman Valadi

Abstract To tackle the problem of the feature selection, the newly introduced algorithm of equilibrium optimization is featured in this paper. The algorithm is inspired from dynamic mass balance inside a control volume. Each agent (particle) updates concentration with best solutions obtained so far leading to equilibrium state, i.e., optimum result. To find effectiveness of the proposed algorithm, it was tested on 13 benchmark datasets from UCI machine learning repository from different domains.

Keywords Equilibrium optimizer · Feature selection · Optimization · Meta-heuristic

1 Introduction

Feature selection simplifies data by reducing its dimensionality and helps select relevant features without impacting the predictive accuracy of the model. Feature selection is essential to address problems like data mining, document classification, bio-metrics, object detection, etc. [1]. Feature selection captures informative attributes and removes noisy and redundant features. Feature selection algorithms include filter, wrapper and ensemble methods [2–21].

The two important features of meta-heuristics are (A) exploration and (B) exploitation. Exploration is the ability to globally search the sample space which is associated with escaping the local optima entrapment. Exploitation is the ability to refine promising solutions which are found by exploration. A good performing model can be developed by maintaining a good balance between these two features. In this work, we have described equilibrium optimizer a newly developed algorithm for selecting attributes.

G. Tikhe · T. Joshi · A. Lahorkar
Centre for Modelling and Simulation, Savitribai Phule Pune University, Pune, India

A. Sane · J. Valadi (✉)
Department of Computer Science, FLAME University, Pune, India

V. Bhateja et al. (eds.), *Data Engineering and Intelligent Computing*,
Advances in Intelligent Systems and Computing 1407,
https://doi.org/10.1007/978-981-16-0171-2_29

2 Equilibrium Optimizer

Equilibrium optimizer originates from the physical laws of nature and is based upon the unsteady state mass balance equation tracking the instantaneous concentration of a nonreactive component. The generic equation of mass balance is identified with a first-order differential equation [7] given by Eq. (1).

$$V\frac{dC}{dt} = QC_{eq} - QC + G \tag{1}$$

where V denotes the control volume, C is the concentration inside the control volume, $V\frac{dC}{dt}$ is the rate of change of control volume, Q is the flow rate, C_{eq} concentration inside the control volume at equilibrium state in which there is no generation and G is the mass generation rate. A steady equilibrium state can be achieved when $V\frac{dC}{dt}$ reaches to zero. Equation (1) can be rearranged to solve for C (concentration in control volume) as a function of t,

$$\frac{dC}{\lambda C_{eq} - \lambda C + \frac{G}{V}} = dt \tag{2}$$

where $\lambda = Q/V$, which is the turn-over rate.

$$\int_{C_0}^{C} \frac{dC}{\lambda C_{eq} - \lambda C + \frac{G}{V}} = \int_{t_0}^{t} dt \tag{3}$$

where C_0 and C are the concentration levels at t_0 and t. Equation (3) can be furthered solved to,

$$C = C_{eq} + \left(C_0 - C_{eq}\right) F + \frac{G}{\lambda V}(1 - F) \tag{4}$$

$$\vec{F} = e^{-\vec{\lambda}(t-t_0)} \tag{5}$$

In EO, the particles represent the solution and the concentration represents particle's position in the algorithm. There are three terms which contribute in updating the particle's concentration (Eq. 4). The first term gives the concentration of the particle at equilibrium state which is the best solutions so far randomly selected from a pool called the equilibrium pool, the second term gives the difference between a particle and its equilibrium state which induces the exploration aspect encouraging global search, and the third term includes the generation rate which approaches the solution in small steps that mostly cover the exploitation aspect. These terms are further explained below.

2.1 Initialization

EO initializes population at random at first to begin with the process of optimization. The initial level of concentration is constructed of the size of number of particles and number of features. The value of initial concentration (C) lies between [0, 1].

2.2 Equilibrium Pool (C_{eq})

The equilibrium state is the convergence state of the algorithm. At the initial stage, there is no knowledge of the equilibrium state. The search pattern is determined with the help of the particles. The pool consists of four best-so-far particle concentrations, and the fifth particle is the mean of previous concentrations. The first four particles help to achieve better exploration, and the average helps to achieve better exploitation. Equilibrium pool vector for 5 particles will be:

$$\vec{C}_{eq,pool} = \left\{ \vec{C}_{eq(1)}, \vec{C}_{eq(2)}, \vec{C}_{eq(3)}, \vec{C}_{eq(4)}, \vec{C}_{eq(ave)} \right\} \tag{6}$$

2.3 Exponential Term (F)

To maintain a balance between exploration and exploitation while updating the concentration of the control volume, the exponential term plays a very important role. The turn-over rate (λ) changes with time in real control volume; hence, λ is considered as a random value between [0, 1].

$$\vec{F} = e^{-\vec{\lambda}(t-t_0)} \tag{7}$$

The exponential term is a function of t and decreases with increase in number of iterations.

$$t = (1 - \frac{Iter}{Max\_iter})^{(a_2 \frac{Iter}{Max\_iter})} \tag{8}$$

where iter and Max_iter are current iteration and maximum iteration, a_2 is the constant value to manage exploitation ability. Higher a_2 promotes exploitation capabilities. In order to avoid convergence at local minima, the process is slowed down with the help of t_0.

$$\vec{t}_0 = \frac{1}{\vec{\lambda}} \ln(-a_1 sign\,(\vec{r} - 0.5)\,[1 - e^{-\vec{\lambda}t}]) + t \tag{9}$$

where a_1 controls the exploration capabilities, r is a random number between [0, 1], sign gives the direction of search [9].

Substituting Eqs. (9) in (7),

$$\vec{F} = a_1 sign\,(\vec{r} - 0.5)\left[e^{-\vec{\lambda}t} - 1\right] \tag{10}$$

2.4 Generation Rate (G)

Generation rate is a term that provides optimal solution using exploitation. Out of many methods of describing generation rate as a function of time, the first-order exponential decay method can be defined as,

$$\vec{G} = \vec{G}_0 e^{-\vec{\lambda}(t - t_0)} = \vec{G}_0 \vec{F} \tag{11}$$

where

$$\overrightarrow{G_0} = \overrightarrow{GCP}\left(\overrightarrow{C_{eq}} - \vec{\lambda}\vec{C}\right) \tag{12}$$

And

$$\overrightarrow{GCP} = \begin{cases} 0.5 r_1 & r_2 \geq GP \\ 0 & r_2 < GP \end{cases} \tag{13}$$

where $\overrightarrow{GCP}$ is defined as the generation rate control parameter, r_1 and r_2 are random numbers between [0, 1]. The probability which decides the contribution of generation rate is defined by generation probability (GP). This tells us whether the specific particle is updated using generation rate. The concentration updating rule for EO is as follows,

$$\vec{C} = \vec{C}_{eq} + \left(\vec{C} - \vec{C}_{eq}\right).\vec{F} + \frac{\vec{G}}{\vec{\lambda} V}\left(1 - \vec{F}\right) \tag{14}$$

3 Proposed Algorithm

The proposed algorithm is inspired from EO [9] which originally uses a continuous search space to be utilized in a binary search algorithm. Particles analogous to FS solution are set in the search space. Their concentration defines their position in the space. The particle population, initial concentration and the fixed parameters are initialized. The concentration is expressed as a vector of 0 or 1 where 0 implies feature is deselected and 1 implies feature is selected. FS is considered as a multi-objective optimization problem due to two different criteria to evaluate the feature subset in consideration: classification accuracy and number of selected features. Hence to encompass both of these objectives, fitness score is used as a measure which is evaluated as follows:

$$F(i) = \frac{Accuracy(i)}{1 + \lambda \cdot n(i)} \tag{15}$$

This score is calculated for the initial population, and the global equilibrium particles are chosen. The concentration is updated to move all the particles towards the global best using Eq. (14). The updating rule assimilates the current equilibrium state (first term), exploration (second term) and exploitation (third term). The GP in Eq. (13) exclusively takes care of exploitation whereas exploration is induced with the help of mutation. We have used Random Forrest as the classifier with random state in order to have reproducible results. Accuracy is determined by with the help of tenfold CV accuracy. The pseudocode for proposed algorithm is as follows:

```
Initialize number of particles (N) in population

Assign values to the parameters a1, a2, GP, λ

while (iteration < max_iteration):

        for each particle in population:
                Initialize concentration vector

                Find classification accuracy for every
                particle

                Evaluate the objective function
                (fitness) using equation (15)

                Based on fitness values, select the fit
                or best-so-far particles (Neq)

                Construct equilibrium concentration
                array using fit particles (Ceq)

                Construct equilibrium pool by using
                equation (6)

                Update the concentration vector with
                rounding using equation (14)

                iteration = iteration + 1
        end for
end while
```

4 Experimental Studies and Results

Effectiveness of the proposed methodology is found by using different datasets from different domains. The datasets were obtained from UCI machine learning repository [22]. The datasets are Breast Cancer, BreastEW, Colon, CongressEW, HeartEW, IonosphereEW, KrvskpEW, Leukemia, Ovarian, Tic-tac-toe, WaveformEW, WineEW, Zoo. These datasets consist of wide variety of dimensions. Table 1 shows the characteristics of the datasets.

Table 1 Characteristics of datasets

Datasets	Number of classes	Number of features	Number of examples
Colon	2	2000	62
Leukemia	2	7129	72
Ovarian	2	15,154	253
Tic-Tac-Toe	2	9	958
WaveformEW	3	40	5000
WineEW	3	13	178
Zoo	7	16	101
Breast Cancer	2	9	699
BreastEW	2	30	569
CongressEW	2	16	435
HeartEW	2	13	270
IonosphereEW	2	34	351
KrvskpEW	2	36	3196

4.1 Parameter Setting

Different parameters which are part of this work are N number of particles, a_1 parameter controlling the exploration ability, a_2 parameter controlling exploitation ability, generation probability (GP) parameter responsible for balance between exploitation and exploration, Neq number of best-so-far solutions. Different values for number of particles (N) used are 25, 35, 50. Values of parameters a_1 and a_2 are set to 2.5 and 1, respectively. Value of GP was set to 0.75. Number of best-so-far solution (N_{eq}) used are 5 and 7. Out of 5 and 7, 5 best-so-far solutions showed promising performance. The weighting parameter λ is tuned for every dataset. The results are averaged over 20 independent runs. Results obtained are compared with [21] and [8]. The methods used by [21] are hybrid grey wolf optimizer with K-nearest neighbors as classifier to obtain accuracy. Approach taken by [8] is binary black hole algorithm with Random Forest classifier (Table 2).

5 Conclusions

The problem of feature selection is addressed in this work using novel Equilibrium Optimization (EO) algorithm. The proposed algorithm is tested using different benchmark datasets from various domains obtained from UCI machine learning repository. The accuracies for majority of datasets were found be better than literature results. Results for datasets like BreastEW, HeartEW, IonosohereEW, Tic-Tac-Toe and WaveformEW are promising. Modifications and further investigations on parameter tuning are the area of future research.

Table 2 Experimental results

Sr. No	Dataset	EO results		Results from literature	
		Accuracy	No. of attributes	Accuracy	No. of attributes
1	Colon	0.9089 ± 0.0147	190.45 ± 54.643	0.9141 ± 0.013	3.4 ± 0.54
2	Leukemia	0.9824 ± 0.0073	466.45 ± 128.77	0.9861 ± 0.012	5.6 ± 2.7
3	Ovarian	0.997 ± 0.0031	1671.85 ± 426.34	0.998 ± 0.0034	2.8 ± 0.44
4	Breast Cancer	0.9614 ± 0.00001	5 ± 0	0.98	4.4
5	BreastEW	0.9782 ± 0.0031	10.45 ± 2.1145	0.97	13.6
6	CongressEW	0.976 ± 0.005	6.7 ± 1.867	0.98	4.4
7	HeartEW	0.8619 ± 0.003	8.65 ± 1.387	0.85	5.8
8	IonosphereEW	0.9572 ± 0.004	12.6 ± 2.349	0.95	13
9	KrvskpEW	0.9815 ± 0.0037	18.05 ± 4.559	0.98	15.8
10	Tic-Tac-Toe	0.8289 ± 0	7.00 ± 0	0.81	5.2
11	WaveformEW	0.8320 ± 0.0075	9.35 ± 1.2680	0.8	14.2
12	WineEW	0.9886 ± 0.001	7.55 ± 0.605	1.00	6
13	Zoo	0.9833 ± 0.0046	5.35 ± 0.022	1.00	6.8

Dataset no. 1, 2, 3 are from [8] and 4–13 are from [21]

References

1. M.A. Khan, S. Rubab, A. Kashif, M.I. Sharif, N. Muhammad, J.H. Shah, … S.C. Satapathy, Lungs cancer classification from CT images: an integrated design of contrast based classical features fusion and selection. Pattern Recogn. Lett. **129**, 77–85(2020)
2. D.N. Le, G.N. Nguyen, V. Bhateja, S.C. Satapathy, Optimizing feature selection in video-based recognition using Max-Min Ant system for the online video contextual advertisement user-oriented system. J. Computat. Sci. **21**, 361–370 (2017)
3. J.H. Holland, *Adaptation in Natural and Artificial Systems* (University of Michigan Press, Ann Arbor, 1975).
4. R. Eberhart, J. Kennedy, A new optimizer using particle swarm theory, in: *Proceedings of Sixth International Symposium on Micro Machine and Human Science, 1995 MHS 95*,1995, p. 39–43
5. E. Rashedi, H. Nezamabadi-pour, S. Saryazdi, GSA: A gravitational search algorithm. Inform. Sci. **179**, 2232–2248 (2009)
6. M. Dorigo, M. Birattari, T. Stutzle, Ant colony optimization. IEEE Comput. Intell. Mag. **1**, 28–39 (2006)
7. S. Kashef, H. Nezamabadi-pour, A new feature selection algorithm based on binary ant colony optimization, in *The 5th Conference on Information and Knowledge Technology, Shiraz* (2013), p. 50–54.
8. E. Pashaei, N. Aydin, Binary black hole algorithm for feature selection and classification on biological data. Appl. Soft Comput. **56**, 94–106 (2017)
9. A. Faramarzi, M. Heidarinejad et al., Equilibrium optimizer: a novel optimization algorithm. Knowl.-Based Syst. **191**, 105190 (2020)
10. R. Storn, K. Price, Differential evolution—a simple and efficient heuristic for global optimization over continuous spaces. J. Glob. Optim. **11**, 341–359 (1997)
11. J.R. Koza, *Genetic Programming* (1992).
12. D. Simon, Biogeography-based optimization. Evolut. Comput. IEEE Trans. **12**, 702–713 (2008)

13. S. Kirkpatrick D. Gelatt, Jr., M.P. Vecchi, Optimization by simulated annealing. Science **220**, 671–680 (1983).
14. O.K. Erol, O.K. EksiErol, I. Eksin, A new optimization method: big bang–big crunch. Adv. Eng. Softw. **37**, 106–111 (2006)
15. A. Hatamlou, Black hole: a new heuristic optimization approach for data clustering. Inf. Sci. (2012)
16. X.-S. Yang, Firefly algorithm, stochastic test functions and design optimisation. Int. J. Bio-Inspired Comput. **2**, 78–84 (2010)
17. X.-S. Yang, A new metaheuristic bat-inspired algorithm, in *Nature Inspired Cooperative Strategies for Optimization* (ed., Springer, 2010). p. 65–74
18. S. Mirjalili, S.M. Mirjalili, A. Lewis, Grey wolf optimizer. Adv. Eng. Softw. **69**, 46–61 (2014)
19. R. Purushothaman, S.P. Rajgopalan, G. Dhandapani, Hybridizing Gray Wolf optimization (GWO) with Grasshopper Optimization Algorithm (GOA) for text feature selection and clustering. Appl. Soft Comput. **96**, 106651 (2020)
20. H. Rao, X. Shi, A. Rodrigue, J. Feng, Y. Xia, M. Elhoseny, X. Yuan, L. Gu, Feature selection based on artificial bee colony and gradient boosting decision tree. Appl. Soft Comput. **74**, 634–642 (2019)
21. Al-Tashi, Kadir, Rais, et al., Binary optimization using hybrid grey wolf optimization for feature selection. IEEE Access **7**, 39496–39508 (2019).
22. UCI Machine Learning Repository, https://archive.ics.uci.edu/ml/datasets.php

A Review on Load Balancing Methods for Cloud Computing Based on Ant Colony Optimization, Honey-Bee and Genetic Algorithm

Trishna Bhowmik, Sanjoy Goswami, and Banani Das

Abstract The load balancing is an important aspect of systematized operations in distributed environments. As cloud computing is emerging expeditiously and clients are claiming more facilities and improved results, load balancing for the cloud environment has become a very fascinating and crucial research area. Load balancing in cloud computing is defined as the method of splitting workloads and other computing properties among different computers, networks or servers in a cloud computing environment. Researchers have proposed many approaches to provide cloud load balancing with the aim to intensify the overall performance of the cloud computing. In this paper, the different cloud load balancing methods with the three most efficient approaches have been investigated. The three approaches that are reviewed in this paper are the ant colony optimization method, the honey-bee algorithm, and genetic algorithm. This research paper summarized the methods for cloud load balancing based on these three mentioned approaches.

Keywords Cloud Computing · Load balancing · Ant colony optimization · Honey bee algorithm · Genetic algorithm

1 Introduction to Cloud Computing

This cloud computing [1], as a term, has been around since the early 2000s, but the concept of computing as a service has been around for much longer as far back in 1960s. Cloud computing helps to control, configure, and access the hardware and software resources remotely. Cloud computing is the method of delivering on-demand computing services from applications to storage. It offers online information storage, infrastructure, and application. This technique makes use of the Internet and remote servers to maintain data and applications. The genuine meaning of cloud is network or Internet, or it is assumed that the cloud is something that is present at

T. Bhowmik (✉) · S. Goswami · B. Das
Assam Downtown University, Guwahati, Assam 788056, India

V. Bhateja et al. (eds.), *Data Engineering and Intelligent Computing*,
Advances in Intelligent Systems and Computing 1407,
https://doi.org/10.1007/978-981-16-0171-2_30

some remote location. Cloud can also provide services over a public and private networks, i.e., WAN, LAN, or VPN.

Cloud computing helps the user to access data easily and efficiently. Some of the popular cloud computing providers are Google, Amazon, and Yahoo. The main disadvantage of this technique is computer capacity which rises dramatically. It also lacks the ability to scale the resources in the event and the necessary resources to support the virtual machine.

According to cloud computing architecture and the demands of clients, the cloud is allotted with some load in it (which might be under loaded or overloaded or balanced). The underloaded and overloaded situations can generate some contrasting types of system failures like power consumption, execution time, machine failure, etc. Therefore, load balancing is essential to solve the above-mentioned problems. Thus, there is a need of load balancing in a cloud computing environment.

The load balancing technology in clouds plays a crucial role in task scheduling. There are many types of loads in the cloud network namely memory load, computation (CPU) load, network load, etc. Load balancing technology is the method to detect overloaded and underloaded hosts and balancing the load among the hosts.

This paper represents an overview of the various load balancing techniques evolved especially to enhance the cloud computing domains. The rest of the paper is arranged as follows. Section 2 provides the challenges of load balancing in cloud computing. Then in Sect. 3, the most popular ant colony optimization-based load methods are reviewed and summarized. In Sect. 4, few of the commonly used honey-bee behavior-inspired load balancing methods are reviewed and summarized. Then Sect. 5 reviewed and listed out the load balancing methods based on the genetic algorithm. And lastly, the conclusion is discussed in Sect. 6 and the references are listed in last section.

2 Load Balancing in Cloud Computing

Load balancing is the process of splitting workloads among different nodes of the network so as to increase the efficiency, reliability, and capacity of the network. It helps to enhance resource use, maximum throughput, minimum response time, and also avoids excess workload of any single node.

Cloud load balancing [2] is the method of distributing workloads and other computing properties in a cloud computing environment. This framework delivers the enterprises to control workload demands or application demands by issuing resources among computers, networks, or servers. It helps enterprises to achieve high-performance levels with lower costs as compared to traditional load balancing technology. It takes the advantage of scalability and agility property of cloud computing to meet the workload demands and to enhance overall availability. Cloud load balancing technology also provides health checks for cloud applications.

Cloud computing provides advantages like cost flexibility and availability of service to the users. Load balancing permits cloud computing to meet up the increasing requirements by abundantly allocating dynamic local workload evenly across all nodes.

Cloud load balancing [3] may also be acquired by an efficient task scheduling mechanism as it helps to improve resource utilization, thereby enhancing the overall performance. Task scheduling is the process of organizing tasks on processors and reduces the completion time of the task.

Task scheduling is a crucial and important part of a cloud computing platform. Task scheduling is used to assign certain tasks to some particular resources at a particular time period. Task scheduling has some issues so different techniques have been proposed to solve the problems of task scheduling. It enhances the efficient utilization of resources and takes less response time so that the execution of the given tasks is performed within a minimum time. The importance of load balancing in cloud computing are: better performance, maintain website traffic, handle sudden traffic burst, flexibility, etc.

2.1 Challenges in Cloud Computing Load Balancing

Before evaluating the load balancing approaches for the cloud computing, there is a necessity to spot the most issues and challenges involved in load balancing in cloud environment.

Spatial Distribution of the Cloud Nodes

It is a challenge to style a load balancing algorithm that may work for spatially distributed nodes. This is often because other factors must be taken under consideration as the rate of the network links among the nodes, the distance between the clients and therefore, the task processing nodes, and also the distances between the nodes involved in providing the service.

Thus, there is a necessity to develop the simplest way to manage load balancing mechanisms among all the spatial distributed nodes while having the ability to effectively tolerate high delays.

Storage/ Replication

A full replication algorithm never takes efficient storage utilization into consideration. This can be because the identical data are stored in all replication nodes. However, partial replication algorithms could save parts of the information sets in each node support each node's capabilities like processing power and capacity.

This might cause better utilization, yet it increases the complexity of the load balancing algorithms.

Algorithm Complexity

In terms of implementation and operations, load balancing algorithms are preferred for less complex. The upper implementation complexity would cause a more complex process that could cause some negative performance issues.

Therefore, load balancing algorithms must be designed within the simplest possible forms.

Point of Failure Controlling

The load balancing and collecting data about various nodes must be designed in an exceedingly way that avoids having one point of failure within the algorithm. Any load balancing algorithm must be designed to beat this challenge. Distributed load balancing algorithms seem to provide a stronger procedure, yet they are far more complex and need more coordination and control to function accurately.

Researchers proposed various load balancing approaches in cloud computing to pander to those challenges. In this paper, the following three handiest approaches are discussed:

- *Ant colony optimization based*
- *Honey-bee behavior based*
- *Genetic algorithm based.*

3 Cloud Load Balancing with Ant Colony Optimization

The technique by which one can resolve computational problems and can find better paths using graphs in less time is called ant colony optimization. Here, the behavior of real ants is compared with those of artificial ants. The mixture of local search algorithms and artificial ants together forms a method that opts for numerous optimization tasks involving various graphs. The rapidly increasing activity in this field has led the companies to adopt the ant colony optimization algorithm. A few of the available load balancing techniques which implement the ant colony optimization methods for cloud computing are summarized in Table 1, and the methods are briefly discussed as follows:

Kun Li, Gaochao Xu, Guangyu Zhao, Yushuang Dong, Dan Wang proposed a research paper[4] in which a load balancing ant colony optimization (LBACO) algorithm is used to find the optimal resource allocation of each task in the dynamic cloud environment. It also balances the whole system load by adapting the dynamic cloud computing system.

Peng Xu, Guimin He, Zhenhao Li, and Zhongbao Zhang proposed a research paper [5] in which an efficient load balancing algorithm for virtual machine allocation based on ant colony optimization is used to achieve the load balancing of multidimensional resources. It solves problems using mathematical models.

In the research paper [6], Zehua Zhang and Xuejie Zhang proposed a load balancing mechanism that is based on the ant colony optimization algorithm and complex network theory in Open Cloud Computing Federation (OCCF).

The research paper [7] proposed by Santanu Dam, Gopa Mandal, Kousik Dasgupta, Paramartha Dutta defines a load balancing algorithm for VMs in the cloud computing algorithm. It is a random search algorithm which is based on the real ant behaviors.

Table 1 Cloud load balancing with ant colony optimization

Paper name	Objectives/goals	Approaches/findings	Future work/drawback
Cloud task scheduling based on Load Balancing Ant Colony Optimization [4]	• An algorithm based on load balancing ant colony optimization (LBACO) algorithm is proposed for cloud task scheduling policy • It mainly focuses to balance the entire system load while trying to minimize the makespan of a given task set	• The LBACO algorithm can be implemented for attaining tasks scheduling with load balancing • Researches have revealed that the proposed LBACO algorithm works usually more efficiently than FCFS (First Come First Serve) and Basic ACO (ant colony optimization)	This system relies on the two subsequent assumptions: • First, all tasks are freelance from one other, i.e., there is no priority management amongst the tasks • The second is that the tasks are computationally intensive, which is not realistic for cloud systems In the future, research can be done to remove these assumptions, and moreover, in order to accommodate the heterogeneous processing of the tasks, the availability vector should be extended to incorporate information about task requirements
An efficient load balancing algorithm for virtual machine allocation based on ant colony optimization [5]	• The main aim of this paper is to maximize the resource utilization of all physical machines using multidimensional resource utilization in the cloud computing environment • Boost physical machine selection approach in order to prohibit the premature convergence or falling into the local optima	• This paper focuses on how to achieve the goal of load balancing of multidimensional resources • Problems are solved using mathematical models • Experiments show that the ACO-based algorithm performs better than the basic ACO technique and other algorithms	Two strategies can be improved in the future: • Improving ACO by proposing the concept of load imbalance degree • Introducing PM selection expectation

(continued)

Table 1 (continued)

Paper name	Objectives/goals	Approaches/findings	Future work/drawback
A load balancing mechanism based on ant colony and complex network theory in Open Cloud Computing Federation [6]	• This paper aims to provide a load balancing technique that is build on ant colony and Complex Network Theory in Open Cloud Computing Federation • This paper enhances many features of the related ant colony algorithms which were proposed to realize load balancing in a distributed system	This acknowledges a load balancing mechanism based on ant colony and complex network theory for OCCF, which aims to cope with the complex and dynamic load balancing problem in OCCF	Further studies can be done by changing the dynamics of the parameters that are used to calculate the probability function of an ant for choosing a neighbor node
An Ant Colony Based Load Balancing Strategy in Cloud Computing [7]	• This paper proposes a technique based on the ant colony optimization algorithm for balancing the load on machines by doing a search operation to the under loaded nodes • It helps to match the tip users' needs by distributing the dynamic workloads to multiple nodes	• This paper gives an approximate calculation for load balancing in cloud computing architecture • The proposed technique functions better than some of the conventional approaches like First Come First Serve (FCFS), Stochastic Hill Climbing (SHC), soft computing approaches like genetic algorithm (GA) and some of the existing ant colony-based techniques	Two assumptions were made for this algorithm: • Fault tolerance issues were not considered • All jobs were predicted with same priority In the future, researches can be done including the fault tolerance and different function variation to calculate the pheromone value

(continued)

Table 1 (continued)

Paper name	Objectives/goals	Approaches/findings	Future work/drawback
Load Balancing of Nodes in Cloud Using Ant Colony Optimization [8]	• This paper proposes an algorithm for the distribution of workloads among different nodes of a cloud by using the ant colony optimization (ACO) algorithm • This technique is the modified strategy of ant colony optimization that is applied for cloud or grid network systems	• The main advantage of using this technique is that it detects the overloaded as well as under loaded nodes of machines so as to perform certain operations on the identified nodes • In this approach, each ant builds its own unique result set then later on a complete solution is produced from it	• In this strategy, the ant uninterruptedly updates a single result set rather than updating their own result set • The solution set is gradually built and continuously improved rather than being compiled only once in a while • The task of each and every ant is exclusive rather than being general and the task depends on the type of the first node that was encountered

Kumar Nishant, Pratik Sharma, Vishal Krishna, Chhavi Gupta, Kuwar Pratap Singh, Nitin, and Ravi Rastogi proposes research paper [8], which is based on ACO for better distribution of workload among the nodes of a cloud computing environment. It helps for the efficient utilization of resources and ensures much lower resource cost.

4 Honey-Bee Behavior-Based Load Balancing Algorithm

The honey-bee load balancing algorithm is fully inspired by the natural honey-bee's behavior. When the tasks are allocated, then those tasks update the other remaining tasks about the virtual machine status in the same way as the honey-bees updates the other bees in the beehive about an abundant food source. This algorithm reduces the overall response time and the processing time of hosts. There are many load balancing methods, which implements the concept of a honey-bee, are proposed by researchers. A few of them are summarized in Table 2 and are discussed briefly as follows:

In this research paper [9], Harshit Gupta and Kalicharan Sahu propose a technique that can execute tasks simultaneously and assigns each of these jobs to a computational resource such as a processor or a computer. This technique generates efficient utilization of processor by balancing out the load and also focuses on many elements such as client, data centers, and distributed servers.

In this research paper [10], Dhinesh Babu L.D., P.Venkata Krishna propose an algorithm that reduces the overall execution time of tasks of the machines as VM's run in parallel so it is not possible to execute the task efficiently. So, the scheduler has to do the scheduling process to utilize the available resources fully. The tasks should be loaded in a well-balanced way in all VM's to ensure the overall utilization of resources.

In this research paper [11], Dr.Sudha Senthilkumar, Dr.K.Brindha, Prof. Rathi.R, Prof. Angulakshmi, Dr. Jothi, YashVardhanThirani propose an optimizing algorithm called the honey-bee foraging algorithm to find an optimal solution to a problem. In this algorithm, an exploitative neighborhood search is done along with a random explorative search. The proposed algorithm has been applied to many processes like multi-objective optimization, neural network training, image processing, etc.

In this research paper [12], Khushbu Zalavadiya and Dinesh Vaghela propose a technique that is based on the priority of the tasks. In this, the tasks having the same priorities are loaded on the VMs, and other tasks are submitted to those VMs having minimum load to achieve better resource utilization. This algorithm makes sure that the loads are well balanced in all VMs.

Table 2 Honey-bee behavior-based load balancing algorithm in cloud computing

Paper name	Objectives/goals	Approaches/findings	Future work
Honey Bee Behavior-Based Load Balancing of Tasks in Cloud Computing [9]	• The aim of this paper is to propose an algorithm called honey-bee behavior-based load balancing (HBB-LB) that aims to achieve a well-balanced load in a virtual machine • It improves average execution time and reduces waiting time of tasks	• In this paper, a flowchart for load balancing in cloud computing environments is proposed which is based on the honey-bee foraging approach • It improves the outturn of process and priority-based balancing that focuses on reducing the quantity of time a task has got to assist on a queue of the VM	• After comparing this algorithm with others, results show that it stands good without increasing additional overheads • In the future, this algorithm can be improved by considering other QoS factors of tasks
Honey bee behavior inspired load balancing of tasks in the cloud computing environments [10]	• This paper aims to achieve a well-balanced load across VMs for maximizing the throughput • This also balances the priorities of the tasks so that the waiting time is minimal	• This algorithm not only balances the load, but also balances the priorities of tasks that have been removed from heavily loaded virtual machines • It reduces the response time of VMs	• This load balancing technique works well for heterogeneous cloud computing systems and is for balancing non-pre-emptive independent tasks • It evaluates precedence as the main QoS parameter • In the future, this work can be extended for workflows load balancing with dependent tasks and can be improved by considering other QoS factors also

(continued)

Table 2 (continued)

Paper name	Objectives/goals	Approaches/findings	Future work
Honey-Bee Foraging Algorithm for Load Balancing in Cloud Computing Optimization [11]	• The aim of this particular research paper is to describe an optimizing algorithm called the honey-bee foraging Algorithm, derived from the natural behavior of honey-bees, to find the ideal solution • This performs an exploitative neighborhood search along with a random explorative search	• In this research paper, the HBFA and its other versions are demonstrated and are implemented to optimize many functions • The findings are contrasted with the findings obtained using other algorithms. The findings conclude that the honey-bee foraging algorithm offers some profit over other optimization techniques based on the problem	• Based on the results, it can be inferred that the results of the new algorithm are significantly better than the results of basic HBFA • Thus, the proposed algorithm performed better than the basic HBFA on higher-dimensional functions such as, Ackley and Griwank
Honey Bee Behavior Load Balancing of Tasks in Cloud Computing [12]	• In this approach, the tasks having similar priorities are allocated to under loaded virtual machine • Produces minimum completion time, amount of waiting time of the tasks in the queue is minimal and achieve better resource utilization	• In this algorithm, priority tasks are removed from the machine and allocated to those machines that have less number of same prioritized tasks as well as the least expected completion time of that task on that machine • It also balances the load of dependent tasks in a pre-emptive manner and helps to produce minimum completion time and better resource utilization	• In the future, it can be modified by considering as many as QoS factors as possible

5 Genetic Algorithm-Based Cloud Load Balancing

A genetic algorithm is a procedure that is inspired by the natural selection theory. It is used for producing high-quality solutions for optimization and problems related to search operations. In this, some individuals called candidate solutions having some optimization problems are progressed toward better solutions. Each candidate solutions is defined by a set of properties that can be changed and represented as an array of bits or binary strings. Researchers have proposed many load balancing methods based on genetic algorithm. Some of them are summarized in Table 3 and are discussed briefly as follows:

In this research paper [13], Kousik Dasgupta, Brototi Mandal, Paramartha Dutta, Jyotsna Kumar Mondal, Santanu Dam propose a technique which is based on the mechanism of natural selection strategy. The main advantage of this technique is that it can handle a huge amount of search space which is applicable to the complex objective function and avoids being trapped into local optimal solutions.

In this research paper [14], Sukhpreet Kaur and Dr. Jyotsna Sengupta propose a technique that balances the load in a cloud computing environment. In this technique, when a new task arrives, it first checks for available machines and if a machine is available, then that task is loaded to that machine and if there is no machine is available, then the task is loaded to a machine whose current task is going to be completed soon.

In this research paper [15], Safwat A. Hamad and Fatma A. Omara propose a task scheduling algorithm for allocating independent tasks to improve its completion time, minimize the execution cost, and increase resource utilization. Here, the task scheduling problem is evaluated as an NP-complete problem. So, the optimizing approaches can be utilized to solve it by referring to the performance parameters.

6 Conclusion

In this paper, we have discussed different load balancing techniques based on ant colony optimization, honey-bee algorithm and genetic algorithm for the cloud computing environment. The techniques reviewed in this paper offer a promising improvement over the conventional load balancing methods for cloud computing.

However, there is still much work to be done to improve the load balancing methods discussed, such as removing the assumptions to make them generalized. In the future, these algorithms can be further extended to improve the response time, finish time, etc., and more dynamic characteristics of VMs can be introduced. A load imbalance degree could be found out to make these algorithms more efficient.

Finally, it can be concluded that this paper summarizes the most promising load balancing methods for the cloud environment along with their limitations. So this study will be helpful to improvise the existing methods in the future.

Table 3 Genetic algorithm-based cloud load balancing

Paper name	Objectives/goals	Approaches/findings	Future work
A Genetic Algorithm (GA) based Load Balancing Strategy for Cloud Computing [13]	• This paper proposes a novel load balancing strategy using a genetic algorithm • The algorithm thrives to balance the load of the cloud infrastructure and tries to minimize • This approach helps to distribute the dynamic workload among multiple nodes to ensure that no single resource is over-utilized or underutilized	• This paper proposes a genetic algorithm-based load balancing strategy for cloud computing that has been developed to provide an efficient utilization of resources in the cloud computing environment	In the future, some variation of the cross-over and selection strategies can be applied to make this algorithm more efficient
Load Balancing using Improved Genetic Algorithm (IGA) in Cloud Computing [14]	• In this paper, an improved genetic algorithm is introduced for balancing the load • The improved genetic algorithm is used in allocating the user's task to the virtual machine • The aim of this algorithm is to maximize resource utilization and to minimize the task execution cost and energy consumption • The proposed algorithm is simulated using the CloudSim toolkit	• The graphical representation of the simulation results shows that the improved genetic algorithm is more efficient than the existing genetic algorithm in terms of several parameters like energy efficiency and cost • The graphical representation of this approach clearly depicts that all the VMs are allocated to the cloudlets in the improved genetic algorithm	In the future, this genetic algorithm can be improved in terms of response time and finish time, etc. Also, it can be implemented in a real-time environment

(continued)

Table 3 (continued)

Paper name	Objectives/goals	Approaches/findings	Future work
Genetic Algorithm Based Task Scheduling Algorithm in Cloud Computing Environment [15]	• This paper proposes a task scheduling algorithm which is based on the genetic algorithm (GA) for allocating and executing an application's tasks • The aim of this algorithm is to minimize the completion time and cost of tasks, and maximize resource utilization • The implementation of this proposed algorithm has been evaluated using the CloudSim toolkit	• The algorithm targets to minimize completion time and cost, and maximize resource utilization • The cost of the proposed TS-GA algorithm is reduced by (4%) and (6%) about the default GA and RR algorithms respectively	For future work, this algorithm can be extended to add possibility dynamic characteristics of VMs through run GA

References

1. Afzal, S., Kavitha, G.: Load balancing in cloud computing-A hierarchical taxonomical classification.
2. Nuaimi,K. A.,Mohamed, N., Nuaimi, M. A.,Al-Jaroodi,J.: A Survey of Load Balancing in Cloud Computing: Challenges and Algorithms, In: Second Symposium on Network Cloud Computing and Applications, 2012, IEEE Xplore: 07 March 2013
3. Chaczko, Z.,Mahadevan, Venkatesh., Aslanzadeh, Shahrzad., Mcdermid, C.: Availability and Load Balancing in Cloud Computing, In: International Conference on Computer and Software Modeling IPCSIT vol.14 (2011)
4. Li, K., Xu, G., Zhao G., Dong, Y.: Cloud Task scheduling based on Load Balancing Ant Colony Optimization, In: Sixth Annual China Grid Conference (2011)
5. Xu, P., He, G., Li, Z., Zhang, Z.: An efficient load balancing algorithm for virtual machine allocation based on ant colony optimization. In: International Journal of Distributed, Sensor Networks, Vol. 14(12), DOI: https://doi.org/10.1177/1550147718793799, (2018)
6. Zhang, Z.,Zhang, X.: A Load Balancing Mechanism Based on Ant Colony and Complex Network Theory in Open Cloud Computing Federation,In: The 2nd International Conference on Industrial Mechatronics and Automation, (2010)
7. Dam, S.: An Ant Colony Based Load Balancing Strategy in Cloud Computing,In: Advanced Computing, Networking and Informatics - Volume 2, Smart Innovation, Systems and Technologies 28, 403 DOI: https://doi.org/10.1007/978-3-319-07350-7_45, © Springer International Publishing Switzerland (2014)
8. Nishant, K., Sharma, P., Krishna, V., Gupta, C., Singh, K. P.,Rastogi, N. R.: Load Balancing of Nodes in Cloud Using Ant Colony Optimization, In: 14th International Conference on Modelling and Simulation (2012)
9. Gupta, H.,Sahu, K.: Honey Bee Behavior Based Load Balancing of Tasks in Cloud Computing, In: International Journal of Science and Research (IJSR) ISSN (Online): 2319–7064 (2012)
10. L.D., D. B., Krishna, P.V.: Honey Bee Behaviour Inspired Load Balancing of Tasks In Cloud Computing Environments, In: Applied Soft Computing 13,pp. 2292–2303, (2013)
11. Senthilkumar, Dr. S., Brindha, Dr. K., R, Prof. R., Prof. A., Y.V.T., Dr. J.:Honey-Bee Foraging Algorithm for Load Balancing in Cloud Computing Optimization, In: International Journal of Engineering Science and Computing, December, Volume 7 Issue No.12.pp. 15840–15844 (2017)
12. Zalavadiya, K., Vaghela, D.: Honey Bee Behavior Load Balancing of Tasks in Cloud Computing, In: International Journal of Computer Applications pp. 0975 – 8887 Volume 139 – No.1, April (2016)
13. Dasgupta, K., Mandal, Brototi., Dutta, P.,Mondal, J. K., Dam, S.: A Genetic Algorithm (GA) based Load Balancing Strategy for Cloud Computing, In: International Conference on Computational Intelligence: Modeling Techniques and Applications (CIMTA) (2013)
14. Kaur, S., Sengupta, Dr. J.: Load Balancing using Improved Genetic Algorithm (IGA) in Cloud Computing, In: International Journal of Advanced Research in Computer Engineering & Technology (IJARCET) Volume 6, Issue 8, ISSN: 2278 – 1323, August (2017)
15. Hamad, S. A., Omara, F. A.: Genetic-Based Task Scheduling Algorithm in Cloud Computing Environment,In: (IJACSA) International Journal of Advanced Computer Science and Applications, Vol. 7, No. 4, (2016).

Adaptive Ensemble Learning with Concept Drift Detection for Intrusion Detection

Deepa Mulimani, Shashikumar G. Totad, Prakashgoud Patil, and Shivananda V. Seeri

Abstract Intrusion Detection Systems (IDS) operate in a dynamic real-time environment and are subject to evolving data streams. Recent IDS use machine learning techniques to rapidly identify new attacks and update themselves. However, proposed techniques in literature do not consider the problem of Concept Drift in which data come from different distributions. This paper proposes a technique of adaptation to Concept Drift using ensemble learning for intrusion detection in a network system. The primary objective is minimizing the number of 'malicious attacks which are marked as genuine' thereby increasing the 'Recall' metric. The technique firstly identifies the change in data distributions and then applies ensemble learning to detect if it is an attack. It is incremental in adapting to the changing data distributions on the go. The experimental dataset was converted into streaming data and input to the model. The proposed technique performed better when compared with logistic regression.

Keywords Intrusion detection systems (IDS) · Concept drift · Ensemble learning · Adaptive XGBoost

1 Introduction

The present world of technology is ever-evolving and increasingly challenging in terms of protecting systems and networks from malicious attacks or intrusions. Intrusion Detection Systems (IDS) are one of the important security technologies

D. Mulimani (✉) · S. G. Totad · P. Patil · S. V. Seeri
KLE Technological University, Hubballi, Karnataka, India

S. G. Totad
e-mail: totad@kletech.ac.in

P. Patil
e-mail: prakashpatil@kletech.ac.in

S. V. Seeri
e-mail: seeri@kletech.ac.in

V. Bhateja et al. (eds.), *Data Engineering and Intelligent Computing*,
Advances in Intelligent Systems and Computing 1407,
https://doi.org/10.1007/978-981-16-0171-2_31

developed for detecting and preventing attacks in a network system [1]. The present IDS are designed using techniques of machine learning toward being smart, robust and defensive. Though there are numerous techniques developed, IDS may fail to address the dynamic nature of the data. These techniques usually train a static classifier notwithstanding the varying characteristics of the dynamic data. Moreover, there would be various new intrusions evolving infinitely. As the static classifier is not trained for the evolving data, it performs poorly. It needs to be retrained to detect new classes of intrusions. Retraining the static classifier proves to be expensive and inefficient. Instead, the IDS classifier could be designed to be incremental and adaptive to the streaming data. Incremental learning [2] provides two approaches—single classifier and ensemble classifier. The ensemble learning approaches are found to be highly accurate, efficient and robust [3, 4]. In literature, there are only a few pieces of research on ensemble learning that deal with varying data in IDS [5, 6].

In dynamic environments, the Concept Drift problem is the phenomenon of data distribution change that affects the learning algorithm's performance. The adversaries may introduce new classes of intrusions. On the other hand, the varying data might introduce new genuine classes. In any case, the IDS has to train itself on any such concept drifts and robustly adapt to the changes. This paper introduces an implementation of using an ensemble technique—Adaptive eXtreme Gradient Boosting [7] that integrates Concept Drift problems to enhance the performance of IDS. The proposed IDS classifier uses ensemble learning in two ways—Push Strategy and Replacement Strategy for learning the statistical characteristics of data. It then tries to identify new classes for being intrusions or genuine. The KDD CUP 99 dataset used in the experiment is converted from a.csv file into streaming data for the input. The technique proposed proves to perform better in comparison with the results of logistic regression in predicting intrusions.

The organization of the paper is as follows—the related work carried out is presented in Sect. 2. In Sect. 3, Concept Drift and Adaptive eXtreme Gradient Boosting algorithm is discussed. Section 4 presents the proposed methodology and experiments, while Sect. 5 discusses the experiment results. Section 6 concludes the work.

2 Related Work

In [8], the authors proposed an experimental framework to detect malicious attacks in the network under the setting of supervised and unsupervised learning. The framework was tested with Support Vector Machines (SVM), k-nearest neighbors, decision trees and multilayer perceptron among supervised learning. K-means clustering and single-linkage clustering were used to evaluate the framework under unsupervised learning. The implementation works on the assumption that the train and test data are from the identical distribution. This assumption, however, suffers from Concept Drift and hence does not perform well under supervised learning. On the contrary, the unsupervised algorithms performed well in case of unseen attacks in the test data.

The authors of [9] pointed out that most of the IDS are not incremental and fail in discriminating false alarms from the true ones. The authors propose intrusion detection using artificial intelligence (AI)-based approach and computational intelligence (CI) methods.

The work presented in [10] is a methodology for IDS that trains multiple classifiers on a set of selected features. It tries to combine feature selection techniques such as PCA, CFS, IGR, and minimum redundancy was used with the classifiers SVM, Naïve Bayesian, decision tree, k-nearest neighbors and neural networks. Every combination of feature selection and classifier works differently on a given dataset and has its pros and cons. Hence, it is difficult to choose a particular combination to devise intrusion detection and prevention. The IGR technique along with k-NN gave the highest accuracy among others.

The authors of [11] proposed Concept Drift Ensemble Incremental Learning for intrusion detection. The Drift Diffusion model using Hoeffding inequality was used to detect Concept Drift and abnormality. The ensemble method used only three classifiers whose learning was combined to detect intrusion. The KDD CUP 99 dataset was used for the experimentation.

The work in [12] proposes Diversity and Transfer-based Ensemble Learning (DTEL). It is a diversity-based ensemble learning that applies Transfer learning (TL) on a selected set of features instead of the entire set. This method proves to be expensive and is complex. Forest Cover type, Poker Hand, Electricity and CTR prediction were the datasets used in the experiments.

3 Concept Drift and Adaptive eXtreme Gradient Boosting Technique

This section presents an overview of the Concept Drift problem and Adaptive eXtreme Gradient Boosting technique (AXGB) suitable for evolving data streams.

3.1 Concept Drift

Stream learning is characterized by the problem of models accessing data only once and processing it on the go. Additionally, the models have to resolve the challenges of Concept Drift [13]. This is an innate problem of evolving data wherein the relationship between the features and targets change potentially.

Concept change refers to the changes in the underlying concept over time [14]. Concept Drift may occur gradually or abruptly between two concepts. The formal definition of Concept Drift is as follows.

In an evolving streaming data of unbounded size, the sequence is in the form $\{X, y\}$ where $X = \{x1, x2, x3, \ldots xn\}$ features for which $y = \{y1, y2, y3, \ldots yn\}$

comprises the labels. A function $y = f(X)$ predicts the labels for unknown X values. The underlying distribution of X is non-stationary. Either the posterior probability $p_t(X|y)$ varies overtime or the class probability $p_t(x)$ changes, while the posterior probability of y (class labels) is not affected.

The change may have a large impact on model's error rate calling for revising the current model. An important challenge for the change detector would be to separate the true changes from false alarms. In an IDS, the analysis of communication logs involves distinguishing between the Concept Drift and real attacks. Learning against an adversary would be the main focus of such systems.

3.2 Adaptive eXtreme Gradient Boosting (AXGB) Technique for Ensemble Learning [7]

The Concept Drift in streaming data can be handled by using incremental and ensemble learning in an adaptive learning setup. An ensemble-based system combines multiple models strategically to resolve a computational problem. Ensemble learning improves the system performance by employing multiple base learners (a combination of classifiers) and combining their outputs to determine the final result. The ensemble methods are popular for improving predictive performance and increasing robustness.

One of the popular algorithms for ensemble learning in supervised learning tasks is the eXtreme Gradient Boosting (XGB) algorithm [15]. XGB implements ensemble through boosting technique using decision trees. To minimize the objective function the ensemble adds new weak learners during training. Generally, a weak learner with less complexity such as trees is considered for this process. The main challenge with XGB is its configuration using multiple hyperparameters. If configured well XGB offers state-of-the-art results.

For a corresponding set of features $X = (x_1, x_2, \ldots, x_n)$ a supervised task predicts the labels $Y = (y_1, y_2, \ldots, y_n)$. An ensemble method predicts $\hat{y}_i$ for a corresponding given input $\vec{x}_i$ by averaging the results in the ensemble. Boosting creates the ensemble E sequentially by adding new members one at a time. For every iteration k, the existing ensemble members evaluate the training data and output the prediction scores $Y^{(k)}$ correspondingly. These predictions further create new ensemble members for E. The predictions are combined using Eq. (1)

$$\hat{Y}^{(k)} = \sum_{k=1}^{K} f_k(X) = \hat{Y}^{(k-1)} + f_k(X) \quad (1)$$

where the number of members in E is given by K. The space for possible base functions is given by F and each $f_k \in F$. The Eq. (2) sums up the predictions for a sample $\hat{y}_i$ for each tree f_k in E and gives the final prediction.

$$\hat{y}_i = \sum_{k=1}^{K} f_t(x_i) \quad (2)$$

In streaming data setting, the data stream can be defined as $A = \{(\overrightarrow{x}_t, y_t)\}| \; t = 1, \ldots, T$ where $T \to \infty$, $\overrightarrow{x}_t$ represents the feature vector and the respective target is given by y_t. AXGB creates an ensemble using incremental strategy. For each base function f_i it creates non-overlapping windows from which it uses the data sub-samples. That is, it doesn't consider the same data for selecting the base function f_i. As it receives new data samples, it creates a buffer $w = (\overrightarrow{x}_t, y_i)$ where $i \in \{1, 2, \ldots, W\}$ with $|w| = W$ samples as the size. AXGB trains a single f_k once w is full. Hence, Eq. (2) can be rewritten as Eq. (3).

$$\hat{Y}^{(k)} = \sum_{k}^{K} f_k(w_k) = \hat{Y}^{(k-1)} + f_k(w_k) \quad (3)$$

This function is determined by the new base function index k of the ensemble. For the first ensemble member f_1 the data in the entire buffer is used. When $k > 1$ the new base function f_k in the ensemble is obtained using the data passed through the $k - 1$ models of the ensemble and their residuals. The learned predictors are updated constantly by AXGB once the ensemble is full. This done using two strategies explained as follows:

- *Push Strategy* (AXGB$_{[p]}$)—In this the ensemble members form a queue. New models join this queue as and when they are created. The older models are discarded from the queue whenever the ensemble is full.
- *Replacement Strategy* (AXGB$_{[r]}$)—Here the new models replace the older models in the ensemble.

A completely new ensemble is got for every K iteration. The AXGB uses a dynamic window that grows exponentially in every iteration until it reaches the maximum size.

The experimental study uses ADWIN [16] for handling Concept Drift. AXGB triggers an ensemble update whenever ADWIN signals Concept Drift. The procedure is implemented as follows

i. AXGB resets the size of w to the minimum defined.
ii. Depending on the strategy chosen, the new members are trained and added. In the Push Strategy, the models are trained and added based on the increasing sizes of the window. In the case of the Replacement Strategy, the index points to the beginning of the ensemble and replaces the older models with the newer ones. In most of the cases, the training of new models does not consider the residuals of the old ones.

4 Experimental Methodology

The working of the proposed technique is shown in Fig. 1, and the detection of Concept Drift or Attack by every estimator is shown in Fig. 2. The proposed model has AXGB as the ensemble classifiers as Estimators 1, 2, ..., *n*, along with ADWIN embedded for drift detection in each one of them. The predictions are aggregated, and the final prediction is given as 'attack' or 'no attack.' In case of Concept Drift for new genuine classes, the ensemble is updated. Also, the model updates itself for every unseen attack. The model distinguishes between the attack and real drift. In the hyperparameter tuning the model is trained using the first 30% of the stream data and by tweaking the hyperparameters with different values. Further, the best-performing models are tested on 70% of the stream data. In data stream processing, the resources memory and time need to be properly managed as there would be infinite samples. The number of ensemble members can be used to get an insight on model complexity and memory usage. The proposed technique is implemented using Python's stream learning framework—scikit-multiflow [17].

The KDD CUP 99 dataset consists of [18] 29,175 records each having 41 features with 2 classes labels—'normal' or 'an attack.' There are over 20 types of attacks captured in the dataset. Every record sequences the TCP packets from source IP to destination IP. The connections are labeled as either normal or as an attack. Each connection record consists of about 100 bytes.

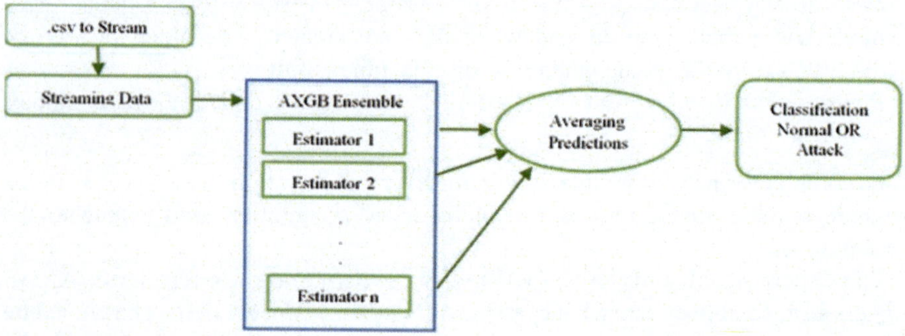

Fig. 1 Working of the proposed technique

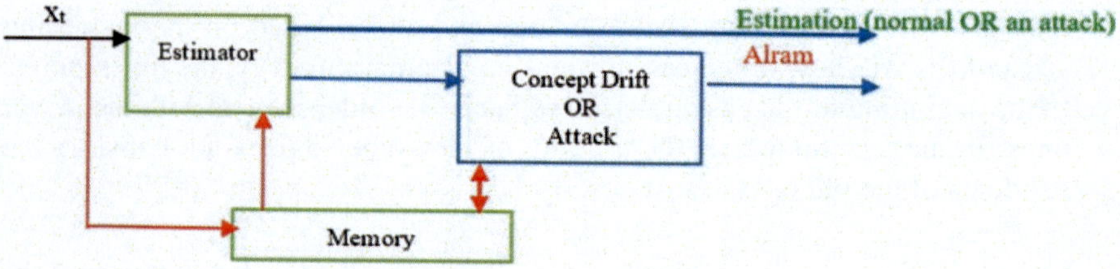

Fig. 2 Working of the estimator to identify Concept Drift or attack

5 Experimental Results

The experimental results presented below include the details of predictive performance, hyperparameter relevance, memory and time usage for $AXGB_{[p]}$ and $AXGB_{[r]}$. The study uses prequential evaluation [19] in which the model predicts for a sample before it is used for training/updating the model. Classification Accuracy and Recall are used to measure the performance. There were multiple tests conducted on various combinations of hyperparameters shown in Table 1. The sample results for two of the different combinations of hyperparameters are given in Table 2. The best results achieved is with learning rate = 0.3, ensemble size = 30, maximum window = 1000.

The estimators are capture change in data better when the maximum window size ranges between 1000 and 2000, and the ensemble size is higher. The different learning rate values did not show much significant change in the accuracy metric. Also, when the ensemble size is small, there is a bigger gap between the observer accuracies, and hence, there is a decrement of Kappa value.

Table 3 gives a comparison of the proposed technique and logistic regression. It can be seen that the proposed approach has better Accuracy and Recall scores. The Recall is the proportion of appropriately predicted attack cases to the actual size of the attack class. Hence, it is an important parameter to evaluate the performance of the algorithm.

Table 1 Range values

Learning rate (LR)	0.01–0.5
Ensemble size (E_size)	5–100
Max depth	1–15
Max win size	500–8000
Min win size	0–16

Table 2 Sample results

LR = 0.01, E_size = 5, Max_depth = 1, Max_win = 8000	AXGBp—Accuracy: 0.8451 AXGBp—Kappa: 0.6834 AXGBr—Accuracy: 0.8451 AXGBr—Kappa: 0.6834
LR = 0.3, E_size = 30, Max_depth = 6, Max_win = 1000	AXGBp—Accuracy: 0.9802 AXGBp—Kappa: 0.9601 AXGBr—Accuracy: 0.9504 AXGBr—Kappa: 0.8998

Table 3 Proposed technique versus logistic regression

	Proposed technique (AXGBp, AXGBr)	Logistic regression
Class 0 (No attack)	Precision: 0.9989	Precision: 0.8879
	Recall: 0.9976	Recall: 0.9076
Class 1 (Attack)	Precision: 1.00	Precision: 0.9201
	Recall: 0.9957	Recall: 0.8629
Accuracy	0.99078	0.90580

6 Conclusion

This paper proposes a network intrusion detection approach of using adaptive ensemble learning with Concept Drift detection in IDS that receives streaming data. The main objective has been to design an IDS classifier that is incremental and incorporates Concept Drift detection in the evolving data. The proposed technique updates the classifier whenever the performance drops thereby, increasing the overall 'Accuracy' and 'Recall.' A major emphasis has been given on 'Recall' as it is an important metric in any IDS that gives an overview of the false alarm rate. The model is tested against logistic regression and is found to perform better. It is also optimized in terms of memory and time usage. A notable finding is that the model performs better with hyperparameter tuning. The proposed technique can be tested on various other benchmark IDS datasets to evaluate its performance in terms of adaptiveness and robustness to drifting attacks.

References

1. A. Zainal, M.A. Maarof, S.M. Shamsuddin, Ensemble classifiers for network intrusion detection system. J. Inf. Assurance Secur. **4,** 217–225 (2009)
2. H. He, S. Chen, K. Li, X. Xu, Incremental learning from stream data. IEEE Trans. Neural Networks **22**(12), 1901–1914 (2011)
3. H. Rajadurai, U. Gandhi, A stacked ensemble learning model for intrusion detection in wireless network. Neural Comput. Appl. (2020). https://doi.org/10.1007/s00521-020-04986-5.E
4. A.A. Aburomman, M. Bin IbneReaz, A survey of intrusion detection systems based on ensemble and hybrid classifiers. Comput. Secur. **65**, 135–152 (2017). https://doi.org/10.1016/j.cose.2016.11.004
5. A. Rasoulifard, A.G. Bafghi, M. Kahani, *Incremental Hybrid Intrusion Detection Using Ensemble of Weak Classifiers* (Springer Berlin Heidelberg, 2008)
6. G. Yin, Y.T. Zhang, Z.N. Li, G.Q. Ren, H.B. Fan, Online fault diagnosis method based on incremental support vector data description and extreme learning machine with incremental output structure. Neurocomputing **128**(5), 224–231 (2014)
7. J. Montiel, R. Mitchell, E. Frank, B. Pfahringer, T. Abdessalem, A. Bifet, Adaptive XGBoost for evolving data streams. Source: arXiv:2005.07353v1 [cs.LG] 15 May 2020
8. P. Laskov, P. Dfussel, C. Schfaferand, K. Rieck, Learning intrusion detection: supervised or unsupervised? Comput. Sci. **3617**, 50–57. https://doi.org/10.1007/11553595_6
9. M. Zamani, M. Movahedi, M. Ebadzadeh, H. Pedram, A danger-based approach to intrusion detection, December 2013, Source: arXiv

10. S.K. Biswas, Intrusion detection using machine learning: a comparison study. Int. J. Pure Appl. Math. **118**(18), 101–114
11. A. Khraisat, I. Gondal, P. Vamplew, et al., Survey of intrusion detection systems: techniques, datasets and challenges. Cybersecurity **2**, 20 (2019). https://doi.org/10.1186/s42400-019-0038-7
12. N. Lu, Y. Sun, H. Liu, S. Li, Intrusion detection system based on evolving rules for wireless sensor networks. Adv. Internet Things Big Data Technol. Smart Hum.-Care Serv. **2018**, Article ID 5948146. https://doi.org/10.1155/2018/5948146
13. M. Althabiti, M. Abdullah, Streaming data classification with concept drift. Biosc. Biotech. Res. Comm. **12**(1) (Spl Issue February) (2019). https://doi.org/10.21786/bbrc/12.1/20
14. Janardan, S. Mehta, Concept drift in streaming data classification: algorithms, platforms and issues. ITQM (2017). https://doi.org/10.1016/j.procs.2017.11.440
15. T. Chen, C. Guestrin, Xgboost: A scalable tree boosting system, in *Proceedings of the 22Nd ACM SIGKDD International Conference on Knowledge Discovery and Data Mining, ser. KDD '16* (ACM, New York, NY, USA, 2016), pp. 785–794
16. A. Bifet, R. Gavalda, Learning from time-changing data with adaptive windowing, in *Proceedings of the 2007 SIAM International Conference on Data Mining*, pp. 443–448 (2007.
17. J. Montiel, J. Read, A. Bifet, T. Abdessalem, Scikit-multiflow: a multi-output streaming framework. J. Mach. Learn. Res. **19**, 1–5 (2018)
18. M. Tavallaee, E. Bagheri, W. Lu, A.A. Ghorbani, A detailed analysis of the KDD CUP 99 data set, in *Proceedings of the 2009 IEEE Symposium on Computational Intelligence in Security and Defense Applications (CISDA 2009)*
19. A.P. Dawid, Present position and potential developments: some personal views: Statistical theory: the prequential approach. J. R. Stat. Soc. Ser. A (General), pp. 278–292 (1984)

An Effective Algorithm to Enumerate Spectrum of Laplacian Matrix for Complete Fuzzy Graphs

A. Amutha, C. Jebakiruba, and M. Davamani Christober

Abstract In this paper, Laplacian spectrum of fuzzy complete graphs is investigated to excerpt and explore the fuzzy complete graph's structure, the pattern it follows and its invariant behaviour towards its fuzzy vertex set and fuzzy edge set numbering. The major shortcoming in using the Laplacian spectrum as a depiction of a graph is its enormous calculation. Finding the Laplacian spectrum comprises various steps, such as enumerating the corresponding degree matrix, adjacency matrix, Laplacian matrix and then computing its spectrum, i.e. eigenvalues. The Laplacian spectrum may change drastically with a very small structural change itself. Even if we go for machine assistance, such as MATLAB, the data that has to be entered physically plays the major role and if any error occurs in entering the data, will make a vast difference in their eigenvalues. This is very complicated and challenging. To calculate this effectively, an algorithm is constructed with polynomial time complexity $\mathcal{O}(n)$.

Keywords Fuzzy graph · Complete fuzzy graph · Laplacian matrix · Laplacian spectrum

Mathematics Subject Classification MSC 05C72 · MSC 94C15 · MSC 68R10

1 Introduction

Graph theory is a flourishing area of mathematics introduced by Leonard Euler in 1736 [1]. Fuzzy sets were deliberated by L.A Zadeh, in 1965 [2]. Later, in 1973, Kaufmann defined fuzzy graph based on Zadeh's insight [3]. Fuzzy sets and relations were developed and protracted as fuzzy graph theory, by Azriel Rosenfeld in 1975 [4]. In 1990, Robert Grone, Russell Merris and V. S. Sunder established the Laplacian spectrum of graphs and gave particular attention to multiplicities of integer

A. Amutha (✉) · M. Davamani Christober
The American College, Madurai, India

C. Jebakiruba
Lady Doak College, Madurai, India

V. Bhateja et al. (eds.), *Data Engineering and Intelligent Computing*,
Advances in Intelligent Systems and Computing 1407,
https://doi.org/10.1007/978-981-16-0171-2_32

eigenvalues and to the effect on the spectrum of various modifications of graphs [5]. Further, in 1991, Mohar, Bojan and Chartrand did a survey of known results about the spectrum of the Laplacian matrix of graphs with special emphasis on the second smallest Laplacian eigenvalue [6]. Mean while few algorithms such as algorithm to compute the supremum of the max-min powers of a map from the Cartesian product of a set to a bounded subset of the real numbers were given by Bhattacharya and Suraweera during 1991 [7]. During 1996, an algorithm for finding the connectedness matrix of a fuzzy graph was enumerated by Tong, Zengxiang and Zheng [8]. Berthold, Michael R and Huber in 1999 derived an algorithm for developing fuzzy graphs [9]. Sarwar, Musavarah and Akram, Muhammad in 2016, developed an algorithm for computing certain metrics in intuitionistic fuzzy graphs [10], and during 2017, they also enumerated an algorithm for computing strength of competition in bipolar fuzzy graphs [11]. Still, fuzzy graph theory is a booming concept with enormous applications in various grounds. This paper, first, recalls the basic definitions required to compute the algorithm. Then, an effective algorithm is introduced to find the Laplacian spectrum (exact values) of fuzzy complete graphs only by using its $\sigma(x)$ values, so as to minimize lots of manual exertion.

2 Basic Concepts

Consider a simple connected graph $G(V, E)$ where V denotes the vertex set of G and E denotes the edge set of G having n number of vertices, m number of edges and $n, m > 1$. Then, a fuzzy graph with V as the underlying set is a pair $G : (\sigma, \mu)$ with n vertices and m edges where $\sigma : V \to [0, 1]$ is a fuzzy subset and $\mu : V \times V \to [0, 1]$ is a fuzzy relation on σ such that $\mu(x, y) \leq \sigma(x) \bigwedge \sigma(y)\ \forall\ x,\ y \in V$, where $\bigwedge$ stands for minimum [4]. A fuzzy graph is said to be complete if $\mu\,(x, y) = \sigma\,(x)\ \Lambda\ \sigma\,(y)\ \forall\, x,\ y \in V$ [12, 13]. For a fuzzy graph, the adjacency matrix $A(G)$ is a square matrix of order n and $A(G) = [a_{ij}]$ where $a_{ij} = \mu(u_i, u_j)$ entry is as the strength of relation between the vertices u_i and u_j [14]. Let u be the vertex of the fuzzy graph and then the degree of u is defined as, $d_G(u) = \sum_{uv \in E(G)} \mu(uv)$ and its degree matrix $D(G)$ is also a square matrix of order n and $D(G) = [d_{ij}]$ where $d_{ij} = \begin{bmatrix} d_G(v_i) & \text{if} i = j \\ 0 & \text{otherwise} \end{bmatrix}$ [14]. The Laplacian matrix $L(G)$ is defined as the difference between the adjacency matrix from the degree matrix of the given fuzzy graph, i.e. $L(G) = D(G) - A(G)$. The spectrum of $L(G)$ denoted by $S(G) = (\lambda_1, \lambda_2, \lambda_3, \ldots, \lambda_n)$ where $\lambda_1 \geq \lambda_2 \geq \lambda_3 \geq \ldots \geq \lambda_n$ are the eigenvalues of $L(G)$. The Laplacian spectrum of a graph is spectrum of its Laplacian matrix.

3 Laplacian Spectrum of a Fuzzy Complete Graph

Consider the following complete fuzzy graph with 11 vertices. By the regular procedure, let us find the spectrum of laplacian matrix of the given complete fuzzy graph G (Fig. 1).

The degree matrix of graph G is,

$$D(G) = \begin{bmatrix} 1 & 0 & 0 & 0 & 0 & 0 & 0 & 0 & 0 & 0 & 0 \\ 0 & 2.5 & 0 & 0 & 0 & 0 & 0 & 0 & 0 & 0 & 0 \\ 0 & 0 & 3.4 & 0 & 0 & 0 & 0 & 0 & 0 & 0 & 0 \\ 0 & 0 & 0 & 1.8 & 0 & 0 & 0 & 0 & 0 & 0 & 0 \\ 0 & 0 & 0 & 0 & 1 & 0 & 0 & 0 & 0 & 0 & 0 \\ 0 & 0 & 0 & 0 & 0 & 3.6 & 0 & 0 & 0 & 0 & 0 \\ 0 & 0 & 0 & 0 & 0 & 0 & 2.5 & 0 & 0 & 0 & 0 \\ 0 & 0 & 0 & 0 & 0 & 0 & 0 & 3.7 & 0 & 0 & 0 \\ 0 & 0 & 0 & 0 & 0 & 0 & 0 & 0 & 3.7 & 0 & 0 \\ 0 & 0 & 0 & 0 & 0 & 0 & 0 & 0 & 0 & 3 & 0 \\ 0 & 0 & 0 & 0 & 0 & 0 & 0 & 0 & 0 & 0 & 3.4 \end{bmatrix}$$

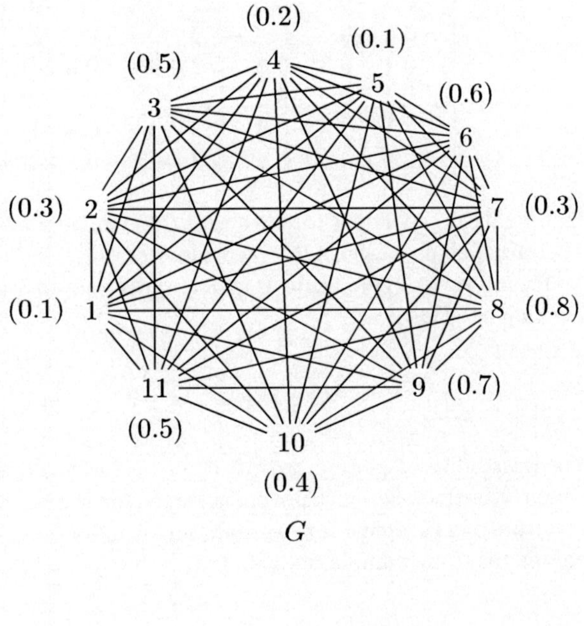

Fig. 1 Complete fuzzy graph

The adjacency matrix of graph G is,

$$A(G) = \begin{bmatrix} 0 & 0.1 & 0.1 & 0.1 & 0.1 & 0.1 & 0.1 & 0.1 & 0.1 & 0.1 & 0.1 \\ 0.1 & 0 & 0.3 & 0.2 & 0.1 & 0.3 & 0.3 & 0.3 & 0.3 & 0.3 & 0.3 \\ 0.1 & 0.3 & 0 & 0.2 & 0.1 & 0.5 & 0.3 & 0.5 & 0.5 & 0.4 & 0.5 \\ 0.1 & 0.2 & 0.2 & 0 & 0.1 & 0.2 & 0.2 & 0.2 & 0.2 & 0.2 & 0.2 \\ 0.1 & 0.1 & 0.1 & 0.1 & 0 & 0.1 & 0.1 & 0.1 & 0.1 & 0.1 & 0.1 \\ 0.1 & 0.3 & 0.5 & 0.2 & 0.1 & 0 & 0.3 & 0.6 & 0.6 & 0.4 & 0.5 \\ 0.1 & 0.3 & 0.3 & 0.2 & 0.1 & 0.3 & 0 & 0.3 & 0.3 & 0.3 & 0.3 \\ 0.1 & 0.3 & 0.5 & 0.2 & 0.1 & 0.6 & 0.3 & 0 & 0.7 & 0.4 & 0.5 \\ 0.1 & 0.3 & 0.5 & 0.2 & 0.1 & 0.6 & 0.3 & 0.7 & 0 & 0.4 & 0.5 \\ 0.1 & 0.3 & 0.4 & 0.2 & 0.1 & 0.4 & 0.3 & 0.4 & 0.4 & 0 & 0.4 \\ 0.1 & 0.3 & 0.5 & 0.2 & 0.1 & 0.5 & 0.3 & 0.5 & 0.5 & 0.4 & 0 \end{bmatrix}$$

Laplacian matrix for the graph G denoted by $L(G) = D(G) - A(G)$ is,

$$L(G) = \begin{bmatrix} 1 & -0.1 & -0.1 & -0.1 & -0.1 & -0.1 & -0.1 & -0.1 & -0.1 & -0.1 & -0.1 \\ -0.1 & 2.5 & -0.3 & -0.2 & -0.1 & -0.3 & -0.3 & -0.3 & -0.3 & -0.3 & -0.3 \\ -0.1 & -0.3 & 3.4 & -0.2 & -0.1 & -0.5 & -0.3 & -0.5 & -0.5 & -0.4 & -0.5 \\ -0.1 & -0.2 & -0.2 & 1.8 & -0.1 & -0.2 & -0.2 & -0.2 & -0.2 & -0.2 & -0.2 \\ -0.1 & -0.1 & -0.1 & -0.1 & 1 & -0.1 & -0.1 & -0.1 & -0.1 & -0.1 & -0.1 \\ -0.1 & -0.3 & -0.5 & -0.2 & -0.1 & 3.6 & -0.3 & -0.6 & -0.6 & -0.4 & -0.5 \\ -0.1 & -0.3 & -0.3 & -0.2 & -0.1 & -0.3 & 2.5 & -0.3 & -0.3 & -0.3 & -0.3 \\ -0.1 & -0.3 & -0.5 & -0.2 & -0.1 & -0.6 & -0.3 & 3.7 & -0.7 & -0.4 & -0.5 \\ -0.1 & -0.3 & -0.5 & -0.2 & -0.1 & -0.6 & -0.3 & -0.7 & 3.7 & -0.4 & -0.5 \\ -0.1 & -0.3 & -0.4 & -0.2 & -0.1 & -0.4 & -0.3 & -0.4 & -0.4 & 3 & -0.4 \\ -0.1 & -0.3 & -0.5 & -0.2 & -0.1 & -0.5 & -0.3 & -0.5 & -0.5 & -0.4 & 3.4 \end{bmatrix}$$

Finding the eigen values for the above $L(G)$ manually will be a very tedious process. Therefore, using MATLAB, eigenvalues are,

Therefore, from Fig. 2, the Laplacian spectrum for the given complete fuzzy graph G using MATLAB is,

$$S(G) = \begin{bmatrix} 4.4 & 4.2 & 3.9 & 3.4 & 2.8 & 2 & 1.1 & 0 \\ 1 & 1 & 2 & 1 & 2 & 1 & 2 & 1 \end{bmatrix}$$

The possibility of getting error in any stage of the above calculation is very high, which will affect the spectrum enormously. In the next section, enumerating laplacian spectrum of a complete fuzzy graph is made effortless by an effective algorithm with polynomial time complexity $\mathcal{O}(n)$.

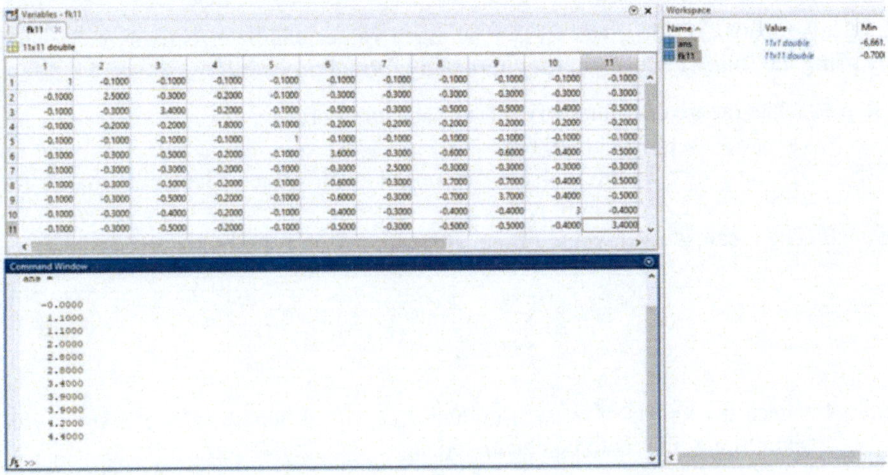

Fig. 2 Laplacian spectrum using MATLAB

4 Main Results

4.1 Interpretation of Algorithm

Before interpreting the algorithm, let us prove the following theorem.

Theorem 4.1.1 *One of the eigen values of the laplacian matrix for any complete fuzzy graph is zero.*

Proof Let $G : (\sigma, \mu)$ be a complete fuzzy graph (CFG) with n vertices and m edges. Let $L(G)$ be the laplacian matrix of the given CFG. We know that, the laplacian matrix of a graph is a symmetric matrix, i.e. $L(G) = L(G)^T$ Also, sum of entries of each row of $L(G)$ equals zero. We know that, if sum of entries of each row of a matrix is zero, then the given matrix is a singular matrix. Therefore, $L(G)$ is a singular matrix. We also know that, if a matrix is singular then one of its eigen value is zero. Therefore, one of the eigen value of the laplacian matrix for any complete fuzzy graph is zero.

4.2 Algorithm

Let $G : (\sigma, \mu)$ be a complete fuzzy graph (CFG) with n vertices. Since, the given CFG has n vertices, there exist n number of eigen values (say)$e_1, e_2, ..., e_n$ for the

Laplacian matrix of the given CFG. The eigen values can be effectively calculated by using the following algorithm without any formal procedure.

Step 1 Write the given values of $\sigma(x)$ in ascending order.
(say) $\sigma_1(x), \sigma_2(x), ..., \sigma_n(x)$

Step 2 By the above Theorem 4.1.1, one of the eigen value of laplacian matrix for CFG is zero. let $e_1 = 0$.

Step 3 The greatest eigen value of the laplacian matrix for CFG is given by

$$e_2 = \sum_{i=1}^{n-2} \sigma_i(x) + [2 \times \sigma_{n-1}(x)]$$

Step 4 Check the values of $\sigma_{n-(i-2)}(x)$and $\sigma_{n-(i-1)}(x)$, where i denotes the corresponding e_i that is to be calculated.

- Case (i) If the values are equal then go to step 6
- Case (ii) If not go to step 5

Step 5 Next eigen value is calculated using the formula

$$e_i = e_{i-1} - [(\sigma_{(n-(i-2))}(x) - \sigma_{(n-(i-1))}(x)) \times (i-1)]$$

where i denotes the corresponding e_i that is to be calculated. After finding an eigen value once again go to step 4 to find the next eigen value

Step 6 Then eigen value also gets repeated, i.e., $e_i = e_{i-1}$. Then go to step 4 to find the next eigen value.

Proof of Correction Let $G : (\sigma, \mu)$ be a complete fuzzy graph (CFG) with n vertices. Let $e_1, e_2, \ldots, e_n$ be the eigen values for the Laplacian matrix of the given CFG. Arrange the given values of $\sigma(x)$ in ascending order
(say) $\sigma_1(x), \sigma_2(x), \ldots, \sigma_n(x)$. Then, by Theorem 4.1.1, $e_1 = 0$. Then,

$$e_2 = \sum_{i=1}^{n-2} \sigma_i(x) + [2 \times \sigma_{n-1}(x)]$$

For $i = 3, 4 \ldots n$ and i now denotes the corresponding e_i that is to be calculated, check if the values of $\sigma_{n-(i-2)}(x) = \sigma_{n-(i-1)}(x)$, then $e_i = e_{i-1}$, else

$$e_i = e_{i-1} - [(\sigma_{(n-(i-2))}(x) - \sigma_{(n-(i-1))}(x)) \times (i-1)]$$

Now, increment the value of i by one and repeat the process till i takes the value n.

4.3 Illustration

Let us use the above algorithm to find the laplacian spectrum of Fig. 1: Complete fuzzy graph with 11 vertices.

Arrange the given $\sigma(x)$ in ascending order $\sigma_1(x) = 0.1$, $\sigma_2(x) = 0.1$, $\sigma_3(x) = 0.2, \sigma_4(x) = 0.3, \sigma_5(x) = 0.3, \sigma_6(x) = 0.4, \sigma_7(x) = 0.5, \sigma_8(x) = 0.5, \sigma_9(x) = 0.6$, $\sigma_{10}(x) = 0.7$,$\sigma_{11}(x) = 0.8$. Here, $n = 11$. Let $e_1, e_2, \ldots, e_{11}$ be the corresponding eigen values of Laplacian matrix of given CFG. Then, by Theorem: 4.1.1, one of the eigen value is zero.
let $\boldsymbol{e_1 = 0}$.

$$e_2 = \sum_{i=1}^{n-2} \sigma_i(x) + [2 \times \sigma_{n-1}(x)]$$
$$= (0.1 + 0.1 + 0.2 + 0.3 + 0.3 + 0.4 + 0.5 + 0.5 + 0.6) + [2 \times 0.7] = 4.4$$

$\therefore \boldsymbol{e_2 = 4.4}$
Now, i=3. Also, $\sigma_{10}(x) \neq \sigma_9(x)$ Therefore, using the formula,

$$e_i = e_{i-1} - [(\sigma_{(n-(i-2))}(x) - \sigma_{(n-(i-1))}(x)) \times (i-1)]$$
$$e_3 = e_2 - [(\sigma_{10}(x) - \sigma_9(x)) \times 2]$$
$$= 4.4 - [(0.7 - 0.6) \times 2] = 4.2$$

$\therefore \boldsymbol{e_3 = 4.2}$
Now, i=4. Also $\sigma_9(x) \neq \sigma_8(x)$. Therefore, using formula,

$$e_4 = e_3 - [(\sigma_9(x) - \sigma_8(x)) \times 3]$$
$$= 4.2 - [(0.6 - 0.5) \times 3]$$
$$= 3.9$$

$\therefore \boldsymbol{e_4 = 3.9}$
Now, $i = 5$. But $\sigma_8(x) = \sigma_7(x)$. Hence, $e_5 = e_4 = 3.9$.
$\therefore \boldsymbol{e_5 = 3.9}$
Now $i = 6$. Also, $\sigma_7(x) \neq \sigma_6(x)$ Therefore, using formula,

$$e_6 = e_5 - [(\sigma_7(x) - \sigma_6(x)) \times 5]$$
$$= 3.9 - [(0.5 - 0.4) \times 5]$$
$$= 3.4$$

$\therefore \boldsymbol{e_6 = 3.4}$

Now $i = 7$. Also, $\sigma_6(x) \neq \sigma_5(x)$ Therefore, using formula,

$$\begin{aligned} e_7 &= e_6 - [(\sigma_6(x) - \sigma_5(x)) \times 6] \\ &= 3.4 - [(0.4 - 0.3) \times 6] \\ &= 2.8 \end{aligned}$$

$\therefore \boldsymbol{e_7 = 2.8}$
Now, $i = 8$. But $\sigma_5(x) = \sigma_4(x)$. Hence, $e_7 = e_8 = 2.8$.
$\therefore \boldsymbol{e_8 = 2.8}$
Now $i = 9$. Also, $\sigma_4(x) \neq \sigma_3(x)$. Therefore, using formula,

$$\begin{aligned} e_9 &= e_8 - [(\sigma_4(x) - \sigma_3(x)) \times 8] \\ &= 2.8 - [(0.3 - 0.2) \times 8] \\ &= 2 \end{aligned}$$

$\therefore \boldsymbol{e_9 = 2}$
Now $i = 10$. Also, $\sigma_3(x) \neq \sigma_2(x)$. Therefore, using formula,

$$\begin{aligned} e_{10} &= e_9 - [(\sigma_3(x) - \sigma_2(x)) \times 9] \\ &= 2 - [(0.2 - 0.1) \times 9] \\ &= 1.1 \end{aligned}$$

$\therefore \boldsymbol{e_{10} = 1.1}$
Now, $i = 11$. But $\sigma_2(x) = \sigma_1(x)$. Hence, $e_{10} = e_{11} = 1.1$
$\therefore \boldsymbol{e_{11} = 1.1}$

Therefore, the Laplacian spectrum of the given CFG is,

$$S(G) = \begin{bmatrix} 4.4 & 4.2 & 3.9 & 3.4 & 2.8 & 2 & 1.1 & 0 \\ 1 & 1 & 2 & 1 & 2 & 1 & 2 & 1 \end{bmatrix}$$

Thus, the eigenvalues obtained through algorithm are exactly same as the one calculated using MATLAB just by considering the values of $\sigma_1(x), \sigma_2(x), \ldots, \sigma_n(x)$ and there is no need for writing the degree matrix, adjacency matrix or the Laplacian matrix.

Preposition 4.3.1 *Let $\sigma_1(x) = \sigma_2(x) = ... = \sigma_n(x) = \sigma(x)(say)$ for a CFG with n vertics. Let $e_1, e_2, ..., e_n$ be the eigen values of the laplacian matrix of CFG. Then, one of the eigen values of the laplacian matrix of CFG will be zero (by Theorem: 4.1.1) ie., $e_1 = 0$ and remaining $(n - 1)$ eigen values $e_2, e_3, \ldots e_n$ are given by, $e_i = \sigma(x) \times n \; \forall \, i = 2$ to n.*

5 Conclusion

The algorithm given in this paper enumerates the exact eigen values of Laplacian matrix for any given fuzzy complete graphs as well as fuzzy complete bipartite graphs with the polynomial time complexity $\mathcal{O}(n)$. Further work is in progress to extend this paper in various types of fuzzy graphs.

References

1. N. Biggs, E.K. Lloyd, R.J. Wilson, Graph Theory, pp. 1736–1936 (Oxford University Press, 1986)
2. L.A. Zadeh, Fuzzy sets. Information and control **8**, 3338–3353 (1965)
3. A. Kaufmann, Introduction a la Theorie des Sous-Ensembles Flous (Masson, Paris, Ch. 2:41–189) (1973)
4. A. Rosenfeld, Fuzzy graphs, in *Fuzzy and Their Applications to Cognitive and Decision Processes*, ed. by L.A. Zadeh, K.S. Fu, K. Tanaka, M. Shirmura, pp. 77–95 (Academic Press, New York, 1975)
5. R. Grone, R. Merris, V.S. Sunder, The laplacian spectrum of a graph. SIAM Journal on matrix analysis and applications **11**(2), 218–238 (1990)
6. B. Mohar, Y. Alavi, G. Chartrand, O.R. Oellermann, The Laplacian spectrum of graphs. Graph Theory, Combinator. Appl. **2**(12), 871–898 (1991)
7. P. Bhattacharya, F. Suraweera, An Algorithm to compute the max-min powers and a property of fuzzy graphs. Patt. Recogn. Lett. **12**, 413–420 (1991)
8. Z. Tong, D. Zheng, An algorithm for finding the connectedness matrix of a fuzzy graph. Congressus Numerant. **120**, 189–192 (1996)
9. M.R. Berthold, K.-P. Huber, Constructing fuzzy graphs from examples. Intell. Data Anal. **3**(1), 37–53 (1999)
10. M. Sarwar, M. Akram, An algorithm for computing certain metrics in intuitionistic fuzzy graphs. J. Intell. Fuzzy Syst. **30**(4), 2405–2416 (2016)
11. M. Sarwar, M. Akram, Certain algorithms for computing strength of competition in bipolar fuzzy graphs. Int. J. Uncertainty, Fuzz. Knowl.-Based Syst. **25**(06), 877–896 (2017)
12. P. Bhattacharya, Some remarks on fuzzy graphs. Pattern Recogn. Lett. **6**(5), 297–302 (1987)
13. T. Al-Hawary, Complete fuzzy graphs. Int. J. Math. Comb. **4**, 26 (2011)
14. K.H. Lee, *First Course on Fuzzy Theory and Applications* (Springer Verlag, Berlin, 2005)

Enhancement of Retinal Images Using Morphological Filters

Archita Johri, Vikrant Bhateja, Deepika Pal, and Babita Pal

Abstract Glaucoma is the second largest cause of permanent blindness all over the world. Glaucoma is marked by change in shape and size of optic disc (OD) and optic cup (OC). Contrast enhancement is necessary to improve the visuality of OD and OC in images for early detection, screening and timely treatment planning of glaucoma. Morphological filters are adapted for the contrast enhancement of the retinal fundus images. Combination of morphological filters such as top-hat transform and bottom-hat transform are considered for enhancement of retinal fundus images. The quality of the improvement of the image can be determined by the use of measure of enhancement (EME). The resultant enhanced images have been obtained with the increased value of EME indicating that enhancement has been done.

Keywords Contrast enhancement · Morphological filter · Measure of enhancement (EME)

1 Introduction

Glaucoma is chronic disease where a loss of vision occurs as a result of change in shape of optic cup and optic disc and high intraocular pressure (IOP). Diagnosis of glaucoma is mainly based on the IOP, medical history of patient's family [1] and change in optic disc structure [2]. Glaucoma suspect will have IOP more than 21 mm Hg [3]. There has been emphasis on glaucoma screening, since most of the times there is an insidious start of the disease (i.e. with no clear start point), and progression may be slow and unnoticed to the patient. In addition to that, there is an alarming stage of the disease in which patients suffer from loss of the visual field, which brings a challenge to the diagnosis and screening techniques [4]. Glaucoma has common symptoms as any other eye disease such as blurred vision, halos

A. Johri · V. Bhateja (✉) · D. Pal · B. Pal
Department of Electronics and Communication Engineering, Shri Ramswaroop Memorial Group of Professional Colleges (SRMGPC), Lucknow 226028, Uttar Pradesh, India

Dr. A.P.J. Abdul, Kalam Technical University, Lucknow, Uttar Pradesh, India

V. Bhateja et al. (eds.), *Data Engineering and Intelligent Computing*,
Advances in Intelligent Systems and Computing 1407,
https://doi.org/10.1007/978-981-16-0171-2_33

round lights, severe eye pin, headache, nausea; hence, its detection is difficult until it has reached its advanced stage. The people of age group 40–80 years are usually affected by glaucoma. Image analysis and processing are of great significance in early detection and treatment planning of glaucoma. Precise detection and accurate analysis of the disease are necessary, for timely treatment to prevent permanent vision loss. Various challenges such as low intensity contrast, unclear boundaries and background noise are associated with retinal images which might significantly affect detection of the region of interest. Therefore, enhancement of retinal images is necessary for the precise localization and segmentation of optic disc (OD) and optic cup (OC). There are various enhancement techniques used earlier for the improvement of the retinal images. Histogram matching is very common method for pre-processing. In this technique, histogram matching has been applied to for normalizing illumination and colour tone. Red channel has been considered for it contains most information for OD [5]. Saliency map detection and Winner-take-all algorithm were deployed by Ghassabi et al. ROI extraction takes place. Then vessel in painting converts ROI into a representation suited for ONH and cup segmentation [6]. Template matching with adaptive template size design approach has been used for OD detection by Yu et al. directional matched filtering and level sets approach has been used for OD localization and segmentation [7]. Vessel removal for OD segmentation, and fast hybrid level set model approach has been used for image segmentation purposes [8]. Combination of various methods is for OD segmentation such as use of nonlinear filtering, edge detection methods, Hough transform [9]. Zhang et al. proposed Otsu's method for image segmentation. In this paper, a histogram analysis is proposed to overcome flaws of Otsu thresholding when Otsu method leads to the failure segmentation when the object size is much less than background [10]. Anusorn et al. proposed pre-processing by the method of optic disc (OD) and optic cup (OC) localization and segmentation. OD had been extracted using the method of edge detection and variational level set approach. OC segmentation has been performed by colour component analysis (CCA) method and threshold level set method. The threshold level set method takes more time for cup boundary detection [11]. Maheshwari et al. extracted correntropy features using empirical wavelet transform (EWT). It was observed that system performance does not depend on the resolution of the image [12]. In order to overcome the existing limitation of large computation time for glaucoma detection, an improved method based on computer-aided diagnosis (CAD) of glaucoma through enhancement of retinal fundus images has been presented. Contrast enhancement based on the combination of morphological filters using disc-type structuring element has been proposed in this paper. The rest of the paper is organized in the following sections: Sect. 2 describes the overview of morphological filters and proposed methodology for contrast enhancement along with image quality assessment by calculating measure of enhancement (EME), Sect. 3 discusses the obtained results and Sect. 4 discusses the conclusion of the work.

2 Proposed Methodology Using Morphological Filters

2.1 *Morphological Filters*

Morphological processing is a collection of nonlinear operations associated different shapes of an image. Morphology is tool for extracting image components useful in further analysis of image [13]. These are implemented on an input image by the combinations of morphological operations. These operations are applied on input images with the structuring element defining the shape of entities in an image. Structuring element are of various shapes such as ball, diamond, line and disc. It is selected according to the shape and size of objects one wants to process in the input image [14]. Commonly used morphological operations are erosion, dilation, opening and closing. Erosion removes pixels on object boundaries and dilation adds pixels on object boundaries [15]. Opening is dilation followed by erosion and closing is erosion followed by dilation [16]. Morphological filtering has wide range of applications in Biomedical Image Processing such as lung cancer detection, tumour detection etc. It helps in improvement of detection, screening and timely treatment of various ophthalmic pathologies. It is simple method for the contrast enhancement of retinal fundus images for the detection and screening of glaucoma.

2.2 *Proposed Morphological Filter for Contrast Enhancement*

The proposed enhancement technique for retinal images for glaucoma detection is computationally simple using morphological operations [17]. Top-hat and bottom-hat transforms with the combination of image subtraction are implemented. Top-hat transform highlights the bright objects on a dark background, and bottom-hat transform highlights the darker regions of the image [18]. Firstly, coloured image is converted into greyscale input image (i) and structuring element (se) is defined. The size and shape of the structuring element are decided to extract the ROI, i.e. optic disc (OD) and optic cup (OC). Disc-shaped flat structuring element of size 10 is suitable for enhancement of red and green channels for extracting OD and OC. Extracted images are distorted, so their sharpening is done and then image is subjected to top-hat ($T(j)$) and bottom-hat (B(j)) transformations defined by Eqs. (1) and (2), respectively. Bottom-hat transform is subtracted from top-hat [19]. Then the resultant image is added to the greyscale (red and green channel) image. Morphological enhancement module has been shown in Fig. 1.

$$T(j) = j - (j \cdot \text{se}) \tag{1}$$

$$B(j) = (j \cdot \text{se}) - j \tag{2}$$

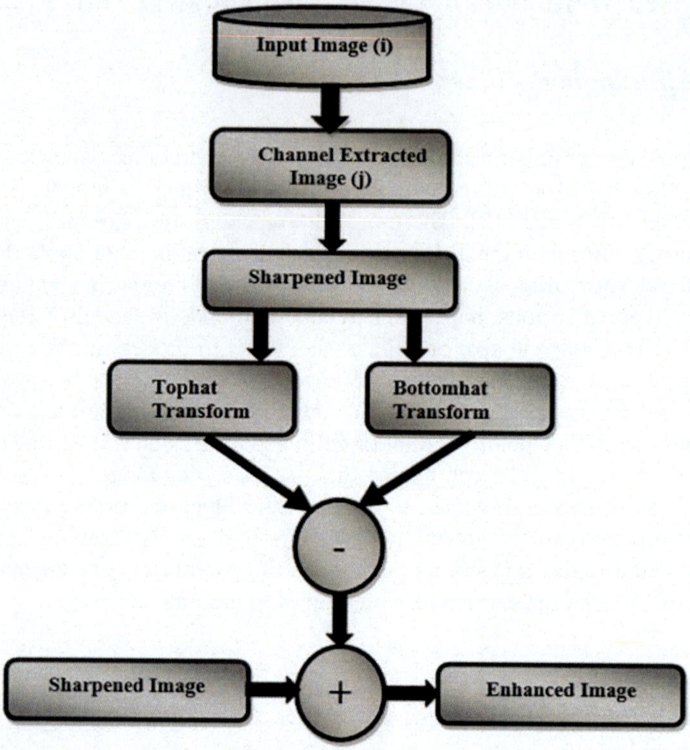

Fig. 1 Morphological contrast enhancement module

$$I = [(T(j) + j) - B(j)] \tag{3}$$

where

- j Sharpened image,
- $T(j)$ Top-hat transform of the sharpened image,
- $B(j)$ Bottom-hat transform of the sharpened image.

2.3 Measure of Enhancement (EME)

It is used to improve the visual appearance of the image for future analysis, detection, segmentation and recognition [20]. Measure of enhancement or measure of improvement (EME) helps to assess the quality of the enhanced image in comparison to the original image. The improvement of an image after the enhancement is very difficult to measure; only by seeing the image, hence EME is used to so determine the quantitative measure of the enhanced image. If the enhanced image has greater EME value than original image and the region of interest which is to be processed further is clearly visible, then the image is considered to be enhanced. A single retinal image can have different EME value by using different enhancement techniques such as EME for the measurement of frequency and time domain. In the proposed methodology, different EME values are obtained for both red channel enhanced image and green channel enhanced image of thc samc input image.

EME or measure of enhancement can be formulated as follows [21]:

$$\mathrm{EME} = \max_{\emptyset \in \{\emptyset\}} x \left(\frac{1}{k_1 k_2} \sum_{l=1}^{k_2} \sum_{k=1}^{k_1} 20 \log \frac{I^{\omega}_{\max;k,l}}{I^{\omega}_{\min;k,l}} \right)$$

3 Simulated Results and Discussions

Image quality assessment of the enhanced image can be easily done by the measure of enhancement (EME). Enhancement of extracted channel by using morphological filtering has been given in Fig. 2. Image quality assessment (IQA) or improvement in image after pre-processing can be done by values of EME. EME of the enhanced image must be greater than the original image for the improvement of the retinal image. EME for the Test_Image#A, Test_Image#B and Test_Image#C has been given in Table 1.

4 Conclusion

Early detection of glaucoma is necessary for its timely treatment; otherwise, it would cause permanent blindness in affected patients. The proposed algorithm for pre-processing or enhancement in retinal images is carried out by morphological filtering. It involves top-hat and bottom-hat transforms by using disc-shaped structuring element. Image quality assessment has been done by the measure of enhancement or measure of improvement (EME) for the enhancement of the retinal images. The proposed algorithm is simple for the implementation of enhancement of retinal images.

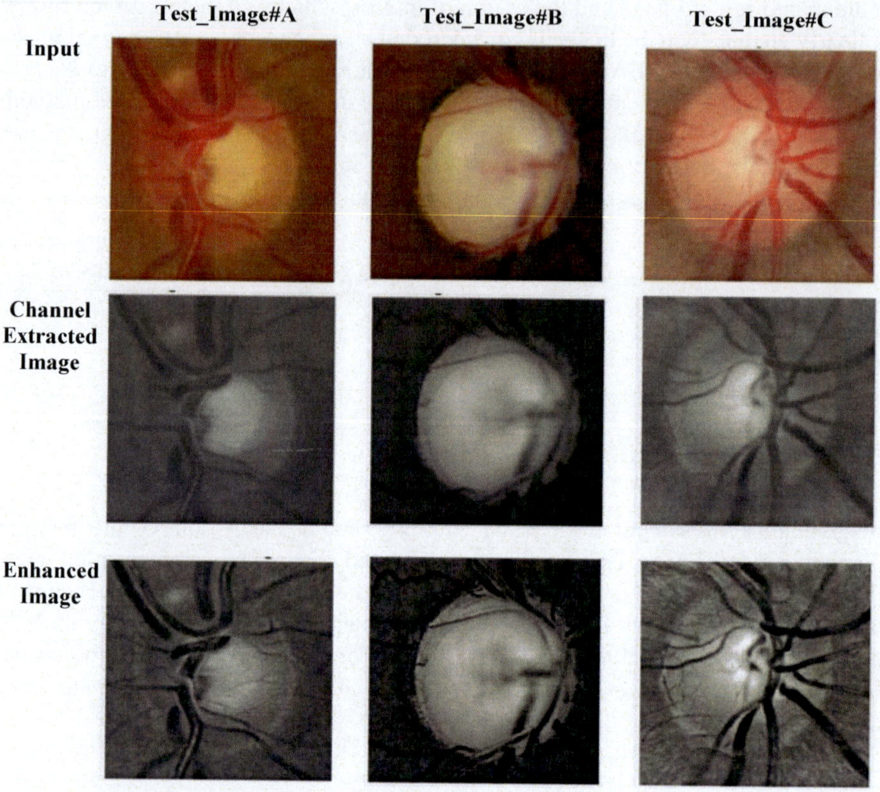

Fig. 2 Enhanced images obtained by using morphological filters

Table 1 Image quality assessment

	EME of extracted image	EME of enhanced image
Test_Image#A	2.65227	6.92571
Test_Image#B	4.00719	10.35056
Test_Image#C	3.15812	8.92906

References

1. S. Kavitha, N. Zebardast, K. Palaniswamy, R. Wojciechowski, E.S. Chan, D.S. Friedman, R. Venkatesh, P.Y. Ramulu, Family history is a strong risk factor for prevalent angle closure in a South Indian population. Ophthalmology **121**(11), 2091–2097 (2014)
2. M.J. Greaney, D.C. Hoffman, D.F. Garway-Heath, M. Nakla, A.L. Coleman, J. Caprioli, Comparison of optic nerve imaging methods to distinguish normal eyes from those with glaucoma. Invest. Ophthalmol. Vis. Sci. **43**(1), 140–145 (2002)
3. S.Y. Shen, T.Y. Wong, P.J. Foster, J.L. Loo, M. Rosman, S.C. Loon, W.L. Wong, S.M. Saw, T. Aung, The prevalence and types of glaucoma in Malay people: the Singapore Malay eye study.

Invest. Ophthalmol. Vis. Sci. **49**(9), 3846–3851 (2008)
4. Glaucoma screening is available at: https://eyewiki.aao.org/Glaucoma_Screening. Last updated: June 2019
5. M.N. Zahoor, M.M. Fraz, Fast optic disc segmentation in retina using polar transform. IEEE Access **5**(7), 12293–12300 (2017)
6. Z. Khassabi, J. Shanbehzadeh, K. Nouri-Mahdavi, A unified optic nerve head and optic cup segmentation using unsupervised neural networks for glaucoma screening, in *Proceedings of 40th International Conference of the IEEE Engineering in Medicine and Biology Society*, Honolulu, HI, USA, pp. 5942–5945, July 2018
7. H. Yu, E.S. Barriga, C. Agurto, S. Echegaray, M.S. Pattichis, W. Bauman, P. Soliz, Fast localization and segmentation of optic disk in retinal images using directional matched filtering and level sets. IEEE Trans. Inf. Technol. Biomed. **16**(4), 644–657 (2012)
8. M.M. Rama Krishnan, O. Faust, Automated glaucoma detection using hybrid feature extraction in retinal fundus images. J. Mech. Med. Biol. **13**(1), 1–23 (2017)
9. H. Yu, S. Barriga, C. Agurtoa, S. Echegaray, M. Pattichis, G. Zamora, W. Bauman, P. Soliz, Fast localization of optic disc and fovea in retinal images for eye disease screening, in *Vision Quest, Biomedical*, Albuquerque, NM, USA 87106
10. J. Zhang, J. Hu, Image segmentation based on 2D Otsu method with histogram analysis, in *International Conference on Computer Science and Software Engineering*, pp. 105–108 (2008)
11. C.B. Anusorn, W. Kongprawechnon, T. Kondo, S. Sintuwong, K. Tungpimolrut, Image processing technique using cup to disc ratio. Thammasat Int. J. Sci. Technol. **18**(1), 22–34 (2013)
12. S. Maheshwari, R.B. Pachori, U.R. Acharya, Automated diagnosis of glaucoma using empirical wavelet transform and correntropy features extracted from fundus images. IEEE J. Biomed. Health Inf. **21**(3), 803–813 (2016)
13. R.C. Gonzalez, R.E. Woods, *Digital Image Processing*, 3rd edn., Chap. 9 (Pearson), pp. 649–710
14. Structuring Element is available at: https://in.mathworks.com/help/images/structuringelements.html
15. V. Bhateja, S. Devi, A novel framework for edge detection of microcalcifications using a non-linear enhancement operator and morphological filter, in *IEEE 3rd International Conference on Electronics Computer Technology (ICECT)*, Kanyakumari, India, vol. 5, pp. 425–430, April 2011
16. A. Srivastava, Alankrita, A. Raj, V. Bhateja, Combination of wavelet transform and morphological filtering for enhancement of magnetic resonance images, in *International Conference on Digital Information Processing and Communications*, Ostrava, Czech Republic, pp. 460–474, 7–9 July 2011
17. V. Bhateja, M. Misra, S. Urooj, Human visual system based unsharp masking for enhancement of mammographic images. J. Comput. Sci. **21**, 387–393 (2017)
18. A.S. Bhadauria, M. Nigam, A. Arya, V. Bhateja, Morphological filtering based enhancement of MRI, in *Proceedings of 2nd International Conference on Computing, Communication and Control Technology (IC^4T)*, Lucknow, (U.P.), India, pp. 54–56, October 2018
19. A.S. Bhadauria, V. Bhateja, M. Nigam, A. Arya, Skull stripping of brain MRI using mathematical morphology, in *Proceedings of 3rd International Conference on Smart Intelligent Computing and Applications*, pp. 775–780 (2020)
20. S.S. Agaian, K.P. Lentz, A.M. Grigoryan, A new measure of image enhancement, pp. 153–163 (2000)
21. S.S. Agaian, K.P. Lentz, A.M. Grigoryan, A new measure of image enhancement, in *International Conference on Signal Processing & Communication*, pp. 1–7, January 2000

Sentiment Analysis Framework for E-Commerce Reviews Using Ensemble Machine Learning Algorithms

N. Dhamayanthi and B. Lavanya

Abstract Sentiment analysis or opinion mining as a research field is gaining importance. In this study, we are focusing on building a sentiment analyzer for Amazon product reviews in multi-domain category using ensemble machine learning algorithms. We have used five machine learning algorithms random forest, extra trees, bagging, AdaBoost, and stochastic gradient boosting for our experiments with four datasets, books, DVD, kitchenware, and electronics categories. We have compared the models to explore which model gives better performance in analyzing sentiments. The result shows that stochastic gradient boosting with 83% accuracy outperforms the other algorithms including the random forest algorithm which is generally considered as the best.

Keywords Sentiment analysis · Text mining · Opinion classification · Natural language processing · Ensemble machine learning algorithms

1 Introduction

We have moved away from measuring customer satisfaction to measuring customer experience. Customers do not hesitate to share their experience about the product they purchase in social media. Hence, it is important for the product manufacturers to understand the sentiment of the people who are buying products using which they can improve the quality of the product.

Opinion from customers will affect the future sales for hotels, video games, and E-commerce [1, 2]. In fields like customer relationship management, mining sentiments through social networks have become an invaluable resource [3]. Hence, opinion mining has become an active research area in recent years due to the practical demands and challenges [4]. New approaches for various domains are required as there are several issues which are yet to be addressed [5]. Survey on text classification elements used in the recent researchers is published by Marcin et al. [5]. Another

N. Dhamayanthi (✉) · B. Lavanya
Department of Computer Science, University of Madras, Chennai, India

V. Bhateja et al. (eds.), *Data Engineering and Intelligent Computing*,
Advances in Intelligent Systems and Computing 1407,
https://doi.org/10.1007/978-981-16-0171-2_34

survey on challenges on sentiment analysis is published by El-Din Mohamed Hussein [6].

Capturing the feedback available and analyzing it for various stakeholders involves text mining and sentiment analysis which is a prominent research area. Assigning the polarity for the review text as positive, negative, or neutral is the basic task in opinion mining [7]. There are many researchers focusing on analyzing and improving sentiment analysis [8–12].

We have used five ensemble machine learning algorithms: random forest, extra trees, bagging, AdaBoost, and stochastic gradient boosting in building a sentiment analysis framework for E-commerce reviews. We have used four multi-domain datasets from Amazon reviews to compare the performance.

We have organized the paper as follows. Section 2 describes the proposed methodology. Details of implementation with datasets description, ensemble machine learning algorithms used are detailed in Sect. 3. Results of the study along with the inference are described in Sect. 4. We then present our conclusions in Sect. 5.

2 Proposed Methodology

We have used ensemble machine learning algorithms described in Sect. 3.2 for building the sentiment analyzer to predict the sentiment of E-commerce reviews.

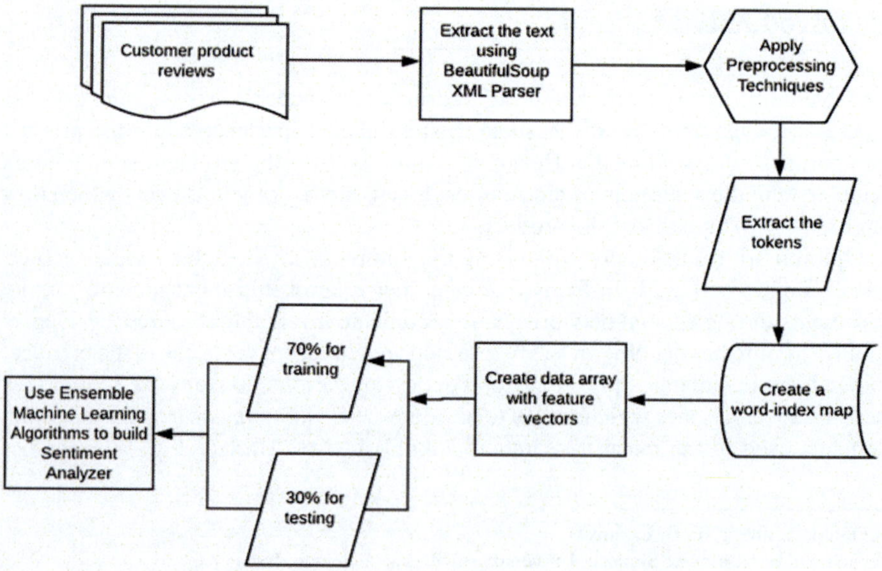

Fig. 1 Proposed methodology

Figure 1 shows the proposed methodology. We have used the datasets described in Sect. 3.1 with 70% for training and 30% for testing. We evaluate and compare the algorithms for their performance on each of the datasets.

We have used natural language toolkit (NLTK) platform and Scikit-learn machine learning library to build the sentiment analyzer. Tokenization, parsing, stemming, and classification tasks are performed in Python.

BeautifulSoup XML parser is used to extract the review text from the dataset. After preprocessing the data, we extract the tokens from the reviews. Two arrays are created to store positive and negative review tokens.

To create a feature vector, we will need two passes, one to determine the vocabulary size and one more to create data vectors. Then we create a dictionary of word index map with the word as key and value, an unique integer ranging from 0 to $V - 1$ where V is the vocabulary size.

To convert each review into feature vectors, we need to turn the sentence into a binary bag of words. If a word appears in a sentence, it gets converted into "1" in feature vector; otherwise, it remains "0". Data array is created to store the feature vectors along with the sentiment polarity values of "1" for positive and "0" for negative reviews.

After splitting the train and test datasets from the data array, we have evaluated the five ensemble machine learning algorithms for their accuracy, precision, recall, and F1-score using four datasets.

3 Implementation

3.1 Datasets from Amazon Product Reviews

The datasets published by Blitzer et al. [13] are used for our research. They have made multi-domain sentiment datasets from Amazon product reviews available for researchers.

We have used Amazon product reviews under four categories, books, DVD, kitchenware, and electronics. Each review is an XML file which contains the fields as given in Table 1. Each dataset contains 2000 reviews with 10 features.

Since the fields except Review_Text is not useful for our research, we have ignored the nine features and extracted Review_Text alone from each of the 2000 reviews. In order to avoid imbalance of datasets, we have used 1000 positive reviews and 1000 negative reviews for each dataset with 70% for training and 30% for testing.

Table 1 Dataset description

Field	Description
Unique ID	Unique identifier for reviews
ASIN	This is an identifier used by Amazon (ASINs) for each product
Product name	Name of the product purchased
Product type	Type of product (Electronics, Apparels, Books, DVDs, etc.)
Rating	Rating for the product (on a scale of 1–5) with 5 being the highest rating
Title	Title for the Review
Date	Date of Review
Reviewer name	Name of the Reviewer
Reviewer location	Location where the Reviewer belongs
Review_Text	This field contains the reviews shared by the Reviewer about the product purchased

3.2 *Ensemble Machine Learning Algorithms*

Five ensemble machine learning algorithms, random forests, extra trees classifier, bagging, AdaBoost, and stochastic gradient boosting were used in our experiments.

Random forest relies on the voting from multiple decision trees for prediction. Though similar to random forest classifier, extra trees classifier differs from the way, and the decision trees are constructed in the forest.

Bagging uses aggregation of prediction from subset trees which are created from training sample chosen randomly with replacement.

AdaBoost chooses the training set based on the accuracy of previous training. The final prediction is based on the weighted average of the weak learners pool created by this ensemble technique.

Stochastic gradient boosting uses training data subsample which is picked at random from the full training dataset without replacement at each iteration. In order to fit the base learner, subsample which is randomly selected is used than the full sample.

4 Results and Discussion

We have used five ensemble classifiers random forest, bagging, extra trees classifier, AdaBoost classifier, and stochastic gradient boosting. Performance metrics of accuracy, precision, recall, and F1-score are compared for these five classifiers for four datasets books, DVD, kitchenware, and electronics.

Table 2 lists the results of 5 classifiers on books dataset. Stochastic gradient boosting classifier has the top performance with 78% accuracy, recall, and F1-score and 79% precision. This is followed by random forest classifier with 76%

Table 2 Performance of ensemble classifiers with books dataset

Ensemble Classifiers	Accuracy	Precision	Recall	F1-score
Random Forest classifier	76	78	76	75
Bagging	72	72	72	72
Extra Trees classifier	71	72	71	71
AdaBoost classifier	73	74	73	73
Stochastic Gradient Boosting	**78**	**79**	**78**	**78**

Bold numbers indicate the highest score

accuracy and recall, 78% precision and 75% F1-score. AdaBoost classifier follows random forest with 74% precision and 73% accuracy, recall and F1-score. Bagging scores 72% on all parameters. Extra trees classifier scores the lowest among the five classifiers with 72% precision and 71% for the rest of the parameters.

Performance results of five classifiers against DVD dataset is given in Table 3. Stochastic gradient boosting has scored 75% on all parameters which is the highest score except precision for which random forest classifier scores the highest with 78%. For the rest of the parameters, random forest scores 74%. Bagging follows random forest with 74% precision and 73% on accuracy, recall, and F1-score. AdaBoost classifier stands 4th place with 72% score on all parameters. Extra trees classifier has the least scores with 72% precision and 69% for the rest.

Table 3 Performance of ensemble classifiers with DVD dataset

Ensemble Classifiers	Accuracy	Precision	Recall	F1-score
Random Forest classifier	74	**78**	74	74
Bagging	73	74	73	73
Extra Trees classifier	69	72	69	69
AdaBoost classifier	72	72	72	72
Stochastic Gradient Boosting	**75**	75	**75**	**75**

Bold numbers indicate the highest score

Table 4 Performance of ensemble classifiers with kitchenware dataset

Ensemble Classifiers	Accuracy	Precision	Recall	F1-score
Random Forest classifier	74	76	74	74
Bagging	73	73	73	73
Extra Trees classifier	70	70	70	70
AdaBoost classifier	73	73	73	73
Stochastic Gradient Boosting	**77**	**77**	**77**	**77**

Bold numbers indicate the highest score

Table 5 Performance of ensemble classifiers with electronics dataset

Ensemble Classifiers	Accuracy	Precision	Recall	F1-score
Random Forest classifier	80	81	80	79
Bagging	71	71	71	71
Extra Trees classifier	77	79	77	77
AdaBoost classifier	74	75	74	74
stochastic Gradient Boosting	**83**	**84**	**84**	**84**

Bold numbers indicate the highest score

Table 4 shows the comparison of ensemble classifiers with kitchenware dataset. Stochastic gradient boosting has the top performance with 77% scores. This is followed by random forest classifier with 74% accuracy, recall, and F1-score and 76% precision. Interestingly, both bagging and AdaBoost classifier have scored 73% on all parameters followed by extra trees classifier with 70% score.

Table 5 lists the performance scores of ensemble classifiers with electronics dataset. As in the other datasets, stochastic gradient boosting tops the performance with 83% accuracy and 84% precision, recall and F1-score followed by random forest with 81% precision, 80% accuracy and recall and 79% F1-score. Surprisingly extra trees classifier comes third with 79% precision and 77% score for the rest of the parameters. AdaBoost classifier has 74% accuracy, recall, and F1-score along with 75% precision. Bagging scores the lowest among all the classifiers with 71% on all parameters.

Figure 2 compares the accuracy parameter for the five ensemble algorithms using four datasets. Highest accuracy of 83% was achieved by stochastic gradient boosting followed by random forest with 80%. Extra trees classifier stands third with 77%

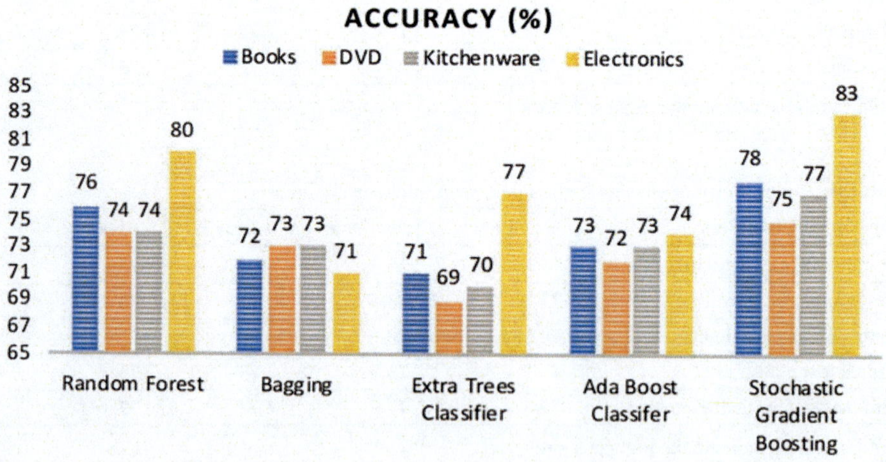

Fig. 2 Comparison of Accuracy parameter

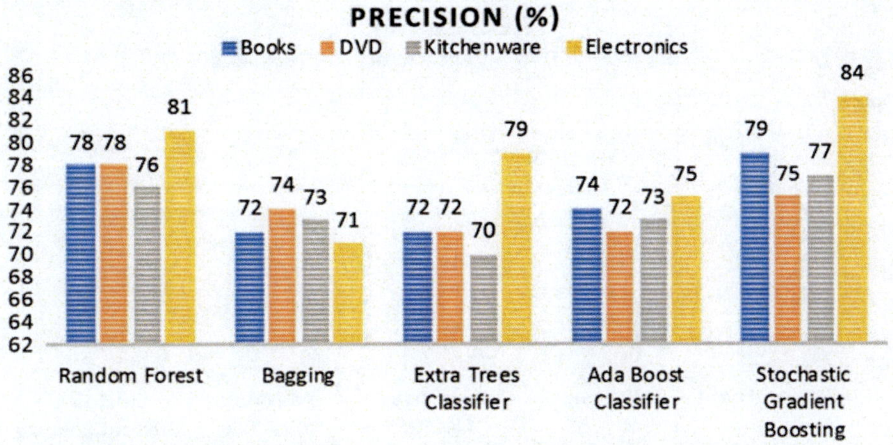

Fig. 3 Comparison of precision parameter

followed by AdaBoost classifier with 74%. While the highest accuracy for stochastic gradient boosting, random forest, extra trees, and AdaBoost were obtained for electronics dataset, bagging achieved the highest accuracy of 73% for both DVD and kitchenware datasets.

The results are similar for precision scores in Fig. 3 as accuracy scores. Stochastic gradient boosting with top precision of 84% followed by random forest with 81% which is followed by extra trees classifier with 79%. AdaBoost has 75% precision followed closely by bagging with 74%. Electronics dataset has the maximum precision scores for all classifiers except bagging in which DVD dataset has the highest precision.

In Fig. 4 also, there is no significant change in performance of ensemble classifiers with respect to recall parameter. Top recall scores of 84% bagged by stochastic gradient boosting followed by random forest with 80%, extra trees with 77%, AdaBoost classifier with 74% and bagging with the least score of 73% for DVD and kitchenware datasets.

Figure 5 depicts the F1-score comparison. Highest score of 84% bagged by stochastic gradient boosting, 79% by random forest, 77% by extra trees classifier, 74% by AdaBoost classifier, and 73% by Bagging.

5 Conclusion

In this paper, we have studied the ensemble machine learning algorithms for building a Sentiment Analysis framework for E-Commerce reviews. We have conducted experiments using Amazon product reviews with multi domain datasets, Books, DVD, Kitchenware and Electronics category in building a sentiment analyzer. Five

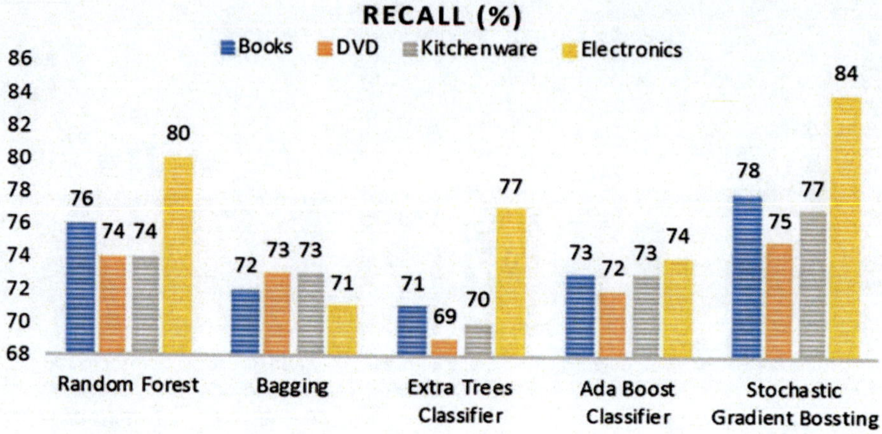

Fig. 4 Comparison of recall parameter

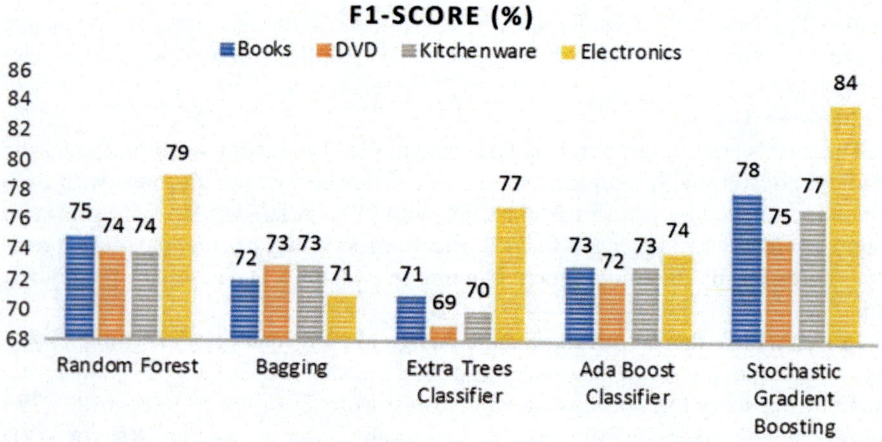

Fig. 5 Comparison of F1-Score parameter

machine learning algorithms random forest, bagging, extra trees classifier, AdaBoost classifier, and stochastic gradient boosting are used for building models for sentiment analysis.

Though random forest classifier is generally considered as the one which gives highest performance, we have seen that stochastic gradient boosting algorithm achieves the highest performance with 83% accuracy and 84% precision, recall and F1-score in comparison with all the other ensemble algorithms.

Stochastic gradient boosting algorithm outperforms random forest with 3% increase in accuracy and precision; 4% increase in Recall; and 5% increase in F1-score.

In future, we would be conducting experiments using datasets from different domains. Exploring other machine learning algorithms and feature selection techniques is also on the cards.

References

1. O. Araque, G. Zhu, C.A. Lelesias, A semantic similarity-based perspective of affect lexicons for sentiment analysis. Knowl.-Based Syst. **165**, 346–359 (2019)
2. F. Zhu, X.M. Zhang, Impact of online consumer reviews on sales: the moderating role of product and consumer characteristics. J. Market. **74**(2), 133–148 (2010)
3. M.M. Mostafa, More than words: social Networks' text mining for consumer brand sentiments. Expert Syst. Appl. **40**, 4241–4251 (2013)
4. S. Sun, C. Luo, J. Chen, A review of natural language processing techniques for opinion mining systems. Inf. Fusion **36**, 10–25 (2017)
5. M.M. Mironczuk, J. Protasiewicz, A recent overview of the state-of-the-art elements of text classification. Expert Syst. Appl. 106 (2018)
6. El-Din Mohamed Hussein, D.M., A survey on sentiment analysis challenges. J. King Saud Univ. Eng. Sci. **30**, 330–338 (2018)
7. A. Kaur, V. Gupta, A survey on sentiment analysis and opinion mining techniques. J. Emerg. Technol. Web Intell. **5**(4) (2013)
8. T. Chen, R. Xu, Y. He, X. Wang, Improving sentiment analysis via sentence type classification using BiLSTM-CRF and CNN. Expert Syst. Appl. **72**, 221–230 (2017)
9. O. Araque, I. Corcurea-Platas, J.F. Sanchez-Rada, C.A. Lelesias, Enhancing deep learning sentiment analysis with ensemble techniques in social applications. Expert Syst. Appl. **77**, 236–246 (2017)
10. M. Kang, J. Ahn, K. Lee, Opinion mining using ensemble text hidden Markov models for text classification. Expert Syst. Appl. **94**, 218–227 (2018)
11. A. Sharaff, A. Soni, Analyzing sentiments of product reviews based on features, in *Proceedings of the 2nd International Conference on Trends in Electronics and Informatics (ICOEI 2018)*
12. Z. Fachrina, D.H. Widyantoro, Aspect-sentiment classification in opinion mining using the combination of rule-based and machine learning, in *2017 International Conference on Data and Software Engineering (ICoDSE 2017)*
13. J. Blitzer, M. Dredze, F. Pereira, Biographies, bollywood, boomboxes and blenders: domain adaptation for sentiment classification, in *Association of Computational Linguistics (ACL 2007)*

Drowsiness Detection Using Viola–Jones Object Detection Algorithm for Real-Time Data

Raghavendra Kumar, Himanshu Rathore, Palak Agrawal, and Prakhar Gupta

Abstract This research paper is proposed a real-time monitoring of drowsiness of the driver, to forestall accidents. With the expansion in population, there has been an incredible increase in road accidents. In India, around 60% of the accidents are caused because of driver fatigue. So, this paper is proposed to detect drowsiness of drivers by capturing video frames of eye closure patterns. This system consists of a small security camera pointing directly into the driver's face. It monitors the eyes of the driver and their closure patterns. The face area and position of eyes are pinpointed by the Viola–Jones object detection algorithm. It uses a 6-coordinate of eye to seek out whether eyes are close or open, using a Haar cascade classifier. When 20 consecutive frames are detected with eyelids aspect ratio less than 0.25, then the system comes to a result that the driver is drowsy and issues an alert.

Keywords Driver drowsiness · Real time · Viola–Jones · Eye aspect ratio

1 Introduction

Drowsiness of drivers is one of the major causes of large numbers of vehicle accidents. The survey of NHTSA in 2017 estimates that 91,000 cases were reported by police officers which involve drowsy drivers. Around 50,000 injuries and 800 deaths took place in these crashes. For reducing these kinds of mishappenings, the drowsiness detection system has been implemented by major automobile companies like Volvo, Mercedes-Benz, Saab, Hyundai, and Nissan which detect the drowsiness behavior of the driver as a precautionary measure. There are several techniques used such as vehicle-based, behavioral and physical measures [1]. Most investigators have used behavioral approaches such as blinking patterns of eye, sleeping eyes, head position, and yawning to detect drowsiness of drivers, as it is an efficient approach [2, 3].

The proposed method prevents or detects drowsiness at the root, which is based on the fixed number of consecutive frames and eye aspect ratio of eyelids. The

R. Kumar · H. Rathore (✉) · P. Agrawal · P. Gupta
Deparment of Information Technology, KIET Group of Institutions, Uttar Pradesh, Ghaziabad 201206, India

V. Bhateja et al. (eds.), *Data Engineering and Intelligent Computing*,
Advances in Intelligent Systems and Computing 1407,
https://doi.org/10.1007/978-981-16-0171-2_35

method uses eye closure patterns, which is the most effective approach for detecting drowsiness. With the help of eyes and its blinking pattern, we are able to find whether the driver is drowsy or not. As eyes are considered to be the most stable feature on face, as compared to other facial features, so systems based on eyes can give the most desirable results. Using Viola–Jones object detection algorithm, we are able to capture the whole face image and determine the position of eyes.

This includes real-time capturing of the image, and generating alert if the driver found drowsy. Here, we first capture the image through a webcam, placed in front of the seat of the driver, then from the image, we extract face and then followed by finding eye. After extracting eyes, we are calculating the average eye aspect ratio to determine the eye state. When the eyes are found to be closed for the fixed number of consecutive frames, then alert is generated.

2 The Proposed Architecture

The real-time drowsiness detection system uses video images recorded by the security camera placed in front of the driver. Video frames are then converted into images using OpenCV. The detection of face and eyes is performed with the help of Viola–Jones object detection algorithm using Haar cascade classifier and Dlib library, from extracted images. Viola–Jones algorithm is an efficient algorithm that can detect the faces and eyes in the most robust way. From 68 facial landmarks, we are able to get 6 coordinates for both the eyes. Using eye aspect ratio (EAR), we can determine whether eyes are open or close. When 20 consecutive image frames are detected with aspect ratio less than 0.25, then the system generates an alert shown in Fig. 1.

Eyes state/behavior is a strong indicator to detect drowsiness. The main focus is on continuously focusing the behavior of eyes with the help of 6 coordinates of eyes, and the eye aspect ratio (EAR).

3 Related Work

Kevan Yuen et al. present a model which is based on cascade convolutional neural network that performs for face detection as well as landmark localization under different occlusion, using AlexNet as a first stage. It includes stacked hourglass as the second stage for detecting faces with a score. However, the second stage found significantly improved accuracy from the stage one. The head position is found using the POSIT algorithm. The AFLW dataset is used to take annotated faces with outline boxes, used for training our face and for occlusion augmentations, SUN2012 dataset detector is used which gives 86.1% accuracy and performs significantly better than its benchmark model [4].

Bhargava Reddy et al. proposed an upgraded deep neural network for drowsiness detection systems, which is designed especially for embedded systems. System

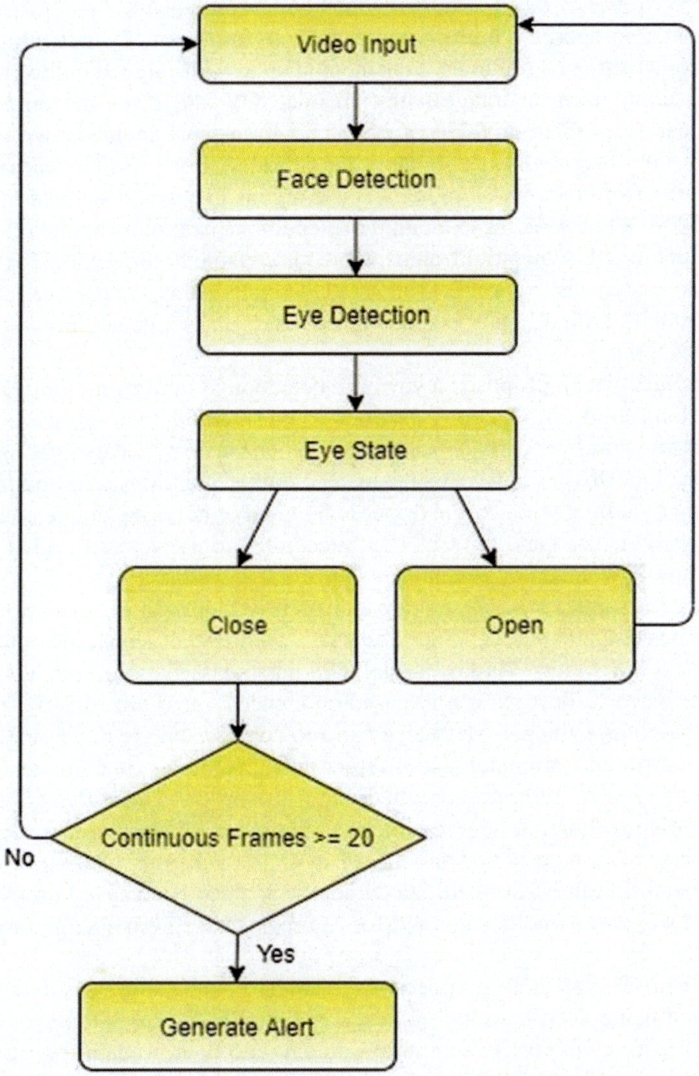

Fig. 1 Drowsiness detection system architecture

takes the driver's face as an input and divides the phase behavior into three phases (drowsiness, yawing, and normal). By using Logitech C920 HD Webcam, custom dataset was collected on their own. The performed experiments were categories into three models which are compressed-2, baseline-2, and baseline-4 models. Here, compressed-2 model served as the best way, as it has the compressed model size as well as less GPU memory usage which gives 89.5% accuracy and is better than its benchmark model like faster RCNN and AlexNet architecture [5].

Jie Lyu et al. proposed a multi-granularity deep framework for detecting drowsiness of drivers using video frames. This method takes well-aligned multi-granularity patches as input from an image using an open CV and gives the output in the form of binary classification (alert or not). This framework includes two important components: multi-granularity convolutional neural network, used to learn the facial representation from the most important sense organs like eye, nose, and mouth and applies LSTM network on these facial representations to explore stable relationship with different length sequential frames, which is capable to differentiate spatial and temporal dependencies. National TsingHua University dataset is used to achieve the accuracy 90.05% with 37 fps speed and build a new dataset named Forward Instant DDD (FI-DDD) [6].

Weiwei Zhang et al. proposed a yawning detection system using a single camera based on a long short-term memory network. In this method, focus is on the mouth of the driver rather than eyes or other parameters. When the driver wears sunglasses, then also fatigue of the driver can be detected using mouth expressions. This model system gives output by telling whether the driver is fatigued or not. The kaggle competition dataset YawDD is used and the CNN architecture GoogLeNet and LSTM network is applied and 87% accuracy is achieved [7].

Mkhuseli Ngxande et al. present a paper on driver drowsiness detection. In this paper, author compares the behavioral features on different techniques of machine learning like SVM, hidden Markov model. This paper describes different ways to find whether the driver is drowsy or not, based on physical measures, driver's behavior, and some based on vehicles. Meta-analysis and comparison are done on around 25 papers with different parameters on different machine learning techniques [8].

Jing-Ming Guo et al. proposed a method for driver drowsiness detection system using hybrid convolutional neural networks. Authors consider the five facial landmarks of face such as eyes and mouth. They experiment this method in different conditions and find the best result. Accuracy of 84.32% is achieved on the ACCV dataset and suggested another method for future work with time skip combination LSTM [2].

Xuan-Phung Huynh et al. proposed a system based on drowsiness detection and implemented using deep learning. A neural network based on 3D convolutional is used to get features in spatial–temporal domain, and then gradient boosting algorithm is used which works as binary classifier. Semi-supervised learning is used to increase overall performance. DDD dataset is a huge and complex dataset consisting of sequences of images. The dataset is divided into five categories: night glasses, bare face, night bare face, glasses, and sunglasses. An efficient tracker is employed by the Haar element face detector, finding the surface area of face from the video that gives an accuracy of 87.46% [9].

Mingfei Sun et al. present idea of taking eye movements (eyelid movement and eyeball movement) for drowsiness.The comparison between three sets (eyelid, eyeball, (eyelid and eyeball)) by using CNN-Net and CNN LSTM-Net on different sizes of video clips (10 s,30 s,60 s,120 s) at 30 fps of 29 subjects dataset. Labelling is done for drowsiness level according to subject current state (not drowsy at all, slightly drowsy, drowsy, significantly drowsy, and extremely drowsy), and every 10 s labeled as average of three labels. Both eyelid and eyeball together give better performance then alone eyelid, alone eyeball gives result below baseline with CNN-Net model on Pearson correlations and mean absolute errors for short period of time [10].

Rateb Jabbar et al. proposed a drowsiness detection system for android by using a neural network in which facial structure of 22 subjects are extracted in the form of images from 30fps AVI video. Subjects have five driving scenarios (with glasses, with sunglasses, without glasses, the night with glasses, and night without glasses). Out of 22 subjects, 18 are used for training and remaining for testing, that classifies as sleepy and non-sleepy. By use of GTX 1080, 7 fps end-to-end real-time speed with 81% accuracy was achieved [11].

Weiwei Zhang et al. suggest the idea of driver yawning detection based on deep convolutional neural network, in which yawning is detected through the mouth. First extract the face using CNN FaceNet model having 5 C-layers and 4 S-layers, then track the nose by the combination of Kalman filter with TLD (redetect tracking on fails). Output passes to the NN_YD in which the mouth divides into two, left and right. It also compares CNN_NoseNet with AdaBoost and WaldBoost, and hence, CNN_NoseNet gives better performance. Overall system gives the accuracy upto 92% [12].

Muhammad Tayab Khan et al. proposed real-time video surveillance system for detecting drowsiness, based on closure pattern of eyelid. As a first step, face is identified by using Viola–Jones object detection algorithm from the image then performs localization and filtration of eyes with Sobel operator and determines the concavity of eyelids. Authors classify the drowsiness of the driver from the curvature of eyelids. Author also designed an alarm generation system. They used the dataset from different papers and achieve the accuracy of 95% [3].

4 Methodology

In this section, we will describe the experimental analysis of the proposed system.

4.1 Video Procurement

Video procurement is performed by using a web camera placed in front of the driver, and capturing the real-time video of the driver while driving. OpenCV provides extensive support for processing live videos. It is also possible to choose the in-built webcam or external camera for capturing video by setting the right parameters. After capturing the video, the next step is video converting into frames. It takes live captured video as an input and converts it into frames, which are processed further. This process took place as follows—first, it grabs a frame from the video, and then the image is decompressed and then retrieved. This retrieved image/frame is used further for face detection.

4.2 Face Detection

The perfect detection of the people's face is an essential process. It uses computer vision technology which is used to detect people's faces from photographs or digital videos, using the Dlib library. It detects faces by using computer vision technology. Dlib is an open source library that provides an excellent software development environment, which supports machine learning. Dlib uses 2 essential components are machine learning tools and linear algebra.

4.3 Eye Detection

Eye is the salient factor to detect driver drowsiness. After the face is detected then we find the position of the eye from the detected face image, by extracting the frontal face feature using python library (i.e., Dlib). Processing time can be reduced by marking the region of interest before detection of the eyes. After detection of eyes using haar cascade, then detect driver drowsiness by calculating eye aspect ratio. Our system is able to detect eyes when the driver is wearing specs, and this is the most challenging part of this system.

Face and eye detection are performed by Viola–Jones object detection algorithm which is described below.

4.4 Viola–Jones Algorithm

Viola–Jones algorithm is the efficient way to detect frontal face features in real time from an image. Haar-like features are used to generate classifier function.

After generation of classifier function, Adaboost algorithm is used for training of the real-time data, for selecting the region of interest. It is an efficient process for searching a small number of good "features" but with great variety, like eyes is the region of interest here [13].

The classifier files (like face.xml, eye.xml) are generated as an output of Viola–Jones algorithm. Face and eyes are recognized using haar cascade classifier files that are face.xml and eye.xml, respectively [14].

4.5 Eye State

The detection of state of eye is the most efficient way to detect drowsiness of the driver. The proposed system defined the state of eye using the aspect ratio of eye.

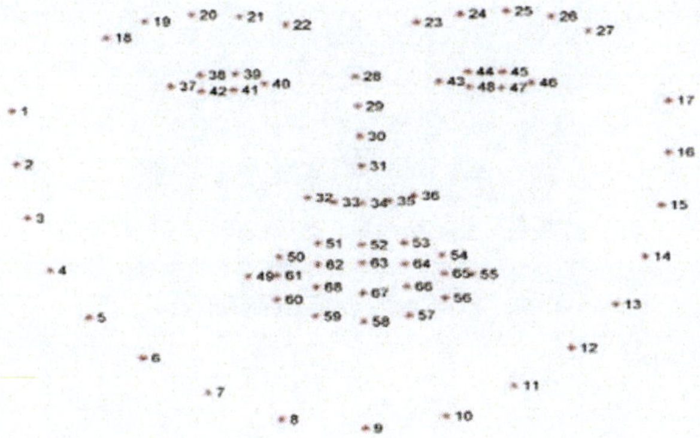

Fig. 2 68 Facial landmarks [15]

To determine the state of eye means eyes either open or close, the proposed method uses a threshold value for classifying the state of eye using eye aspect ratio (EAR).

4.5.1 Calculation of EAR

After detecting eyes, 68-facial landmarks are located by using the python Dlib library for each eye frame shown in Fig. 2. Each eye consists of six located landmarks. These six coordinates are sufficient to track the movement of eyelids at different states.

To find the gap between upper and lower eyelid, the Euclidean distance formula is used to calculate distance.

$$D(m, n) = \sqrt{(m_1 - m_2)^2 + (n_1 - n_2)^2} \tag{1}$$

where m_1, n_1 and m_2, n_2 are x-, y-coordinates of pixels. For example, if we define Euclidean distance between p_2 and p_6 pixel:

$$p_2 - p_6 = D(m, n) \tag{2}$$

Eye aspect ratio (EAR) is formed to get the relation between the gap of eyelids and breadth of the eye. If the aspect ratio is almost zero, then eyes are at a closed state. So, from Eqs. (1) and (2), we get [16]:

$$\mathbf{EAR} = \frac{\|p_2 - p_6\| + \|p_3 - p_5\|}{2\|p_1 - p_4\|} \tag{3}$$

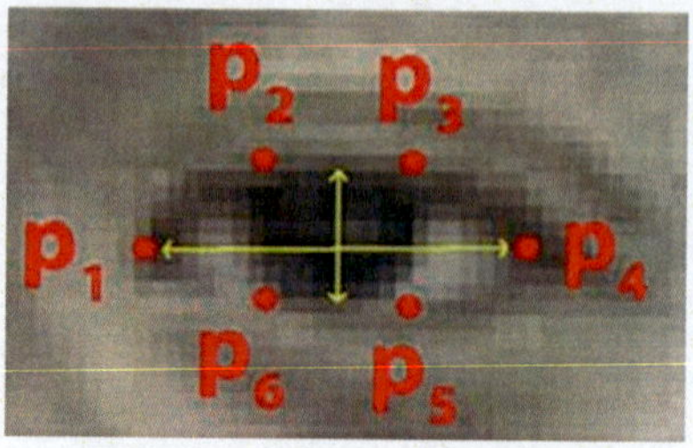

Fig. 3 Open eye [15]

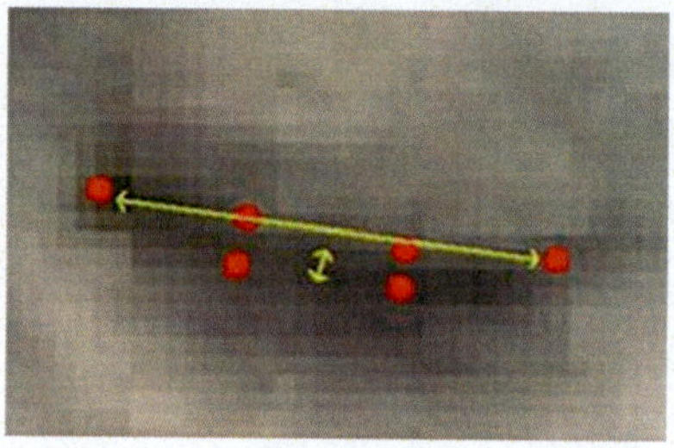

Fig. 4 Closed eye [15]

Average of EAR values is taken for both the left and right eye of a frame. The threshold value is defined below for classify the state of eye:

$$f(\text{EAR}) = \begin{cases} \text{Eyes Open}, & \text{EAR} \geq 0.25 \\ \text{Eyes Close}, & \text{EAR} < 0.25 \end{cases} \tag{4}$$

where $f(\text{Ear})$ is the classification function for defining the state of eye. For the classification of eye state as open shown in Fig. 3, eye aspect ratio would be greater than or equal to the threshold value, i.e., 0.25. If the eye aspect ratio is less than 0.25, then state of eye is considered as close shown in Fig. 4.

4.6 Drowsiness Detection

After detecting the position of the eye from the face image, then we find the Euclidean distance between the pixels. We are considering a 6 (x–y) coordinate system for finding the Euclidean distance between pixels. After finding the Euclidean distance, then we find the aspect ratio for both the eyes. If the average aspect ratio is less than 0.25 for the 20 consecutive frames, the system issues the alert message to the driver. Thus, our system detects that the driver is drowsy and issues the alert.

(a) (b) (c)

Fig. 5 Eye states: **a** Open eyes with Ear >0.25. **b** Partially closed eye with Ear <0.25 and frame count <20. **c** Close eyes with Ear <0.25 and frame count ≥20

Figure 5a shows the eyes having Eye aspect ratio greater than 0.25, and it classifies as open eyes. Hence, drowsiness is not detected. Figure 5b describes the eyes with EAR less than 0.25 and consecutive frames count less than 20. So, no alert is generated for drowsiness. Figure 5c describes the close eyes with EAR less than 0.25 in more than or equal to 20 consecutive frames. So, drowsiness is detected and alert is generated.

Whole experiment was performed using Dlib library and Open CV on Jupyter notebook which was installed on the local machine (with webcam and i5 processor).

5 Result

Many researchers use several techniques for detecting drowsiness of the driver considering behavioral measures like mouth, head movement, lips, and eyes. The proposed system uses eye state for detecting drowsiness of the driver. The proposed system classifies the eye state in terms of open and close based on the eye aspect ratio (EAR). The graph shown depicts different EAR values of various positions of eyes with respect to frames.

In Fig. 6, EAR values for various positions of eyes are shown. The graph shown with sky blue color having eye aspect ratio greater than and equal to 0.25, classifies the eyes as open. Similarly, the graph shown with orange color having EAR value less than 0.25 classifies the eyes as close. Figure 7 shows the EAR values for different eye closure patterns having eye aspect ratio less than 0.25 (Fig. 8).

If we get 20 consecutive frames with eye aspect ratio less than 0.25, then the system generates an alert for drowsiness. For example, Fig. 9 we can see that from 70 to 117 consecutive frames having EAR value less than 0.25, the system will continuously generate drowsiness alert after 90 frames to 117 frames.

Table 1 shows various possible scenarios for different eye aspect ratios with different frame counts. According to them, the system decides whether the driver drowsy or not.

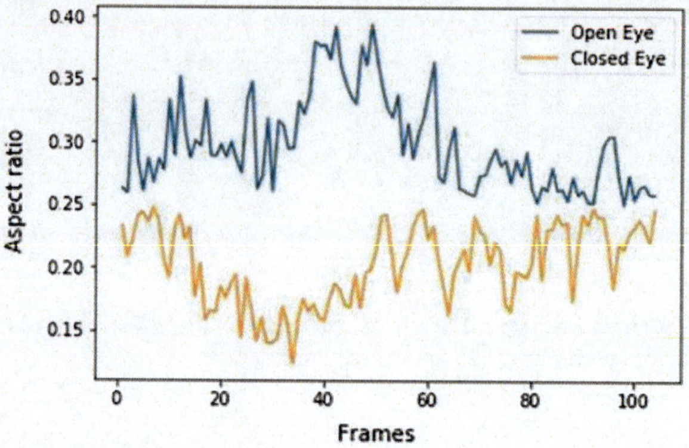

Fig. 6 EAR values for open and close eye

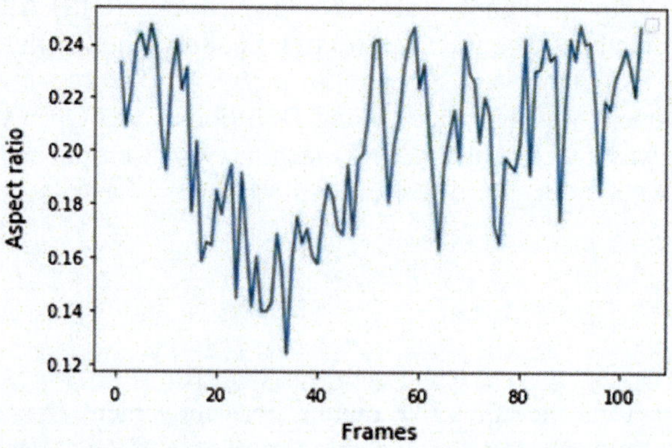

Fig. 7 EAR values graph for closed eye state

6 Conclusion

Automobile driver's faces are detected using frontal faces of humans and eyes from digital images or videos, using Dlib library. Our proposed system is based on real-time monitoring of eye closure patterns which gives efficient results, and it is based on Viola–Jones algorithm which is the backbone for drowsiness detection. This system detects drowsiness on naked eyes, as well as with people wearing spectacles. When 20 consecutive frames are detected with eyelids aspect ratio less than 0.25, the system comes to a result that the driver is drowsy and issues an alert. When this kind of system is implemented in automobiles help in reducing the risk of possible accidents that can occur due to drowsiness.

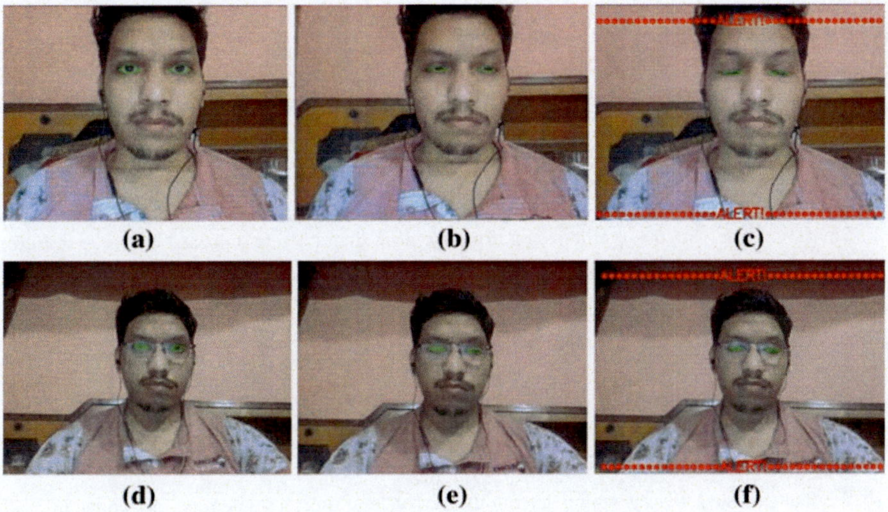

Fig. 8 Different eye positions without specs: **a** Open eyes, **b** partially close eyes, **c** close eyes, Different eye positions with specs: **d** Open eyes, **e** partially close eyes, **f** close eyes

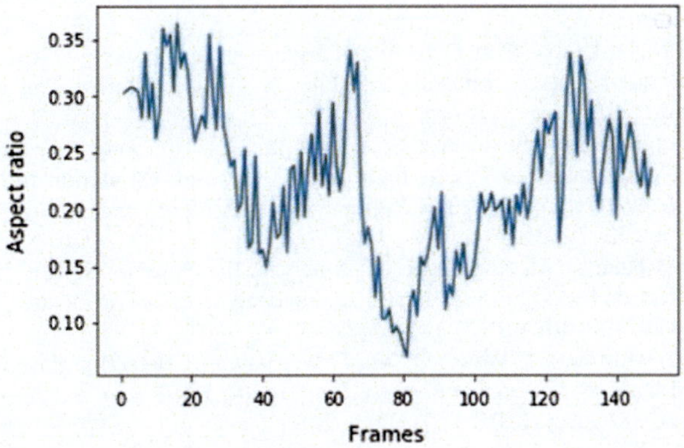

Fig. 9 EAR values for real-time eye monitoring

Table 1 Conditions for drowsiness alert generation

Serial number	Eye aspect ratio (EAR)	Count of consecutive frames	Drowsiness detected
1	<0.25	Less than 20	NO
2	<0.25	At least 20	YES
3	≥0.25	Less than 20	NO
4	≥0.25	At least 20	NO

References

1. A. Sahayadhas, K. Sundaraj, M. Murugappan, Detecting driver drowsiness based on sensors: a review. Sensors **12**(12), 16937–16953 (2012)
2. J.M. Guo, H. Markoni, Driver drowsiness detection using hybrid convolutional neural network and long short-term memory. Multimed. Tools Appl., 1–29 (2018)
3. M. Tayab Khan, H. Anwar, F. Ullah, A. Ur Rehman, R. Ullah, A. Iqbal, B.H. Lee, K.S. Kwak, Smart real-time video surveillance platform for drowsiness detection based on eyelid closure. Wirel. Commun. Mobile Comput., 2019
4. K. Yuen, S. Martin, M.M. Trivedi, Looking at faces in a vehicle: a deep CNN based approach and evaluation, in *2016 IEEE 19th International Conference on Intelligent Transportation Systems (ITSC)* (IEEE, 2016), pp. 649–654
5. B. Reddy, Y. H. Kim, S. Yun, C. Seo, J. Jang, Real-time driver drowsiness detection for embedded systems using model compression of deep neural networks, in *Proceedings of the IEEE Conference on Computer Vision and Pattern Recognition Workshops*, pp. 121–128 (2017)
6. J. Lyu, Z. Yuan, D. Chen, Long-term multi-granularity deep framework for driver drowsiness detection. arXiv preprint arXiv:1801.02325 (2018)
7. W. Zhang, J. Su, Driver yawning detection based on long short term memory networks, in *2017 IEEE Symposium Series on Computational Intelligence (SSCI)* (IEEE, 2017), pp. 1–5
8. M. Ngxande, J.R. Tapamo, M. Burke, Driver drowsiness detection using behavioral measures and machine learning techniques: a review of state-of-art techniques in *2017 Pattern Recognition Association of South Africa and Robotics and Mechatronics (PRASA-RobMech)* (IEEE, 2017), pp. 156–161
9. X.P. Huynh, S.M. Park, Y.G. Kim, Detection of driver drowsiness using 3D deep neural network and semi-supervised gradient boosting machine, in *Asian Conference on Computer Vision* (Springer, Cham, 2016), pp. 134–145
10. M. Sun, M. Tsujikawa, Y. Onishi, X. Ma, A. Nishino, S. Hashimoto, A neural-network-based investigation of eye-related movements for accurate drowsiness estimation, in *2018 40th Annual International Conference of the IEEE Engineering in Medicine and Biology Society (EMBC)* (IEEE, 2018), pp. 5207–5210
11. R. Jabbar, K. Al-Khalifa, M. Kharbeche, W. Alhajyaseen, M. Jafari, S. Jiang, Real-time driver drowsiness detection for android application using deep neural networks techniques. Procedia Comput. Sci. **130**, 400–407 (2018)
12. W. Zhang, Y.L. Murphey, T. Wang, Q. Xu, Driver yawning detection based on deep convolutional neural learning and robust nose tracking, in *2015 International Joint Conference on Neural Networks (IJCNN)* (IEEE, (2015), pp. 1–8
13. G. Bradski, A. Kaehler, *Learning OpenCV: Computer Vision with the OpenCV Library*, O'Reilly Media, Inc. (2008)
14. P. Viola, M. Jones, Rapid object detection using a boosted cascade of simple features, in *Proceedings of the IEEE Computer Society Conference on Computer Vision and Pattern Recognition* (2001)
15. T. Soukupová, J. Cech, Eye blink detection using facial landmarks, in *21st Computer Vision Winter Workshop*, Rimske Toplice, Slovenia (2016)
16. C. Yashwanth, J.S. Kirar, Driver's drowsiness detection, in *TENCON 2019–2019 IEEE Region 10 Conference (TENCON)* (IEEE, 2019), pp. 1622–1625

Application of Association Rule Mining-Based Attribute Value Generation in Music Composition

Shankar B. Naik and Salman Khan

Abstract The aim of the study in this article is to generate musical compositions based on sequence generation using frequent itemset mining and association rule analysis. The algorithm creates a set of rules by scanning the notes sequences of existing compositions. A new composition is created by applying the rules in generating a new musical note in it. The algorithm proposed was implemented to generate new composition on Indian classical music raga *Bhoopali* which were presented to an expert in Hindustani classical music.

Keywords Association rule · Frequent itemset · Indian classical music · Raga · Support threshold

1 Introduction

The problem of composing musical notes involves the generation of a sequence of notes. Be it for a musical instruments or vocal, these sequences can be viewed as sequence of values, generation of which is guided by certain rules.

In Indian classical music, these rules could be in the form of the *raagas* applicable to the composition which limit the composition notes and their combinations strictly to the ones allowed according to the *raaga*.

The study in this article aims to propose an association-based algorithm to generate the sequence of values and apply it to the process of composition of music based on Indian classical music.

S. B. Naik (✉)
Directorate of Higher Education, Govt. of Goa, Alto-Porvorim, Goa, India

S. Khan
IGNOU, Goa, India

V. Bhateja et al. (eds.), *Data Engineering and Intelligent Computing*,
Advances in Intelligent Systems and Computing 1407,
https://doi.org/10.1007/978-981-16-0171-2_36

2 Motivation and Related Work

2.1 Motivation

The notion of creation of a new composition out of existing compositions of the same *raga* set as the motivation for this study. Creation of a new composition is not merely based on the mixtures of sequences from existing compositions but by applying frequent itemset and association rule mining techniques to the existing compositions to identify the frequent and non-frequent patterns of notes and use this knowledge to arrange musical notes to form a new composition.

2.2 Related Works

A similar approach was proposed in which musical composition is generated by using the attribute value generation technique which is based on association rule mining method [1]. The approach has a set of rules where each rule is an relation between two musical notes allowed as per the rules of the applicable *raga*. A new note is added to the composition based on the rule applicable to the current last note. The set of rules are updated manually based on the *raga* upon which the composition is made.

The algorithm proposed in this study finds patterns in the form of rules in existing compositions. The algorithm scans the composition to find out rules which are pairs of notes such that the first note is followed by the second note in an already existing musical composition. Along with each rule, a frequency measure called *support* which is the total number of times the pattern is observed in the existing compositions, is associated.

The process of identification of patterns and their supports from existing compositions is based on the concept of frequent itemset mining (market basket analysis) introduced by Agrawal [2]. It aimed at mining frequent itemsets from transactional databases using the apriori algorithm. The apriori algorithm mines frequent itemsets by generation of candidate itemsets which are large in number and require large memory and processing time [3–5].

Han et al. proposed the FP-growth algorithm [6] which mines frequent itemsets using a frequent pattern tree called FP-tree. The algorithm Eclat proposed in [7] uses a vertical database format to generate frequent itemsets in datasets. The algorithms proposed in [8] and [9] are single-pass and incremental and mine frequent itemsets by scanning the database only once.

The concept of association rule mining is used to find two-note patterns from the existing compositions. The existing compositions which are sequences of musical notes are examined to find out rules between two notes. The algorithm described in [10] finds associations between clusters of values a single attribute. The algorithm described in [11] focuses on generating associations between clusters of values across two attributes. Both the papers generate associations between attribute value clusters. The study in this article is applicable to single attribute value clusters only.

3 Problem Statement

Let a $C = n_1.n_2.n_3 \ldots n_m$ be a sequence of m notes, where m is the size of the composition and each n_i representing a musical note. Let $V = \{v_1, v_2, \ldots\}$ be the set of values such that each $n_i \in V$. Set V is a set of notes allowed in the composition. Let $R = \{r_1, r_2, \ldots\}$ be the set of rules which are applicable in generation of a new composition. A rule r_i denotes as $v_a \rightarrow v_b$ implies that the value v_a in the composition can be followed by the value v_b.

3.1 The Problem Statement

Given a set of compositions $C_1, C_2, C_3, \ldots$, a set of notes V the problem is to generate a new composition C as a sequence of notes n_i's such that each $n_i \in V$. A note can exist multiple times in a composition.

4 The Proposed Approach

4.1 Value Prediction

Let n_{new} be the new note to that to be generated in a composition C. Let the last note in composition C be n_{old}. The value of n_{new} is generated by scanning the set of rules R. All rules with n_{old} on their left hand side are identified in R. Out of these, one rule is selected based on one of the criteria describes in the Sect. 4.3. The note at the right side of the selected rule is assigned to n_{new} using expression 1,

$$n_{\text{new}} = v_o / \exists v_o \rightarrow v_n \in R \tag{1}$$

where v_o is the last value in the composition C.

4.2 Creation of Rules

The rules are generated by scanning the existing composition. For an existing composition, a rule is a pair of notes denoted as $n_a \rightarrow n_b$ such that bot n_a and n_b together exist in at least one of the compositions in a way that the note n_a is followed by the note n_b. The same rule may exist multiple times in the compositions considered. This occurrences are counted and stored a support value of the rule.

4.3 Rule Selection

The rules can be selected based on the number of their occurrences in the existing compositions which can be done by analyzing their support values. If the strategy is to select rules which have high support values or low support values, then the composition generated may have the same subsequence repeated in it. This may lose the creative aspect in the process of composition. Hence, we propose that a hybrid strategy in which for different notes the rules are selected either with high support or with low support. Rules can also be selected by human intervention which involves study and analysis of the rules generated. Rule selection can also employ a random strategy in which the rules are randomly selected.

5 Experimental Study

The algorithm was implemented in C++ on a system with Windows 7 OS, 4 GB RAM and i3 processor. The input to the algorithm is a composition based on Indian classical music raga *Bhoopali* which is a pentatonic scale. It uses 5 notes sa, re, ga, pa and dha in ascending and descending scale [12, 13]. The input composition is $sa\ ga\ re\ sa\ dha\ sa\ re\ ga\ re\ ga\ pa\ ga\ re\ sa\ ga\ re\ sa\ dha\ sa\ re\ ga\ re\ ga\ pa\ ga\ re\ sa$ which is one of the simplest and the first composition introduced to beginners of music learners. Upon the execution of the proposed algorithm, the rules generated are shown in Table 1.

Table 1 Set of rules

Rule	Support	Rule	Support
$sa \rightarrow ga$	2	$sa \rightarrow re$	2
$ga \rightarrow re$	6	$re \rightarrow ga$	4
$re \rightarrow sa$	5	$ga \rightarrow pa$	2
$sa \rightarrow dha$	2	$pa \rightarrow ga$	2
$dha \rightarrow sa$	2		

Table 2 Composition generated

Composition	Selection approach
$sa - re - sa - re - sa - ga - re - sa - dha - sa$	High support
$sa - ga - re - sa - re - ga - re - ga - re - ga$	Random selection

Table 3 Composition ratings

Composition	Rating
$sa - re - sa - re - sa - ga - re - sa - dha - sa$	Good
$sa - ga - re - sa - re - ga - re - ga - re - ga$	Melodious

5.1 Output

Based on the rules, two new compositions of 10 beats were generated as shown in Table 2. The first composition was generating by choosing rules having high support for a choice, while the second composition was generated by selecting the rules randomly.

The generated composition was presented to an expert in Indian classical vocals. The composition was analyzed and was rated as melodious out of the following options:

1. Sounding off
2. Good
3. Melodious
4. Rare.

The rating of the above compositions are given in table 3

6 Conclusion

The aim of the study in this article is to generate musical compositions based on sequence generation using frequent itemset mining and association rule analysis. The algorithm creates a set of rules by scanning the notes sequences of existing compositions. A new composition is created by applying the rules in generating a new musical note in it.

The algorithm proposed was implemented to generate new composition on Indian classical music raga *Bhoopali* which were presented to an expert in Hindustani classical music.

The proposed algorithm considers patterns of only two musical notes while generating the rules which have only one musical notes on their right hand side. This does not allow to capture the melodies of the existing input compositions. The future study will consider rules having a sequence of notes on their right hand side instead of a single note. The experiment was done using a single input composition. In future, the algorithm will be experimented considering multiple compositions as inputs.

References

1. S.B. Naik, Attribute value generation using association rule mining, in *2019 2nd International Conference on Intelligent Computing, Instrumentation and Control Technologies (ICICICT)*, vol. 1, pp. 768–769 (IEEE, 2019)
2. R. Agrawal, T. Imieliński, A. Swami, Mining association rules between sets of items in large databases, in *Acm sigmod record*, vol. 22, pp. 207–216 (ACM, 1993)
3. B. Babcock, S. Babu, M. Datar, R. Motwani, J. Widom, Models and issues in data stream systems, in *Proceedings of the twenty-first ACM SIGMOD-SIGACT-SIGART symposium on Principles of database systems*, pp. 1–16, (ACM, 2002)
4. J.H. Chang, W.S. Lee, Finding frequent itemsets over online data streams. Inf. Soft Technol. **48**(7), 606–618 (2006)
5. S.B. Naik, Mining associations rules between attribute value clusters, in *Advances in Artificial Intelligence and Data Engineering*, pp. 909–917, Springer
6. J. Han, J. Pei, Y. Yin, R. Mao, Mining frequent patterns without candidate generation: a frequent-pattern tree approach. Data Mining Knowl. Discov. **8**(1), 53–87 (2004)
7. J. Han, H. Cheng, D. Xin, X. Yan, Frequent pattern mining: current status and future directions. Data Mining Knowl. Discov. **15**(1), 55–86 (2007)
8. S.B. Naik, J.D. Pawar, An efficient incremental algorithm to mine closed frequent itemsets over data streams, in *Proceedings of the 19th International Conference on Management of Data*, COMAD '13, (Mumbai, India, India), pp. 117–120 (Computer Society of India, 2013)
9. S.B. Naik, J.D. Pawar, A quick algorithm for incremental mining closed frequent itemsets over data streams, in *Proceedings of the Second ACM IKDD Conference on Data Sciences*, CoDS '15, (New York, NY, USA), pp. 126–127 (ACM, 2015)
10. S.B. Naik, J.D. Pawar, Clustering attribute values in transitional data streams, in *2017 International Conference on Computing, Communication and Automation (ICCCA)*, pp. 58–62 (IEEE, 2017)
11. S.B. Naik, J.D. Pawar, Mining association rules between values across attributes in data streams, in *2017 International Conference on Computational Intelligence in Data Science (ICCIDS)*, pp. 1–6 (IEEE, 2017)
12. W.M. Anderson, K.O. Rao, Music of south Asia: India. Multicultural Perspect. Music. Educ. **3**, 105 (2011)
13. P. Chordia, A. Rae, Raag recognition using pitch-class and pitch-class dyad distributions, in ISMIR, pp. 431–436 (Citeseer, 2007)

Pre-processing of Electroencephalography Signals Using Stationary Wavelet Transform-Enhanced Fixed-Point Fast-ICA

Aashi Rastogi and Vikrant Bhateja

Abstract The electroencephalography (EEG) signals are biomedical signals that are acquired from the scalp of a person using electrodes. These EEG signals are the recordings of the neural activities of the brain as observed in a person. These signals are highly susceptible to artifacts such as physiological and external artifacts as well as noises due to random processes occurring in nature. These undesired distortions occurring in the signal can disturb the original signal. Therefore, it is significant to pre-process the EEG signals. The proposed work uses stationary wavelet transform (SWT)-enhanced fixed-point fast-ICA technique in order to eliminate the artifacts or noises occurring in the signal. The fidelity of the pre-processed signal is further assessed at different levels of additive white Gaussian noise (AWGN) using peak signal-to-noise ratio (PSNR).

Keywords Electroencephalography (EEG) signals · Artifacts · Additive white Gaussian noise (AWGN) · Noise filtration · Signal fidelity assessment

1 Introduction

Electroencephalography (EEG) signals define the electrical activity of the brain recorded using electrodes from the scalp of a person. These recorded neural or electrical activities can be used to reflect the status of the brain [1]. The discovery of EEG signals gave great impetus to the study of brain and its functions, and it could also be used to diagnose any form of abnormalities or irregularities associated with the brain along with any form of change in the physiological state and monitor the state of attention in a person. EEG signals have broadly been used in clinical studies to study sleep patterns, aging effects, monitor the state of attention as well as diagnose

A. Rastogi
Department of Electronics and Communication Engineering, Shri Ramswaroop Memorial Group of Professional Colleges (SRMGPC), Lucknow, Uttar Pradesh 226028,, India

V. Bhateja (✉)
Dr. A.P.J. Abdul Kalam Technical University, Lucknow, Uttar Pradesh, India

V. Bhateja et al. (eds.), *Data Engineering and Intelligent Computing*,
Advances in Intelligent Systems and Computing 1407,
https://doi.org/10.1007/978-981-16-0171-2_37

and study mental disorders such as dementia, attention-deficit/hyperactivity disorder (ADHD), epileptic and non-epileptic seizures, etc. They also find applications in brain–computer interface (BCI), neuromarketing and biometrics [2]. The acquisition and study of EEG signals have always intrigued the interest of researchers. EEG signals can be used for different purposes with their largest constraint being noise. These signals are highly susceptible to noises due to random processes occurring in nature as well as several artifacts due to the physiological or external sources. The artifacts are termed as physiological if their source of occurrence is the subject's body itself such as ocular, muscular and cardiac, whereas they are said to be external if the source of their origin is external such as physical movement of the subject's body, artifacts arising due to transmission line and electrodes [3]. Researchers have all along tried to develop a novel technique for the denoising of EEG signals in order to extract true and relevant information from them. The removal of these unwanted disturbances occurring in the signal is highly significant for the fact that it can interfere with the signal, thereby transforming the information that is being conveyed by them [4, 5]. Therefore, several pre-processing techniques have been devised in order to pre-process the EEG signals for effective removal of noise. Okinomou et al. [6] presented a method of spike enhancement for the removal of artifacts present in the EEG signals using Kalman filters. Even though a considerable rise in the signal-to-noise ratio (SNR) of the EEG signals was observed, the process proved to be time consuming. Correa et al. [7] introduced the use of cascade of three adaptive filters instead of simple FIR filters in order to remove noise due to transmission line, ocular and cardiac artifacts present in the EEG signal, but it was observed that the certain parameters affected the stability of the system. Similarly, Hamadicharef et al. [8] made use of multiple filter banks consisting of nine bandpass IIR filters (Chebyshev Type II filter), but since EEG signals are nonlinear and non-stationary signals, the filter banks were unable to remove the artifacts overlapping the pure EEG signal. Asaduzzaman et al. [9] presented the use of discrete wavelet transform (DWT) using Daubechies and orthogonal Meyer wavelet family along with thresholding for effective removal of artifacts present in the EEG signals of healthy and epileptic patients. Root mean square (RMS) was further employed in order to assess the efficiency of the employed method. Mishra and Singla [10] employed fixed-point fast-ICA algorithm for the elimination of artifacts from EEG signals. Similarly, Hu et al. (2016) [11] employed fast-ICA combined with matched filters for the removal of ocular artifacts present in the EEG signal. Ngoc et al. [12] presented the use of low-pass and high-pass filters with cutoff frequencies of 40 Hz and 0.5 Hz, respectively, for the removal of power line and baseline noise present in the acquired EEG signal as well as stationary wavelet transform (SWT) in order to eliminate ocular artifacts recorded along with the pure EEG signal. SWT is known to efficiently eliminate the artifacts present in the signal without distorting the original signal. Since ICA-based methods could not completely distinguish between random noise occurring in nature and EEG signals, Issa and Juhaz [13] introduced an approach that used a combination of wavelets and ICA also known as wavelet-enhanced ICA (wICA). In order to remove artifacts and noise present in the EEG signals, the proposed work employs SWT-enhanced fixed-point fast-ICA algorithm. For effective removal of noise present in the signal, fixed-point fast-ICA algorithm is used to eliminate artifacts, whereas further SWT was employed in order to remove any form of remaining noise along with AWGN externally added to the observed EEG signal. The proposed paper has been organized as follows: Sect. 2 gives an overview of the proposed methodology used for de noising of EEG signals, techniques used for the purpose, i.e., fixed-point fast-ICA algorithm and SWT and signal fidelity assessment using PSNR. Section 3 enlists the results evaluated on application of the proposed methodology over EEG signals. The conclusion of the proposed work has been summed up in Sect. 4 of the paper.

2 Proposed Methodology for Filtering of EEG Signals

We have already studied in the above section that EEG signals are highly susceptible to noises be it artifacts or noise due to random processes occurring in nature. These random processes occurring in nature are generally Gaussian. Therefore, additive white Gaussian noise (AWGN) has been used as an alternative to represent noise due to random processes. Since these noisy signals can easily interfere with the original signal and distort the quality of the signal, it is essential to pre-process the EEG signals. These noises are difficult to interpret using naked eye; therefore, the proposed work employs a technique known as SWT-enhanced fixed-point fast-ICA algorithm. The proposed method is a combination of fixed-point fast-ICA algorithm and stationary wavelet transform (SWT).

The proposed work was performed over a dataset of EEG signals publicly available on "Kaggle" by the title "EEG Data for mental Attention State Detection." It is a collection of 34 experiments and a total of 25 h EEG recording fetched from five participants involved in a low-intensity control task using 14 electrodes [14]. The process first of all adds AWGN to the observed EEG signal $s(t)$ in order to assess the proposed method for the removal of noise due to random processes. The mixed EEG signal $x(t)$ has been given below in Eq. (2.1) where $n(t)$ represents AWGN.

$$x(t) = n(t) + s(t) \tag{2.1}$$

Now, the mixed EEG signals consist of not only artifacts but also AWGN. The process initially uses fixed-point fast-ICA algorithm in order to eliminate artifacts present in the EEG signal. Since fixed-point fast-ICA algorithm is only applicable to mixed signals that are multiplicative in nature, SWT was employed. Then, SWT is employed in order to eliminate the remaining artifacts (present as a result of the shortcomings of the ICA algorithms) as well as AWGN. After the elimination of noises, PSNR of the filtered and noisy EEG signals is evaluated in order to assess the fidelity of the signal and the viability of the proposed technique. The block diagram of the proposed noise filtering model of EEG signals is given below in Fig. 1.

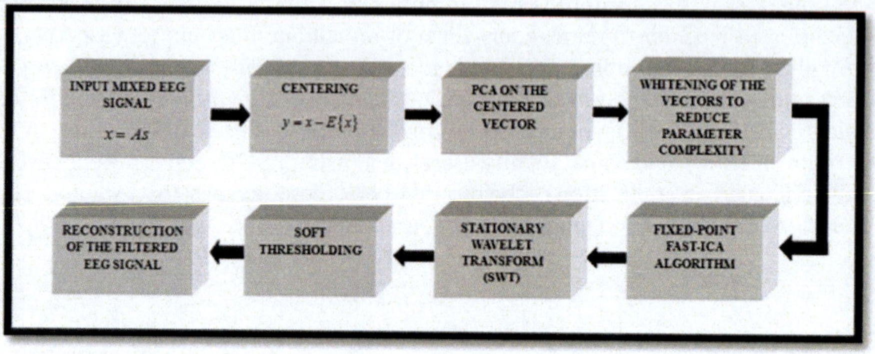

Fig. 1 Block diagram of the proposed noise filtering model of EEG signals

2.1 Pre-processing Using Fixed-Point Fast-ICA Algorithm

ICA algorithms are a family of algorithms that are used for the demixing of source signals from the observed, mixed or noisy signals by computation of independent components using course time recorded information [15]. The proposed work employed the fast-ICA algorithm that is one of the most widely used algorithms of the ICA family. There are two types of fast-ICA algorithms, namely deflation-based fast-ICA algorithm and symmetrical fast-ICA algorithm. The deflation-based fast-ICA algorithm is based on sequential source separation. It is used to extract one source at a time, whereas symmetrical fast-ICA algorithm extracts all the sources simultaneously [16]. The deflation-based fast-ICA algorithm has been proposed in the study for the purpose of pre-processing of EEG signals. The pre-processing of EEG signals using this technique depends upon the initial input to the algorithm and its convergence to a fixed value, i.e., demixing vector. The algorithm is said to converge when the iterate lies in the specified range of fixed point [17]. ICA decomposition is said to be accomplished only when the number of independent components retrieved is same as the number of channels employed for recording of EEG signals [15].

Initially, the observed EEG signals are fed as input to the algorithm. They can be represented as shown in Eq. (2.2) given below where x denotes the vector of observed or mixed EEG signal, A denotes the unknown mixing matrix and s is the unknown source signal that is mutually independent and utmost one of them is Gaussian [16].

$$x = As \tag{2.2}$$

The main purpose of fast-ICA algorithm is to recover the source signal s with the help of observed vector x and estimation of the inverse of mixing matrix A. It is to be noted that neither the mixing matrix A nor source signal s is known, and none of them is identifiable either [17]. As shown in Fig. 1, certain pre-processing steps are needed to be employed before fast-ICA algorithm. Firstly, centering is employed in order to subtract the mean vector from the EEG data. Secondly, principal component analysis (PCA) is statistical method that is employed to evaluate the eigenvectors and diagonal eigenvalues of the centered covariance matrix. This is further used to whiten the data in order to reduce the mixing matrix into an orthogonal one. This reduces the complexity of the algorithm [18]. After simplifying the EEG data for ICA estimation by using certain pre-processing algorithms, the data is now employed to the fixed-point fast-ICA algorithm. Here, fixed point refers to a computational unit or we can say a weight vector w that can be updated with each iteration or learning [18]. The algorithm is said to converge when the iterate lies in the specified range of fixed point. The fast-ICA algorithm may converge to different fixed values rather than the demixing vector. Therefore, in order to prevent false convergence of a vector, negentropy is employed as a measure to maximize non-Gaussianity [17]. If the evaluated normalized value of weight vector w^+ converges with the initial weight vector w, the old and newly normalized weight vector values should point in the same direction or their dot product should almost be equal to 1. Then, they are said to have converged. Finally, the filtered and reconstructed EEG signal can be given by Eq. (2.3), where W is the demixing matrix and $s_i = w_i^{\mathrm{T}} z$ is a component of matrix y [18].

$$s = Wz \tag{2.3}$$

2.2 *Pre-processing Using Stationary Wavelet Transform (SWT)*

In order to enhance the performance of the denoising model of EEG signal and for the removal of noises present as a result of additive random processes present in the signal in the form of AWGN, SWT is used along with the fixed-point fast-ICA technique. The wavelets are capable of disclosing information such as noise present in the signal, signal trend and, discontinuities. Several researches have proved that they provide better filter or denoising quality when employed to a biomedical signal [19, 20].

SWT is an improvised version of wavelet transform. It can overcome the limitations of DWT and is capable of decomposing time-invariant signals without losing any information. It decomposes a signal y into two sets of coefficients, namely approximation and detailed coefficients. These coefficients are obtained by convolving the input signal y with the low-pass filters for approximation coefficients and high-pass filters for detail coefficients. Further, the detail coefficients are retained, whereas the approximation coefficients are used for further decomposition using the same scheme of convolving it with up-sampled low-pass and high-pass filters from the previous step to produce another set of approximation and detail coefficients. Since there is no down-sampling to be employed in SWT, the length of approximation and detail coefficients is same as the length of the signal [21–23].

The approximation and detailed coefficients can be computed by Eqs. (2.4) and (2.5), respectively, where $cA_{j,k}$ is the approximation coefficient, $cD_{j,k}$ is the detail coefficient and j and k are the number of levels of decomposition and position, respectively [23].

$$cA_{j,k} = \sum_{i=1}^{n} cA_{j-1,k+2^{j}(n)} f(n) \tag{2.4}$$

$$cD_{j,k} = \sum_{i=1}^{n} cD_{j-1,k+2^{j}(n)} h(n) \tag{2.5}$$

The proposed methodology employed the use of Coiflet wavelet family of order 5 and the level of decomposition being 5 [24]. The signal was further broken into distinct frequency resolutions, and soft thresholding was employed to it. Thresholding technique is used to segregate noise from the signal. The decomposed signal was then reconstructed using inverse stationary wavelet transform (ISWT). ISWT provides the denoised EEG signal [25].

2.3 *Signal Fidelity Assessment Using Peak Signal-To-Noise Ratio (PSNR)*

After filtering of EEG signals, it is important to assess the quality of the signal to inspect the signal for efficient noise removal. Therefore, to assess the fidelity of the pre-processed or filtered EEG signal, peak signal-to-noise ratio (PSNR) has been used in the proposed methodology. It is the ratio of maximum possible power *peakval* of a signal and the mean square error *MSE*. It is expressed in logarithmic decibel scale and can be computed as shown in Eq. (2.6). MSE computes the sum of squared errors between the filtered and the noisy EEG signal. The lower the MSE, the lower is the error in the filtered signal. It can be computed as shown in Eq. (2.7) where u represents the filtered signal, v represents the noisy signal and m and n represent the dimensions of the signal's matrix. The larger the PSNR, it represents better noise removal from the original or noisy signal [23, 26, 27].

$$\text{PSNR} = 10\log_{10}\left(\text{peakval}^2/\text{MSE}\right) \tag{2.6}$$

$$\text{MSE} = \frac{1}{mn}\sum_{m=0}^{m-1}\sum_{n=0}^{n-1} ||u(i, j) - v(i, j)||^2 \tag{2.7}$$

3 Results and Discussions

In order to analyze the EEG signal for effective noise removal, PSNR between noisy and filtered EEG signal was evaluated. The process was performed at varying levels of AWGN externally added to the observed EEG signal. Here, the observed EEG signal produced on addition of AWGN has been termed as noisy EEG signal. A total of five EEG signals were tested at three different levels of AWGN, namely low-level, medium-level and high-level AWGN. Firstly, the PSNR was evaluated between observed and noisy EEG signals which has been termed as PSNR of noisy EEG signal. Later, after pre-processing of noisy EEG signals, the PSNR between noisy and filtered EEG signal was evaluated that has been termed as PSNR of filtered EEG signal. Both the PSNR values were then compared in order to assess the signal and the process employed for effective noise removal. EEG signal #2 was observed for its PSNR value at different levels of AWGN as shown in Fig. 2.

If the PSNR of the filtered EEG signal is evaluated to be greater than that of the PSNR of noisy EEG signal, it proves that noise has been eliminated from the noisy EEG signal. As observed from Fig. 2 given above, it can be seen that Fig. 2a, b represents the plot of noisy EEG signal #2 and filtered EEG signal #2 at low-level AWGN of 50 dB along with their PSNR values equal to 49.0017 dB and 60.4834 dB, respectively, for noisy and filtered EEG signal. Similarly, Fig. 2c, d represent the noisy and filtered plot of EEG signal #2 at medium-level AWGN of 20 dB along with their PSNR values equal to 19.0012 dB and 27.8865 dB, respectively. Since, for all the cases, PSNR of filtered EEG signal has been observed to be greater than that of the noisy EEG signal, it supports the fact that noise has been eliminated from the noisy EEG signal. The PSNR results of the experiment performed over five EEG signals at different levels of AWGN are enlisted below in Table 1.

If we closely observe Table 1, it can be seen that for every signal, the PSNR of filtered signal obtained is greater than the PSNR of noisy EEG signal. This proves the employed technique to be viable and effective for elimination of noise present in the observed EEG signal.

4 Conclusion

The aim of the proposed work was to denoise the EEG signals from artifacts and random noises occurring in them while acquisition. The noise produced due to random processes occurring in nature was assumed to be Gaussian for simplicity and thus replaced by AWGN. Here, the AWGN at varying levels was added to the observed EEG signal for accurate analysis of the process of noise removal employed. The filtering or pre-processing of EEG signals was performed using SWT-enhanced fixed-point fast-ICA algorithm. Further, the EEG signals were assessed at different levels of AWGN using PSNR. The results were found to be encouraging and proved that the process was viable for noise removal. It can prove to be highly beneficial

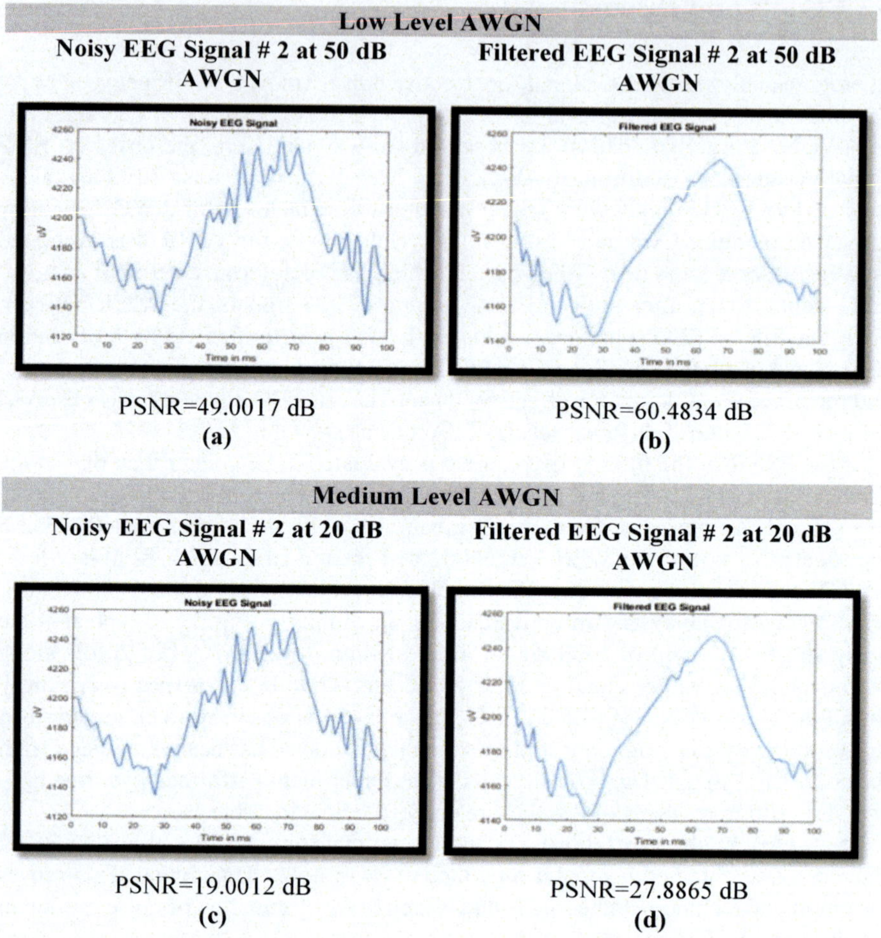

Fig. 2 Simulation results for EEG signal filtering at different levels of AWGN

Table 1 Calculation of PSNR during filtering of EEG signals at varying levels of AWGN

AWGN level	Low-level AWGN (50 dB)	Medium-level AWGN (20 dB)
Noisy EEG signal #1	49.0044	18.9981
Filtered EEG signal #1	53.7236	24.1880
Noisy EEG signal #2	49.0017	19.0012
Filtered EEG signal #2	60.4834	27.8865

with respect to future prospects not only in the field of clinical studies but also BCI, biometric applications, etc., as it can enhance the quality of the EEG signal.

5 Future Scope of the Research

In general, the research deals with the removal of noise be it in the form of physiological or external artifacts or any form of random Gaussian noise present in the EEG signal. Even though the proposed pre-processing technique, i.e., SWT-enhanced fixed-point fast-ICA has returned satisfactory results, it is still uncertain that the noise present in the signal has been eliminated completely. The future prospect may include developing a technique that can assess the amount of artifact or random noise initially present in the observed or acquired EEG signal. This can help verify the end result of pre-processing defining the amount of noise eliminated using a certain technique. Since the technique needs to be applied on real-time systems, other than accuracy, it also needs to be swift so that it can help provide relevant results in time.

References

1. J.G. Webster, *Medical Instrumentation Application and Design*, 4th edn. (Wiley, New York, 2010)
2. S. Sanei, J.A. Chambers, *EEG Signal Processing*, 1st edn. (Wiley, UK, 2007).
3. A. Tandle, N. Jog, P. D'Cunha, M. Chheta, Classification of artifacts in EEG signal recordings and EOG artifact removal using EOG subtraction. Commun. Appl. Electron. **4**, 12–19 (2016)
4. A. Lay-Ekuakille, P. Vergallo, G. Griffo, S. Urooj, V. Bhateja, F. Conversano, S. Casciaro, A. Trabacca, Multidimensional analysis of EEG Features using advanced spectral estimates for diagnosis accuracy, in *Proceedings of 2013 IEEE International Symposium on Medical Measurements and Applications (MeMeA)*, Gatineau, QC, Canada, May 2013, pp. 237–240
5. A. Lay-Ekuakille, G. Griffo, F. Conversano, S. Casciaro, A. Massaro, V. Bhateja, F. Spano, EEG signal processing and acquisition for detecting abnormalities via bio-implantable devices, in *Proceedings of 2016 IEEE International Symposium on Medical Measurements and Applications (MeMeA)*, Benevento, Italy, May 2016, pp. 1–5
6. V.P. Okinomou, A.T. Tzallas, D.I. Fotiadis, A Kalman filter based methodology for EEG spike enhancement. J. Comput. Methods Programs Biomed. **85**(2), 101–108 (2007)
7. A.G. Correa, E. Laciar, H.D. Patino, M.E. Valentinuzzi, Artifact removal from EEG signals using adaptive filters in cascade. J. Phys. Conf. Ser. **90**(1), 1–10 (2007)
8. B. Hamadicharef, H. Zhang, C. Guan, C. Wang, K. Phua, K. Tee, K. Ang, Learning EEG based spectral-spatial patterns for attention level measurement, in *Proceedings of IEEE International Symposium on Circuits and Systems*, Taipei, Taiwan, May 2009, pp. 1465–1468
9. K. Asaduzzaman, M.B.I. Reaz, F. Yasin, K.S. Sim, M.S. Hussain, *Advances in Computational Biology*, Chapter 65 (Springer, New York, 2010), pp. 593–599
10. P. Mishra, S.K. Singla, Artifact removal from biosignal using fixed point ICA algorithm for pre-processing in biometric recognition. J. Meas. Sci. Rev. **13**(1), 7–11 (2013)
11. B. Hu, X. Li, S. Sun, M. Ratcliffe, Attention recognition in EEG-based affective learning research using CFS+KNN algorithm. IEEE/ACM Trans. Comput. Biol. Bioinf. **15**(1), 38–45 (2016)

12. P.P. Ngoc, V.D. Hai, N.C. Bach, P.V. Binh, EEG signal analysis and artifact removal by wavelet transform, in *Proceedings of 5th International Conference on Biomedical Engineering*, Vietnam (2015), pp. 179–183
13. M.F. Issa, Z. Juhaz, Improved EOG Artifact Removal using Wavelet enhanced Independent Component Analysis. Available online at: https://doi.org/10.3390/brainsci9120355. Last accessed on: Dec 2019
14. C. Aci, M. Kaya, Y. Mischenko, Distinguishing mental attention states of humans via an EEG-based passive BCI using machine learning methods. J. Expert Syst. Appl. **134**, 153–166 (2019)
15. J. Onton, M. Westerfield, J. Townsend, S. Makig, Imaging human EEG dynamics using independent component analysis. J. Neurosci. Biobehav. Rev. **30**(6), 808–822 (2006)
16. A. Hyvarinen, E. Oja, Independent component analysis: algorithms and applications. J. Neural Networks **13**(4), 411–430 (2000)
17. T. Wei, A study of the fixed points and spurious solutions of the deflation-based FastICA algorithm. J. Neural Comput. Appl. **28**(1) (2017)
18. E. Bingham, A. Hyvarinen, A fast fixed-point algorithm for independent component analysis. Int. J. Neural Syst. **10**(1), 1–8 (2000)
19. A.A. Jaber, R. Bicker, Development of a condition monitoring algorithm for industrial robots based on artificial intelligence and signal processing techniques. Int. J. Electr. Comput. Eng. (IJECE) **8**(2), 996–1009 (2018)
20. V. Bhateja, S. Urooj, R. Mehrotra, R. Verma, A. Lay-Ekuakille, A composite wavelets and morphology approach for ECG noise filtering, in *Proceedings of International Conference on Pattern Recognition and Machine Intelligence*, Heidelberg (2013), pp. 361–366
21. G.P. Nason, B.W. Silverman, *The Stationary Wavelet Transform and Some Statistical Applications* (Springer, New York, 1995), pp. 281–299
22. M. Alfaouri, K. Daqrouq, ECG signal de-noising by wavelet transform thresholding. Am. J. Appl. Sci. **5**(3), 276–281 (2008)
23. P.A. Babu, K.V.S.V.R. Prasad, Removal of ocular artifacts from EEG signals using adaptive threshold PCA and wavelet transforms, in *Proceedings of 2011 International Conference on communication Systems and Network Technologies*, Katra, Jammu (2011), pp. 572–575
24. A. Srivastava, V. Bhateja, A. Shankar, A. Taquee, On analysis of suitable wavelet family for processing of cough signals, in *Proceedings of Frontiers in Intelligent Computing: Theory and Applications*, Singapore (Springer, Berlin, 2020), pp. 194–200
25. A. Mishra, V. Bhateja, A. Gupta, A. Mishra, Noise removal in EEG signals using SWT-ICA combinational approach, in *Proceedings of Smart Intelligent Computing and Smart Applications*, Singapore, November 2018, pp. 217–224
26. Z. Wang, A.C. Bovik, Mean squared error: love it or leave it? A new look at signal fidelity measures. IEEE Signal Process. Mag. **26**(1), 98–117 (2009)
27. K. Sriniasan, J. Dauwels, M.R. Reddy, Multichannel EEG compression: wavelet-based image and volumetric coding approach. IEEE J. Biomed. Health Inf. **17**(1), 113–120 (2013)

Performance Evaluation of the Supervised Machine Learning Algorithms Using R

Pratibha Maurya and Nidhi Srivastava

Abstract Today, big data occupies a significant place in each one's life. Various supervised and unsupervised machine learning algorithms exist which can be applied on this big data to get the desired information. A wide application of the classification algorithms is seen in analyzing the data of medicine or health where it has been useful in predicting the disease based on the given features. In this paper, the Cleveland database from the dataset of heart disease publicly available on UCI machine learning Web site was taken, and three different classification algorithms were applied on it to predict the presence or absence of heart disease in patients. The performance of the three well-known algorithms—support vector machine, decision tree and random forest—has been compared based on various metrics accuracy, sensitivity, and specificity and error rate. The experiment has been performed on R language, a programming language which is widely used for data visualization, and the result shows that random forest algorithm outperforms the rest of the two algorithms in successfully classifying healthy and unhealthy patients. For this experiment, fourteen different features like age, resting electrocardiographic results, resting blood pressure, chest pain type, etc., of an individual have been considered.

Keywords Big data · Decision tree · R · Random forest · Supervised algorithm · Support vector machine

1 Introduction

Data in today's world is growing rapidly. The data is produced from various resources like finance, education, marketing, health, transportation, logistics, government, etc. With the advent of social media, this data is further multiplying by leaps and bound.

P. Maurya · N. Srivastava (✉)
AIIT, Amity University, Lucknow, Uttar Pradesh, India
e-mail: nsrivatava2@lko.amity.edu

P. Maurya
e-mail: pmaurya@lko.amity.edu

V. Bhateja et al. (eds.), *Data Engineering and Intelligent Computing*,
Advances in Intelligent Systems and Computing 1407,
https://doi.org/10.1007/978-981-16-0171-2_38

The huge data is categorized by namely 5V's. The 5V's stands for Volume, Value, Velocity, Veracity and Variety. This big data can be unstructured, structured or semi-structured. Using the traditional data processing techniques, it is difficult to analyze the big data. Better tools and techniques have come up with time for analysis of this data. There is also a need for management of this data using different approaches and technologies. If handled carefully and thoughtfully, this data can be of great use, and significant and important information can be extracted from this data. But the extraction of information from this huge data is very complex and faces many challenges. One of the challenges it faces is collection and integration of this limited hardware processing. Another challenge being faced is that the data so obtained cannot be used as such. It needs to be cleaned of irrelevant data and noises [1, 2]. Various techniques have been designed to handle this data in proper way and to find out desired information from this gathered data. Machine learning techniques are one way of handling this data. This paper is structured as follows. Section 1 gives the introduction. Section 2 of the paper gives a brief overview about the machine learning. Section 3 lists and explains the three classification algorithms which have been used to conduct the experiment. In Section 4, an experiment has been conducted to show the comparison between the three algorithms—support vector machine (SVM), decision tree and random forest. Section 5 shows the results and discussion, and lastly, Section 6 gives the conclusion.

2 Machine Learning

Machine learning (ML) is a way of teaching machines. The objective of ML is to uncover information and make sensible and smart decisions. In ML, the data is provided to the machines, and by applying various techniques, the machine can extract the desired information or recognize the pattern. The machine tends to learn and enhance its knowledge from data with which it is fed. Sometimes, the result is not as desired, and then, the machine relearns and works on it. All this is done with little or almost no human mediation. Application of ML varies and is used in many areas like medicine, military, recognition systems, control systems, etc. [3].

ML algorithms are generally classified as supervised and unsupervised. However, now, two more types of classification have been included—semi-supervised and reinforcement. In unsupervised learning, there is no labeled data, and there is no supervisor. The machine must learn on its own based on the patterns available. Unsupervised learning can be classified into clustering and association. Clustering is in which clusters are formed. All data which is of same type is grouped together to form one cluster. Similarly, other groups are formed which contain data of similar types based on certain property. In association, certain rules are formed depending on the significant sections of the data [4]. In supervised learning, a machine is taught by using labeled data. A distribution of class labels is shown in the model designed based on the predictor attributes. It can be classified into regression and classification.

In regression, the output is continuous variable, whereas in classification, the output is discrete variable [5, 6].

3 Types of Classifications

Machine classification is a supervised ML approach. The machine learns from the data provided and classifies the data based on certain property. In fact, the data is classified based on the corresponding class labels. Training of data and testing of data are done in classification to create the model. There are many classification algorithms available to handle datasets. Below, we describe the three algorithms—SVM, decision tree and random forest—which we have used in our experiment for the classification of the data.

3.1 Support Vector Machine (SVM)

SVM is widely used supervised machine learning algorithm. SVM is based on statistical learning theory and came up in the mid-1990s [7]. It is used for regression and classification method. It has been successfully used in text classification, handwritten digit recognition, object recognition, etc.

Advantages of SVM are that it is very accurate and is very effective in high-dimensional spaces. While training in this algorithm, only a subgroup of the points is used. This makes it memory efficient. It works well with small datasets, and many kernel functions are also used in this algorithm so that better decisions can be made. Disadvantages are that it takes too much time in training the data and does not work well with datasets which are large in size [8, 9].

3.2 Decision Tree

It is a tree-based classifier. But the tree is an upside-down tree where the root signifies the top node and the branches or sub-nodes show its children. In this, if we have rows and all the values that particular row, also known as attributes, are tested for. In each row, the attribute is tested which is shown by internal node. The test outcome is represented by branch, and lastly, terminal/leaf node shows the class label. This class label shows the decision taken once all the attributes of the row have been processed. As compared to other classifiers, decision trees perform equally or better than other classifiers [10]. It is the most used supervised learning algorithm.

As already mentioned before, the ML algorithm has two phases of training the dataset and then testing the dataset. These two phases are also there in decision tree. In the training phase, the already existing occurrences are used to build the decision

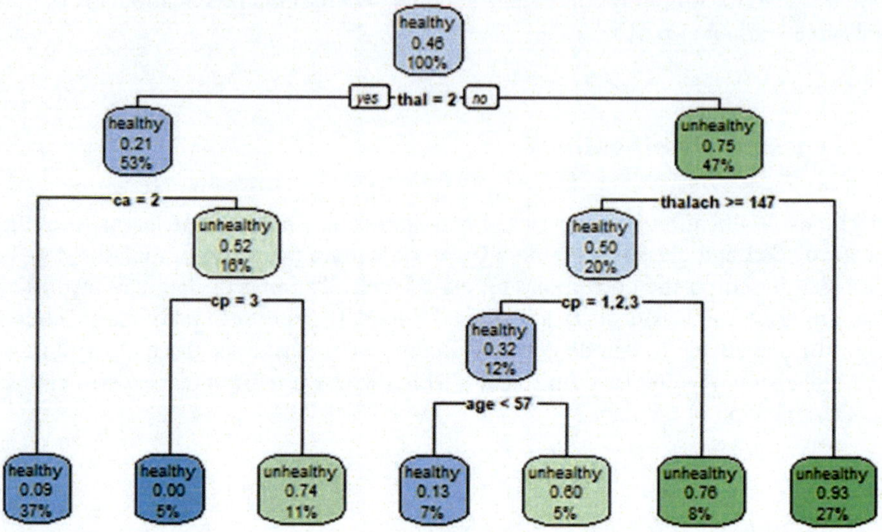

Fig. 1 Sample decision tree built on the Cleveland dataset

tree, and in the testing phase, the new occurrences are classified based on the model created from earlier training phase [11].

Figure 1 represents the decision tree built on the dataset that has been used in this study. 'rpart' library in R is used to build the decision tree for classifying between healthy and unhealthy classes.

3.3 *Random Forest*

This ML algorithm can be used for both regression and classification problems. It is a collection of decision trees. In short, many decision trees are taken together to form random forest. In this, random features are selected to build numerous decision trees. The final output is produced by the decision obtained from voting of various decision trees. Random forest uses bagging (bootstrap aggregation) for selection of the features. These features are randomly selected at each decision split. Decision trees have high variance, but with the use of this bagging, which is an ensemble method, the variance gets reduced. The training set used is called bootstraps, and testing is done on out-of-the-bag bootstraps. The prediction in random forest is thus dependent on group of decision trees and not one. The advantages of random forest lie in this random selection as it enhances the prediction power and thus yields in greater productivity and efficiency. It is best used for large datasets. Random forest easily deals with the problem of overfitting as it deals with more than one decision tree. However, random forest is best suited for regression as compared to classification [12–14].

4 Dataset and Feature Engineering

In this paper, we used the Cleveland database [15] from the dataset of heart disease publicly available on UCI machine learning Web site. Our study uses 14 attributes from this dataset about age, sex, resting electrocardiographic results, resting blood pressure, chest pain type, etc.

At the data cleaning stage, we first start by naming the unnamed columns after the names that were listed on the UCI Web site. We then replace all missing values by 'NA' which are later imputed by using data imputation techniques. The attributes that can be factorized into fixed levels have been converted to factors. The attribute sex which has the values 0 and 1 is substituted by 0 indicating 'female' and 1 indicating 'sex.' Also, the attribute heart disease having values 0–4 has been replaced by 0 indicating 'healthy' and 1's by 'unhealthy' patient. These substitutions are in accordance with the information provided on the Web site. The heat map in Fig. 2 shows the correlation between different features.

In the correlation heat map, red color shows positive correlation among the attributes, while pale or white means nearly no correlation and blue color represents negative correlation.

After feature engineering, the dataset is divided into two parts: the training data and the test data in the ratio of 70/30. These datasets are balanced by having observations for both healthy and unhealthy patients for training and testing our models. Also, the random samples have been selected in the training dataset and are entirely different from the test dataset. The problem statement is to distinguish presence (unhealthy) from the absence (healthy) of heart disease in a patient.

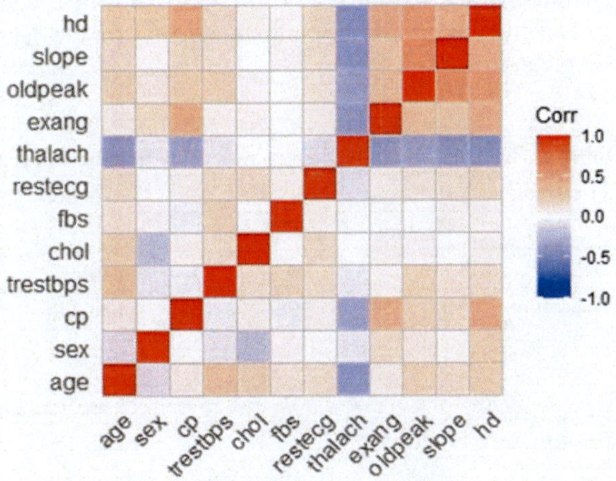

Fig. 2 Correlation heat map between features

5 Results and Discussion

The three classification algorithms random forest, decision trees and support vector machine have been implemented using R language. To compare various machine learning algorithms, accuracy, sensitivity and specificity have been computed. Also, confusion matrix, receiver operating characteristic (ROC) curves and precision–recall curves and have been generated for each algorithm. Care has been taken to ensure that there is no overlap between training and test datasets. The heat disease has been predicted using all the remaining 13 variables in the dataset. The formula so obtained is used as an input for random Forest, svm and rpart functions used to build random forest, SVM and decision trees, respectively.

The random forest is developed by aggregating trees. The default value for number of tree (i.e., ntree) is 500. The variables randomly selected as candidates at each split (i.e., mtry) are sq. root (14) where 14 is number of features in the model.

To test our random forest model, we check the predictions made by the model on both training and test datasets. For this purpose, we use predict function and confusion matrix. A confusion matrix can be very useful in visualizing the efficiency of various classifiers and thereby providing insight about the misclassifications done by the algorithms (Table 1).

The classifier works well by classifying all the values correctly in the training data. For test dataset, the model does some misclassifications, by predicting eight healthy patients as unhealthy while seven unhealthy patients as healthy.

The SVM classifier uses radial basis function as a kernel to make predictions on heart disease. Again, all the 13 attributes have been used for predictions. 120 support vectors have been used to create the hyperplane by the model during learning phase. The predictions and the confusion matrix for the model are given in Table 2.

The predictions and the confusion matrix for the decision tree on the same datasets are given in Table 3.

Table 1 Random forest confusion matrix using training and test datasets

Prediction	Reference		Prediction	Reference	
	Healthy	Unhealthy		Healthy	Unhealthy
Healthy	113	0	Healthy	44	8
Unhealthy	0	98	Unhealthy	7	33

Table 2 Support vector machine confusion matrix using training and test datasets

Prediction	Reference		Prediction	Reference	
	Healthy	Unhealthy		Healthy	Unhealthy
Healthy	107	17	Healthy	44	12
Unhealthy	6	81	Unhealthy	7	29

Table 3 Decision tree confusion matrix using training and test datasets

Prediction	Reference		Prediction	Reference	
	Healthy	Unhealthy		Healthy	Unhealthy
Healthy	95	9	Healthy	35	9
Unhealthy	18	89	Unhealthy	16	32

Using the confusion matrices on test dataset, we now compute the three metrics for evaluating the performance of three machine learning algorithms.

Accuracy: Random forest has accuracy of 83% which is higher than 79 and 72% for support vector machine and decision trees. Higher accuracy clearly indicates that random forest can make a greater number of correct predictions by diagnosing a person having heart disease or not.

Sensitivity: It is calculated as (true positive)/(true positive + false negative). These values are determined using confusion matrix. Sensitivity recognizes what proportion of patients with heart disease can be correctly labeled as having heart disease.

Specificity: It is calculated as (true negative)/(true negative + false positive). This value recognizes what proportion of healthy patients can be correctly labeled as being healthy.

The random forest has higher sensitivity and specificity value as compared with the values obtained by SVM and decision trees. This indicates that if random forest predicts a patient as having no heart disease, then there is a higher probability that one does not have as compared to predictions made by the other two classifiers. Similarly, the specificity value of random forest being higher indicates that if a patient is diagnosed as having heart disease, then it can be said with higher confidence that the patient is unhealthy as compared to the predictions made by decision trees and SVM machine algorithms (Table 4).

Figure 3 demonstrates and compares the ROC curves for every machine learning algorithm that was used in this research. The area under each curve, i.e., AUC value, for the three classifiers can be calculated from the below ROC curve (Fig. 4).

Sometimes, classifiers may have high precision values but low recall or vice versa. In such cases, identifying best classifiers may be an issue. The $F1$ score calculated as harmonic mean of these values can be a clear indicator of a good classifier. To have high $F1$ score, the classifier needs to have both high precision and recall.

Table 4 Prediction accuracy, sensitivity and specificity for three classifiers

	Random forest	Support vector machine	Decision tree
Accuracy	0.837	0.793	0.728
Sensitivity	0.872	0.862	0.686
Specificity	0.804	0.707	0.780

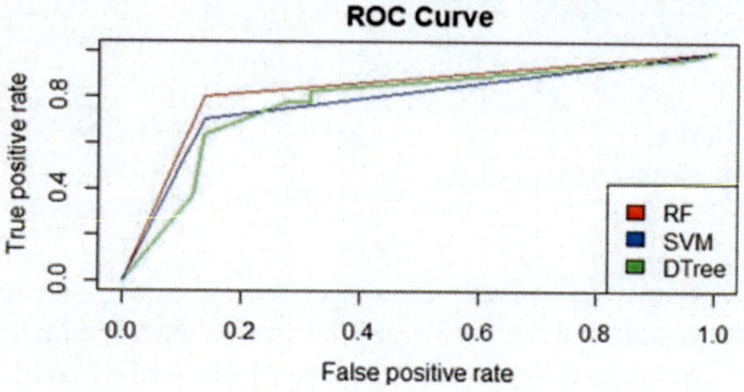

Fig. 3 Classification of three algorithms using ROC curve

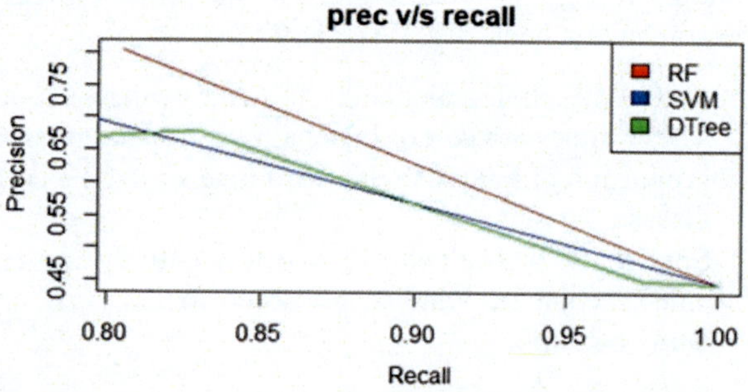

Fig. 4 Precision–recall curve for three classifiers

Using the precision–recall graph, $F1$ score as calculated are:

The higher values of AUC and $F1$ score as obtained from Tables 5 and 6 support our claim that random forest can more accurately predict the presence of heart disease in patients.

Next, we perform tenfold cross-validation on the dataset and calculate the test error rate for all the three classifiers. The random forest has the minimum error rate among the three classifiers (Table 7).

Table 5 AUC values for the three classifiers

	Random forest	Support vector machine	Decision trees
AUC	0.834	0.785	0.782

Table 6 $F1$ score for the three classifiers

	Random forest	Support vector machine	Decision trees
$F1$ score	0.858	0.821	0.736

Table 7 Error rates for three classifiers using tenfold cross-validation

	Random forest	Support vector machine	Decision trees
Error rate	0.152	0.168	0.0170

6 Conclusion

The paper is an attempt to compare the three supervised learning classifiers, i.e., random forest, support vector machine and decision trees, for classifying Cleveland database by doing the implementations in the open-source software R. On validating the performance of the three classifiers from the classification results obtained using various metrics, it can be suggested that random forest has more prediction accuracy than the other two classifiers under study. In future, we aim to implement other unsupervised algorithms to compare the efficiency of the two different classes of classification algorithms.

References

1. A. Oussous, F.Z. Benjellun, A.A. Lahcen, S. Belfkih, Big data technologies: a survey. J. King Saud Univ. Comput. Inf. Sci. **30**(4), 431–448 (2018)
2. P.P. Shinde, K.S. Oza, R.K. Kamat, Big data predictive analysis: using R analytical tool, in *2017 International Conference on I-SMAC (IoT in Social, Mobile, Analytics and Cloud) (I-SMAC)*, Palladam (2017), pp. 839–842
3. A. Dey, Machine learning algorithms: a review. Int. J. Comput. Sci. Inf. Technol. **7**(3), 1174–1179 (2016)
4. M. Alloghani, D. Al-Jumeily, J. Mustafina, A. Hussain, A.J. Aljaaf, A systematic review on supervised and unsupervised machine learning algorithms for data science, in *Supervised and Unsupervised Learning for Data Science* (Springer, Berlin, 2020), pp. 3–21
5. S.B. Kotsiantis, I.D. Zaharakis, P.E. Pintelas, Machine learning: a review of classification and combining techniques. Artif. Intell. Rev. **26**, 159–190 (2006)
6. O. Simeone, A very brief introduction to machine learning with applications to communication systems. IEEE Trans. Cogn. Commun. Networking **4**(4), 648–664 (2018)
7. Y. Yang, J. Li, Y. Yang, The research of the fast SVM classifier method, in *2015 12th International Computer Conference on Wavelet Active Media Technology and Information Processing (ICCWAMTIP)*, Chengdu (2015), pp. 121–124
8. M.A. Hearst, S.T. Dumais, E. Osuna, J. Platt, B. Scholkopf, Support vector machines. IEEE Intell. Syst. Appl. **13**(4), 18–28 (1998)
9. S. Tong, D. Koller, Support vector machine active learning with applications to text classification. J. Mach. Learn. Res. 45–66 (2001)
10. H. Sharma, S. Kumar, A survey on decision tree algorithms of classification in data mining. Int. J. Sci. Res. (IJSR) **5**(4) (2016)
11. M.V. Datla, Bench marking of classification algorithms: decision trees and random forests—a case study using R, in *2015 International Conference on Trends in Automation, Communications and Computing Technology (I-TACT-15)*, Bangalore (IEEE, 2015), pp. 1–7
12. B.R. Patel, K.K. Rana, A survey on decision tree algorithm for classification. IJEDR **2**(1) (2014)
13. Y. Song, Y. Lu, Decision tree methods: applications for classification and prediction. Shanghai Arch. Psychiatry **27**(2), 130–135 (2015)

14. A. Sarica, A. Cerasa, A. Quattrone, Random forest algorithm for the classification of neuroimaging data in Alzheimer's disease: a systematic review. Front. Aging Neurosci. (2017)
15. Machine Learning Databases. https://archive.ics.uci.edu/ml/machine-learning-databases/heart-disease/

COVID-19 and SARS Virus Predictions from Chest X-ray Images Using a Deep Learning Model

Dadi Ramesh, Shashikala, A. Harshavardhan, and Dandugudum Mahesh

Abstract The COVID-19, novel coronavirus, is a pandemic and contaminated disease and it is aggressively increasing day by day from the first occurrence. Till now, within three months, a total of more than twenty lacks people affected with viruses, and over 50 thousand people have died. The virus is mainly attacking the respiratory system of the human body. The cases were increasing exponentially from day by day. And the diagnostic test of COVID-19 is to detect the presence of viral proteins in a sample from a person's respiratory tract. But this process of testing is taking more time, and it is cost effective. For the prediction of COVID-19, we proposed a deep learning-based convolution neural network model using a chest X-ray of humans. The CNN model is trained on the dataset of COVID-19 X-ray and normal human X-ray; so it will able to identify the density of pneumonia in the human body to detect the COVID-19 virus.

Keywords COVID-19 · X-ray · Convolution neural network · Deep learning

1 Introduction

The COVID-19 currently outbreaks in more than 200 countries, and the number of cases increases exponentially and the testing kits are limited, and the gap between the total number of people and number of tests is huge and time consuming for testing is very high therefore there is need of changing the testing and treating methodology. The present testing method for COVID-19 [1, 2] is the reverse transcriptase-polymerase chain reaction (RT-PCR) [3, 4] test, which has to be a labor-intensive process. The COVID-19 virus is entering into the human body from the nose or mouth. It will reach to lungs within four days in 14 days. Therefore, the symptoms

D. Ramesh (✉)
Member Center for Artificial Intelligence and Deep Learning, Computer Science and Engineering, S R Engineering College, Warangal, India

Shashikala · A. Harshavardhan · D. Mahesh
Computer Science and Engineering, S R Engineering College, Warangal, India
e-mail: mahesh_d@srecwarangal.ac.in

V. Bhateja et al. (eds.), *Data Engineering and Intelligent Computing*,
Advances in Intelligent Systems and Computing 1407,
https://doi.org/10.1007/978-981-16-0171-2_39

of COVID-19 are fever and cough, stuffy and runny nose, sore throat, etc., and in the peripheral and posterior lungs on CT images, combined with healthy or decreased white blood cells and history of epidemic exposure, are highly suspected of having 2019 Novel Coronavirus (2019-nCoV) [4, 5] pneumonia.

In recent days, artificial intelligence plays a significant role in the health sectors like brain tumor detection, early detection of a cancer cell from CT scan, and identification pneumonia from chest X-ray, etc. So, we are proposing an AI-based deep neural network to detect COVID-19 from chest X-ray [6, 7]. The dataset was used from Kaggle competitions. In this paper section, two will explain related work, and Sect. 2 will contain our proposed work with model explanation, and Sect. 3 will have experiment results with analysis and Sect. 4 summarizes the conclusion and feature work.

Song et al. [8] confirmed that the COVID-19 virus was found in the CT scan chest X-ray. According to their report, the infection is observed in all patients and accounted for 87% of lesions, and the CT findings combined with predominant distribution in a posterior and peripheral part of the lungs were uncommon in other viral pneumonia. And from the observations, patients' lungs showed that the thickness of pneumonia is increasing daily from fourth day to later. And they grouped all chest images into two groups. The first group is having pictures of affected persons before four days. And the second group contains images of affected persons after four days.

Table 1 shows in detail the observation of chest images [9] of lungs with cell count and distribution of virus in the lungs.

Ibrahim et al. [10] proposed a deep recurrent neural network to forecast coronavirus spread across the globe. The architecture they implemented sequence to sequence model with three components: (1) LSTM, (2) self-attention, and (3) variational autoencoder graph. The LSTM recurrent neural network is playing a significant role in forecasting time series. It works on fixed and time scale parameters using the sigmoid activation function. LSTM will find each cell's behavior based on time and based on the dimensions of different countries. The next in the architecture is the self-attention mechanism; it will take additional features about the country and their inputs based on it will clear data what to forget and what to consider. The third step is the variational autoencoder graph; it will work based on weights of geographical locations and prepares an adjacent weight matrix from these it will predict outbreaks on COVID-19.

Karim et al. [11] proposed a deep neural network to predict COVID-19 based on chest X-ray. They used a ResNet deep learning [12–14] model for classifying the images into positive and negative. And they implemented the CNN model for classification and visualization for four classes like a regular, bacterial, non-COVID-19, and COVID-19 virus. The dataset they used is from kaggle.com, which has three versions of datasets, and they used 3.0 for training the CNN model. ResNet's performance compared to other models like vgg16, vgg19, and DenseNet got good results. And the accuracy they got is 0.85 in the prediction of COVID-19 from X-ray images.

Table 1 CT images findings in 51 patients with COVID-19 pneumonia

Finding	No. of patients	No. of lesions*
Unilateral lung	7(14)	40 (3)
Bilateral lung	44 (86)	1284 (97)
One lobe	4 (8)	12 (0.9)
Two lobes	8 (16)	64 (5)
Three lobes	6 (12)	54 (4)
Four lobes	12 (24)	276 (21)
Five lobes	20 (39)	91 8 (69)
Involved lung zones		
Upper lobes	43 (84)	483 (37)
Middle lobe	30 (59)	136 (10)
Lower lobes	46 (90)	703 (53)
Predominant distribution		
Anterior	9 (18)	145 (11)
Posterior	41 (80)	1179 (89)
Peripheral	44 (86)	1198 (91)
Central	5 (10)	42 (3)
Both central and peripheral	1 (2)	84 (6)

* Number of chest CT lung abnormalities

2 Proposed Approach

2.1 Dataset

We used coronahack-chest-X-ray-dataset of all versions is from Kaggle and GitHub [15]. The complete dataset contains 5910 images which contain 27% of normal person chest X-ray images, it means healthy persons, and 1417 images of virus (COVID-19)-affected chest X-ray, and 47% of images are bacteria-affected chest X-ray images. And remaining images are pneumonia-affected images like ARDS and SARS virus. All the images are labeled with bacteria, viruses, and healthy with a unique number.

2.2 Preprocessing of Data

The dataset contains three categories of images, and all chest X-ray images are in different dimensions and have a different label, so preprocessing [16, 17] is required for a raw dataset. The dataset is majorly divided into two classes. And all bacteria X-ray images are removed from the dataset. We have only 3000 + X-ray images

Table 2 Coronahack dataset after preprocessing

Type	Number of images	Label
COVID-19 and SARS virus images	1665	0
Normal images	1583	1

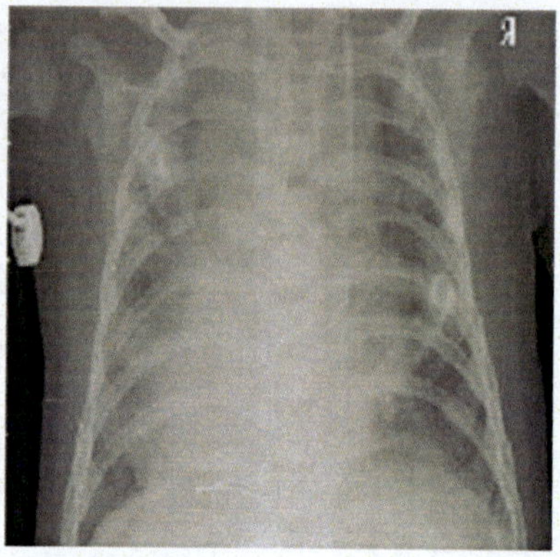

Fig. 1 COVID-19 chest X-ray after preprocessing

from that we separated healthy and virus labeled images as shown in Table 2; the virus images include COVID-19, SARS, and ARDS. For the preprocessing of the coronahack dataset, we used OpenCV, a Python library. And we resized all images to a constant size that is 200 * 200 and assigned label 0 for virus images, 1 for standard images. And all images are converted to grayscale images as shown in Fig. 1. And we turned all images into a 2D array using the Numpy Python library.

2.3 Implementation

We implemented a new convolution neural network for predicting COVID-19 from chest X-ray. We applied this model in a Kaggle notebook. The CNN model [18] consists of five layers in that three layers are for feature extraction. Each feature extraction and learning layer will have conv2d unit, max polling of size 2 * 2, and batch normalization with axis 1and the activation function is ReLU. The ReLU activation function is faster than other activation functions, and the gradient is always high.

And reaming two layers are the dense layers. The first dense layer activation function is ReLU, and the dropout rate is 0.5 that at every iteration, iteration of 0.5% of units is randomly dropped from the network to reduce the over fitting. And the second dense layer has the same dropout rate, but the activation function is sigmoid. The sigmoid function is used for binary classification of output. The first dense layer

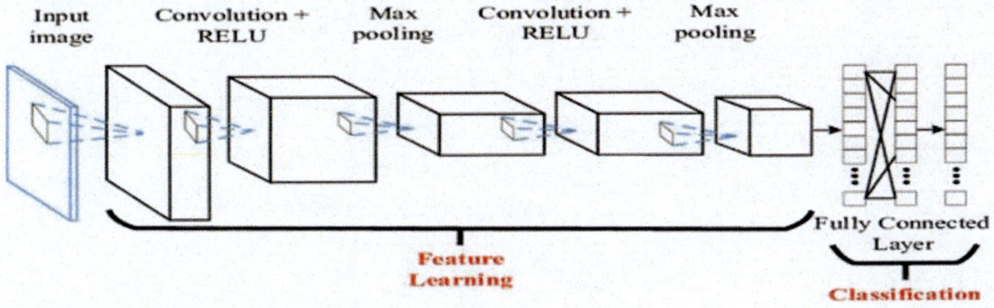

Fig. 2 Block diagram CNN

will reduce the number of units to 64. And the second dense layer will reduce the units to 1 the complete block diagram is shown in Fig. 2.

And in our model, we used Adam optimizer, and loss calculated with binary cross-entropy, which will calculate the probability of prediction. Hear $P(y)$ is the predicted probability for all N points.

$$H_p(q) = -\frac{1}{N}\sum_{i=1}^{N} y_i \cdot \log(p(y_i)) + (1 - y_i) \cdot \log(1 - p(y_i))$$

2.4 *Training*

The total dataset is divided into two parts: training and testing, in the ratio of 80:20 from the complete dataset, which is 3248 images for testing and training. And the learning rate is set to 0.0001, and the model trained for 15 epochs. The average time for each epoch is 11 s. At each epoch, we calculated loss and accuracy. The Kaggle GPU was used for training the model.

Model Evaluation
The model trained for 15 epochs, and at every epoch, the accuracy is improved starting from 0.54 to reach to 0.84. And model loss is reduced at every epoch starting from 0.7 and reached to 0.34. Figures 3 and 4 will show the model performance. And Figs. 5 and 6 will show model validation accuracy and loss.

3 Result Analysis

Compared to some predefined standard models combinations like vgg-16, DenseNet-161and other models, our model performance, precision, and recall are better—the comparison of accuracy and recall are shown in Table 3. And in Fig. 7, a sample

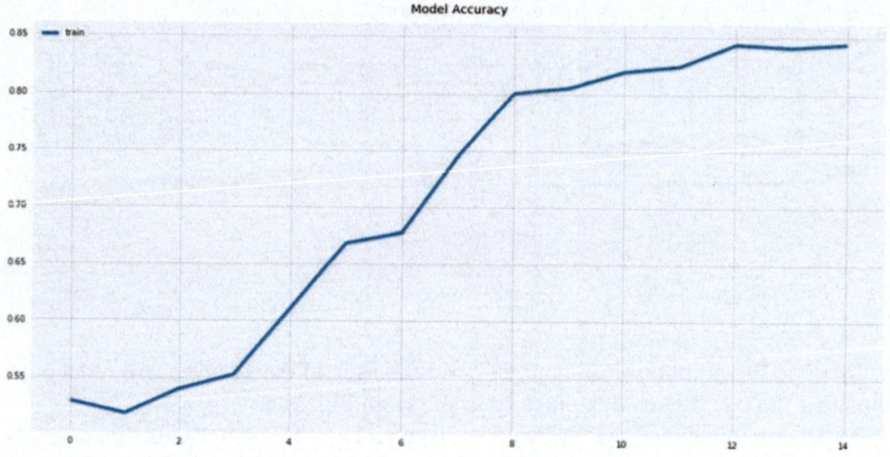

Fig. 3 Model accuracy

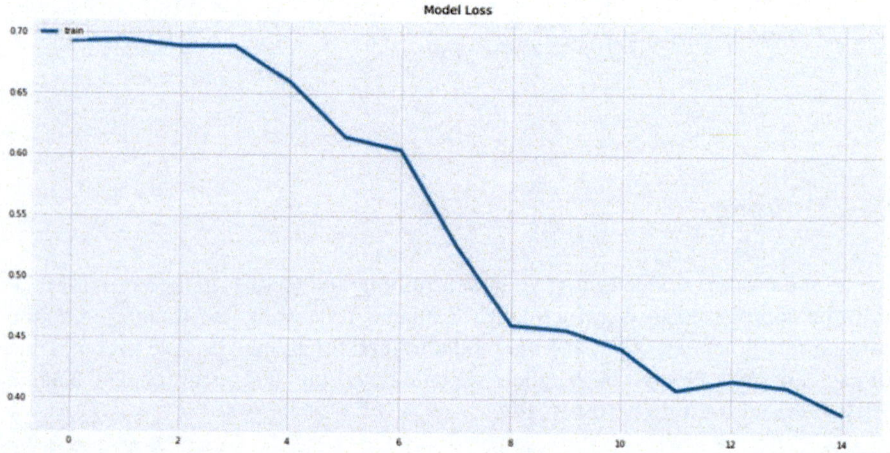

Fig. 4 Model loss

of 25 images is given for prediction, and our model mispredicted 10% of images predicted. The average patient chest X-ray was predicted as COVID-19 and SARS, and one image with COVID-19 and SARS was predicted as usual.

Precision: The precision is the value when dividing the total number correctly predicted images by the total number of correctly predicted and false predictions. The precision value of our model is 0.972.

Recall: The recall was the value when we divided the total number correctly predicted images by the total number of positive samples. The recall value of our model is 0.82. And from precision and recall, we can say precision is higher than recall that

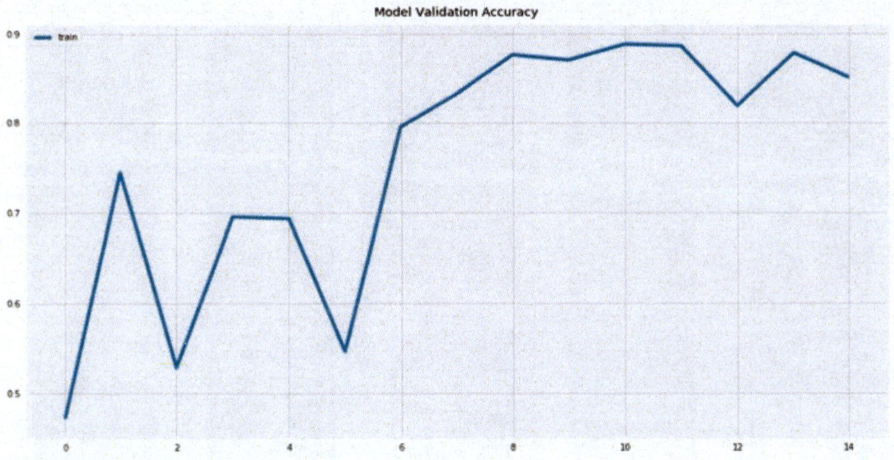

Fig. 5 Model validation accuracy

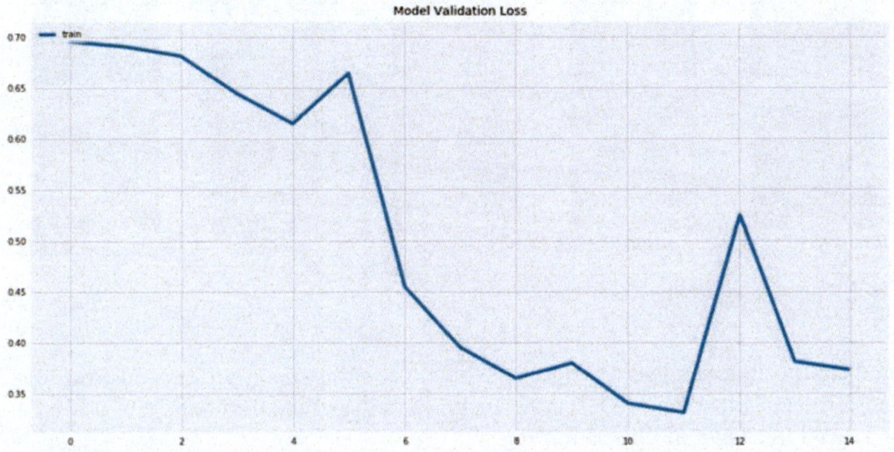

Fig. 6 Model validation loss

Table 3 Comparison of precision and recall

Model	Precision	Recall	F1-score
VGG-19 + DenseNet-161	0.905	0.913	0.909
VGG-19 + ResNet-18	0.926	0.917	0.925
Our model	0.97288	0.834	0.897

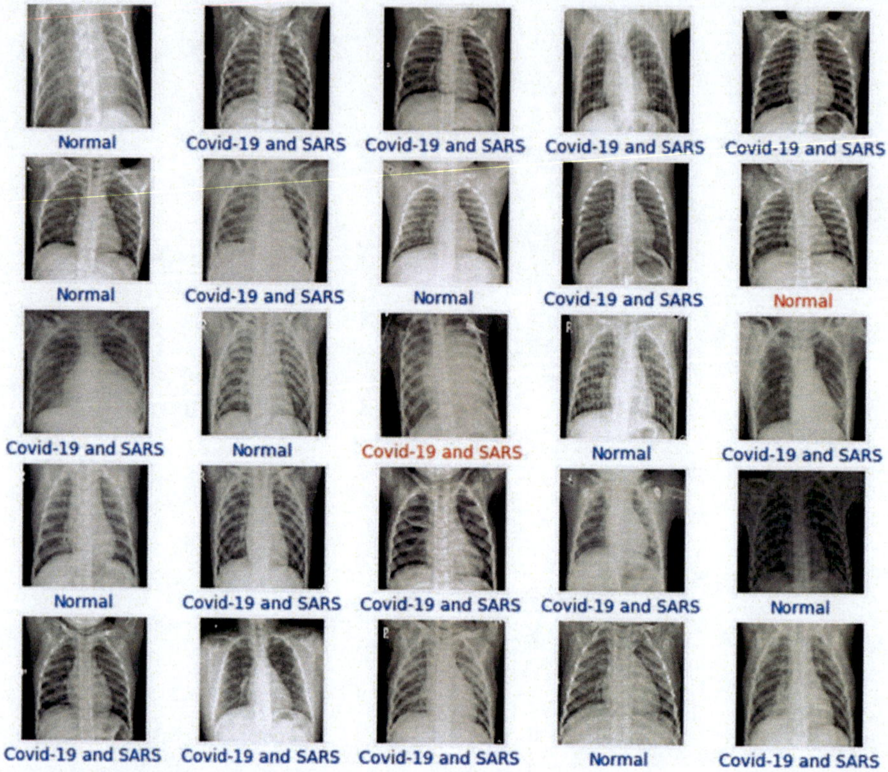

Fig. 7 Sample prediction

predicted positive samples are highly positive. The comparison between false positive and correct positive rate (ROC curve) is shown in Fig. 8 and the contrast of precision and recall as shown in illustration Fig. 9

Confusion Matrix
A confusion matrix is a table that shows the accuracy of the model on prediction. The total number of correct and incorrect predictions summarized in a table. In Fig. 10, an overview of a confusion matrix is shown and in Fig. 11 overall prediction of our model as represented.

4 Conclusion

Despite the combination of some standard models of deep learning for image processing to classify the images or prediction, the evaluation results can show that performance and accuracy are excellent without any combinations. By having more

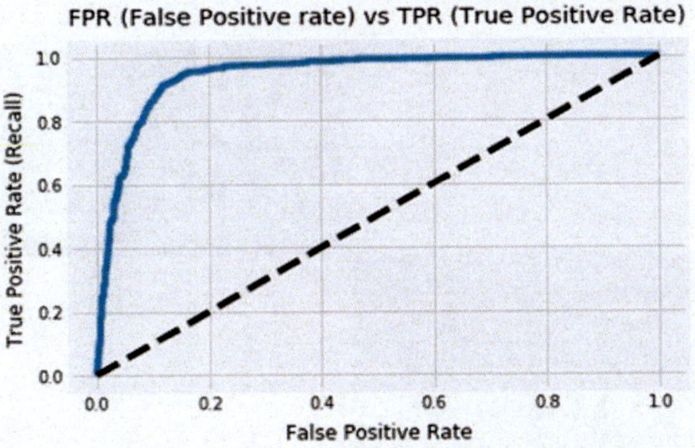

Fig. 8 ROC curve

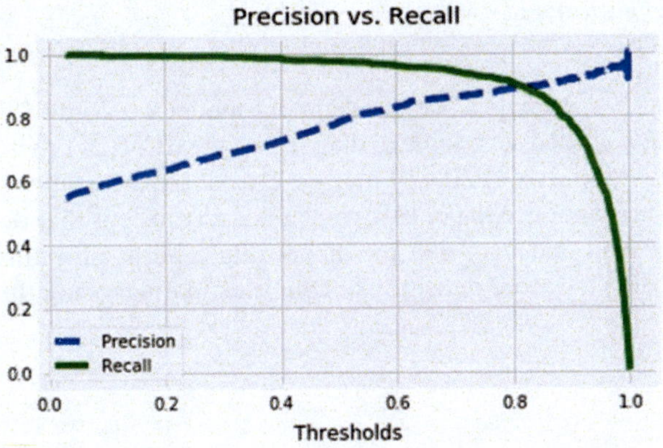

Fig. 9 Precision vs. recall

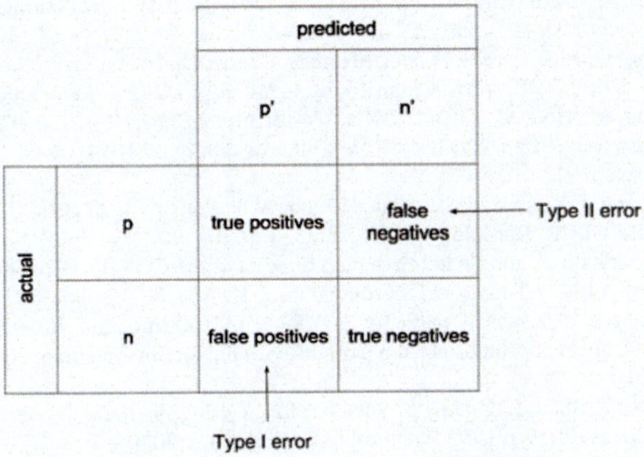

Fig. 10 Sample confusion matrix

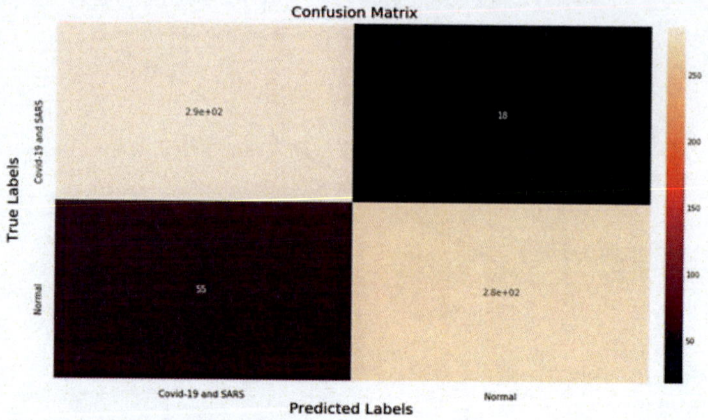

Fig. 11 Confusion matrix for our model

COVID-19 chest X-ray images, we can train our model with only COVID-19 virus images only then it will be accessible to diagnosing the COVID-19 virus in future we train our model with only COVID-19 images. By our experiment, we can conclude that deep learning cannot replace, but we can heartily accept that deep learning is useful and consistent and very fast for diagnosing reports, and this type of technology is beneficial for more density like countries in diagnosing the bulk number of patients.

References

1. C. Huang et al., Clinical features of patients infected with 2019 novel coronavirus in Wuhan, China. The Lancet **395**, 497–506 (2020)
2. M.A.A. Al-qaness, A.A. Ewees, H. Fan, M. Abd, E. Aziz, Optimization method for forecasting confirmed cases of COVID-19 in China 19. https://doi.org/10.3390/jcm9030674
3. T. Ai, Z. Yang, H. Hou, C. Zhan, L. Xia, Correlation of chest CT and RT-PCR testing in coronavirus disease 2019 (COVID-19) in China: a report of 1014 cases. Radiology **2020**, 200642 (2020)
4. Y. Fang, H. Zhang, J. Xie, M. Lin, L. Ying, P. Pang, W. Ji, Sensitivity of chest CT for COVID-19: comparison to RT-PCR. Radiology **2020**, 200432 (2020)
5. E. Botha, W. Dednam, A simple iterative map forecast of the COVID-19 pandemic 1–6 (2020)
6. E. Soares, P. Angelov, S. Biaso, M.H. Froes, Daniel, K. Abe, SARS-CoV-2 CT-scan dataset: a large dataset of real patients CT scans for SARS-CoV-2 identification, in medRxiv
7. N. Zhu, et al., A novel coronavirus from patients with pneumonia in china, 2019. New Engl. J. Medicine (2020)
8. F. Song, N. Shi, F. Shan, Z. Zhang, J. Shen, H. Lu, Y. Ling, Y. Jiang, Y. Shi, Emerging 2019 Novel Coronavirus (2019-nCoV) Pneumonia, in Radiology 2020
9. W. Kong, P.P. Agarwal, Chest imaging appearance of covid-19 infection. Radiol. Cardiothorac. Imaging **2**, e200028 (2020)
10. M.R. Ibrahim, J. Haworth, A. Lipani, N. Aslam, T. Cheng, N. Christie, Variational-LSTM Autoencoder to forecast the spread of coronavirus across the globe, in medRxiv 2020

11. M. Rezaul Karim, T. DAűhmen, D. Rebholz-Schuhmann, S. Decker, et al., DeepCOVIDExplainer: explainable COVID-19 predictions based on Chest X-ray images, in eess (2020)
12. I. Goodfellow, Y. Bengio, A. Courville, *Deep Learning, Adaptive Computation and Machine Learning Series* (The MIT Press, Cambridge, 2017)
13. A. Narin, C. Kaya, Z. Pamuk, Automatic detection of coronavirus disease using X-ray images and deep convolutional neural networks. arXiv:2003.10849 (2020)
14. S.N. Pasha et al., Variation analysis of artificial intelligence, machine learning and advantages of deep architectures. Int. J. Adv. Sci. Technol. **28**(17), 488–495 (2019)
15. J.P. Cohen, P. Morrison, L. Dao, COVID-19 image data collection. arXiv 2003.11597 (2020). https://github.com/ieee8023/covid-chestxraydataset
16. A. Harshavardhan, S. Babu, et al., An improved brain tumor segmentation and classification method using SVM with various kernels. J. Int. Pharma. Res. **46**(2), 489–495 (2019)
17. A. Harshavardhan, S. Babu, et al., Brain Tumor Segmentation Methods—A Survey. Special Issue 11-Special Issue, July 2017, pp. 240–45
18. L. Wang, A. Wong, COVID-Net: a tailored deep convolutional neural network design for detection of COVID-19 cases from chest X-ray images. arXiv:2003.09871 (2020)

Vehicle-Type Classification Using Customized Fuzzy Convolutional Neural Network

D. T. Mane, P. B. Kumbharkar, P. S. Dhotre, and Santosh Borde

Abstract Vehicle-type classification is fast becoming a popular domain of interest due to its various application areas ranging from traffic surveillance to autonomous navigation. In this paper, customized fuzzy convolutional neural network (CFCNN) is proposed for the classification of vehicle types by incorporating a customized convolutional neural network (CCNN) with a fuzzy hypersphere neural network (FHSNN). The two pivotal factors for the improvement of the CFCNN learning algorithm are: first, the ability to extract dominant hidden features from vehicle front raw images using CCNN and second, fuzzy hypersphere membership function used for pattern classification. The proposed model's performance evaluated using the BIT-vehicle standard dataset, which consists of high-resolution 9850 images of various five types of vehicles, and the testing accuracy obtained 94.26%, which is exceptionally better in comparison to the existing algorithms.

Keywords Convolutional neural network · Vehicle-type classifications · Fuzzy hypersphere neural network · Deep learning

1 Introduction

Intelligent transport system application intends to give new administrations identifying with various methods of transports and traffic managing and allows users to be better knowledgeable and make more secure, more organized, and smarter utilization of transport systems. One of the essential components of the intelligent transport system is vehicle-type classification. This field has a wide scope of applications, crossing from traffic flow statistics to smart parking method. Current methods mainly incorporate an ultrasonic, camera, and magnetic. As there is the widespread use of traffic scrutiny cameras these days, image-based techniques garner significant attention. Till date, many picture-based techniques have been proposed,

D. T. Mane (✉) · P. B. Kumbharkar · P. S. Dhotre · S. Borde
JSPM's Rajarshi Shahu College of Engineering, Pune, Maharashtra 411033, India

V. Bhateja et al. (eds.), *Data Engineering and Intelligent Computing*,
Advances in Intelligent Systems and Computing 1407,
https://doi.org/10.1007/978-981-16-0171-2_40

and they can be arranged into: model-based techniques and appearance-based techniques. The most important objective of model-based techniques is to calculate the height, width, and length parameters of the vehicle and use that to retrieve the three-dimensional model view of the vehicle. On the other hand, appearance-based methods are based on extracting features representing the vehicle appearance to classify the vehicles. Majority of the vehicle classification methods involve side-view images of the vehicle. However, in the current scenario, many traffic surveillance cameras are installed at various places, making it easier to capture vehicle frontal-view images. Thus, models involving vehicle frontal-view images require comparatively less effort. Petrovic and Cootes [1] facilitated the modeling of vehicle appearance, making use of vehicle frontal-view images by extracting various features like edge orientation, locally normalized gradients, Sobel edge response, Harris corner response, and direct normalized gradients. Zhang [2] combined the Gabor transform feature with the PHOG characteristic to correspond to the vehicle class pattern and also put forth a go down classifier system to identify vehicle types. A voting algorithm is proposed by Negri et al. [3] revolving around sloping contour positions for the detection of different types of vehicle vehicles. Psyllos et al. [4] incorporated the SIFT features in his approach to identify the manufacturer, model, and vehicle logo. Dong et al. [5] proposed method that can consequently learn dominant features in favor of the vehicle without the need for inputting hand-labeled features manually. Classification of the type of vehicle is done via a softmax regression. The unsupervised sparse filtering is used to give training to the network. Nevertheless, the above-mentioned methods use numerous features which might not be discerning enough in complex scenarios. Peng et al. [6] invented a robust vehicle-type classification using adaptive multiclass PCA. Zhen Dong et al. [7] have presented a novel semi-supervised CNN technique for classification of types of the vehicle. They applied a Laplacian filter on unlabeled data, and softmax classifier applied on labeled data. Bai et al. [8] proposed a deformable-based framework using a support vector machine (SVM) for vehicle detection and classification. Roecker et al. [9] presented a CNN-based automatic vehicle-type classification model with frontal perspective low-resolution images. Unlike conventional techniques of using the handcrafted characteristics of features, CCNN is capable to automatically get familiar with numerous phases of invariable hidden dominant features in support of the specific applications [10, 11]. And due to supervised clustering, a fuzzy hypersphere neural network (FHSNN) is very much efficient in handling the different pattern classification problems [12, 13]. In this paper, a CFCNN method is presented for vehicle-type classifications of frontal vehicle images that combine the benefits of two CCNN and FHSNN techniques. The performance of the model is evaluated on BIT-vehicle standard dataset, which contains high-resolution 9850 frontal five types of vehicle images. Some of each class BIT samples represented in Fig. 1. CFCNN created ideal outcomes over CNN because of the following reasons:

- CFCNN deals with the balanced and irregular allocation of data examples, while conventional CNN is more effective on balanced data.

Fig. 1 BIT-vehicle dataset

- In CFCNN, only representative cluster centroids are used for vehicle pattern classification at the time of testing, while on account of CNN, the correlation is made with all the examples.

An overview of the paper is as follows: Learning algorithm and CFCNN system architecture is described in Sect. 2. Experimental results and performance evaluation is explained in Sect. 3, and finally, Sect. 4 depicts the conclusion of the proposed work.

2 CFCNN Architecture

The strong point of CNN is that it automatically takes out unknown dominant hidden features and makes use of it to efficiently detect the patterns which are present into input pattern data samples [14]. In conventional classifiers like support vector machine with kernel trick, logistic regression, multilayer neural networks, Naive Bayes algorithms, etc., require the dominant features for their training, which is a very complex preprocess task in the classification task. To produce an appropriate outcome, require the proper dominant hidden features and many times as a human also cannot find such hidden features properly. Due to this overhead problem, traditional classifiers could not produce efficient outcomes. Moreover, FHSNN is a technique which effectively handles small datasets and also very much useful to reduce overall complexity in classification and different clustering tasks [13, 15]. Due to the supervised fuzzy clustering, it effectively handles pattern classification tasks, but it requires the dominant features of each pattern. Merging the strength points of together these techniques outcomes in an amalgam method, i.e., customized fuzzy CNN (CFCNN). Therefore, in this paper, we proposed CFCNN technique which incorporating the strength of both FHSNN and CNNN techniques. The two pivotal factors for the improvement of the CFCNN learning algorithm are: first, the ability to extract dominant hidden features from vehicle front raw images using CCNN and second, fuzzy hypersphere membership function is used for pattern classification. The general CFCNN system model architecture is represented in Fig. 2.

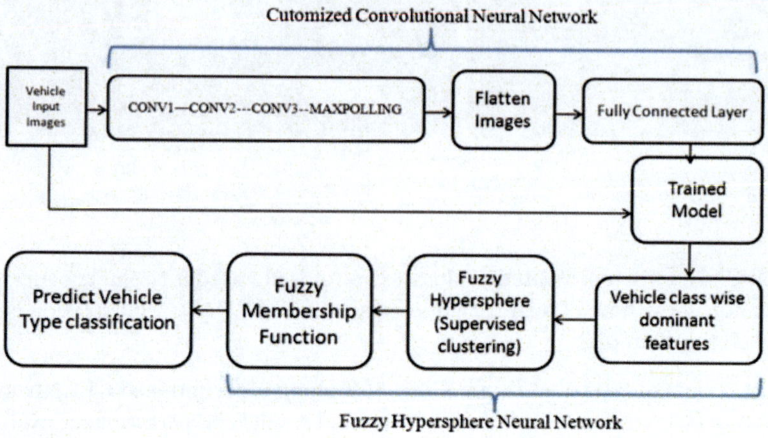

Fig. 2 CFCNN system architecture

2.1 Customized CNN

In neural networks, a single layer has only two layers, the input layer and the output layer. It takes input with weights and passes the summation/net value through the activation function. The summation value is calculated by input with weights, and the summation value is passed by different activation functions as per problem requirement, i.e., function (net).

$$\text{function}(\text{output}) = \text{function}(\text{net}) = \frac{1}{1 + \mathrm{e}^{-\text{net}}} \tag{1}$$

$$\text{net} = I_1 w_1 + I_2 w_2 + I_3 w_3 + \cdots + I_n w_n + I_b w_b \tag{2}$$

Multilayer perceptron (MLP) has more than two layers that contain multiple combinations of perceptrons. The expansion of MLP is a convolutional neural network (CNN), which has two additional layers for extract hidden features from inputs data, i.e., convolutional layer (CL) and pooling layer (PL). CL and PL are responsible for extracting dominant hidden features, and MLP is used for reducing the overall model error [14]. Aside from the fundamental architecture of CNN, several distinctive researches have proposed variation in traditional CNN [11, 16]. Here, customizes the traditional CNN architecture in this way, so it is capable of finding out unknown hidden features against the input vehicle samples without constraining the number of layers, instead of optimizing the number of layers for requirement of vehicle pattern classification task. This feature extractor CCNN architecture is shown in Fig. 3.

The input for CCNN is a frontal vehicle image of size 1 * 128 * 128, and eight CLs are used with rectifier linear units (ReLU). To handle the overfitting, the max PL is used along with two strides after the CL. Then, the PL output of the pooling layer

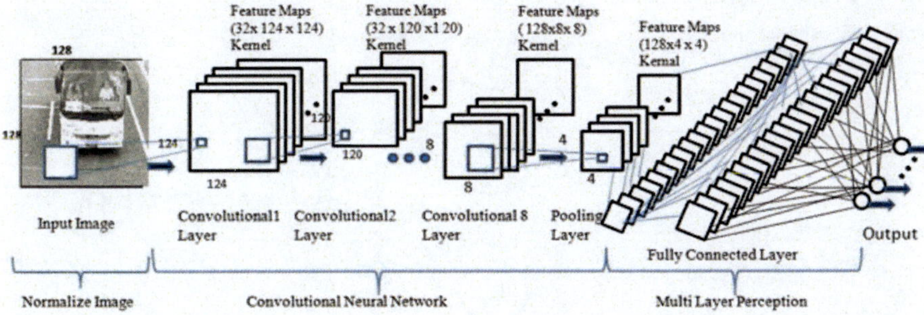

Fig. 3 Feature extractor CCNN architecture

128 * 4 * 4 is compressed and passed through MLP to produce dominant features. Shared filter weights are useful to discover a dominant feature pattern inside the image, similar to diagonal, vertical, polygonal curves, etc. For, the input I layer and jth hidden perceptron the output is.

$$\text{fun}(\boldsymbol{X} * \boldsymbol{W})_{i,j} = \text{fun}\left(\text{bais} + \sum_{a=0}^{4} \cdot \sum_{b=0}^{4} W_{a,m}.\boldsymbol{X}_{i+a,j+b}\right) \tag{3}$$

The trained model weights are saved, which will be utilizing to find the hidden patterns of all vehicles. Once these dominant features from each class of vehicle images are produced, they will further be feed into a FHSNN instead of the softmax activation function.

2.2 *FHSNN Architecture*

Extracted dominant features from CCNN passed through the FHSNN. Here, class-wise vehicle cluster is generated by using the concept of supervised clustering in FHSNN [12]. Fuzzy hypersphere clusters are determined through its fuzzy membership function, cluster radius, and clusters centroid [15]. The technique utilizes similarity measures and labels for every cluster, where the inclusion by a neuron relies upon the correlation of the input class pattern by the class learning pattern. In case of failure, a fresh hypersphere is formed. This shows the way to the building various hypersphere toward separate pattern as per its measurement of data dimension. The FHSNN system architecture is represented in Fig. 4. Multidimensional N patterns are the inputs in the input layer, and it is not an actual part of the computation; it just forwards the input information to the next hidden layer. Based on distance measurement functions, Jth nodes of fuzzy hypersphere are generated in the hidden layer. The output layer represents the creation of clusters by concern label and its centroid.

Fig. 4 FHSNN architecture

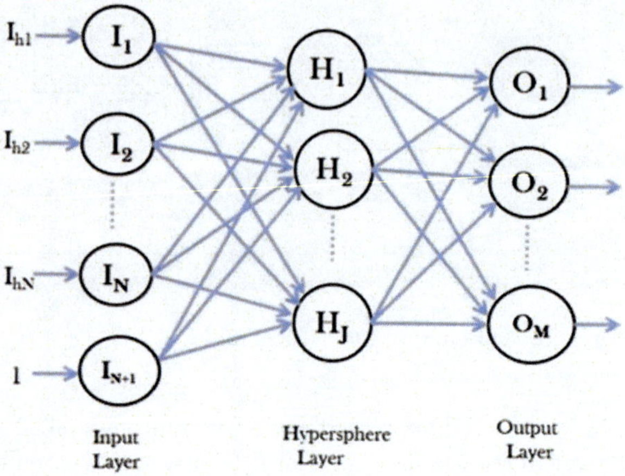

2.3 *Learning CFCNN Algorithm*

In this section, learning CFCNN algorithm steps are illustrated alongside its short form, which is illustrated in Table 1.

Table 1 CFCNN learning algorithm notations

Notation	Meaning	Notation	Meaning
Fea_area	Receptive field of CNN	α	Learning rate
Output	Output of the process	w	Weights
Fea_data	Feature vector data from CNN	€	Expansion factor for distance
Class_count	Total classes of fuzzy cluster	mem	Membership function
Distance	Distance vector of input from centroid	c_class	Centroid of each class present in cluster
Visited	Vector to check if every input point has class predicted	overlap	Check if cluster overlaps with other class cluster
Radius_class	Radius of each class present in cluster	class	Vector in which class of each input data is stored

Step I: Take an input image and pass through the Convolutional layer. Randomly initialize the weights and bias and calculate the part of the image through which the receptive field of the CNN would pass depending upon the stride considered.

Output = sum (w * fea_area) + b

Step II: Pass the filter of CNN over that part of the image and sum the element multiplication between the filter and receptive field. Output=sum (layer_slice*wt)+bi Step III: Repeat step I and II till the whole image is not passed through the filter

Output = output>0

Step IV: Flatten all the negative output of the output after the CNN filter passes to zero. Output = max (fea_area)

Step V: Calculate the part of the image through which the receptive field of the max pooling would pass depending upon the stride.

Step VI: Extract the maximum element of that area

Step VII: Repeat step V and VI till the whole feature is not passed through the max pooling layer.

Step VIII: Calculate the error depending upon the output predicted by the CNN.

Step IX: With respect to loss, compute the error partial derivative with respect to weights.

Step X: Update the layer weights with the partial derivative of the error using the formula. w-= α *dw

Step XI: Output the weights of the image

Step XII: Perform Step I to XI for all image

Step XIII: For j=0 till fea_data do // Fuzzy Hypersphere Neural Network
For i=0 till class_count do
Calculate Distance using the formula distance[k]=(fea_data[j]-c_class[i])^2
k=k+1, Sort distance
For k=0,z = 0 till distance, class_count do
if(distance[k]<radius_class[z])
Visited[j]=true, class[j]=k
End For
if(visited[j] not true)
For k=0 till class_count do
if(distance[k]<= €)
radius_class[k]=distance[k]
For l=0 till class_count do
if(l!=k)
if(c_class[k]- c_class[l]<=radius_class[k]+radius_class[l]) do

```
        overlap=true, radius_class[k]=radius_old[k]
      Else
      Update visited = True and class = k, radius_class[k]=distance[k], Break
      if(overlap=true) do
         class_count=class_count+1,  c_class[class_count-1]=fea_data[j]
         radius_class[class_count]=0
    For m=0 to class_count do
      if(m!=(class_count-1))
      if(c_class[class_count-1]-c_class[m]<radius_class[m])
       mem=max(radius_class,distance for class_count)
       radius_class[m]=c_class[class_count-1]-c_class[m], visited[m]=False
Step XIV: Return Class Predicted
Step XV: End
```

3 Experimental Results

BIT vehicle dataset comprises of 9850 samples. It consists of different images of five types of vehicles. The existing FHNN algorithm is efficient for pattern classification, but on the downside, it takes as an input the image consisting of dominant features. BIT dataset consists of high-resolution raw images, and due to similar appearance of different vehicle types, extracting handcrafted features involves very complex processing. Hence, all the 9850 sample images were processed using the CCNN algorithm, and then the model was trained using these images. Post-processing, we obtained 1024 features pertaining to each sample. Now, these samples were fed as an input to the FHSNN algorithm. Out of the complete 9850 examples, 7850 examples were utilized for preparing training, and the left behind 1970 examples were utilized for testing. While training, the radius of the hypersphere was set to 6, and the expansion factor was set to 8. After applying MHSNN learning rules, we obtained different clusters corresponding to bus, minivan, Sedan, SUV, and truck vehicle types, i.e., 324, 505, 882, 417, and 465, respectively. 2D visualization representation of hypersphere clusters and traditional CNN verses CFCNN results is represented in Fig. 5. The accuracy obtained while training the model was almost 100%. At the time of testing, the membership function was used to derive the hypersphere membership for every testing pattern. The individual accuracy of each vehicle class is represented in Table 2.

The performance confusion matrix parameter for the presented method is shown in Fig. 6. The testing accuracy obtained was 94.26%, which is exceptionally better in comparison to the existing algorithms. The comparative accuracy obtained by the proposed model with all existing algorithm is depicted in Table 3.

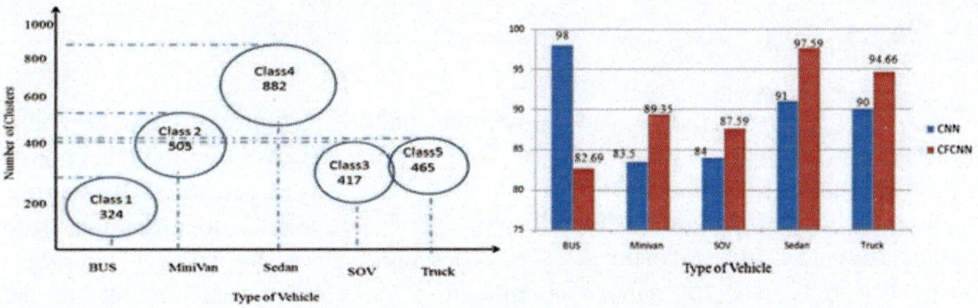

Fig. 5 Classwise cluster hypersphere and individual accuracy CNN versus CFCNN

Table 2 Individual accuracy of each class of vehicle

Training 7880 samples + testing 1970 samples						
Vehicle	Bus	Minivan	Sedan	SUV	Truck	Avg. (%)
Clusters	324	505	882	417	465	
CNN	98	83.5	91	84	90	88
CFCNN	82.69	89.35	87.59	97.59	94.66	94.26

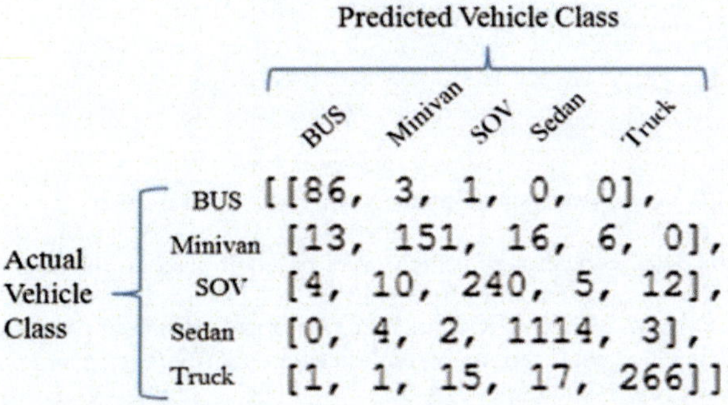

Fig. 6 Confusion matrix

Table 3 Comparative analysis proposed approach with existing systems

Year and Ref	Method	Accuracy
2004 [1]	Rigid structure approach	82.7
2011 [4]	Probabilistic neural network	78.3
2012 [6]	PCA with self-clustering	87.6
2014 [5]	Convolutional neural network (CNN)	88.8
2015 [7]	Laplacian filter + CNN	89.4
2017 [8]	Support vector machine	91.08
2018 [9]	CNN	93.6
Proposed	CCNN + FHSNN	94.26

4 Conclusion

This paper proposes CFCNN, for a proficient pattern recognition technique. The acquired outcomes are evidence for that the proposed CFCNN method has enough for getting better performance classifications of vehicle types because of two significant learning protocols: first, the ability to extract dominant hidden features from vehicle front raw images using CCNN and second, fuzzy hypersphere membership function is used for pattern classification. Amalgam-based CFCNN model combines the benefit of CCNN and FHSNN, as two are amazing artificial intelligence methods in the pattern classification tasks. The training time is moreover very much decreased because of the utilization of a GPU alongside the CPU. Performance evaluation is done using the BIT-vehicle standard dataset, and the testing accuracy obtained 94.26%, which is exceptionally better in comparison to the existing algorithms. Therefore, the authors conclude that CFCNN is capable of delivering superior outcomes, and it very well may be particularly helpful in various pattern recognition problems. Due to noisy and uncertainty in the similarity a type of vehicle, the proposed model sometimes demonstrates inappropriate outcomes which can be more enhanced in future.

References

1. V. Petrovic, T. Cootes, Analysis of features for rigid structure vehicle type recognition, in BMVC, pp 1–10 (2004)
2. B. Zhang, Reliable classification of vehicle types based on cascade classifier ensembles. IEEE Trans. Intell. Transp. Syst. **14**(1), 322–332 (2013)
3. P. Negri, X. Clady, M. Milgram, R. Poulenard, An oriented-contour point based voting algorithm for vehicle type classification, in *18th International Conference on Pattern Recognition (ICPR'06)*, Hong Kong (2006), pp. 574–577
4. A. Psyllos, C.N. Anagnostopoulos, E. Kayafas, Vehicle model recognition from frontal view image measurements. Comput. Stand. Interfaces **33**(2), 142–151 (2011)
5. Z. Dong, M. Pei, Y. He, T. Liu, Y. Dong, Y. Jia, Vehicle type classification using unsupervised convolutional neural network, in *22nd International Conference on Pattern Recognition*, Stockholm (2014), pp. 172–177
6. Y. Peng, J. S. Jin, S. Luo, M. Xu, Y. Cui, Vehicle type classification using pca with self-clustering, in *IEEE International Conference on Multimedia and Expo Workshops* (2012), pp. 384–389
7. Z. Dong et al., Vehicle type classification using a semisupervised convolutional neural network. IEEE Trans. Intell. Transp. Syst. **16**, 2247–2256 (2015)
8. S. Bai, et al., Classify vehicles in traffic scene images with deformable part-based models. Mach. Vis. Appl. **29**, 393–403 (2017)
9. M.N. Roecker, Y.M.G. Costa, J.L.R. Almeida, G.H.G. Mat-sushita, Automatic vehicle type classification with convolutional neural networks, in *IEEE 25th International Conference on Systems, Signals and Image Processing (IWSSIP)* (2018), pp. 1–5
10. D.T. Mane, U.V. Kulkarni, A survey on supervised convolutional neural network and its major applications. Int. J. Rough Sets Data Anal. **4**, 71–82 (2017)
11. D.T. Mane, U.V. Kulkarni, Visualizing and understanding customized convolutional neural network for recognition of handwritten Marathi numerals. Procedia Comput. Sci. **132**, 1123–1137 (2018)

12. D.T. Mane, U.V. Kulkarni, Modified fuzzy hypersphere neural network for pattern classification using supervised clustering. Procedia Comput. Sci. **143**, 295–302 (2018)
13. D.T. Mane, U.V. Kulkarni, A novel fuzzy convolutional neural network for recognition of handwritten Marathi numerals. Int. J. High Perform. Comput. Netw. **15**, 158–169 (2019)
14. Y. LeCun, L. Bottou, Y. Bengio, P. Haffner, Gradient-based learning applied to document recognition. Proc. IEEE **86**(11), 2278–2324 (1998)
15. U.V. Kulkarni, D.D. Doye, T.R. Sontakke, General fuzzy hypersphere neural, in *International Joint Conference on Neural Network*, vol. 3, Honolulu, HI, USA (2002), pp. 2369–2374
16. A. Khan, et al., A survey of the recent architectures of deep convolutional neural networks. Artif. Intell. Rev. 1–62 (2020)

Investigation of Student Performance with Contextual Factors Using Association Rules in Higher Educational Institutions (HEIs)

Subhashini Sailesh Bhaskaran

Abstract Association rules provide an important clue on how hidden knowledge and patterns can be extracted and the attributes extracted are associated when big data is mined by means of a particular data mining algorithm. The context of HEIs, where big data mining is uncommon, there is a need to know how association rules could be utilized to discover hidden knowledge and forecast student performance based on time-to-degree and cumulative grade point average (CGPA) and also make decisions. This research has demonstrated this by taking the example of the dataset of 337 graduated students of a university in Bahrain belonging to the Bachelor's degree in Accounting and Finance. Two types of data mining concepts were applied, regression and a priori algorithm. Regression results showed that there is a statistically substantial link between time-to-degree and course-difficulty and not CGPA. Similarly, association rules showed the link between time-to-degree, course-taking pattern, no. of courses and course-difficulty pattern but not CGPA. The association rules generated provided the basis to predict optimum time-to-degree by controlling the no. of courses, course-taking pattern, and course-taking difficulty. The findings of this research provide an important window of opportunity for HEIs to enhance student performance using data mining techniques. This investigation has contributed theoretically showing the application of data mining theory using apriori algorithm and regression, for HEI big data mining where student factors and organizational factors are said to be fairly different from that of the HEIs where research on big data is seen to be a well-known activity.

Keywords HEIs · Data mining · Association rules · Time-to-degree · Student performance · Contextual

S. S. Bhaskaran (✉)
Multimedia Science, College of IT, Ahlia University, Manama, Bahrain
e-mail: sbhaskaran@ahlia.edu.bh

V. Bhateja et al. (eds.), *Data Engineering and Intelligent Computing*,
Advances in Intelligent Systems and Computing 1407,
https://doi.org/10.1007/978-981-16-0171-2_41

1 Introduction

Improving decision making is a continuous process in organizations including HEIs who strive hard to improve student experience. HEIs are constantly concerned with how student experience in terms of learning and performance could be improved. Learning and performance of students have been evaluated in a no. of ways in the HEIs. Examples include cumulative grade point average (CGPA), semester grade point average (SGPA), letter grades (e.g., A+, A, B+, B, C, P, and F), and percentage of results as transcripts. However, there are certain other aspects that have been hardly used by HEIs in improving student learning and performance, for instance time-to-degree. According to MacNeill et al. [1], there is a growing interest among researchers to understand, investigate, and analyze education data due to the potential it has in supporting business intelligence applications, that help in decision making. In addition, literature points out that hidden information and knowledge in student datasets need to be considered while making business decisions in HEIs as they have the potential to affect the accuracy of decision making [2]. Course-taking pattern is an example of hidden information and knowledge that reside in the student dataset and how it impacts the decisions related to learning and performance of students is not well discussed in the extant literature [3]. Adding to the course-taking pattern is the importance of contextual factors in decision making that are also considered to be hidden in the education dataset (student dataset). Contextual attributes are seldom extracted from the dataset to be included alongside course-taking pattern in decision making [4]. Examples of contextual attributes that could have serious implications to decision making in higher education but hidden in the educational dataset include course-difficulty, course weight age, no. of courses registered in a semester, and no. of courses offered in a semester along with repetition of courses. While some research outcomes have suggested that there is a need to gain knowledge on how course-taking patterns along with contextual attributes determine time-to-degree, semester GPA (SGPA) and cumulative GPA (CGPA) (student performance attributes), knowledge about the association between the student performance attributes on one side and course-taking pattern along with contextual attributes on the other could reveal useful rules and patterns relating them. Association rules and patterns have immense potential in enabling accurate decision making and predicting future performance using past data [5]. So as to discover the association rules this research has examined the use of data mining concepts [6] to education data, which are usually large.

1.1 Research Problem

While student dataset and its analysis can be used to study SGPA, CGPA and time-to-degree on one side and their relationship to course-taking pattern on the other, literature shows that such studies need to take into account contextual factors mentioned above, to enable a more accurate understanding of the relationship. Such an understanding is expected to be highly useful in accurate decision making by managers in HEIs with regard to improving teaching and learning although hardly any study has investigated the impact of contextualized course-taking patterns on student performance attributes. Considering the usefulness of the understanding of the relationship between student performance attributes, time-to-degree and contextual factors in accurate decision making in HEIs that follow American system and the volume of data that need to be analyzed, this article examines how data mining, an electronic method, can be used to discover the association between SGPA, time-to-degree, CGPA, students' course-taking pattern in HEIs, and contextual factors. Applying analytic software to handle large volume of student data to discover hidden knowledge by applying knowledge discovery and data mining principles is a major challenge not discussed well in the literature. Research on how to face this challenge successfully is considered as important by some and useful research outcomes produced in this area could significantly alter the ability of HEIs to make accurate data-driven decisions. Some of the important data-driven decisions that could be facilitated by an investigation into the association between CGPA and contextual attribute are: categorizing students on the unearthed contextualized course-taking patterns that describe time-to-degree, SGPA, and CGPA; predict course-taking pattern of students and the discovered relationship between time-to-degree, CGPA, SGPA and contextual factors; optimization of time-to-degree on student categorization and forecast. How to discover the association rule using appropriate data mining technique is the main objective of the research using which the above-mentioned decisions can be taken in HEIs. In order to achieve this, it is pertinent to explain about the background information about data mining, which is provided next.

2 Research Context

The research was conducted in a Higher Education Institution in Bahrain taking the example of the Bachelor's degree in Accounting and Finance (BSAF) where the no. of students who have graduated was identified as 337 no. The main factors investigated were student performance factors like CGPA, SGPA, time-to-degree and the hidden factors like course-taking patterns, course-difficulty, course weightage, withdrawals, course repetition, and summer sessions and obvious attributes like semester and no. of courses. Although literature shows that many research papers have been published to study the student performance using data mining [7], it is also seen that such research outcomes are often segregated and not unified and have only covered certain academic attributes. Thakar et al. [7] argue that a no. of factors that act in a significant role to predict student performance that are not cognitive in nature such as skills and attitudes have not been studied yet using data mining technique. Again Thakar et al. [7] claim that there is a need to measure, monitor, and infer many factors that could help in prediction of student performance and there is a necessity to use specific type of data mining techniques. These arguments provide support to this research. As hardly any study could be found to use data mining techniques in the literature that have been applied to student data in HEIs to predict the students' time-to-degree having relevance to the course-taking pattern and the GPA, the outcome of this research is expected to bridge this gap. In this context, this study has set the main objective to uncover patterns of registered courses of students and their association to their GPA and time-to-degree in addition to reveal how contextual factors (course-taking patterns, course-difficulty, course weightage, withdrawals, and repetition of courses). This knowledge is new and could help the HEIs in a no. of decision-making processes including but not limited to student advising, deciding on the no. of students to be taken in a particular stream, deciding on which courses to be offered thereby optimizing the resources and improving student performance in terms of GPA, time-to-degree, and knowledge on the set of courses to be registered in any particular semester.

3 Methodology

This research has taken the case of university in Bahrain and students who graduated through the BSAF program between the years 2009 and 2015. The students have graduated completing 132 credits which are equivalent to 44 courses with a minimum CGPA of 2.0 and usually completed in about 4 years of study were examined. Full-time students have registered in a minimum of 4 courses per semester but not exceeding 6. The maximum CGPA that could be scored is 4. All students who have completed their time-to-degree within 6 years were collected and a separate student dataset was created. The no. of students covered was 337. Only those students who registered at least 4 courses and maximum of 6 courses per semester were considered in the experiment. No part time students were considered. Then the attributes defined in Table 1 were computed to create data as per the scales given in the table.

Next a comprehensive table detailing all the attributes under investigation in this research was prepared which provide detail including definition, formulas used, quantity, scale point, coding used, and cited authors.

After tabulating the details of the attributes, the following steps were identified as recommended by Chapman et al. [8] that led to the generation the association rules.

Step I: Business understanding

Step II: Data understanding

Step III: Data preparation

Step IVa: Modeling using regression

Step IVb: Modeling using association

Step V: Evaluation

Step VI: Deployment

Table 1 Features related to student that were used in the research

No.	Attribute	Description	Formula
1	Student id	Student identification number	
2	CGPA	Cumulative GPA of student at graduation	Sum of all gradpoints scored/ total number of credits
3	len	Time to degree in terms of Semesters	Graduated semester - Join semester / 2
4	C1difficulty	difficulty for Course no.1	$Difficulty_c = \frac{\sum_{t \in BE_c} \sum_{j=1}^{m_t} G_{j,t} * W_t}{\sum_{t \in BE_c} W_t * m_t}$ where c, current course; t, course equivalent to the current one; BE_c, Set of equivalence courses for course c; m_t, Total number of students in course t; $G_{j,t}$ Grade of the jth student in course t; W_t Number of credits of course t.
5	C2difficulty	difficulty for Course no.2	
6	C3difficulty	difficulty for Course no.3	
7	C4difficulty	difficulty for Course no.4	
8	C5difficulty	difficulty for Course no.5	
9	C6difficulty	difficulty for Course no.6	
10	No. of courses	No. of courses registered	Number of courses registered in the semester
11	SGPA	Semester Grade Point Average	Sum of all gradpoints scored in the semester/number of credits in semester

Table 2 Association rules

1. c6difficulty=Difficult 47 ==> noofcourses=6 47 <conf:(1)> lift:(4.74) lev:(0.11) [37] conv:(37.08)
2. c5difficulty=Difficult c6difficulty=Difficult 37 ==> noofcourses=6 37 <conf:(1)> lift:(4.74) lev:(0.09) [29] conv:(29.19)
3. c4difficulty=Difficult c6difficulty=Difficult 36 ==> noofcourses=6 36 <conf:(1)> lift:(4.74) lev:(0.09) [28] conv:(28.4)
4. len=4.5 c2difficulty=Average noofcourses=5 37 ==> c5difficulty=Difficult 35 <conf:(0.95)> lift:(1.53) lev:(0.04) [12] conv:(4.71)
5. c4difficulty=Average 44 ==> c3difficulty=Difficult 41 <conf:(0.93)> lift:(1.25) lev:(0.02) [8] conv:(2.79)
6. len=4.5 c2difficulty=Average c5difficulty=Difficult 41 ==> c4difficulty=Difficult 38 <conf:(0.93)> lift:(1.24) lev:(0.02) [7] conv:(2.6)
7. len=4.5 c2difficulty=Average noofcourses=5 37 ==> c4difficulty=Difficult 34 <conf:(0.92)> lift:(1.23) lev:(0.02) [6] conv:(2.35)
8. len=4.5 c2difficulty=Average 46 ==> c4difficulty=Difficult 42 <conf:(0.91)> lift:(1.22) lev:(0.02) [7] conv:(2.34)
9. len=4.5 c4difficulty=Difficult noofcourses=5 44 ==> c5difficulty=Difficult 40 <conf:(0.91)> lift:(1.47) lev:(0.04) [12] conv:(3.36)
10. len=4.5 c2difficulty=Average c4difficulty=Difficult 42 ==> c5difficulty=Difficult 38 <conf:(0.9)> lift:(1.46) lev:(0.04) [12] conv:(3.21)
11. c1difficulty=Average c3difficulty=Average 41 ==> c4difficulty=Difficult 37 <conf:(0.9)> lift:(1.21) lev:(0.02) [6] conv:(2.08)

Step VII: Decision making.

Step I: Business understanding: Predicts the optimum time-to-degree considering course-taking patterns, CGPA, and contextual factors namely no. of courses and course-difficulty to find relationship between them.

Step II: Data understanding: This step concerns with data collection from student information system of a university in Bahrain. It was attained using SQL queries. The outcome was a dataset which can be mined to deal with the business problem. The dataset was checked for quality problems like incomplete values and extreme values.

Step III: Data preparation: This stage comprised of the choice of records from student dataset, data cleaning, creating, assimilating, and arranging the data. The final dataset consisted of partial data on registered courses, details of semester, semester GPA, course-difficulty, and time-to-degree. Using the formula suggested by Zainudin [9] course-difficulty was calculated. The size of dataset contained 337.

Step IVa: Modeling using regression: SPSS version 21 was used to regress the following equations:

$$\text{Time-to-degree} = u0 + \alpha 1\ \text{difficulty}(\text{course-difficulty of the courses per year}) \\ + \alpha 2\text{no. of courses}(\text{per year}) + \text{e}1 \tag{1}$$

$$\text{CGPA} = u1 + \alpha 3\ \text{difficulty}(\text{course-difficulty of the courses per year}) \\ + \text{no. of courses}(\text{per year})\alpha 4 + e2 \tag{2}$$

Thus, the regression results shows that it is course-difficulty that affects time-to-degree. Then Eq. (1) can be rewritten as:

$$\text{Time-to-degree} = 0.625 + 0.5\ \text{difficulty} \\ (\text{course-difficulty of the courses per year}) + 0.139 \tag{3}$$

A closely resembling activity was finished as to condition (2), which did not give any noteworthy outcome demonstrating that CGPA is not measurably connected to either the no. of courses taken by the understudies in a year or course-difficulty. The elucidation of condition (3) will be given in the assessment step (stage 5).

Step IVb Modeling Using Association Rules:

This step shows the use of modeling algorithm that enabled choice of modeling technique, generation of test design, creating a model, and valuation of the model. The modeling technique was used to find the connection between course-difficulty, time-to-degree, no. of courses registered, and CGPA. In the test design a test, training and validation datasets were generated. The model was generated based on the following factors: CGPA $\leq$4, time-to-degree $\geq$3.5 but $\leq$6, program under consideration—BSAF; students who joined in 2011, first semester were examined and course-difficulty was fixed as <1. The final model generated showed significant association between the four above-mentioned attributes.

Association rule was applied to find the association between the four attributes under examination. Weka 3.7.11 was used. The result of association rule generated by Weka is provided in the below table.

After obtaining the result of regression and association rule mining from Steps 4a and 4b, respectively, the next step taken was to evaluate the results which is provided next.

Evaluation: This step involved the sum up of assessed results to achieve business goals. In this research, the business goals are to forecast the optimal time-to-degree using CGPA, course-taking patterns, and contextual factors like course-difficulty and no. of courses and to group courses and students with the help of discovered information. Two steps were involved in evaluation. Evaluation of the regressed results is the first step and the second step was to evaluate the association rules.

Regression result evaluation: From Eq. (3) it was found that there was statistical and significant relationship between course-difficulty and time-to-degree. There was no statistical and significant relationship between no. of courses and time-to-degree. Likewise it was also found that there was no statistical and significant relationship between CGPA and course-difficulty or no. of courses. An essential characteristic is that regression only offered information related to statistical relationship among time-to-degree and course-difficulty, but not on unseen knowledge about contextual factors or course-taking pattern. Therefore, regression only is not sufficient to make judgment as apart from providing statistically important relationship between the attributes, however it does not provide results to classify students or courses and establish the best possible time-to-degree or CGPA. Association rule mining is likely to offer this knowledge which is evaluated next.

Association rule evaluation: Evaluation of the results obtained using apriori algorithm involved the scrutiny of various association rules that were discovered (see Fig. 1). From (Fig. 1), it is shown that there is a link between course-difficulty level, course-taking pattern, CGPA, time-to-degree and no. of courses. Based on the literature, confidence value equal to or exceeding 0.9 is considered. Accordingly the report in Table V 11 rules was regarded as important.

A key finding was that a particular course pattern of a student in a particular semester had course-difficulty pattern leading to a particular time-to-degree. For instance, student id Stud1 in a particular semester (e.g., 3) had courses ranging from 1 to 5 (1 ACCT 301, 2 ECON 121, 3 ENGL 202, 4 FINC 421 and 5 STAT 202) with the course-difficulty identified for each course (e.g., EASY, AVERAGE, VERY EASY, DIFFICULT, VERY DIFFICULT) the time-to-degree was determined as 4.5.

Although this time-to-degree could be improved by changing the course-difficulty pattern, for instance:

Association rule:

1. If c6difficulty=Difficult then noofcourses = 6
2. If c5difficulty=Difficult and c6difficulty=Difficult then noofcourses=6
3. If c4difficulty=Difficult and c6difficulty=Difficult then noofcourses=6
4. If len=4.5 and c2difficulty=Average and noofcourses=5 then c5difficulty=Difficult
5. If c4difficulty=Average then c3difficulty=Difficult
6. If len=4.5 and c2difficulty=Average and c5difficulty=Difficult then c4difficulty=Difficult 38
7. If len=4.5 and c2difficulty=Average and noofcourses=5 then c4difficulty=Difficult
8. If len=4.5 and c2difficulty=Average then c4difficulty=Difficult
9. If len=4.5 and c4difficulty=Difficult and noofcourses=5 then c5difficulty=Difficult
10. If len=4.5 and c2difficulty=Average and c4difficulty=Difficult then c5difficulty=Difficult
11. If c1difficulty=Average and c3difficulty=Average then c4difficulty=Difficult

Fig. 1 Association rules

If time-to-degree = 4.5 and difficulty of course 2 = average and difficulty of course 4 = difficult then difficulty of course 5 = difficult.

If course-taking pattern is 1 ACCT 301, 2 ECON 121, 3 ENGL 202, 4 FINC 421, and 5 STAT 202 with course-difficulty pattern EASY, AVERAGE, VERY EASY, DIFFICULT, VERY DIFFICULT and no. of courses 5, based on this rule time-to-degree is 4.5. This shows that when there is modification done to the course-difficulty of a course then there will be change in the time-to-degree also. Similarly, changes to no. of courses affect the time-to-degree. This is verified using the subsequent illustration.

The below illustration shows the use of association rule to forecast time-to-degree and categorizes students and courses which can attain best time-to-degree based on their course-taking pattern. Further essential point that requires to be pointed out here is that no association rule was produced while mining that gave a link between course-taking pattern, CGPA, course-difficulty, and no. of courses.

On comparison of the results got by using association rules and regression, the subsequent conclusion could be made. Regression as an assessment method cannot provide a hint to classify students or courses by means of the students' course-taking pattern while association rule provides. Nonetheless, regression gave a clear initiative regarding the statistically significant relationship that subsist between course-difficulty and time-to-degree which showed that time-to-degree is reliant on course-difficulty which behaves as the independent factor.

Moreover, regression showed the extent of variance that take place in the dependent factor as a result of alterations happening in the independent factor. From Eq. (3) it can be seen that β1 is equal to 0.5 which implies that when a change occurs in course-difficulty then 50% of variance in time-to-degree is explained due to a change in course-difficulty. This was not possible to ascertain using association rule. However, regarding CGPA, regression and association rule mining generated similar results. That is to say regression demonstrated that there is no relationship between

CGPA, no. of courses, and course-difficulty that is statistically significant. Likewise association rule did not generate any relationship between course-difficulty, no. of courses, and CGPA. This is an important finding. Further, dissimilar from regression in which dependent and independent attributes can be shown to forecast, association rules do not offer any such an investigation. This subsequently leaves a choice to the user with difficulty in choosing factors and controls them either as determined or determinant. Thus assessment of the findings reveal that equally regression and association rules offer a more complete understanding of predicting time-to-degree using data mining and applied in executive decisions.

Rule	Description	Implication
If len=4.5 and c2difficulty=Average and c4difficulty=Difficult then c5difficulty=Difficult	According to this rule time to degree is 4.5 if course-taking pattern is 1 ACCT 301, 2 ECON 121, 3 ENGL 202, 4 FINC 421 and 5 STAT 202 with course difficulty pattern EASY, AVERAGE, VERY DIFFICULT,DIFFICULT, DIFFICULT and number of courses 5.	If course-taking pattern is altered to 1 ACCT 301, 2 ECON 121, 3 PHOT 101, 4 FINC 421 and 5 STAT 202 with course difficulty pattern EASY, AVERAGE, VERY EASY,DIFFICULT, DIFFICULT and number of courses 5 then the time to degree becomes 4.

Deployment: The deployment would be executing a recursive data mining process across the HEI so that the HEI is able to make use of the association rules generated in this research. That is to say that the HEI could mine the data of past students who have graduated and advise the future students regarding the no. of courses to be taken by a student based on the course-difficulty. The adviser need not pay heed to the CGPA scored by the student which is a major area of concern in the current advising activities in the HEIs. In addition, using the association rules HEIs can take decisions to offer specific courses of a particular difficulty level in semesters so that students can register in a specific no. of courses with a particular pattern defined by the association rule that would optimize their time-to-degree. The HEIs can also categorize students based on the course-taking pattern and no. of courses and focus on their performance.

Decision making: This research aims to help HEIs in taking data-driven decisions, i.e., categorizing students based on the discovered contextualized course-taking patterns that establish CGPA and time-to-degree; predict course-taking patterns and the discovered relationship between CGPA, time-to-degree, and contextual attributes; and optimization of time-to-degree on student categorization and forecast.

Example

Rule	Decision-making
If len=4.5 and c2difficulty=Average and noofcourses=5 then c5difficulty=Difficult	In order to achieve time to degree 4.5 when course 2 difficulty is average then the decision on course 5 difficulty has to be Difficult
If len=4.5 and c2difficulty=Average and noofcourses=5 then c4difficulty=Difficult	In order to achieve time to degree 4.5 when course 2 difficulty is average and course 4 difficulty is Difficult then the decision on number of courses to be registered is 5

4 Conclusion, Limitations, and Future Work

The results of this research clearly show that the objective set has been achieved. An example of big dataset of a HEI has been mined using association rule technique and regression. Findings reveal that the model created by regression algorithm has produced an association between course-difficulty and time-to-degree that is statistically significant but not no. of courses students have enrolled. Further regression findings confirm that CGPA cannot be forecasted using no. of courses or course-difficulty in which students have registered. The association rules demonstrated the strategy by which a particular course-difficulty level enrolled by a student in a semester could be modified and modify the course-taking pattern and forecast the time-to-degree. Findings demonstrate that course-difficulty might not be a major characteristic that forecasts the time-to-degree. As an illustration when the no. of courses of a student was incremented with course whose difficulty level was known as difficult, time-to-degree was found to reduce.

Further, a significant finding appeared. Regression demonstrated causal association between course-difficulty and time-to-degree and to how much proportion a variation in time-to-degree can be clarified by a variation in course-difficulty although hidden patterns could not be extracted. On the contrary association rule generated hidden course pattern registered by students and their course-difficulty, but was not able to explain which factor is contingent and which one is independent leading to the intricacy in finding out the level to which the linked factors could differ and be associated as dependent and independent factors. Unexpectedly the findings of both association rules and regression showed that CGPA is not a major student performance factor that could be made optimal using course-difficulty pattern, course-taking pattern, and no. of courses, that seems to be to some extent strange. The regression model offered a means to link the factors as dependent and independent factors while association rules offer the relationship between the factors that can be differed to forecast the time-to-degree. Judgments can be made by means of this knowledge as shown in the illustrations.

This investigation has contributed in terms of generating both association rule and regression models in HEIs that can be very helpful in improving the student performance and as well create the foundation for upcoming research on this topic. This investigation has contributed theoretically showing the application of data mining theory using apriori algorithm and regression, for HEI big data mining where student factors and organizational factors are said to be fairly different from that of the HEIs where research on big data is seen to be a well-known activity.

As observed in many research results, this examination additionally has a few downsides; e.g., this examination utilized just one program to wrap up on the viability of information mining procedures in HEI decision making. Generally programs have contrasting classes of courses and understudies and fluctuating course-trouble levels. Subsequently summing up the consequences of this examination is precarious dependent on one program, more projects should be broke down to sum up the outcomes. Also, as CGPA is not a factor of understudy execution that is influenced by course-trouble, course-taking example, and no. of courses, there may be other logical components shrouded that should be removed to discover in the event that they are connected to CGPA or how they impact CGPA, for example, course weightage or student potential. Such an endeavor can create an alternate outcome. These confinements become potential points for future research.

References

1. S. MacNeill, L.M. Campbell, M. Hawksey, Analytics for education. J. Interact. Media Edu. (1) Art. 7. (2014). http://doi.org/10.5334/2014-07
2. V.A. Lotkowski, S.B. Robbins, R.J. Noeth, The Role of Academic and Non-Academic Factors in Improving College Retention, Iowa City, IA: ACT Policy Report (2004)
3. Z.J. Kovacic, Early prediction of student success: mining student enrollment data. s.l., Proceedings of Informing Science & IT Education Conference (2010)
4. G. Vert, A. Chennamaneni, S.S. Iyengar, Potential application of contextual information processing to data mining. Las Vegas Nevada, USA, in *Proceedings of the 2010 International Conference on Information & Knowledge Engineering*, IKE 2010, 12–15 July 2010 (2010)
5. B.D. Gerardo, J. Lee, J. Lee, M. Park, M. Lee, The association rule algorithm with missing data in data mining, in *Proceedings of Computational Science and its Applications*—ICCSA 2004, pp. 97–105.
6. U. Fayyad, G. Piatetsky-Shapiro, P. Smyth, R. Uthurusamy, *Advances in Knowledge Discovery and Data Mining* (MIT Press, 1996)
7. P. Thakar, Performance analysis and prediction in educational data mining: a research travelogue (2015)
8. P. Chapman, J. Clinton, R. Kerber, T. Khabaza, CRISPDM 1.0 step-by-step data mining guide. Technical report, CRISP-DM, s.l.: CRISP-DM (2000)
9. A. Zainudin, A handbook on SEM: structural equation model-ling using amos graphics (4th ed.). Kelantan, University Technology MARA Press (2012).
10. C. Vialardi, J. Chue, J.P. Peche, G. Alvarado, A data mining approach to guide students through the enrollment process based on academic performance. User Model. User—Adapt. Interact. **21**, 217–248 (2011)

11. R. Agrawal, T. Imieliński, A. Swami, Mining association rules between sets of items in large databases, in *Acm sigmod record,* vol. 22, no. 2. ACM (1993), pp. 207–216
12. N.R. Aljohani, H.C. Davis, Learning analytics in mobile and ubiquitous learning environments, in *Proceedings of 11th World Conference on Mobile and Contextual Learning (mLearn2012)*, Helsinki, Finland, 16–18 Oct 2012 (2012)

Feature Engineering by a Brain-Computer Progressively Interactive Simulated Annealing on Random Forest for Health Data Analytics

Gayathri Nagarajan and L. D. Dhinesh Babu

Abstract Health care generates voluminous data and establishing predictive models on such voluminous data is not only challenging in terms of accuracy but also in terms of computational complexity. Feature engineering is a hot research topic that has gained significance over the recent years. Feature engineering is the process of using domain knowledge to understand the features, combine them or discard them for efficient feature selection. We propose a new feature selection framework by incorporating domain knowledge to guide simulated annealing (SA). Our approach uses the classification accuracy of random forest as a positive feedback for SA. The results prove that our proposed framework of feature engineering reduces the dimensionality to a greater extent in a reasonable amount of time without any adverse effect in the accuracy of the model.

Keywords Feature engineering · Health data analytics · Simulated annealing · Random forest

1 Introduction

Feature selection is the process of identifying the important features that contribute to prediction and discarding the remaining features to reduce the computational complexity without creating any adverse effect on the accuracy of prediction. Though wrapper methods are found to be effective in yielding an optimal set of features, the computational cost is their biggest drawback [1]. Hence, most of the wrapper approaches use evolutionary algorithms by considering the feature selection problem as a multi-objective optimization problem. Multi-objective optimization problems are generally solved by achieving Pareto-optimal solutions in a definite amount of time. The objectives generally considered in feature selection problem are maximizing the accuracy of prediction, minimizing the number of selected features, and

G. Nagarajan · L. D. Dhinesh Babu (✉)
School of Information Technology and Engineering,
VIT university, Vellore, Tamil Nadu 632014, India

V. Bhateja et al. (eds.), *Data Engineering and Intelligent Computing*,
Advances in Intelligent Systems and Computing 1407,
https://doi.org/10.1007/978-981-16-0171-2_42

minimizing the computational time. There is always a trade-off between the objectives and evolutionary algorithms are proved to provide satisfactory solutions. Our proposed framework user interactive simulated annealing on random forest (UISARF) introduces the concept of feature engineering in developing a feature selection framework on health datasets. We incorporate domain knowledge and decision maker's(DM) preferences in the feature selection process using evolutionary algorithm. User interaction in evolutionary algorithms helps to guide the search in a particular direction according to the DM 's preferences so that a satisfactory solution is obtained in a reasonable amount of time. There are always scenarios in real time where a satisfactory solution in a considerable amount of time is much more important and required than the best solution obtained at the cost of computational time [2]. Our approach uses symmetric uncertainty(SU) to select the features to be fed as input to SA and uses the accuracy of random forests as the fitness function. SU is an unbiased correlation measure based on entropy and can also predict nonlinear dependencies [3]. User interactive SA reduces the computational complexity by repeatedly evaluating the objectives after few iterations of SA and incorporating DM's preferences to guide further search in a particular direction so that a satisfactory solution is achieved in a reasonable amount of time. The rest of the paper is organized as follows—Sect. 2 throws light on multi-objective optimization, interactive approaches, simulated annealing and random forests, Sect. 3 explains our proposed approach, Sect. 4 discusses the experimental framework and results, Sect. 5 winds up with conclusion.

2 Preliminaries

Multi-objective optimization problems aim to achieve the best possible values for different objectives. Interactive approaches are popular in multi-objective optimization problems owing to their ability to lower the computational cost by generating a small amount of Pareto-optimal solutions depending on the DM's preferences. They are proved to be better in terms of flexibility, adaptability and computational cost [4]. The interactive methods differ depending on the kind of information asked from the DM. They are classified into apriori, posteriori and progressively interactive approaches. Progressively interactive approaches are found to be effective as the search is narrowed down to a limited number of decision calls [5]. Few examples of DM's interactive approaches on evolutionary algorithms include [6, 7].

SA is an effective optimization technique based on the concept of 'annealing' in thermodynamics. SA starts the search in a randomized state and the neighboring states are accepted based on the energy level and bad moves are also accepted with a probability based on the temperature of the system. The acceptance probability for the bad moves decreases with the temperature and Boltzmann distribution is used to decide the acceptance probability of the bad moves. The states represent the solutions of the optimization problem, energy is the fitness function and the temperature is the control parameter. SA is a stochastic search method that proved to be a stable model insensitive to variations in optimization parameters and is also capable of capturing complex nonlinear relationships [8]. Meiri and Zahavi [8], Maurizio et al. [9] are few examples using SA in feature selection.

Random forest (RF) is an ensemble bagging technique and is considered to be an efficient algorithm in predictive tasks. Though RF is a good classification technique, it is difficult for the traditional parameter values to achieve optimal performance [10]. Methods like grid search, exhaustive search and orthogonal selection are used to optimize the parameters but they suffer from computational complexity. Researchers are working on combining random forests with intelligent algorithms like GA, PSO, etc., to improve its performance on imbalanced datasets, for parameter optimization and feature selection [10]. Wen and Wang [11], Freiman [12] are few examples that combine random forests with evolutionary algorithms for feature selection.

3 Proposed Feature Selection Framework—UISARF

The search in normal SA and user interactive SA is shown in Fig. 1. Simulated annealing starts the search for a Pareto-optimal solution from an initial random solution. It continues to search along the neighboring regions as determined by the acceptance probability of the previous solutions. User interactive simulated annealing helps to obtain a satisfactory solution with reduced computational cost when compared with the normal simulated annealing. It works as follows; normal SA is executed for a fewer number of iterations n, where n is chosen to be between 25 and 50 usually, in order to be both effective and computationally not expensive. Some solutions accepted by SA in n number of iterations may be rejected by DM, and on the other hand, those that are rejected by SA may be accepted by DM based on his/her preferences. Each solution is evaluated for a new fitness value according to the weights assigned to the objectives by DM. In order to ensure that the DM's preferences are not ending up in local optima or is restricted within a specific region, a turbulence component is added to the threshold set by the DM or subtracted from the same

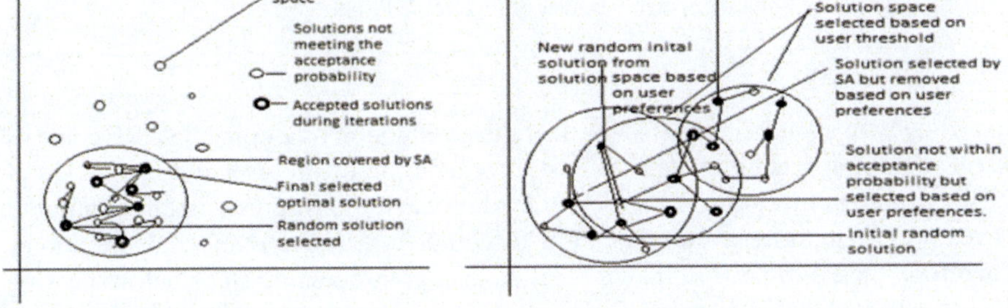

Fig. 1 Sample search in **a** normal simulated annealing, **b** user interactive simulated annealing

if required. New solution space is constructed based on the solutions accepted by the DM. A new feature set is constructed from the selected solution space. Feature engineering concept is incorporated into this constructed feature set and the DM can add or discard features accordingly. A new random initial solution is selected from the constructed feature set space and SA is executed with n number of iterations. The output is presented to the DM in the visualization form and DM continues to choose his preferred solution space and continue with the iterations until he is satisfied with the solution. The proposed feature selection framework uses interactive simulated

---

Algorithm 1 Pseudocode of UISARF

Input = DS[m,n] where m is no. of observations, n is no. of features
Step 1 : Features such as ID, gene description are all eliminated
$DS[m, n]$ is reduced to $DS[m, n1]$
Step 2 : Feature with missing value more than 30% discarded.
$DS[m, n1]$ is reduced to $DS[m, n2]$
Step 3 : Normalization of features.
$Normalized(x) = \frac{x-min(x)}{max(x)-min(x)}$
$DS[m, n2]$ is normalized to NDS[m,n2]
Step 4 : Calculate feature importance of each feature using SU
$varImp(f_i) = SU(f_i, y)$
Step 5 : Get threshold t_1 from DM and select the features accordingly
$f(selected) = f_i \epsilon n2 \forall f_i \geq t_1$
$NDS[m, n2]$ is reduced to $NDS[m, n3]$
Step 6 : Invoke SA
$SA(NDS[m, n3])$ is reduced to $NDS[m, n4]$
Step 7 : Get w1 and w2 – weights for the objectives from the DM.
Step 8 : Calculate the new fitness value
$Fitness(new) = w1(acc) + w2(1/k)$
Step 9 : Get t2 from DM. Turbulence component added/removed.
if ($t2 \geq Q1$ and $\leq Q3$), $t2 = t2$ else $t2 = t2(+/-)tc$
Step 10 : Select the solution space
Solutions(selected) = $Solution_i \epsilon$ Solutionspace(SA) $\forall Solution_i \geq t2$
Step 11 : Construct the new feature set space
$NDS[m, n4]$ is increased or decreased to $NDS[m, n4a]$
Step 12 : Input the dataset with new feature set to SA
$SA(NDS[m, n4a])$ is reduced to $NDS[m, n5]$
Step 13 : If solution from Step 12 is satisfactory, stop else go to step Step 8
Output:- Satisfactory Solution S_i with number of features k.

---

annealing with a random forest classifier as the evaluation function. UISARF consists of few data preprocessing steps before the execution of the user interactive simulated annealing algorithm. Execution of evolutionary algorithms with larger number of features sometimes even takes days to complete a fewer number of iterations. Therefore, there is a need to include certain data preprocessing steps before passing them as input to the evolutionary algorithms. The algorithm for UISARF is shown in Algorithm 1.

4 Experimental Framework

Table 1 shows the dataset description. The datasets are taken from open-source data repository UCI and Kaggle. The datasets are examples of classification tasks. Figure 2 shows a sample histogram representing the distribution of feature importance values for dataset1 and dataset4, respectively. DM assigns weights for the objectives. In this case study, DM inputs 0.7 for maximizing accuracy and 0.3 for minimizing the number of features. The new fitness value is calculated for each solution and its distribution is shown to the DM. DM inputs the threshold value. Solutions with fitness value equal to or greater than the threshold are selected. A plot of the solution space is drawn and shown as visualization for the DM. DM can add/remove solutions in the solution space according to his/her preferences. The fitness value distribution and the solution space for first, second and fifth iterations of UISARF on dataset 1 is shown in Figs. 3, 4 and 5, respectively. The size of the solution space for different iterations of UISARF is shown in Table 2 for all the six datasets. It is evident that the dimensionality is reduced to a greater extent on successive iterations as DM's preferences are incorporated in the direction of the search. The final results (satisfactory solution) are also displayed. The results prove that the dimensionality is

Table 1 Dataset description

S. No	Dataset	Number of features	Number of observations
1	CancerRNA	20,533	801
2	Breast cancer gene	1927	133
3	Leukemia	7131	72
4	Cancer disease	32	569
5	Disease prediction	133	4920
6	Hospital readmission prediction	373	55,136

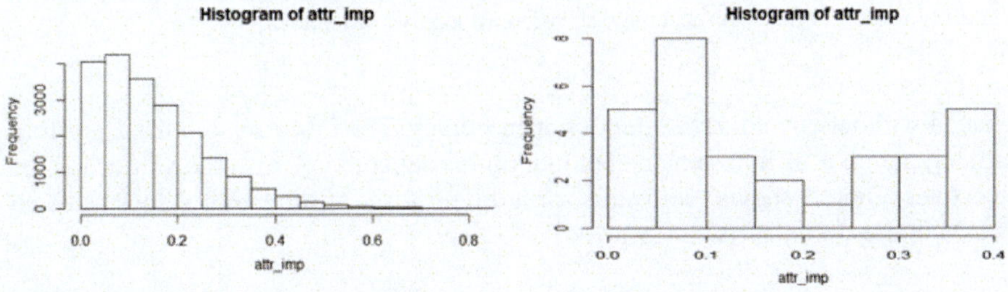

Fig. 2 Histogram represents the distribution of feature importance values for dataset 1 and dataset 4, respectively. In this case study, DM selects the threshold as 0.2 for dataset 1 and 0.1 for dataset 4

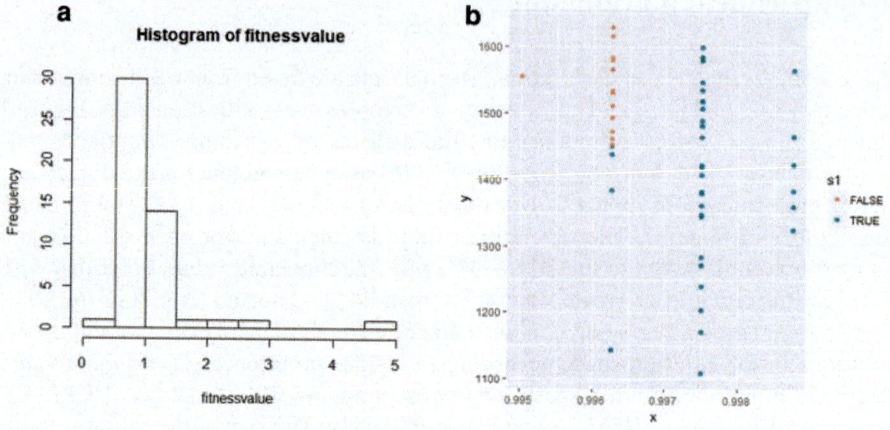

Fig. 3 Results of first iteration of UISARF—**a** DM selected threshold value—0.7, **b** selected solutions in blue. X represents accuracy, Y represents number of features

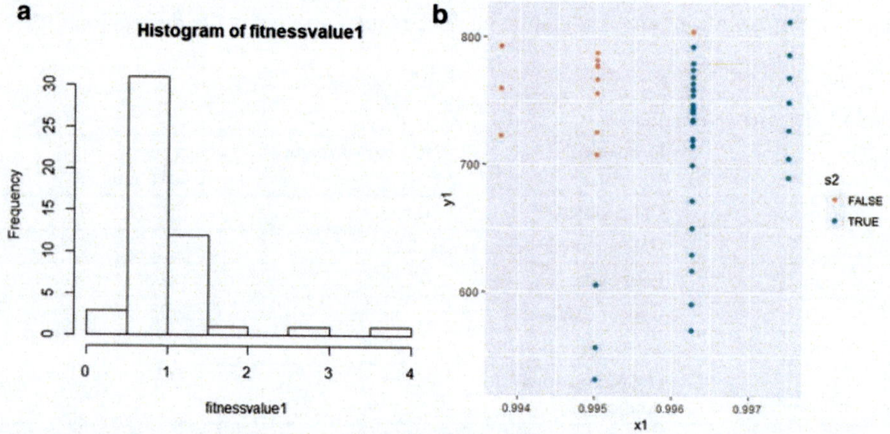

Fig. 4 Results of second iteration of UISARF—**a** DM selected threshold value—0.8, **b** selected solutions in blue. X represents accuracy, Y represents number of features

reduced without any adverse effect on its accuracy. The DM can stop the algorithm with any number of iterations depending on his satisfaction with the values for the objectives. Table 3 shows the results for a different set of threshold values selected by DM in the six datasets.

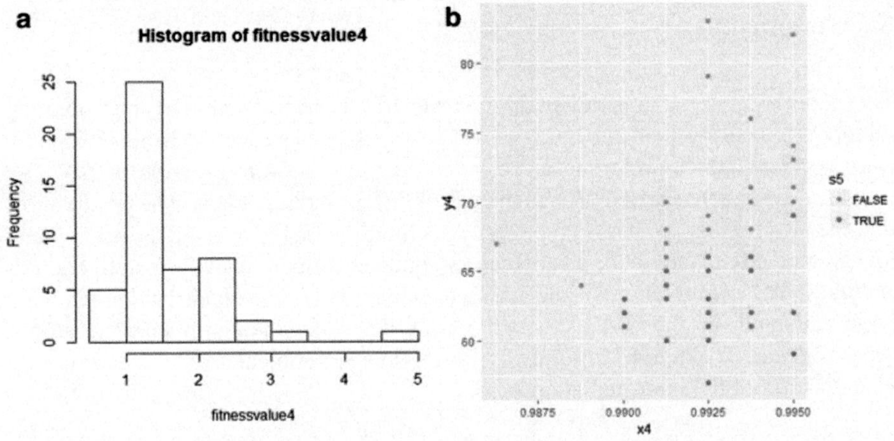

Fig. 5 Results of fifth iteration of UISARF—**a** DM selected threshold value—1.3, **b** selected solutions by DM in blue. X represents accuracy, Y represents number of features

Table 2 Size of solution space

Step	Dataset1	Dataset2	Dataset3	Dataset4	Dataset5	Dataset6
Filtering metrics	20,265	1926	7128	31	103	278
UISARF(Iteration1)	7192	557	213	17	52	241
UISARF(Iteration2)	1481	166	62	13	41	197
UISARF(Iteration3)	809	96	35	8	28	164
UISARF(Iteration4)	340	37		6		95
UISARF(Iteration5)	188	16				47
UISARF(Iteration6)	61					
Accuracy	99.6	96.3	87.8	96.3	95.6	81.2
Dimensionality reduced	61	16	35	6	28	47
Computational time	9 h	13.07 min	19 min	10.3 min	13 min	15 min

Table 3 Results for different sets of threshold values

Dataset	No. of iterations	DM threshold values	Accuracy	No. of features	Time
1	3	0.9, 1.2,1.4	98.7	287	6.3 h
2	4	0.9, 0.7, 0.9, 1.3, 1.4	95.7	32	9.3 min
3	2	0.8, 1.1	85.4	31	12 min
4	2	1.4, 1.6	95.6	7	7.9 min
5	3	0.9,1.3,1.5	96	35	12 min
6	4	0.7,1.2,1.4,1.6	78	42	12 min

5 Conclusion

This paper proposes a new framework UISARF for feature selection by incorporating feature engineering concept on SA. The search space is reduced in our proposed framework as DM's preferences are included in the search direction of SA. Our proposed approach is tested on six datasets of different dimensions. The results show that the proposed approach is an effective technique that reduces the dimensionality to a greater extent and yields a satisfactory solution without any adverse effect on its accuracy. The computational time is also reasonable in our proposed approach but it largely depends on the DM's satisfaction that determines the number of iterations. This work can be extended in future by developing a graphical user interface tool for visualization of the intermediate results.

References

1. S.-W. Lin, Z.-J. Lee, S.-C. Chen, T.-Y. Tseng, Parameter determination of support vector machine and feature selection using simulated annealing approach. Appl Soft Comput **8**(4), 1505–1512 (2008)
2. M. Richter, G. Maier, R. Gruna, Langle, T., J. Beyerer, Feature selection with a budget, in *World Congress on Electrical Engineering and Computer Systems and Science (EECSS)* (2016) pp. 104–112
3. M. Garcia-Torres, F. Gomez-Vela, B. Melian-Batista, J.M. Moreno-Vega, High-dimensional feature selection via feature grouping: a variable Neighborhood Search approach. Inf. Sci. **326**, 102–118 (2016)
4. S. Zhong, Y. Chen, J. Zhou, Y. Liu, An interactive satisficing approach for multi-objective optimization with uncertain parameters. J. Intell. Manu. **28**(3), 535–547 (2017)
5. A. Sinha, P. Korhonen, J. Wallenius, K. Deb, An interactive evolutionary multi-objective optimization algorithm with a limited number of decision maker calls. Eur J Oper Res **233**(3), 674–688 (2014)
6. L. Cruz-Reyes, E. Fernandez, P. Sanchez, C.A. Coello, C. Gomez, Incorporation of implicit decision-maker preferences in multi-objective evolutionary optimization using a multi-criteria classification method. Appl. Soft Comput. **50**, 48–57 (2017)
7. J. Hettenhausen, A. Lewis, Sanaz Mostaghim, Interactive multi-objective particle swarm optimization with heatmap-visualization-based user interface. Eng. Opt. **42**(2), 119–139 (2010)
8. R. Meiri, J. Zahavi, Using simulated annealing to optimize the feature selection problem in marketing applications. Eur. J. Operat. Res. **171**(3), 842–858 (2006)
9. M. Filippone, F. Masulli, S. Rovetta, Supervised classification and gene selection using simulated annealing. In: IEEE International Joint Conference on Neural Network Proceedings, pp. 3566–3571 (IEEE, 2006)
10. L. Ma, Suohai Fan, CURE-SMOTE algorithm and hybrid algorithm for feature selection and parameter optimization based on random forests. BMC Bioinform **18**(1), 1–18 (2017)

11. W. Xiong, C. Wang, A hybrid improved ant colony optimization and random forests feature selection method for microarray data, in *Fifth International Joint Conference on INC*, IMS and IDC, pp. 559–563 (IEEE, 2009)
12. M.H. Freiman, Using random forests and simulated annealing to predict probabilities of election to the Baseball Hall of Fame, J. Quant. Analy. Sports **6**(2) (2010)

Quantum Cryptography and Blockchain System: Fast and Secured Digital Communication System

C. Nandni and S. Jahnavi

Abstract Brief introduction about quantum cryptography and blockchain system: Fast and secured digital communication network-based applications followed by its background and motivation are explained. Short review on the idea is enclosed with aiming network-based application for data exchange with quantum blockchain technology. The main motivation about this quantum blockchain technology in data communication is to enable fast and reliable communication channel in future reverence. In order to avoid data theft in exchange of data and fast mode of exchanging the data communicated over the Internet, quantum cryptography is fused with blockchain technology

Keywords Quantum cryptography · Blockchain technology · Quantum key distribution system.

1 Blockchain Technology for Data Security

With the aim of providing security while performing online transaction, the sensitive data is transmitted over the network without involving any third-party software. In line with this, exchange of data through blockchain technology to the intended person, the information is shared among two parties. One the sender who wants to share the data and the other is the intended person as shown in Fig. 1. A prototype of fast and secure communication of the data over the Internet is achieved by performing quantum cryptography followed by passing the data using blockchain technology that ensure fast and reliable exchange of data.

In a blockchain every user in the network will be aware of the public key of other user like an email address. But only user has the knowledge about his private key

C. Nandni · S. Jahnavi (✉)
Computer Science and Engineering Department, Dayananda Sagar Academyof Technology and Management, Bangalore, Karnataka, India

C. Nandni
e-mail: hodcse@dsatm.edu.in

V. Bhateja et al. (eds.), *Data Engineering and Intelligent Computing*,
Advances in Intelligent Systems and Computing 1407,
https://doi.org/10.1007/978-981-16-0171-2_43

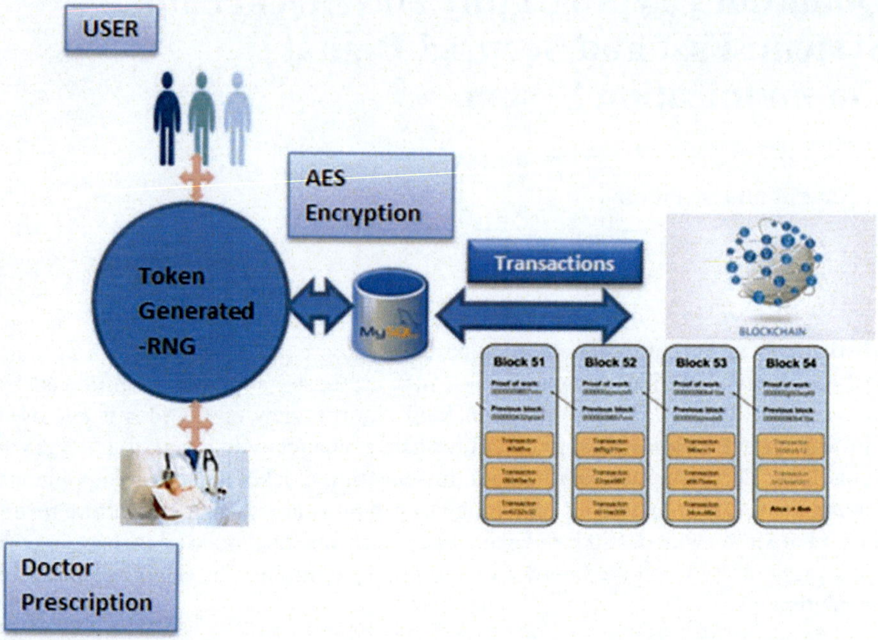

Fig. 1 Enhanced security model using blockchain technology for data privacy in medical system

which is not shared in the public domain like password. The digital signature is performed by using an encryption algorithm and Khans private key. Public bitcoin uses SHA256 algorithm. Ethereum, one of the famous cryptocurrency uses Ethash algorithm. These transactions are validated before considering as a block. The validation of blocks is done by a character called minors and in order to perform this, minors have to solve a complex mathematical problem. So the minors on validating the blocks it is added to the blockchain for which the miners are rewarded with bitcoins. This process is called proof of work [1]. The process which minor performs to add the blocks to blockchain is called mining.

2 Considering Blockchain Technology for Data Privacy in Medical System as an Enhanced Security Model

Maintenance of patient record when the patient visits the clinic in timely manner is major problem. If the doctor is unaware of the patient history, the medication could be a repetition. So it is better to save in a centralized database. But in communicating securing records is very important since it should not be altered, should reach the required person like doctor and the patient. So model is created using blockchain

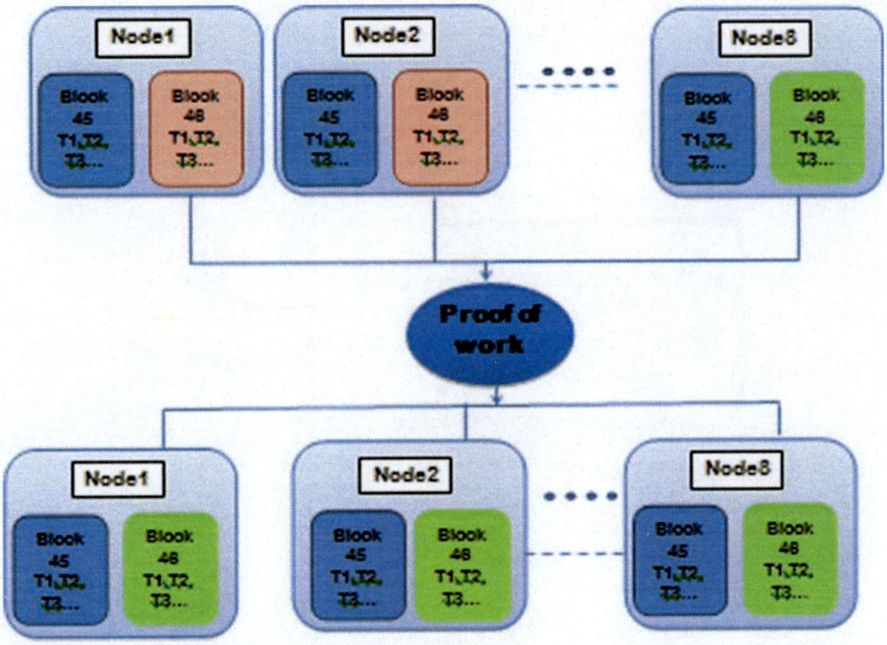

Fig. 2 Blockchain design structure

applied in healthcare system [2], that provides high security compared to communicated records in the network as shown in Fig. 1. Referring this network as a distributed registry, data is stored on each node operating in each of the individual networks as shown in Fig. 2.

2.1 *Implementation of Blockchain*

The blockchain holds absolute information. Small chunks called blocks are shown in Fig. 4, which are logically structured, that are centralized, distributed, and immutable database. Each block $b_i > 0$ is connected to next preceding block $b_i - 1$ with hash function $H(b_i - 1)$ calculated as shown in Fig. 3 [3]. Any changes made in $b_i - 1$ cause an unacceptable hash in b_i and successive blocks. The first block does not have predecessor. To assure integrity of block and data it contains, it is digitally signed. A blockchain can be used to store arbitrary data. In its simplest form, a block b_i consists of following steps:

- A cryptographic hash of the previous block, i.e., hi: $= h(b_i - 1)$.
- Payload (pi), which is stored in this block. (e.g., transactions in bitcoin).
- Digital signature (si) of the block data, i.e., si $= \sigma$ sk (hi|pi), signed with secret keys k of creator of the block. Public key pk that verifies the signature.

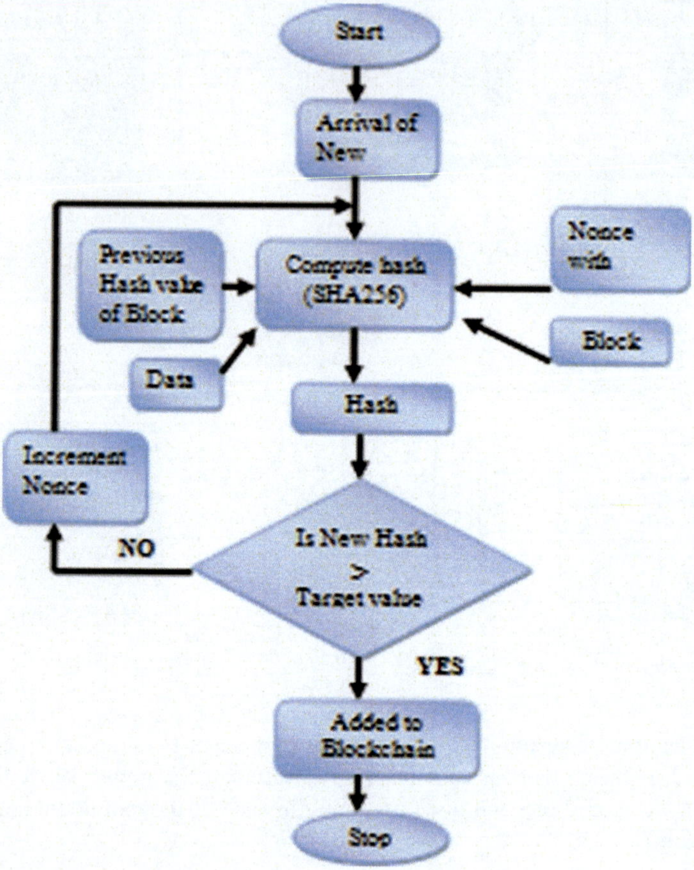

Fig. 3 Blockchain implementation with proof of work

```
mysql> use cam;
Database changed
mysql> select * from token;
+--------+-----+------+-----------------+--------------+------------+
| tok_id | uid | name | token           | query        | qdate      |
+--------+-----+------+-----------------+--------------+------------+
|     48 | 15  | 1858 | HMW3W:undefined | cold         | 2020-05-24 |
|     49 | 15  | 7125 | 3ST2V:HMW3W     | Stomach ache | 2020-05-24 |
|     50 | 15  | 7366 | FTE4S:3ST2V     | Fever        | 2020-05-24 |
+--------+-----+------+-----------------+--------------+------------+
3 rows in set (0.00 sec)

mysql>
```

Fig. 4 MySQL command line showing chain of blocks

2.2 Database Security with AES Encryption

Collection of data with organized form provides information which is considered valuable and important corporate resource called database. Therefore, database information with patient medication history, blockchain tokens are stored and retrieved securely using AES encryption algorithm [4] as shown in Fig. 7.

Advanced Encryption Standard (AES) and symmetric encryption algorithm work with logic of substitution permutation and others operations involving shuffling bits around (permutations). It is a byte operation than bits. So, 128 bits plaintext is considered as 16 bytes, which are arranged in 4 row 4 columns for matrix operation. The process uses round keys for encryption following successive steps as shown in Fig. 6:

1. Deriving round keys.
2. Initializing state array.
3. Initial round key addition to state array (considering 4 keys: W0, W1, W2, W3).
4. Performing 9 rounds of operation, manipulating state array called intermediate array.
5. 10th and final round of state array manipulation.
6. Considering state array as ciphertext.

Round Keys: 128-bit long chipper key is used in encryption process which is removed from secret master key before it arrives at AES encryption held by authentication server, which is then later used to generate 128-bit round key combined with data in the encryption process. 11 keys are used before the round begins, as shown in Table 1 and initial value of the state array in Table 2.

To alter the state array, every round of encryption continuously follows below steps:

Table 1 Round key array

Round#	32-bit word	32-bit word	32-bit word	32-bit word
Round#1	W.4	W.5	W.6	W.7
Round#2	W.8	W.9	W.10	W.11
Round#3	W.12	W.13	W.14	W.15
Round#4	W.16	W.17	W.18	W.19
Round#5	W.20	W.21	W.22	W.23
Round#6	W.24	W.25	W.26	W.27
Round#7	W.28	W.29	W.30	W.31
Round#8	W.32	W.33	W.34	W.35
Round#9	W.36	W.37	W.38	W.39
Round#10	W.40	W.41	W.42	W.43

Table 2 Round key array

$S_{1,1}$	$S_{1,2}$	S_{13}	$S_{1,4}$
$S_{2,1}$	$S_{2,2}$	$S_{2,3}$	$S_{2,4}$
$S_{3,1}$	$S_{3,2}$	$S_{3,3}$	$S_{3,4}$
$S_{4,1}$	$S_{4,2}$	$S_{4,3}$	$S_{4,4}$

1. Sub-bytes: its simple substitution operation converts every byte into an altered value. It has table of 256 values for substitution. It works with state array; each byte is an index into the 256-byte value substitution table to replace, which results in a new state array after the replacement with 256-byte value from the table, considering inverse substitution table original contents can be drawn back.
2. Shift rows: Operation performed to each row of state array, where the row is rotated to the right with a fixed number of bytes as shown:
   - 1st row rotated by 0 bytes, 2nd Row rotated by 1 byte, and so on.
3. Mix columns: The state array column is processed separately to produce a new column by replacing the old one using matrix multiplication. For instance, each state array column C_0 to C_3 replaces new column by performing matrix multiplication.

$$\begin{pmatrix} C_0' \\ C_1' \\ C_2' \\ C_3' \end{pmatrix} = \begin{pmatrix} 02\ 01\ 05\ 03 \\ 03\ 02\ 01\ 05 \\ 05\ 03\ 02\ 01 \\ 01\ 05\ 03\ 02 \end{pmatrix} \bullet \begin{pmatrix} C_0 \\ C_1 \\ C_2 \\ C_3 \end{pmatrix}$$

The new column is computed as follows:

$$C_0' = 02 * C_0 + 01 * C_1 + 05 * C_2 + 03 * C_3$$
$$C_1' = 03 * C_0 + 02 * C_1 + 01 * C_2 + 05 * C_3$$
$$C_2' = 05 * C_0 + 03 * C_1 + 02 * C_2 + 01 * C_3$$
$$C_3' = 01 * C_0 + 05 * C_1 + 03 * C_2 + 02 * C_3$$

4. About XOR round key:

The XOR operation in this context takes the existing state array after mixing column operation is carried out and performs XOR with suitable round key and replacing sate array with result. This is done before the rounds starts and once per round, considering each round key. Operations are done in the order listed for first 9 rounds. In 10th round sub-byte, shift row and XOR round key operations are carried out as shown in Fig. 6.

2.3 Digital Token Generation

Data security is imperative to protect privacy of patient details and medication history. So in this view tokenization is precise key in considering medical record transactions. Before sending out the reports, the validity of the token generated is examined by the storage server which is needed while accessing the reports that is linked to the previous token as shown in Fig. 2. Cryptography uses integer key but in cases it could be a randomly generated called random number generator (RNG) or pseudorandom number generator (PRNG) [5]. PRNG uses system entropy to seed the data which yields better results, since it is initial conditions and much more difficult to make a guess by attacker since it is encrypted even when record is accessed by doctor as shown in Fig. 5. Following steps show the functioning of pseudorandom key generation algorithm (Figs. 6 and 7).

Name	Email	Mobile	Query
wc9xOLbffFdRnR27M1C+7Q==	2blBiLpvCJVWiHUbempzDA==	ixYBF53RtWmGjBvPUj6VbA==	Cold
wc9xOLbffFdRnR27M1C+7Q==	2blBiLpvCJVWiHUbempzDA==	ixYBF53RtWmGjBvPUj6VbA==	Strep Throat
wc9xOLbffFdRnR27M1C+7Q==	2blBiLpvCJVWiHUbempzDA==	ixYBF53RtWmGjBvPUj6VbA==	Pinkeye
wgCWj5RHdtxxGoeMw8Cjdw==	rzXV/LGK3e2niRlNwWLKHg==	F4KL6IElPljWj60HpgXshw==	Influenza

Fig. 5 User details are encrypted except query for doctor

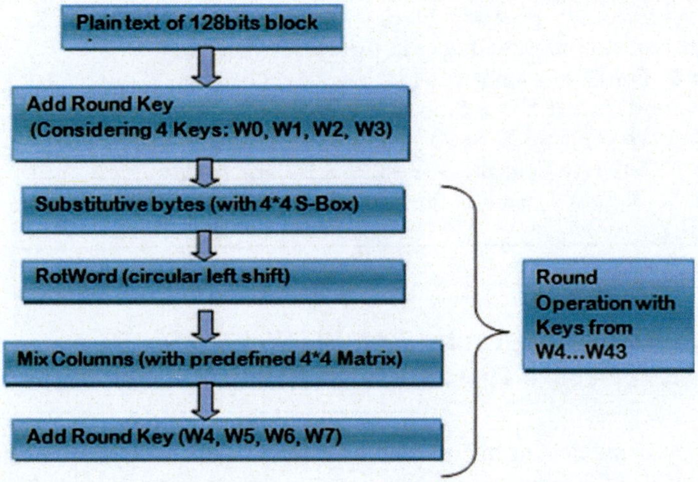

Fig. 6 Flowchart for XOR round key operation

Fig. 7 Data encrypted in database

Algorithm of Pseudo Random Key Generation

Step 1: Pseudo-random bytes generation using seed key generator and adding user's Seed (U) if any
Step 2: Set 192 bit key (K) and Seed which is last 8 bytes, from the step 1 outcome
Step 3: Set current date and time as 64 bit representation called as C
Step 4: Generate block Y0=R(S,K,D) where R is the Random Number Generation (RNG) Algorithm, updating S accordingly as per algorithm.
Step 5: Carryout Continuo's RNG tests such that if Y0=L or Y0=P(Previously generate block) stop and notify failure. Set L=Y0, P=Y0 and save in thread save in thread safe memory.
Step 6: For G = N until G equal zero do: Generate block Y=R(S,K,D) of 64bit, update S if Y = P Stop and notify failure, Set B=lesser of G and 8. Output B byes from X. Set G=G-B and P=Y.
Step 7: Generate final Block Y_f=R(S,K,D) and set Y=Y_f
Step 8: Retain L and P for next following use. Reset the parameters K,S,D,Y.

---

3 Quantum Cryptography with Blockchain for Fast and More Secure Transmission of Data

In the current rising field of Internet and machine-to-machine (M2M) technologies, physical objects of diverse technologies are connected together for intelligent decision making using Internet of things (IoT). The massive data generated or captured by ATM transaction, online bank application having highly confidential and valuable

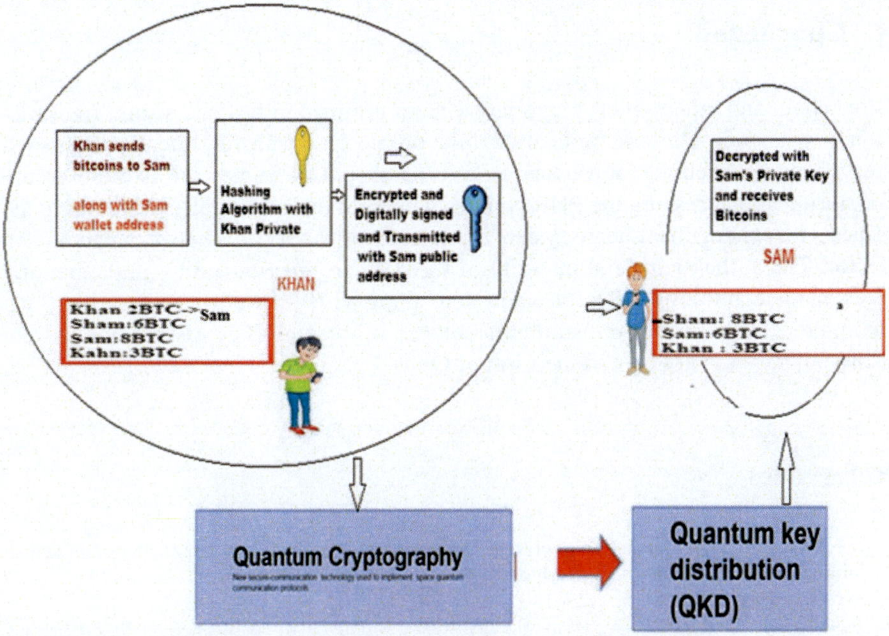

Fig. 8 Quantum cryptography with blockchain for secure transmission of data

information are placed in cloud. Biometrics give many more methods in high-end secure applications using natural, user-friendly, and fast authentication.

Cryptography is used to ensure secrecy and message authenticity. Since sometimes cryptographic key will be lengthy, difficulty in remembering, protecting confidentiality is the significant issues to be dealt with. That could be solved by the integrating of biometrics with quantum cryptography. The blockchain technique is combined with quantum cryptography to ensure high level of security for cloud applications by using biometric features used for authentication as shown in Fig. 8. This research of quantum key distribution (QKD) [6], a data exchange technology grounded in the principles of quantum physics, is available to the energy sector, preventing unauthorized access to critical energy infrastructure data.

A milestone in the field of quantum information science that encodes information using the photon properties, so called polarization, is securely used to distribute encryption keys between two communicating parties [7] (a transmitter and a receiver). Quantum cryptography with blockchain for biometric information exchange promises to give high secure way of protecting the critical data.

4 Conclusion

Blockchain and quantum cryptography system promise in fast and secure transmission of data over network applications like online fund transfer, bitcoins, transmission of medical reports, and many more. The proposed framework is more secure for storing and accessing the PHR with easy use and understanding, overcoming the issues with existing healthcare system in transferring the records between patient and doctor. The authenticity and integrity of the data are maintained by implementing blockchain technology in healthcare system along with quantum cryptography. Such that medical records are transmitted as blocks in quantum key distribution system using polarization concept in the form of Qbits.

References

1. R.F.S. Gemeliarana, Evaluation of proof of work (pow) Blockchains security network on Selfish mining, in *2018 International Seminar on Research of Information Technology and Intelligent Systems (ISRITI)* (2018), pp. 126–130
2. A.A. Omar, S. Rahman, A. Basu, S. Kiyomoto, MediBchain: a blockchain based privacy preserving platform for healthcare data, in *The 10th International Conference on Security, Privacy and Anonymity in Computation, Communication and Storage* (2017), pp. 534–543. https://doi.org/10.1007/978-3-319-72395-2_49
3. F. Knirsch, A. Unterweger, D. Engel, Implementing a blockchain from scratch: why, how, and what we learned. EURASIP J. Inf. Secur. **2019**, 2–14 (2019). https://doi.org/10.1186/s13635-019-0085-3
4. T.M. Zaw, M. Thant, S.V. Bezzateev, Database security with AES encryption, elliptic curve encryption and signature, in *Wave Electronics and its Application in Information and Telecommunication Systems*. IEEE (2019), p. 6. https://doi.org/10.1109/WECONF.2019.8840125
5. C.-H. Murillo, L. Cardoza-Avendaño, et al., A novel pseudorandom number generator based on pseudo randomly enhanced logistic map. Nonlinear Dyn. **87**, 407–425 (2016). https://doi.org/10.1007/s11071-016-3051-3
6. N. Sasirekha, M. Hemalatha, Quantum cryptography using quantum key distribution and its applications. Int. J. Eng. Adv. Technol. (IJEAT) **3**(4) (2014). ISSN: 2249-8958
7. C. Bennett, G. Brassard, Quantum cryptography: public key distribution and coin tossing, in *Proceedings of IEEE International Conference on Computers, Systems and Signal Processing*, vol. 560, I (2014), pp. 7–11. https://doi.org/10.1016/j.tcs.2014.05.025

An Efficient Framework to Bifurcate Healthy and Diseased Vegeatables and Fruits Using Multimodal Approach

T. M. Rajesh, S. G. Shaila, Lavanya B. Koppal, and Nyrika Bhargavaram Renuka

Abstract The importance of agriculture is known to mankind at the start of early human civilization, whereby farming and use of domestic animals created food supplies to people which helped cities to develop. Due to immense growth of population and industrialization, the demand for fruits and vegetables is more than its supply. Considering the wastage factor of modern agriculture, early detection of diseased fruits and vegetables by image processing and its techniques can hugely benefit in preventing rotting and its spread to neighbouring crops. Generally, these diseased fruits and vegetables are plucked by manual inspection. The same work is done by the autonomous robot with pattern and machine learning-based detection and recognition of diseased fruits and vegetables which covers much greater distance in a short period of time. The objective is divided into three stages. Firstly, after some basic preprocessing steps, the local and global features of diseases in fruits and vegetables are extracted using well-known algorithms of pattern recognition in feature extraction phase. Then the techniques like supervised and unsupervised learning are used to categorize them effectively and stored the same in training database. Then secondly, validating is done using k-NN classification and k-NN regression which is a type of instance-based learning. Lastly, after identification of the diseased fruit or vegetable, the robot plucks the target infected object and this cycle continues until it reaches the end of the agricultural land. This is beneficial for people like farmers, gardeners, home makers who have little or no information about the diseased crops which are growing and this invention saves their time and health by increasing the yield of healthy fruits and vegetables. Instead of having an expert agronomist to find out and study the disease every time, the robot can identify it by placing certain algorithms

T. M. Rajesh (✉) · S. G. Shaila · L. B. Koppal · N. B. Renuka
Computer Science and Engineering, Dayananda Sagar University, Bengaluru 560102, India
e-mail: rajesh-cse@dsu.edu.in

S. G. Shaila
e-mail: shaila-cse@dsu.edu.in

L. B. Koppal
e-mail: lavanya-cse@dsu.edu.in

V. Bhateja et al. (eds.), *Data Engineering and Intelligent Computing*,
Advances in Intelligent Systems and Computing 1407,
https://doi.org/10.1007/978-981-16-0171-2_44

for disease detection. This also reduces the cost for people who cannot invest a large amount money and time for such work.

Keywords DIP · Pattern recognition · Feature extraction · Clustering model · PCA · Obstacle avoidance

1 Introduction

Since agriculture is the main source of income for India's economy, it is commonly known as the backbone of Indian economy. India is the second largest production unit for growing fruits and vegetables worldwide. Due to this, a large per cent of people directly (farming) or indirectly (doing business with these commodities) dependents on agriculture. In recent years, due to rapid industrialization and modern techniques, the production of quality fruits and vegetables has decreased due to various reasons. In this work, we plan to tackle and give solution for one of the main reasons that causes less production and affected yield of fruits and vegetables. At certain epidemic out lash, lots of fruits and vegetables have been subjected to diseases due to bacterial and fungal growth. This makes the fruits and vegetables unfit for consumption. These if not identified on an earlier stage, the infected growth spreads to the surrounding plants and affects the ratio of increasing in growth of fruits and vegetables. This directly and drastically affects the production yield of fruits and vegetables that bring downfall for both the farmer's income and degrades the quality of agriculture lands which is a wasteful produce. This can be prevented by earlier detection of such diseased commodities which should be ripped off from the plant thereby saving the other fruits and vegetables as well. This is currently being done manually by farmers or gardeners in most part of India, which is not always accurate and precise; also it is unsafe and time-consuming process for the farmers. Due to intense physical and mental strain involved during the manual process of plucking, lots of these people have found and shifted for better job opportunities. This however does not solve the problem faced by large agricultural farms that have to invest more for skilled labourers. In this era, a lot of research work has been continuously developed in different domains and is broadening its approach. Agriculture is one such field where researchers all over the world are coming up with effective algorithms to tackle real-life problems using various approaches. Camera-based image capturing is most commonly used to get the raw input images of fruits and vegetables which is otherwise termed as image acquisition. After acquiring the images, some basic pre-processing techniques are used to enhance the image. So that it will help in getting the accurate results for feature extraction phase. Since agricultural farms have more than one kind of fruits and vegetables grown, we first need to overcome from this problem by enforcing unsupervised learning where we can differentiate each of them into different category. We focus on cluster analysis method using k-means clustering which partitions the data set based on its size, colour, appearance and features. For example, a ripe tomato is small, red and oval in appearance whereas a capsicum is comparatively large, green and cylindrical in shape, based on these features we can put the two in different datasets. This will further help us to analyse and identify diseased produce for the individual fruits and vegetables. After the fruits and vegetables have been segregated, the robot will now move towards the individual fruit or vegetable plant and begin to analyse it. We use the PCA algorithm to extract the features which helps to identify the diseased fruit or vegetable set. Using SVM classifier, we can classify the diseased fruits and vegetables. For example, we can detect black spots on tomatoes and from learning we can classify them as a fungal disease which is commonly called as black spot or speck disease. Similarly, even for diseased capsicums, such black and brown spots occur which is also a fungal disease, termed as Anthracnose which on a long period of time left will discolour the vegetable. Even in vegetables such as cabbage, Alternaria, a type of fungus pathogen that causes allergies in humans can be detected by this technique along with the help of feature extraction and pattern recognition. After collecting the diseased and healthy datasets, we now use supervised learning to train the robot to differentiate between healthy and diseased fruits and vegetables. After the learning process is finished, the robot is now capable of identifying the disease on its own and then produces a response. Now the final stage to solve the problem taken is by segregating the diseased fruits and vegetables that is done by the automated system. The proposed model of the system is to first identify the diseased fruit or vegetable and then it produces a response to move the arm. The moment of the arm is triggered only if a diseased produce is present and then with the help of its end effectors plucks the target object successfully. In addition to this, we also integrated obstacle avoidance logic, whereby the robot is able to overcome and continue to do its work normally if any obstacle is present in its path, which is one of the ideal conditions for a farm. With the help of this novel model, we can save a huge amount of investment and also helps to increase the yield and to grow healthy fruits and vegetables. By doing so we are not only overcoming the problem of growing the diseased commodities but also prevent the other fruits and vegetables from spreading the disease. This also helps to prevent epidemic outburst of such diseases which affects a huge area of land. Even though fruits and vegetables plucking robots are invented, their collection of produce does not identify the diseased produce and when harvesting is completed the disease is passed on to the other collected fruits and vegetables which leads to huge amount of loss and wastage. Hence to overcome this, we have proposed the novel model that overcomes all these suggested problems in a long run.

2 Background

As technology collides with science, a lot of impossible creations have been emerging and being implemented. Researchers have been implementing and improvising new algorithms that help us get the optimal solutions for real-time problems in a small period of time. Image processing is one such field that is being developed rapidly in all domains. In the domain of agriculture, researchers have tried to develop real-time solutions such as creating and building algorithms that help to replace human involvement in the fields. In 2012, the approach to detect the diseased left was implemented by finding the difference between the healthy leaf and diseased leaf. The algorithm to analyse and compare the leaves was done by the use of Otsu threshold and k-means clustering which is a type of unsupervised learning. Here, Mrunalini R. Badnakhe et al. extracted the features of the diseased leaf by using feature extraction and used digital analysis for crop colour as changes in colour determine the health of the crop. The values of feature extraction are then plotted on a graph which enables easy detection of the infected parts. It was then stated that extraction of edges, background and foreground and background ratio and differences for the Otsu threshold is more compared to the same extracted vales for k-means clustering. Hence, we can conclude that even though k-means clustering takes less extracted values, the clarity of clusters is more accurate and precise than other methods [1]. In 2016, Sushil R. Kamlapurkar used three steps such as pre-processing, training a model and identification of diseased parts for the identification of diseased leaf. The grape plant is given more concentration in this paper. Then the principles of segmentation were used such as similarity and discontinuity to group similar neighbouring pixels into groups with predefined criteria and to extract regions of different properties such as colour, intensity and texture, respectively, for feature extraction phase. The paper stated that it will benefit people having less knowledge about the occurring plant diseases. So, the proposed system was able to detect the disease as soon as it appeared on the leaf but to a certain extent only [2]. Since supervised learning is used, referring to the teaching–learning-based optimization algorithm for solving complex constrained optimization problems is used. Venkata Rao et al. stated that this algorithm provides a teaching–learning class room experience which does not require any algorithm-specific control parameters. Here, the investigated algorithm has been tested its performance on 35 constrained benchmark functions with different characteristics and compared the working of this algorithm to other optimization algorithms which solve various optimization problems dealing with the industrial environment. It was then observed that the algorithm still needs fine tuning of the common controlling parameters. Also, after including the elitism, many functions produced better results than without it. So, this algorithm can be used for optimizations involving large number of variables and objectives [3]. To obtain even more fast and accurate results, these days researchers have included artificial intelligence using its subdomain of neural networks to identify diseased areas in the field of agriculture. In 2016, Sladojevic et al. then improvised the existing system by implementing deep convolution networks for the detection of plant diseases in leaf. In this paper, the authors have implemented the use of convolutional neural network to create a model that could recognize up to 13 different types of plant diseases shown of leaves. Then implemented this on Caffe, which is a framework for deep learning using 3000 original images that were available on the Internet and proved the model was working to distinguish between a healthy leaf and 13 diseased leaves. Then it later was extended to more than 30,000 images after applying appropriate transformations to those images. Having able to achieve precision between 91 and 98% for separate classes, leads to an overall approximation of 96.3%. Due to the modern lifestyle, fast and accurate results are needed in all fields [4]. Al-Hiar et al. added two more steps to the existing model. Those steps are to identify most green coloured pixels then they are marked based on specific threshold values to compute using Otsu's method, where mostly green pixels are masked. Addition to this, the pixels with pixels with zero red, green and blue values along with the pixels of infected cluster boundaries were completely removed. So, the built models were able to increase the overall accuracy from 89.5 to 94% while comparing from the results of the paper they stated in their research paper. The accuracy of classification was also increased from 92.7 to 94.67% while comparing the results from the paper the authors stated to improve the previously used technique [5]. Even the convergence curve of the neural network in learning phase was better in the authors proposed study compared to the previous ones which was implemented in MATLAB under Windows XP environment. Therefore, concluding that this approach was 19% speed up of the overall approach of the paper they stated to improve, thereby tested this approach for five different diseases. Now, this raised a concern for the diseases that affects the plant on its fruits. Since a wide range of study has been implemented in the domain of disease detection in leaves, a novel way for detection and classification of diseased for fruits has been implemented much later as well [6]. In 2017, Thakur et al. suggested that using image segmentation and k-means clustering technique can be used to identify and detect the diseased on fruits using feature extraction on its vectors for diagnosis of disease morphology. The authors discussed various techniques and approaches existing and how effectively it was combined and redesigned to create a faster model. The Naïve Bayes classifier was created which is more precise and relatively faster in terms of implementation. The treatment was also stated for the specified disease [7]. Tamanna Jena et al. proposed an elitist TLBO method for identification and verification of plant diseases. [8]. Rajesh et al. proposed a methodology for automatic data acquisition and spot disease identification [9].

2.1 Analysis on Different Types of Diseases Found in Agricultural Fields and the Work Done in the Agricultural Pathology Domain

The main reasons for the cause of rotting and black spots arise due to fungal and bacterial attack. This epidemic out blast is passed on to another healthy vegetable or fruit by insects, wind, water, etc., which spreads infection to the entire crop ultimately leading to its death. The fungi spread its infection through its spores whereas the bacteria can propagate with the help of the above transportation agents. Although the diseases for fruits and vegetables are described under several symptom types, the difference is not clear and there are many names for the same kind. The following are the list of diseases caused by both fungi and bacteria and symptoms on vegetables and fruits:

1. **Bacterial Speck**—This disease is caused by the bacteria Psedomonas syringae. It is caused by contaminated seeds or transplants. The spots on fruit are very small and do not penetrate deep into the fruit tissue. The spots can be raised, flat or hollow of varied size. The spots range in colour from brown to black. On green unripen fruits the spots often have darker green haloes while on ripe fruits the spots can have slight yellow haloes.
2. **Powdery Mildew**—This disease is caused by the fungi Podosphaera, Sphaerotheca or Erysiphe nectar. These fungi generally do not require moist conditions to release spores and the spores to germinate and infect the host plant. So, they can infect in warm areas as well. The disease can be recognized by the white to grey powdery mycelium and spore growth. This affects even the fruits. The fungi grow as thin layers of mycelium like patches on the surface of the parts affected by the plant. As the vegetables and fruit mature, the discolouration of fruits is observed like in grapes, pumpkins, apples, pears, etc. The fruit and vegetables may also become dwarfed.
3. **Anthracnose**—This disease is caused by the fungi of species Colletotrichum. The symptoms on vegetables and ripe fruits are small, circular spots that can increase up to half an inch of diameter in size. The centres of the spots become blackish. As the spots become larger, they group together to form large blotches which usually start to drip out. The spots may become numerous in severe cases. This can be caused to plants like tomatoes, melons, etc.
4. **Damping-Off**—Majorly this disease is caused by the fungus and oomycetes, among which one of them is Pythium debaryanum Hesse. It kills seedlings of most plants before they emerge or soon after which results in plant collapse. The fungus attacks the seedlings at or near the surface of the soil. Moisture level and depth of planting play a major role in occurrence of this disease. This can be caused to vegetables and fruits such as cabbage and berries.
5. **Powdery Mildew**—It is a fungal disease which is responsible for white patches most commonly on cucumbers, pumpkins and zucchini. It also affects peas and melons. This disease affects ideally in warm weather and can cause destruction in dry and hot seasons. The superficial white or ash grey coating of mycelium grows which turn into spores in plants like grapes, cucumbers, pumpkins, etc.
6. **Tuber Diseases**—It is a bacterial disease which infects superficial skin, such as common scab and powdery scab as in the case of potato and sweet potato. The symptoms of common scab are manifested on the surface of the tuber which are in the form of cork-like lesions. The damaged skin has rough, cracked skin with scab like spots which may lead to black welts. Powdery scab includes small lesions in the early phases of the disease and later turn into powdery mass.
7. **Wilt Diseases**—It is a fungal disease that causes wilting and death of most vegetables and makes it unfit for consumption. It is caused by the bacteria Erwinia tracheiphila which affects cucumber, pumpkin, guards, muskmelon, etc. Once the plant is affected by the bacteria, it spreads through the vessels and

reaches the main stem and the entire plant becomes dry and dies. Initial symptoms include the drying of single leaves and smaller stems. Later symptoms are drying up of stem and entire plants and disease.

8. **Downy Mildew**—Initially the symptoms are large, angular, yellow areas visible on the upper surface. As lesions become matured, they expand quickly and turn brown. The below surface of affected leaves appears water soaked. A purple-brown mould becomes apparent. Small spores shaped can be seen among the mould. In disease favourable conditions, downy mildew will spread quickly, destroying leaf tissue without infecting stems in plants like cucumber, gourd, soybeans, spinach, etc.
9. **Crown Galls**—It is a disease caused by bacteria soil-borne bacterium Argo bacterium tumefaciens which causes abnormal growth or galls on plants which weakens and stunt the growth of the plant such as barberry and cauliflower.
10. **Bacterial Canker**—This disease is caused by the bacteria. Symptoms on vegetables first appear as round, small and raised spots. These spots then gradually increase in size and have brown centre with white halo in plants like mango, tomatoes, etc.
11. **Alternaria Spot Disease**—It is a type of fungi. Alternaria are the major plant pathogens. Causes early blight of potato, leaf spot disease in cabbage and can infest many other plants (Table 1).

3 Proposed Methodology

The work of the proposed model is to identify the diseased fruits and vegetables which in turn initiate a response for the end manipulator of the arm to pluck the diseased commodities. With the help of image processing and machine learning techniques, we are able to achieve the accurate results. By doing so, we hereby help a lot of farmers, gardeners and homemakers who have agricultural lands to grow and produce a healthier yield of fruits and vegetables.

3.1 Basic Pre-processing Steps of Digital Image Processing

Conversion to grey scale, resizing and denoising—After image acquisition, the RGB image needs to be converted into grey scale image for further analysis. This is an important step because we can extract and manipulate more information from a grey scale image, after which we need to eliminate the unwanted and wasted backgrounds that are not the region of interest. This helps to reduce the computation time and complexity involved to analyse the image. In image acquisition and transition stage, there will be external sources (noise) that affect its quality which has to be removed or else it would add unnecessary weights to the problem. These can be removed by noise removal techniques (Table 2).

Table 1 Analysis on different algorithms and approach used by various researchers on different types of diseased commodities

S. No.	Disease analysis	Approach/algorithms used	Results
1	Recognition and classification of the diseases in infected Crop. (Mrunalini R. Badnakhe and Prashant R. Deshmukh)	Otsu threshold and K-means clustering	K-means clustering was 91% as claimed by the authors
2	Leaf infected with powdery mildew and downy mildew. (Sushil R. Kamlapurkar)	Pre-processing, training and identification	About an average of 93.4% was obtained
3	Solving complex constrained optimization problems. (R. Venkata Rao and Vivek Patel)	Teaching–learning-based optimization algorithm	Between 55–59% accuracy achieved
4	Diseases found in pear, peach, apple and grapevine. (Srdjan Sladojevic, Marko Arsenovic,, Andras Anderla, Dubravko Culibrk and Darko Stefanovic)	Deep convolutional networks	An overall accuracy of 96.3% was recorded by authors
5	Plant stem that is infected with the white mould disease. (H. Al-Hiary, S. Bani-Ahmad, M. Reyalat, M. Braik and Z. ALRahamne)	K-means clustering and feature extraction	Achieved 20% speedup over the approach proposed mentioned by authors
6	Fruit diseases are based on the naked eye observation by experts. (Rupam Thakur, Priyanka Mehta)	ANN, Naïve Bayes classification, K-means clustering	Estimated 94% precision claimed by the authors
7	Data acquisition where pre-processing of images. (Rajesh T. M., Kavyashree Dalawai, Pradeep N.)	K-mean, PCA and ICA algorithms along with basis of histogram of gradient	Proved to have an accuracy of 97.9% as mentioned by authors
8	Spot detection in leaves. (Tamanna Jena, T. M. Rajesh and Mallanagouda Patil)	Elitist TLBO for identification and verification of plant diseases	Found to have an overall accuracy about 95%

3.2 *Feature Extraction for Training Phase*

The pre-processed output image is now used in this step for further analysis. First, using k-mean clustering, the fruits and vegetables are identified and grouped into two independent clusters "*K*". Now this subset is subjected to our novel algorithm which in turn gives a count of objects as "kn" present in the individual cluster "*K*".

Table 2 Different phases of pre-processing stages in digital image processing

S. No.	Pre-processing steps	Input	Output
1	**Input image**: Image will be acquired by taking the picture using digital camera and stored in the digital format		
2	**Grey image:** The color image will be converted to grey colour image to ease out the difficulties in feature extraction level. The simple command to convert color image to grey colour image in MATALB is "rgb2gray"		
3	**Resizing:** If the images are of different size, the algorithm will not give efficient result. So in order to maintain the same aspect ratio, resizing will be done. In our work, all images are resized to 150X150. Simple command to resize the image in MATLAB is "imresize"		
4	**Denoising**: At times while capturing the image, due to noise present on the lens or while converting the images into digital format there are chances of images getting noise. In our work, we have used salt and pepper noise and median filter to denoise the image [10, 11] Simple command to add noise in MATLAB is "imnoise". Simple command to use median filter in MATLAB is "medfilt2"		

Once again by applying the same algorithm, the identification and extraction of the diseased part of the fruit or vegetable take place and this is further classified into two groups known as diseased and healthy subset which is represented as "k_{nn}".

3.2.1 Proposed Novel Algorithm to Extract Features in Training Phase

1. The pre-processed image is subjected to K-mean clustering. To differentiate the objects as fruits/ vegetables based on their well-defined characteristics such as size, colour and shape
2. The identified objects are grouped as different clusters as shown in Eq. (1).

$$\mathrm{K} = \mathrm{f(u, v)} \sum_{1}^{n} (g(x1, y1)g(x2, y2) + \cdots g(xn, yn)) \tag{1}$$

where K is the number of clusters identified in the image f(u, v) with respect to fruits and vegetables.

3. Each cluster "*K*" is subjected to further analysis to identify the number of objects "*k*" present in each cluster using *k*-NN algorithm is shown in Eq. (2).

$$k_n = f(x, y) = \sum_{1}^{n}(a0 + a1 + \cdots an) \tag{2}$$

4. The obtained result is again subjected to K-means algorithm for classification to identify the diseased part of the fruit or vegetable, which is shown in Eq. (3).

$$k_{nn} = f(xn, yn) = \sum_{1}^{n}(a0 + a1 + \cdots an)\sum_{1}^{n}(b0 + b1 + \cdots bn) \tag{3}$$

5. The results obtained from Eq. (3) are the different types of diseases found in vegetables or fruits and these identified diseases will be given the unique ID number in the training phase.

The common diseases occurring in fruits and vegetables are due to fungal and bacterial infections. So, we have taken black spot disease in tomato, Anthracnose disease in capsicum and Alternaria disease in cabbage.

3.3 *Feature Extraction for Testing Phase*

Now the learning phase is implemented by computing for the same stages all over again. First, using k-mean clustering, the fruits and vegetables are identified and grouped into two independent clusters "*K*". Now this subset is subjected to *k*-NN algorithm which in turn gives a count of objects as "k_n" present in the individual cluster "*K*". Once again by applying the same algorithm, the identification and extraction of the diseased part of the fruit or vegetable takes place and this is further classified into two groups known as diseased and healthy subset which is represented as "k_{nn}".

3.4 *Detecting the Spot Diseases Using Proposed Methodology Extraction for Testing Phase*

3.4.1 Colour Detection Using K-Means Clustering

$$s(x) = \pi_l \eta\left(x/\mu_i \sum_{j}\right) + \pi_i \eta\left(y/\mu_j \sum_{j}\right) \tag{4}$$

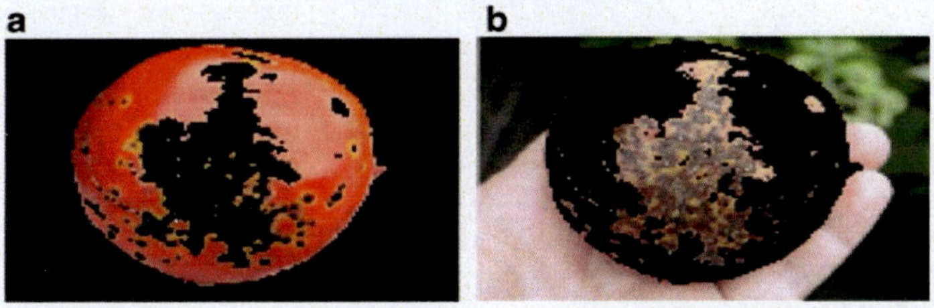

Fig. 1 **a** Foreground extraction (non-diseased) **b** Background extraction (diseased)

In the above equation, $S(x)$ is nothing but the diseased fruit/vegetable image information retrieved. It is combination of various Gaussians values where "i" is called mixing coefficient where $\eta(x/\mu_i, \Sigma_i)$ is foreground parameter and $\eta(y/\mu_j, \Sigma_j)$ is the background parameter. The foreground and background parameters bifurcation are shown in the Fig. 1a, b. Each $\eta(x,y/\mu_{i,j}, \Sigma_{i,j})$ is known as the colour components of the combination. The mixture is expressed in Eq. 4. A GMM is parameterized by two types of parameters, the component weight/mean and component variances/co-variances. For a GMM with K components, the kth component has a mean of μ_i and variance of vector Σ_i for the and a mean of vector μ_j and covariance matrix of Σ_j for the background extraction case these mixture parameters weights/mean and variances/co-variances are defined as π, the summation of GMM equation is given in the equation.

3.4.2 Spot Detection Using PCA

Principal component analysis (PCA) concept is widely used to decrease the size and dimensions of data. It is used to extract a set of feature vectors. In PCA, all important parameters are called as principal components and these principal components satisfy the condition of identification of RoI. PCA helps to convert correlated variables to lesser values of uncorrelated variables called principal components [12, 13]. The first principal component expresses the inconsistency in the data and the next set of expresses remaining values. If first principal component could not represent the prominent features, then next principal components are selected until best prominent feature of data is extracted. To match the working procedure of PCA, almost 13 features of diseased fruits/vegetables have been taken to identify the diseased spots. 13 features such as contrast of the image, correlation, energy, homogeneity, mean, standard deviation, entropy RMS, variance, smoothness have been considered. Standard algorithm used for PCA.

1. Whole image has been considered for m+1 features as shown in Eq. (5).

$$S(x) = \sum (f_1, f_2, f_3, \ldots f_{13}) \tag{5}$$

where f_1 to f_{13} are the number of features considered.

2. For the whole image, mean has been computed for each and every feature as shown in Eq. (6).
   $A(x) = [f_1, f_2, f_3, \ldots f_{13}]$ is matrix

$$\hat{A} = [f_n] \tag{6}$$

   Mean of an image
3. Covariance matrix is computed for the whole image as shown in Eq. (7).

$$C(x, y) = \frac{1}{n-1} \sum (x_i - x)(y_i - y) \tag{7}$$

4. Eigenvectors and eigenvalues have been computed for the whole image as shown in Eq. (8).

$$\mathrm{E(A - \lambda v) = o} \tag{8}$$

   E is eigenvector, A is mean matrix and λ is the eigenvalue of v vector.

4 Experimental Results and Analysis

We have created a database that stores all the picture frames of different kinds of fruits and vegetables. This raw image is used as the input for disease identification. The database basically has two different kinds of image sets which represent the healthy and the diseased commodities. For sample, we have taken three different fruits and vegetables which are tomato, capsicum and cabbage. Figure 2 is the small sample of diseased commodities from the database collection of 5000–10,000 pictures taken from the Internet.

4.1 Detection and Identification of Diseased Tomato

One of the sample pictures of the tomato is taken from the database and subjected to the model built. The diseased identified here is spot disease which is commonly seen in tomatoes due to fungus. The table below shows the two cluster which help in detection of the diseased tomato (Figs. 3, 4 and 5).

Fig. 2 Dataset of fruits and vegetable images for spot identification

Phase1: The healthy part of tomato is extracted. Background = Healthy pixels (color) Foreground = Spot(disease) Pixels (Black) Cluster 1: Healthy pixels
Phase2: The diseased part tomato is detected. Background = Healthy pixels (Black) Foreground = Spot(disease) Pixels (Color) Cluster 2: Spot(disease) Pixels

Fig. 3 Cluster analysis of diseased tomato using K-means clustering algorithm

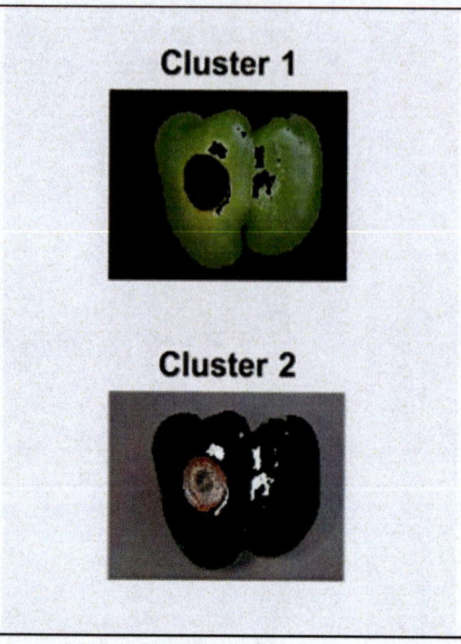	Phase1: The healthy part of capsicum is extracted. Background = Healthy pixels (color) Foreground = Spot(disease) Pixels (Black) Cluster 1: Healthy pixels
	Phase2: The diseased part capsicum is detected. Background = Healthy pixels (Black) Foreground = Spot(disease) Pixels(Color) Cluster 2: Spot(disease) Pixels

Fig. 4 Cluster analysis of diseased capsicum using K-means clustering algorithm

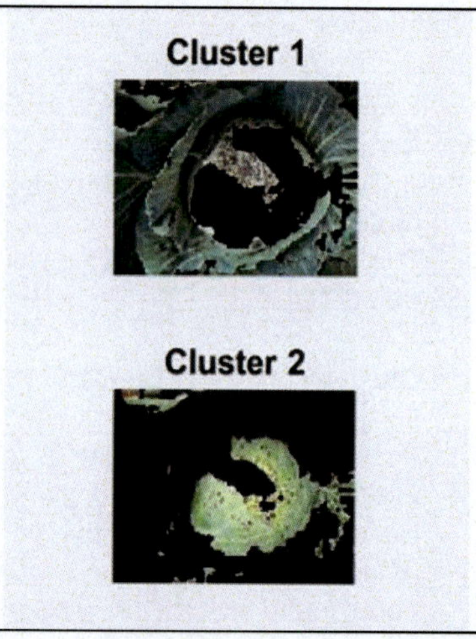	Phase1: The healthy part of cabbage is extracted. Background = Healthy pixels (color) Foreground = Spot(disease) Pixels (Black) Cluster 1: Healthy pixels
	Phase2: The diseased part cabbage is detected. Background = Healthy pixels (Black) Foreground = Spot(disease) Pixels(Color) Cluster 2: Spot(disease) Pixels

Fig. 5 Cluster analysis of diseased cabbage using K-means clustering algorithm

4.2 *Detection and Identification of Diseased Capsicum*

One of the sample pictures of the capsicum is taken from the database and subjected to the model built. The diseased identified here is Anthracnose which is commonly seen in capsicum due to fungus. The table below shows the two cluster which help in detection of the diseased capsicum (Tables 3, 4 and 5).

Table 3 Cluster analysis of diseased tomato using K-means clustering algorithm

	Input	Actual output
Case 1	Cluster 1	Healthy Tomato portion in the background
Case 2	Cluster 2	Diseased Tomato portion in the foreground

Table 4 Cluster Analysis of diseased capsicum using K-means clustering algorithm

	Input	Actual output
Case 1	Cluster 1	Healthy capsicum portion in the background
Case 2	Cluster 2	Diseased capsicum portion in the foreground

Table 5 Cluster Analysis of diseased cabbage using K-means clustering algorithm

	Input	Actual Output
Case 1	Cluster 1	Healthy cabbage portion in the background
Case 2	Cluster 2	Diseased cabbage portion in the foreground

4.3 Detection and Identification of Diseased Cabbage

One of the sample pictures of the cabbage is taken from the database and subjected to the model built. The diseased identified here is Alternaria which is commonly seen in cabbage due to fungus. The table below shows the two cluster which help in detection of the diseased cabbage.

The obtained results are accurate and precise. The diseased cause in each fruit or vegetable is identified and mentioned in the command window. For the arm movement we have modeled a small robot that moves its end effectors due to the response produced by the diseased commodity. The microcontroller signals the robotic arm to move and pluck the diseased commodity. This microcontroller signals the arm based on the command of the program to detect the disease. If the commodity is healthy the arm moves over the next commodity and does not pluck it Fig. 6.

5 Conclusion

This paper mainly discussed about the disease detection and identification of diseased fruits and vegetables which is highly beneficial in the modern era of agriculture. This model helps the farmers, gardeners and homemakers to easily identify the diseased that is cause due to infected growth of fungal and bacterial parasites.

Fig. 6 Robot the moves its end effectors based on the response from diseased commodity

References

1. M.R. Badnakhe, P.R. Deshmukh, Infected leaf analysis and comparison by Otsu threshold and k-means clustering. Int. J. Adv. Res. Comput. Sci. Soft. Eng. **2**(3) (2012). ISSN: 2277 128X
2. S.R. Kamlapurkar, Detection of plant leaf disease using image processing approach. Int. J. Sci. Res. Publ. **6**(2), 73 (February 2016). ISSN 2250-3153
3. R. Venkata Rao, V. Patel An elitist teaching-learning-based optimization algorithm for solving complex constrained optimization problems. Int. J. Ind. Eng. Comput. **3**, 535–560 (2012) (© 2012 Growing Science Ltd.)
4. S. Sladojevic, M. Arsenovic, A. Anderla, D. Culibrk, D. Stefanovic, Deep neural networks based recognition of plant diseases by leaf image classification, *Hindawi Publishing Corporation Computational Intelligence and Neuroscience* volume 2016, Article ID 3289801, 11 pages, (2016)
5. V. Narayanan, B. Parsi, Center symmetric local descriptors for image classification. Int. J. Nat. Comput. Res. (IJNCR) **7**(4), 56–70 (2018)
6. H. Al-Hiary, S. Bani-Ahmad, M. Reyalat, M. Braik and Z. ALRahamneh, Fast and accurate detection and classification of plant diseases. Int. J. Comput. Appl. (0975-8887) **17**(1) (March 2011)
7. R. Thakur, P. Mehta, Bayesian classifier based advanced fruits disease detection. © 2017 IJEDR **5**(3) (2017). ISSN: 2321-9939
8. T. Jena, T.M. Rajesh, M. Patil, Elitist TLBO for identification and verification of plant diseases. 2019. https://doi.org/10.1007/978-981-13-6569-0_3. Springer
9. T.M. Rajesh, Dalawai, K., N. Pradeep, Automatic data acquisition and spot disease identification system in plants pathology domain: agricultural intelligence system in plant pathology domain. https://doi.org/10.4018/978-1-5225-9632-5.ch006.igi. Global
10. V. Bhateja, A. Verma, K. Rastogi, C. Malhotra, S.C. Satapathy, Performance improvement of decision median filter for suppression of salt and pepper noise, in *Advances in Signal Processing and Intelligent Recognition Systems. Advances in Intelligent Systems and Computing ed by* S. Thampi, A. Gelbukh, J. Gelbukh, vol. 264. (Springer, Cham 2014). https://doi.org/10.1007/978-3-319-04960-1_26
11. K. Shukla, V. Bhateja, R. L. Verma, M.S. Alam, An improved directional weighted median filter for restoration of images corrupted with high density impulse noise, in *2014 International Conference on Reliability Optimization and Information Technology (ICROIT)* (Faridabad, 2014), pp. 506–511. https://doi.org/10.1109/icroit.2014.6798376

12. V. Himanshi, A.K. Bhateja, A. Sahu, An improved medical image fusion approach using PCA and complex wavelets,in *2014 International conference on medical imaging, m-Health and emerging communication systems (MedCom)* (Greater Noida, 2014), pp. 442–447. https://doi.org/10.1109/medcom.2014.7006049
13. Krishn, A., Bhateja, V., S.A. Himanshi, PCA based medical image fusion in ridgelet domain, in *Proceedings of the 3rd International Conference on Frontiers of Intelligent Computing: Theory and Applications (FICTA) 2014. Advances in Intelligent Systems and Computing* ed. by S. Satapathy, B. Biswal S. Udgata, J. Mandal, vol. 328 (Springer, Cham, 2015). https://doi.org/10.1007/978-3-319-12012-6_52

Water Management System Using Machine Learning

Harsh Jain, Mohit Buch, and Preeja Babu

Abstract ARIMA model, unlike conventional models, is capable of self-learning data features layer by layer in supervised or unsupervised manner and creating a data-driven model with the given dataset. A detailed analysis along with automation in system would help reduce wastage of resources. This report presents an integrated data-driven modeling framework that couples machine learning with a well-developed IoT device. The integrated framework enables users to monitor water levels as well as forecast usage patterns using an Android application. We aim at unraveling usage patterns, and correlation of various factors would help in better analyzing and predicting the consumption pattern of the most valuable resource-potable water. Water consumption in India is rising rapidly due to the ever-increasing population and hence necessary precautions need to be taken in order to regulate the water flow in different parts of the country and minimize wastage.

Keywords Machine learning · ARIMA · IoT device · Water tank · Android

1 Introduction

Currently drinking water is very prized for all the humans and provisioning of sufficient water quality and quantity is one of the major challenges major cities and districts face in sustaining an economic growth. In recent times, water levels are very low and water in the lakes is going down. So, it is too important to find the solution for water monitoring and control system. IoT is a solution which has been used to tackle problems involving automation of water control and management systems. In recent days, development in computing and electronics technologies has triggered Internet of things technology. Internet of things can be described as the network of

H. Jain (✉) · M. Buch · P. Babu
Department of Information Technology, Mukesh Patel School of Technology Management and Engineering, NMIMS Deemed-to-be-University, Mumbai, India

P. Babu
e-mail: preeja.babu@nmims.edu

V. Bhateja et al. (eds.), *Data Engineering and Intelligent Computing*,
Advances in Intelligent Systems and Computing 1407,
https://doi.org/10.1007/978-981-16-0171-2_45

electronics devices communicating among them by the help of a controller. It is a collection of devices that work together in order to serve human tasks in an efficient manner and combines computational power to send data about the environments.

According to statistics, the wastage of water due to overflow of tanks, running taps etc., has a major impact on the water scarcity. Currently, the manual monitoring and management of water filling in the tanks, the over-exaggeration of demand in a locality account for scarcity in a more water-demanding locality.

Using machine learning provides great opportunity for innovations and new business of smart monitoring, data acquisition and management, and to explore the power of deep analysis and prediction. A truly smart water management system in order to achieve the best outcomes of water management and monitoring is required to be "smart" at each of its steps.

In this paper, firstly we provide the overview of a water tank and report the preliminary investigation of artificial neural network, one of the breakthroughs in deep learning for extracting the intelligence from data, with a potential of predictive analysis which will serve as a data-driven engine of smart water management system.

2 Related Work

Shanmugam et al. [1] presented Automated Water Tank Overflow Control Unit Integrated with Mobile Application which comprised of (1) 8051 microcontroller which acts as a brain to the system, (2) a Bluetooth device for communication between mobile application and the system, and (3) a three-level sensing terminal which is fixed inside the tank and operates on three levels (a) FULL, (b) MEDIUM, and (c) LOW. The moisture of the tank is checked using the three-level sensor and the water motor is switched ON or OFF according to the moisture level presence which is predefined in the requirements. The system also has a manual ON or OFF switch. The merit of the system presented was conservation of water resource and the manual effort reduction. However, drawback of such a system includes the human proximity required by the Bluetooth device to operate and lack of analysis or data collection.

Haesung Tak et al. [2] proposed a Water Tank Monitoring And Visualization System Using Smart-Phones using sensors and server. The water tank comprises of three components: tank, pumps, and valves. The tank sensor data takes into account the water level, temperature, and pressure. The valve sensor monitors the valve status-open/close. Pump sensor monitors the various actions of pumps such as pump run, pump start, pump stop, and pump fail. The visualization of the data was limited to the modularity of each component and recurring issues with water-level accuracy remained. The author was unable to quantify values together as a unit. The system proposed provides only real-time monitoring and does not provide any analysis of the events that occur and utilize the same for predicting or forecasting.

3 Overview of Water Tank System

3.1 Components

The developed system comprises of an IoT device which is placed inside the water tank. Various components of the system are: -

1. Float-type level sensor—These sensors are waterproof and submersed in the tank at different levels. They have digital value and correspondingly give output as a binary 0 or 1 based on switch being ON/OFF. Switch ON indicates the water level has crossed the sensor, subsequently CLOSING the switch in the sensor. Switch OFF indicates the water level is below the sensor and subsequently the switch is OPEN.
2. Vibration sensor—It displays digital value of 1/0 based on if the sensor senses any vibration or it does not. The sensitivity of the vibration sensor can be adjusted.
3. Temperature sensor—This sensor is an "integrated circuit sensor." The yield voltage is linearly proportional. It is not suitable for remote applications and directly measures in Celsius.
4. Current transformer—is a non-invasive sensor used to measure the current flowing through a component and displays analog value in volts.
5. Power source—The PCB board requires power source to operate and connect different modules
6. Microcontroller—Operates the different sensors by using the codes provided to gather the data and transfer it
7. Server—The raw data is collected on the server where it is then preprocessed (Fig. 1).

3.2 Working

The float-type level sensors are placed inside the water tank at uniform distance covering water levels of 20, 40, 60, 80, and 100%. These levels allow us to quantify water level without knowing the dimensions of the tank which would otherwise be difficult.

Vibration sensor is placed on either the inlet or an outlet pipe. Since water flowing through a pipe produces vibrations on the pipe, it allows to ensure the free flow of water and assures us there is no blockage within the pipe which could hamper the flow.

Current transformer is a non-invasive sensor wrapped around the wires of the water-pump. Since the inlet pipe requires a water pump to act against the force of gravity to fill the tanker, use of current transformer allows us to monitoring the current utilization of the water pump. This would further allow us to monitor the health of the pump by discovering cases when the pump works below the specified current

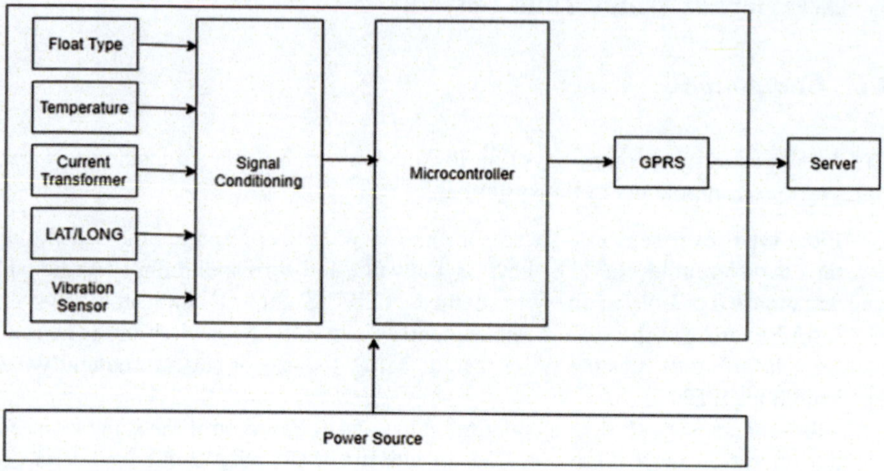

Fig. 1 Component model

rating which indicates the water is being pumped very slowly in the storage tank or in case the pumps draw heavy current which could potentially damage the pump.

Temperature sensor is placed above the tank to monitor the temperature of the environment. Since placing the temperature inside a closed system provides no useful data, it serves as a useful parameter in a water management system (Fig. 2).

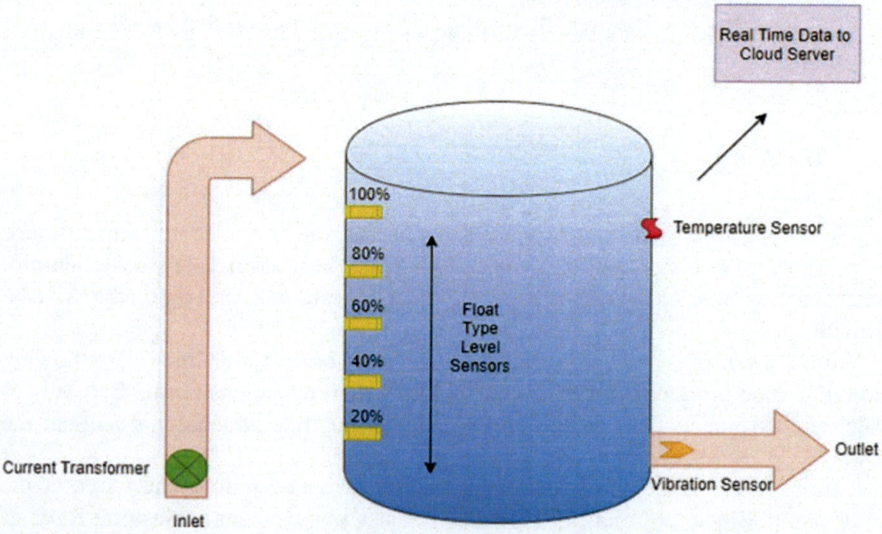

Fig. 2 Tank layout

3.3 System Flow

The analysis of the data collected from the water tank sensors is used to implement machine learning. The system flow, which is configured to implement this system, is shown in Fig. 3.

As shown in Fig. 3, the device communicates with the server and sends in the data collected by the sensors. Input the data in the machine learning model which is divided in two parts—the ARIMA model which smoothens the curve for the water level to give us an accurate water level reading. The need for smoothening the curve is because the float-level sensors provide the accuracy in values 20, 40, 60, 80, 100% and by using the ARIMA model we can increase the accuracy.

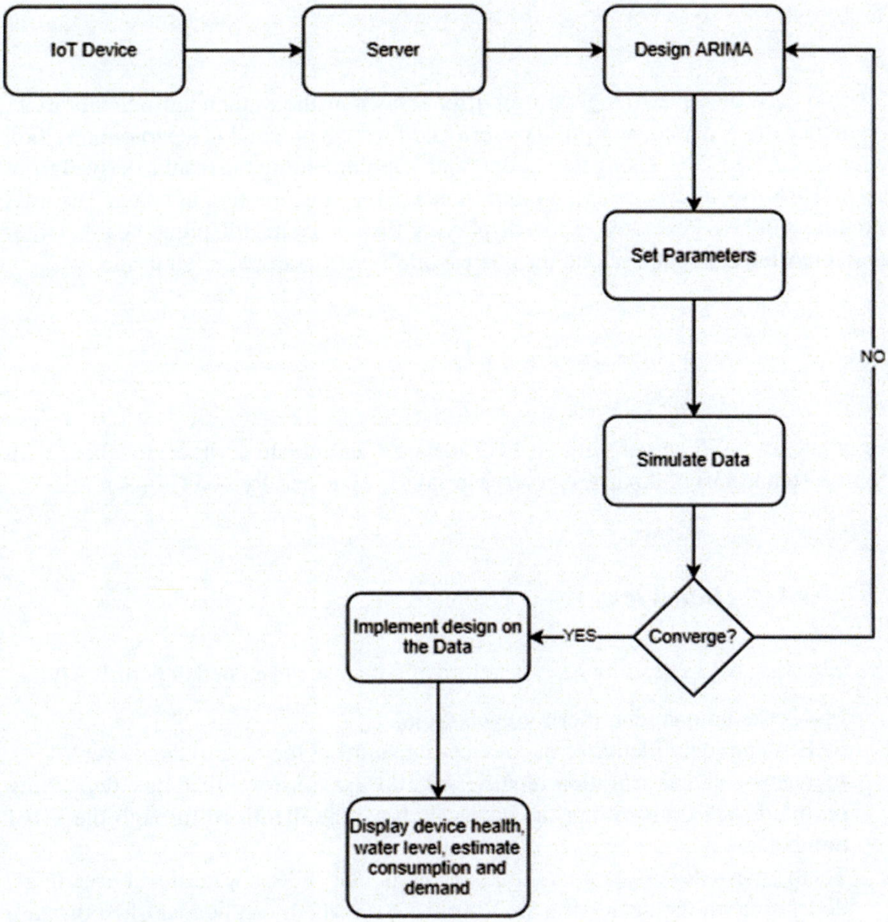

Fig. 3 System process flow

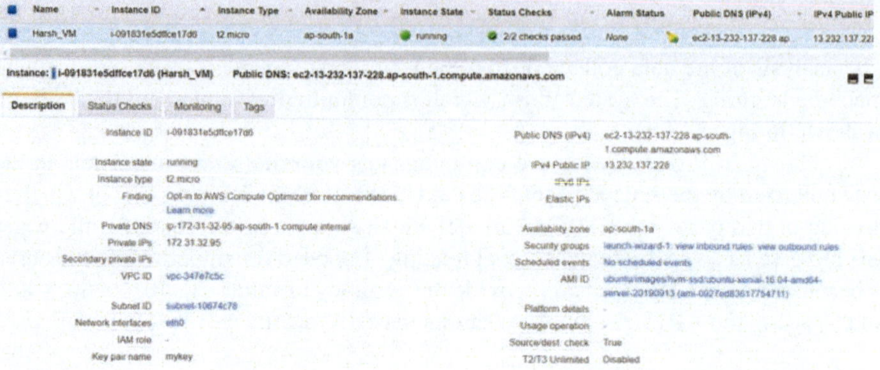

Fig. 4 AWS

From the water level, temperature, current drawn by the water pump and vibration, we deduce the water consumption pattern and forecast demand in a particular region based on LAT/LONG. Using the model, with supplementing the results derived from the ARIMA model, we are able to find trends in the consumption of water. The final outcome will be a generated report displaying the device health, pump health, water level, consumption pattern, and the corresponding demand based on usage.

3.4 Server

We create an AWS server with an EC2 instance and create a virtual machine with Ubuntu 16 and install Apache Server, MariaDB, PHP, and Python (Fig. 4).

3.5 Server Data View

The following figure describes the structure of database developed in phpMyAdmin.

1. Id—is the unique identifier for each tuple.
2. IMEI—uniquely identifies the device transmitting the data to the server.
3. Io_status—shows vibration readings or 100% water-level readings, depending on the device transmitting the data which can be identified through the IMEI number.
4. Temp_adc—shows temperature readings or 80% water-level readings, depending on the device transmitting the data which can be identified through the IMEI number.

←T→				id ▲ 1	imei	io_status	temp_adc	temp_adc2	temp_adc3	temp_adc4	timestamp
☐	Edit	Copy	Delete	13701	869696043614481	0	27	5	0	0	0000-00-00 00:00:00
☐	Edit	Copy	Delete	13702	869696043558829	0	0	1	1	1	0000-00-00 00:00:00
☐	Edit	Copy	Delete	13703	869696043614481	0	27	5	0	0	0000-00-00 00:00:00
☐	Edit	Copy	Delete	13704	869696043558829	0	0	1	1	1	0000-00-00 00:00:00
☐	Edit	Copy	Delete	13705	869696043614481	1	27	5	0	0	0000-00-00 00:00:00
☐	Edit	Copy	Delete	13706	869696043558829	0	0	1	1	1	0000-00-00 00:00:00
☐	Edit	Copy	Delete	13707	869696043614481	1	27	5	0	0	0000-00-00 00:00:00
☐	Edit	Copy	Delete	13708	869696043558829	0	0	0	1	1	0000-00-00 00:00:00
☐	Edit	Copy	Delete	13709	869696043614481	0	27	5	0	0	0000-00-00 00:00:00
☐	Edit	Copy	Delete	13710	869696043558829	0	0	0	0	1	0000-00-00 00:00:00
☐	Edit	Copy	Delete	13711	869696043614481	1	27	5	0	0	0000-00-00 00:00:00
☐	Edit	Copy	Delete	13712	869696043558829	0	0	0	0	1	0000-00-00 00:00:00
☐	Edit	Copy	Delete	13713	869696043614481	0	27	5	0	0	0000-00-00 00:00:00
☐	Edit	Copy	Delete	13714	869696043558829	0	0	0	0	1	0000-00-00 00:00:00
☐	Edit	Copy	Delete	13715	869696043614481	1	27	5	0	0	0000-00-00 00:00:00
☐	Edit	Copy	Delete	13716	869696043558829	0	0	0	0	1	0000-00-00 00:00:00
☐	Edit	Copy	Delete	13717	869696043614481	1	27	5	0	0	0000-00-00 00:00:00
☐	Edit	Copy	Delete	13718	869696043558829	0	0	0	0	1	0000-00-00 00:00:00
☐	Edit	Copy	Delete	13719	869696043614481	0	27	5	0	0	0000-00-00 00:00:00
☐	Edit	Copy	Delete	13720	869696043558829	0	0	0	0	1	0000-00-00 00:00:00

Fig. 5 phpMyAdmin

5. Temp_adc2—shows current readings or 60% water-level readings, depending on the device transmitting the data which can be identified through the IMEI number.
6. Temp_adc3—shows 40% water-level reading.
7. Temp_adc4—shows 20% water-level reading.
8. Timestamp—to add timestamp to the entry (Fig. 5).

As seen in the above figure, two different devices are transmitting data: -

1. The device with IMEI number 869696043558829 has float-level sensors connected to it indicating readings at 20%, 40%, 60%, 80%, and 100%, respectively, corresponding to temp_adc4, temp_adc3, temp_adc2, temp_adc2, and IO_status.
2. The device with IMEI number 869696043558829 has vibration sensor reading at IO_status, Temperature readings at temp_adc, and current readings at temp_adc2.

The data here displayed is the data generated in a tanker for a span of 38 days.

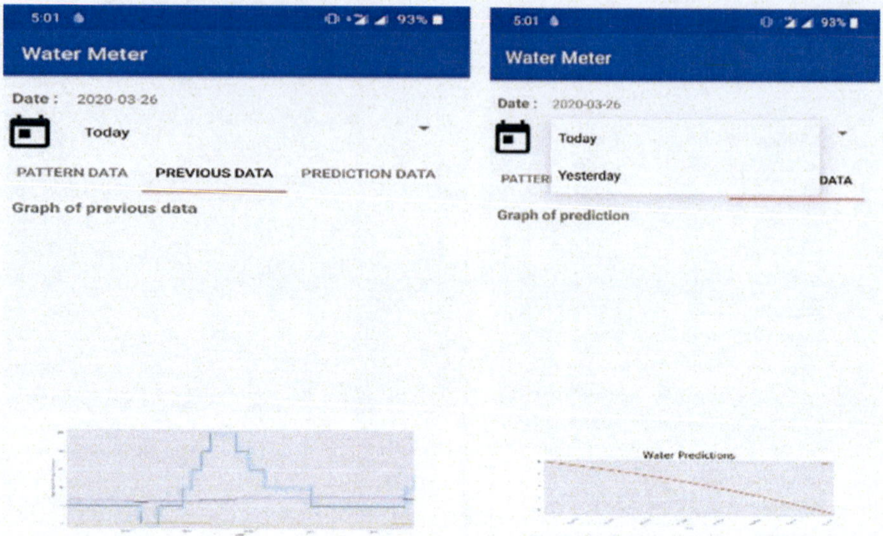

Fig. 6 Android application

3.6 Android

We display the results and the graphs on the Android application. The data can be generated for the current date or the previous date. Real-time data is fetched from the server and the ARIMA model.

The Android application consists of a login page and a home page (Fig. 6).

3.7 ARIMA Model and Machine Learning Algorithm

An ARIMA model is a class of statistical models for analyzing and forecasting time series data.

It explicitly caters to a suite of standard structures in time series data, and as such provides a simple yet powerful method for making skillful time series forecasts.

ARIMA is an acronym that stands for autoregressive Integrated Moving Average. It is a generalization of the simpler autoregressive moving average and adds the notion of integration.

This acronym is descriptive, capturing the key aspects of the model itself. Briefly, they are:

AR: Autoregression. A model that uses the dependent relationship between an observation and some number of lagged observations.

```
mod_pr_post_vol = auto_arima(df_comp.capacity[ann_2:end_date], exogenous = df_comp[['capacity']][ann_2:end_date],
                             m = 5, max_p = 5, max_q = 5)
```

Fig. 7 ARIMA model

I: Integrated. The use of differencing of raw observations (e.g., subtracting an observation from an observation at the previous time step) in order to make the time series stationary.

MA: Moving average. A model that uses the dependency between an observation and a residual error from a moving average model applied to lagged observations.

Each of these components are explicitly specified in the model as a parameter. A standard notation is used of ARIMA (p,d,q) where the parameters are substituted with integer values to quickly indicate the specific ARIMA model being used (Fig. 7).

The parameters of the ARIMA model are defined as follows:

p: The number of lag observations included in the model is also called the lag order.

d: The number of times that the raw observations are differenced, also called the degree of differencing.

q: The size of the moving average window is also called the order of moving average.

A linear regression model is constructed including the specified number and type of terms, and the data is prepared by a degree of differencing in order to make it stationary, i.e., to remove trend and seasonal structures that negatively affect the regression model.

A value of 0 can be used for a parameter, which indicates to not use that element of the model. This way, the ARIMA model can be configured to perform the function of an ARIMA model, and even a simple AR, I, or MA model.

Adopting an ARIMA model for a time series assumes that the underlying process that generated the observations is an ARIMA process. This may seem obvious but helps to motivate the need to confirm the assumptions of the model in the raw observations and in the residual errors of forecasts from the model.

4 Results and Discussion

4.1 Machine Learning Analysis Graph

The graph shows the water-level consumption pattern. The blue line represents water level, yellow represents the temperature readings, and the green line represents the current level, respectively. We observe that at the start of the day the water begins to get pumped in the water tank and so the blue curve in the graph increases. The curve begins to go down as the tank water gets consumed by evening. The temperature

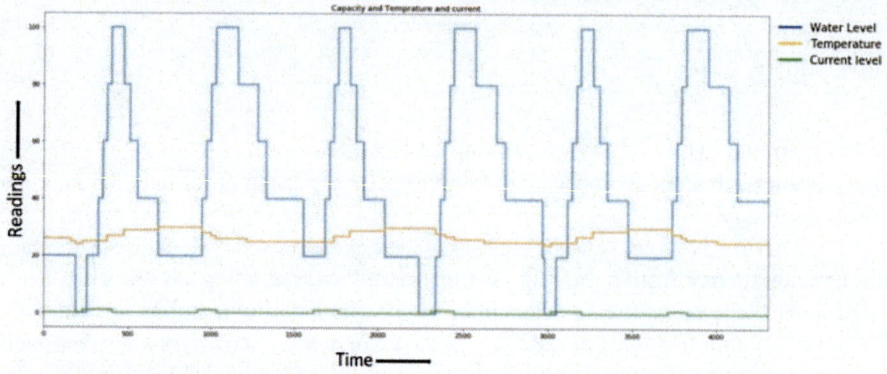

Fig. 8 Pattern generated in the span of 30 days

readings take peak during the afternoons and the current readings take peak when the water is being pumped into the tank (Fig 8).

4.2 *Real-Time Data Versus Prediction*

Here, we can see at what rate the water in been consumed. Based on the current rate of consumption, we can develop a slope which tells by what time the entire resource will get finished. Based upon this, the user can perform the following task:

1. Can be prepared to order for water tankers knowing how much quantity to order.
2. Put a stay on ordering as there is sufficient amount of water already available (Fig. 9).

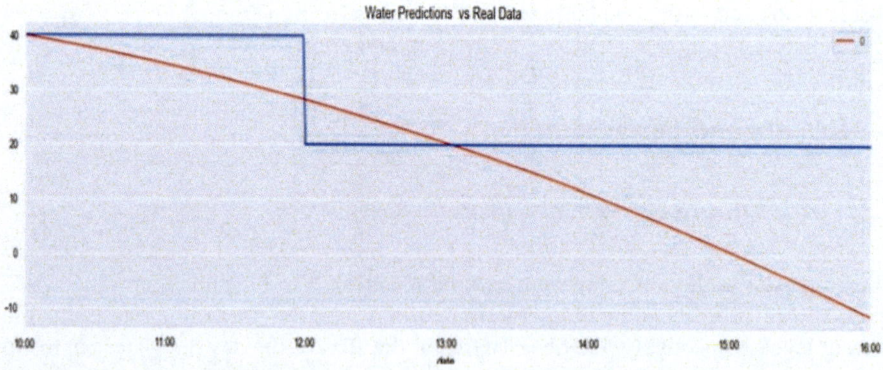

Fig. 9 Water prediction vs real data

5 Conclusion

This work brings in the accuracy of a sophisticated technology at an economical cost which not only provides real-time monitoring but also a detailed analysis of the data for a better interpretation. The automation that the system brings not only reduces the human errors but also reduces the overall operational cost, making it much easier and convenient to manage. Earlier the system was restricted to proximity of the user to the system or use of sophisticated sensors such as infrared or ultrasonic sensors which are difficult to place due to the alignment of sensors playing the key role in accuracy. These sensors are not only costly but also delicate, causing a major barrier in developing a cost-effective system.

We use ARIMA to generate accurate water levels and to generate consumption patterns and forecast demand. Accuracy and optimization of the machine learning algorithms require a large dataset to provide better results. Therefore, in order to achieve maximum accuracy, a growing and continuous data is required.

The major merit of the system includes:

1. The conservation of water as a resource
2. Reduction in manual effort and human errors
3. Can be used with any liquid storage tank with minimal changes.

6 Future Works

In future works, visualization techniques can be used to enhance readability of the system. There is a possibility to expand the scope whereby there could be an option added wherein the organization can also put the sensors on moving water tanks and keep track of the data received. Also, there could be an upgrade to the mobile app user interface where values from additional sensors could be displayed. This will make the application more and more useful increasing our customer base appealing to a larger audience.

References

1. G. Shanmugam, P. Pranathi, K. Sravya, Automated water tank overflow control unit integrated with mobile application. Int. J. Inf. Sci. Comput. **9**, 10–12 (2015). https://doi.org/10.18000/ijisac.50155
2. H. Tak, D. Kwon, H.-G. Cho, Water tank monitoring and visualization system using smartphones. Int. J. Mach. Learn. Comput. **3**(1), (2013)
3. M. Almasarweh, S. Alwadi, ARIMA model in predicting banking stock market data. Mod. Appl. Sci. **12**, 309 (2018). https://doi.org/10.5539/mas.v12n11p309
4. Z. Wu, M. El-Maghraby, S. Pathak, Applications of deep learning for smart water networks. Procedia Eng. **119**(479–485), 2015 (2015). https://doi.org/10.1016/j.proeng.2015.08.870
5. V. Madhavireddy, B. Koteswarrao, Smart water quality monitoring system using Iot technology. Int. J. Eng. Technol. (UAE). **7**, 636–639 (2018) https://doi.org/10.14419/ijet.v7i4.36.24214

6. M. Suresh, U. Muthukumar, J. Chandapillai, A novel smart water-meter based on IoT and smartphone app for city distribution management, pp. 1–5 (2017). https://doi.org/10.1109/TENCONSpring.2017.8070088

Convolution Neural Network-Based CAPTCHA Recognition for Indic Languages

Shaikh Sahil Ahmed and Kumar M Anand

Abstract CAPTCHAs (completely automated public turing test to tell computers and humans apart) are automated turing tests to identify whether the user is a human or not. There are several types of CAPTCHAs available; among them, most of the CAPTCHAs are based on characters. These are some distortions of character images and symbols identifiable by most human beings but are difficult for machines like a computer to recognize. CAPTCHAs are used for many security reasons in various sectors. We can see that existing alphanumeric CAPTCHAs are still difficult to recognize by humans also. In comparison with the traditional English CAPTCHAs, Indic languages like Hindi and Bengali are difficult to recognize. For this purpose, we use deep learning models like convolution neural networks to recognize CAPTCHAs automatically. In this paper, we have used three Indic languages, and two of them are northern language, one is Hindi, the other is Bengali, and one southern language Tamil as Indic CAPTCHAs. We generated the dataset with the help of the Pillow library for all three languages. After experiments, we observed that the model has reasonable accuracy for the Bengali language, Hindi language, and for the Tamil language.

Keywords CAPTCHA · Recognition · Image processing · Convolution neural networks

1 Introduction

CAPTCHAs are now used on the Internet on different Web sites to differentiate humans from the bots to deny various spammers that can cause any breakage to security issues, preventing bots from accessing or performing transactions on the site or any other violations. There are many websites in which there may be CAPTCHAs of different Indian languages. It is more difficult for humans to recognize CAPTCHAs

S. Sahil Ahmed (✉) · K. M. Anand
National Institute of Technology Karnataka, Surathkal, India
e-mail: m_anandkumar@nitk.edu.in

V. Bhateja et al. (eds.), *Data Engineering and Intelligent Computing*,
Advances in Intelligent Systems and Computing 1407,
https://doi.org/10.1007/978-981-16-0171-2_46

of these languages. For this, we have developed an automatic recognizer for three languages. From the northern part, we have used Hindi and Bengali, and from the southern part, we have used Tamil. We have seen that most of the CAPTCHAs are in English, but when it comes to Hindi, Bengali, and Tamil CAPTCHAs, it is difficult to recognize automatically. For that purpose, convolution neural networks are used to identify the CAPTCHAs from images. This can be used for checking the reading proficiency of humans.

So, we can say that CAPTCHA is an image that consists of letters and digits having some distortions, and the user is called to write the letters that are shown in the CAPTCHA. There are various types of CAPTCHA, like Audio, 3D, Drag and Drop, math solution, etc., as the user who gives a correct response for solving CAPTCHA is assumed to be human otherwise considered a bot.

2 Literature Review

Ruti Gafni [1] investigated the dissimilarity between various CAPTCHA tests and inspected how they influence user experience, success rate, and response time. They performed tests on different types of CAPTCHAs such as text-based CAPTCHA, audio-based CAPTCHA, arithmetic operation-based CAPTCHA, picture-based CAPTCHA, and game-based CAPTCHA. They concluded all five types of tests that influenced user experience at a certain level.

Anju Bala and Baljit Singh Saini [2] described the categorization of CAPTCHAs and their application areas and rules for generating a CAPTCHA. It has used three methods. The first is visual method, the second one is optical character recognition, and the third one is the non-optical character recognition visual method. Various existing CAPTCHA strategies give the details for working of CAPTCHA.

Sushma Yalamanchili, M.Kameswara Rao [3] introduced a substructure for a text-based CAPTCHA based on the Devanagari script to distinguish the reading proficiency between humans and computer programs. They have used DevaDB, Query Generator, Obfuscato, DevaGUI, and Match Response as their framework. They concluded that DevaCAPTCHA is highly useful and is easy for users to respond.

Shafiya Afzal Sheikh , Tariq Banday [4] reviewed existing CAPTCHA scripts and Indian regional Web sites in terms of their accessibility, helpfulness, and multilingual support. They reported the design of CAPTCHAs on Hindi, Urdu, Punjabi, and the English language, which gives rise to CAPTCHA testing in Web sites. They concluded that random image patterns and the use of multiple fonts make it more secure against bots.

Kailash Atal, Ashish Arora, Devendra Singh Sachan [5] used an integrated OCR-reCAPTCHA system where OCR digitizes documents with high accuracy for Devanagri script that learns from the human response. They concluded the accuracy of character on single font using OCR.

Kumar and Gokul [6] covered different types of CAPTCHAs, attacks on CAPTCHAs such as basic CAPTCHAs-decode attacks, and bypassing attacks of

CAPTCHAs. They considered Sanskrit, which in Devanagari. They used the script's database that consists of large data of Devanagari script samples and used module obfuscation for distorting and adding noise to it. They explained various types of CAPTCHAs that are commonly used in sites and further ideas that can carry out for higher security.

3 CAPTCHA Generation

We have used the Pillow library [12] to generate random CAPTCHAs with the help of different fonts taken sidewise. We imported modules like Image, ImageFont, and ImageDraw from PIL. From the main function, we will call a model that will select the random text of length four and stored it in a variable. After that, a function is called that takes the text and fits it in the fixed-size image drawn by ImageDraw in background. Random fonts are chosen to distort the CAPTCHAs. Then, CAPTCHA file names are saved with the same name as CAPTCHA in the image with an extension ".png" in the desired directory having name "generated captcha images". 30,000 CAPTCHAs are generated for every language for single and multiple fonts. Four-letter CAPTCHA images are shown in Fig. 1.

3.1 Dataset Distribution

With a total of 30,000 datasets for each language, 75% of the dataset is used for the training purpose, 15% is used for testing, and the remaining 10% is used for validation as shown in Table 1. We used the dataset based on the extracted letters for all the three languages.

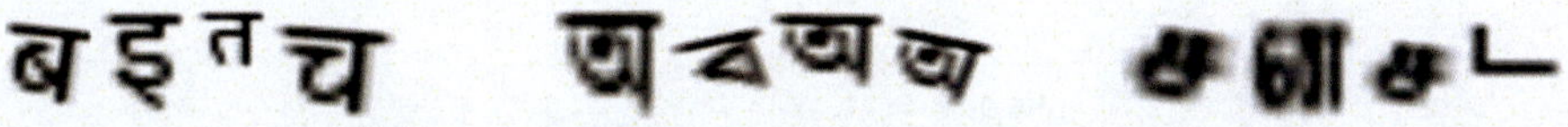

Fig. 1 Four-letter CAPTCHA images for three languages

Table 1 Training, testing, and validation distribution of CAPTCHAs

Language	Training	Testing	Validation	Total
Bengali	22,500	4500	3000	30,000
Hindi	22,500	4500	3000	30,000
Tamil	22,500	4500	3000	30,000

4 Methodology

We build the network using the OpenCV [13] package used as a popular framework for computer vision and image processing. Our training input (created CAPTCHA images) is given to deep neural networks and getting its expected output, as shown in Fig. 2. We have also used the TensorFlow library for a deep learning framework.

As new random CAPTCHAs are generated, they are given a file name as the text written on CAPTCHA. Later on, the letters are extracted to different directories with the help of Python script for letter extraction.

4.1 Letter Extraction Algorithm

The image is loaded and converted it into grayscale. After that, extra padding is added around the image. Now, a threshold of the image (i.e., convert it into pure black and white) and note the contours (i.e., continuous blobs of pixels) of the image. Four contours are looped, and the letters are extracted. Get the rectangle that holds the contour and compare the height and the width of the contour to detect the letters that are joined as a single quantity. If the ratio of width to height is greater than 1.25, then break it in half into two-letter regions; otherwise, it is a normal letter by itself. If more or less than four-letter CAPTCHAs are found, then CAPTCHA images are skipped; otherwise, we will continue through the program. Get the directory to save the image and write the letter image to the file. The letters are extracted and stored in the given path. Figures 3, 4 and 5 show some of the extracted letters of Bengali, Hindi, and Tamil.

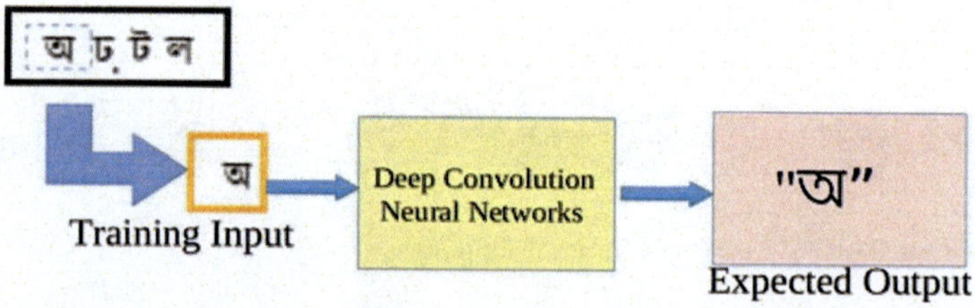

Fig. 2 Recognizing a single letter at a time

Fig. 3 Extracted single letters of Bengali CAPTCHA

Fig. 4 Extracted single letters of Hindi CAPTCHA

Fig. 5 Extracted single letters of Tamil CAPTCHA

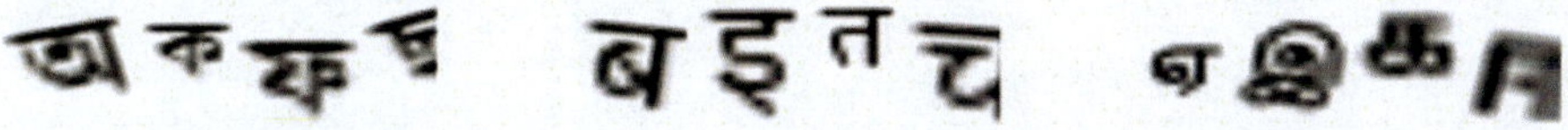

Fig. 6 Skipped CAPTCHA images

We will process the dataset directory and extract it into separate letters into a different directory. The following subsection explains the letter extraction.

The CAPTCHAs that were not well enough to go for training were skipped as shown in Fig. 6, and the remaining CAPTCHAs were extracted into letters that will be used for our training. The total number of letters remaining after extraction that is used for training (75%), testing (15%), and validation (10%) in each language is visualized into two categories as follows:

1. Using a single font.
2. Using multiple fonts.

The dataset distribution with filtering the erroneous split is shown in Table 2 and 3.

The flow of architecture having convolution layer with max pooling and fully connected layer with one hidden fully connected layers and one output fully connected layer and decode prediction from a binary number to a letter as shown in Fig. 7.

Table 2 Using single font

Language	Training	Testing	Validation	Total
Bengali	22,974	4595	3063	30,632
Hindi	15,384	3077	2051	20,512
Tamil	14,085	2817	1878	18,780

Table 3 Using multiple font

Language	Training	Testing	Validation	Total
Bengali	20,091	4018	2679	26,788
Hindi	32,871	6574	4383	43,828
Tamil	11,931	2386	1591	15,908

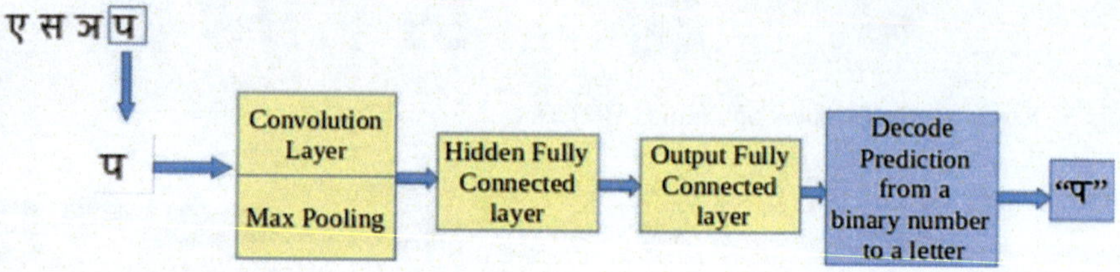

Fig. 7 Structure of convolution neural network

4.2 *CAPTCHA Identification*

Input images are loaded and converted it into grayscale. Letters are resized such that it fits in a 20 × 20 pixel box, and a third-channel dimension is added to the image. Now, get the folder name where the CAPTCHAs were extracted. The letter image is labeled and is added to the training data, and the raw pixel intensities are scaled to the range [0,1]. The data used are split into train and test sets. Now, the labels are converted into one-hot encodings. Further, we built the neural network with the first layer of convolution with max pooling, the second layer of convolution with max pooling, a hidden layer with 500 nodes, and an output layer with 62 nodes compatible with the given input letter. Keras is used to construct the TensorFlow model behind the scenes and train the neural network, and then the trained model is saved to the disk.

Here, we used the convolution neural network with two convolution layers and two hidden fully connected layers, as shown in Fig. 8. In the design, we visualized some

Fig. 8 CNN structure with input, filter sizes, hidden nodes, and output nodes

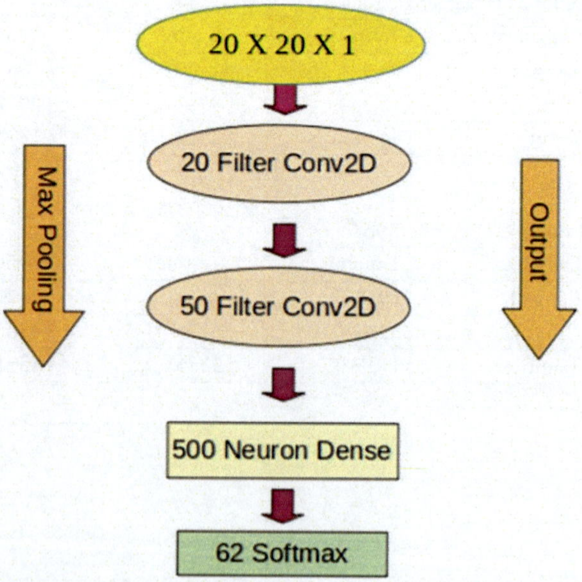

factors like layers, filter sizes, number of parameters, etc. After training the model, we will now solve the CAPTCHAs with the help of a neural network that helps in the prediction of letter images. We will take some random four-digit CAPTCHA for all the three languages and will solve the CAPTCHAs with the model.

Suppose that we have some $N * N$ square neuron layer, which is followed by our convolutional layer. If we use an m×m filter ω, our convolutional layer output will be of size $(N - m + 1) \times (N - m + 1)$. To compute the pre-nonlinearity input to some unit p_{ij}^{ℓ} in our layer, we need to sum up the contributions (weighted by the filter components) from the previous layer cells as shown in Eq. 1.

$$p_{ij}^{\ell} = \sum_{a=0}^{m-1}\sum_{b=0}^{m-1} \omega_{ab} y_{(i+a)(j+b)}^{\ell-1} \tag{1}$$

5 Experimental Results and Analysis

In our experiment, we have gone through thirty thousand CAPTCHAs having the length of four letters. We selected some random CAPTCHAs and tested them under our model. Figure 7 shows the CAPTCHA recognition framework. We have extracted the letters to recognize the CAPTCHAs more efficiently. To visualize the functionalities for CAPTCHAs, we further found the accuracy and other parameters dividing it into four cases:

Case 1: Using a single font and filtering the dataset (i.e., skipping CAPTCHAs for bad training data).
Case 2: Using a single font and without filtering the dataset (i.e., without skipping CAPTCHAs).
Case 3: Using five multiple fonts and filtering the dataset.
Case 4: Using five multiple fonts and without filtering the dataset.

While training the dataset with ten epochs, we observed that the testing and validation accuracy for the three languages as shown in Table 4.

Table 4 Test and validation accuracy for four cases

Cases	Hindi		Bengali		Tamil	
	Validation	Test	Validation	Test	Validation	Test
Case 1	1.0	1.0	1.0	1.0	1.0	1.0
Case 2	0.655	0.656	0.807	0.811	0.471	0.468
Case 3	1.0	1.0	0.983	0.975	1.0	0.999
Case 4	0.780	0.782	0.704	0.698	0.438	0.444

Table 5 Same font and filtering the erroneous split

Language	Micro-average	Macro-average	Weighted average	Samples average
Hindi	1.0	1.0	1.0	1.0
Bengali	1.0	1.0	1.0	1.0
Tamil	1.0	1.0	1.0	1.0

Table 6 Same font and without filtering the erroneous split

Language	Micro-average	Macro-average	Weighted average	Samples average
Hindi	0.71	0.70	0.70	0.63
Bengali	0.86	0.86	0.86	0.80
Tamil	0.52	0.52	0.52	0.44

Table 7 Multiple font and filtering the erroneous split

Language	Micro-average	Macro-average	Weighted average	Samples average
Hindi	1.0	1.0	1.0	1.0
Bengali	0.97	0.96	0.97	0.97
Tamil	1.0	1.0	1.0	1.0

Table 8 Multiple font and without filtering the erroneous split

Language	Micro-average	Macro-average	Weighted average	Samples average
Hindi	0.85	0.85	0.85	0.78
Bengali	0.76	0.76	0.76	0.67
Tamil	0.43	0.41	0.41	0.32

Based on all the four cases, we calculated the F1-score for each language based on various parameters like micro-average, macro-average, weighted average, and samples average, as shown in Tables 5, 6, 7, and 8. As shown in Table 5, in case of using four multiple fonts with filtering the dataset, Hindi with test accuracy of 1.0 is better as compared to the Bengali having a test accuracy of 0.975 and Tamil having 0.999. In the case of using five multiple fonts and without filtering the dataset, we can observe that Tamil has more segmentation errors as compared to the other two languages. Test accuracy for Hindi is 0.782, Bengali is 0.698 and lowest for Tamil, i.e., 0.444. In this Hindi is giving good results as compared to the other two. In the case of single font and filtering the dataset, all languages have good accuracy, but taking single font without filtering the dataset, Bengali with 0.811 is giving better results as compared to the other two languages, Hindi with a test accuracy of 0.656 and Tamil with 0.468.

Here we can observe that the Bengali language has good F1-score as compared to Hindi and Tamil language.

5.1 Error Analysis

We found out the letters having the least recall in a particular language.

1. Single fonts and without filtering the erroneous split: In the case of the Hindi language, the lctter "NYa" has the least recall of 0.60. In the case of Bengali language, letters "e", "u", "chh", "sh" have the least recall of 0.77. In the case of the Tamil language, the letter "j" has the least recall of 0.37.
2. Multiple fonts and without filtering the erroneous split: In the case of Hindi language, letters "i" and "Va" have the least recall of 0.73. In the case of the Bengali language, the letter "gh" has the least recall of 0.58. In the case of the Tamil language, the letter "nGa" has the least recall of 0.03.

6 Conclusion and Future Work

CAPTCHAs are one of the methods that distinguish the machines from human beings in a network environment. In this paper, CAPTCHA recognition is based on convolution neural networks, which helps in efficient recognition of Hindi and Bengali CAPTCHA. In future work, some other Indic languages for CAPTCHA recognition will be added. Handwriting recognition and testing of Indic CAPTCHAs like Telugu, Malayalam, etc., are another future research activity.

References

1. R. Gafni, I. Nagar, CAPTCHA–security affecting user experience. Issues Inform. Sci. Inf. Technol. **13**, 063–077 (2016)
2. B.S. Saini, A. Bala, A review of bot protection using CAPTCHA for web security. IOSR J. Comput. Eng. 36–42 (2013)
3. S. Yalamanchili, M. Kameswara Rao.: A framework for devanagari script-based captcha. Int. J. Adv. Inf. Technol. **1**(4), 47–57 (2011)
4. M. Tariq Banday, S. Afzal Sheikh, Design of CAPTCHA script for Indian Regional Websites, SSCC 2013, in *International Symposium on Security in computing and Communications*, pp. 98–109
5. K. Atal, A. Arora, D.S. Sachan, P.K. Bora, reCAPTCHA assisted OCR for Devanagiri Texts (2015)
6. A.A. Jecinth Kumar , A.R. Guru Gokul, A study on CAPTCHAs the challenge response test. Int. J. Latest Trends Eng. Technol. Special Issue April-2018, pp. 005–011
7. L.V. Ahn, M. Blum, N.J. Hopper, J. Longford, Captcha: using hard AI problems for security, in *Advances in Cryptology–EUROCRYPT*, vol. 2656 (Springer, 2003), pp. 294–311
8. A. Krizhevsky, I. Sutskever, G.E. Hinton, ImageNet classification with deep convolutional neural networks. Commun. Acm **60**(2) (2012)
9. K. Chellapilla , K. Larson , P. Simard, M. Czerwinski, Computers beat humans at single character recognition in reading based human interaction proofs, in *Proceedings of the Third Conference on E-Mail and Anti-Spam 2005*

10. J. Yan, A.S. El Ahmad, A low-cost attack on a microsoft CAPTCHA, in *15th ACM Conference on Computer and Communications Security (CCS'08)* (ACM Press, Virginia, USA, 2008), pp. 543–554
11. S.Y. Huang, Y.K. Lee, G. Bell, Z.H. Ou, An efficient segmentation algorithm for CAPTCHAs with line cluttering and character warping. Multimedia Tools Appl. **48**, 267–289 (2010)
12. https://pypi.org/project/Pillow/
13. https://pypi.org/project/opencv-python/
14. G. Mori, J. Malik, Recognizing objects in adversarial clutter: Breaking a visual CAPTCHA, Computer Vision and Pattern Recognition, in *Proceeding IEEE Computer Society Conference*, vol. 1 (IEEE, 2003), pp. I–I

Automatic Weapon Detection in Video Using Deep Learning

B. Madhushree, K. N. Sowmya, and H. R. Chennamma

Abstract In recent years, terror attacks have become more common in public places like schools, banks, religious places and transportation system throughout the world. Automatic video surveillance techniques at these places aid the security personnel for identifying threats. Convolutional neural network-based approaches have been very successful in image/video cataloging and object recognition. In this work, Alex-net is used to identify terror weapons from the video feeds and classify them to aid further investigation by security personnel.

Keywords Weapon detection · Deep learning · Alex-net · CNN · Video surveillance · SPP · RBM

1 Introduction

Nowadays, most of the crimes are carried out by utilizing hand weapons like gun, bazooka, knife and tanker by terrorists to create social disturbance. In order to mitigate such activities, identification of the culprits is a societal need. Human limitation in supervision has lead mankind to look for alternatives. In order to effectively address this problem, surveillance systems are adopted by security/government agencies to track and monitor crime.

Automatic armament detection demands a set of computer vision tasks that consists of object recognition, localization and classification as shown in Fig. 1. During image classification, bounding boxes over multiple objects of interest in a video/image frame distinguish each of them and allot a class label. Objects like weapons and arm recognition problem are solved by adopting machine learning and deep learning techniques in recent years [1–5]. Deep learning approach has much

B. Madhushree · H. R. Chennamma (✉)
JSS Science and Technology University, Mysuru, India
e-mail: hrchennamma@jssstuniv.in

K. N. Sowmya
JSS Academy of Technical Education, Bengaluru, India

V. Bhateja et al. (eds.), *Data Engineering and Intelligent Computing*,
Advances in Intelligent Systems and Computing 1407,
https://doi.org/10.1007/978-981-16-0171-2_47

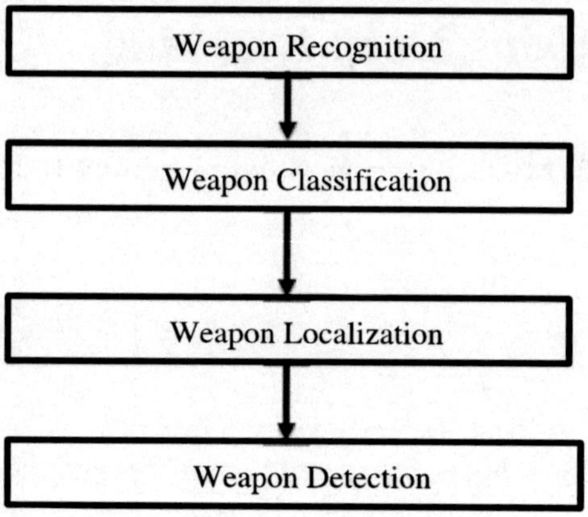

Fig. 1 General architecture of automatic surveillance for weapon detection system

advantage over machine learning when trained with enormous data. It can be trained for unstructured data, and it performs feature engineering by its own in an incremental approach. In deep learning, a large number of data can be labeled which is tedious in machine learning and also provide accurate results.

Machine learning is suitable for simple problems and does not help to solve complex problems because there are few thousand data points used for analysis. Deep learning is introduced for large-scale data and is suitable for complex calculation. Deep neural network (DNN) is one of the artificial neural networks (ANN) having many layers, and learning takes place in between the input and the output layers. The probability of every output is measured by the network while moving through the layers.

Deep learning can be implemented with different types of models like convolutional neural network (CNN)-based methods, restricted Boltzmann machines (RBM)-based methods, autoencoder-based methods and sparse coding-based methods depending on the problem type. Among the deep learning approaches, autoencoder is the simplest and easy method to implement. Autoencoder distinguishes few data vectors with respect to other, whereas RBM are productive, and it considers inbuilt features, and with the united distribution, it is able to produce new data efficiently. Sparse coding is found to be concise where the representation of input is of high level. Sparse coding is simpler in training when compared to RBM approach. Nowadays, CNN-based methods are used in most of the image recognition problems since human supervision is not required to learn important features. With respect to performance, CNN is the best neural network for picture recognition task. Multiple types of CNN models are proposed in the literature based on the problem statement and type of object recognition involved like LeNet, Alex-net, ResNets, GoogleNet, SPPNet and VGG. Clarifai platform is used in the enterprise field to manage large amount of data related to enterprise. Spatial pyramid pooling (SPP) is mainly used for isolation, classification, enumeration and identification that are used in laboratory section. VGG-based methods are used for face recognition.

Alex-net is used in this work since it consists of pre-trained model where it is pre-trained for more than 1000 classes. It cuts down the training time for new class and is easy to train within a short period of time.

2 Related Work

The conception of automatic image understanding from video for open security applications is remarkable, and it is investigated extensively. Amanjeet and Lakhwinder [6] have proposed a framework for weapon identification using SIFT and SURF features. In olden days, the primitive and conventional section in firearm recognition centers used to recognize disguised weapons through x-beam or millimetric wave scanned pictures. In [7, 8], the belongings of passenger in airports were monitored effectively by utilizing the mixture of different feature identifier and abstractor, like simple density descriptors, pattern matching and border detection approaches. In order to detect weapon effectively, complex strategies involving classifiers with boosting are also used. The feasibility of all these strategies has limitation due to the non-availability of the specific machines at all public places due to various circumstances like cost, institutional importance, etc. These systems cannot detect non-metal weapons since they depend on metal recognition.

In [9], Rohit et al. detected gun in RGB pictures using Harris interest points and FREAK points. This technique is incapable of identifying different guns in a similar scene. Rohit et al. [10] used the SURF descriptors for gun detection in RGB pictures. Color-based segmentation is done using k-means to eliminate the unrelated object. This technique is time consuming to implement in real time. Nadir et al. [11] have used bag-of-words surveillance system (BoWSS) in order to detect the weapons in surveillance systems. SIFT features are extracted and clustered using k-means. Histogram approach is adopted to count occurrence of the clusters that are fed to SVM for training purposes. It is slow and cannot be used in open regions. Ren et al. [12] introduce region proposal network (RPN) for accurate region proposal generation. RPN is trained end to end to generate high-quality region proposal that is utilized by fast R-CNN for detection. RPN and fast R-CNN are combined into single network by sharing convolutional features. RPN improves the region proposal quality that leads to increase in accuracy of object detection. Pedro et al. [13] discuss about object detection using deformable part models. Deformable part models use discriminative model for classification and is capable of identifying many object classes. Pedro et al.[14] adopt pictorial structures for object recognition using statistical approach for estimation. The desired objects are represented and identified by parts based on qualitative description. Each part is responsible for detection of objects. Once features are extracted, they are attached to each other by deformable connections. The limitation of their approach is due to error that occurs in recognizing individual parts. Ross et al. [15] use R-CNN detection algorithm for detection of objects. R-CNN is used to extract region proposals that are fed into CNN for feature extraction, and then, each region is classified using SVM. R-CNN requires large number of region proposals and requires more time to train. Jasper et al. [16] proposed selective search for object recognition. It helps to group similar regions based on size, color, and texture and shape compatibility. It is fast and achieves good accuracy but expensive since it needs to search for object in thousands of windows even for small image size.

Context-oriented object recognition and its detection are in much focus in recent years. In the proposed system described in the next section, we adopt Alex-net model for weapon detection to facilitate video forensics.

3 Proposed Work

In forensic investigation, weapon detection and classification in frames of the video is a challenging task depending on the video size and type. Automatic weapon detection in surveillance video involves computer vision tasks like weapon recognition, localization in frames, classification and weapon detection. In this work, Alex-net model is used to detect weapons shown in Fig. 2.

Alex-net has five convolutional layers and three fully connected layers along with a softmax layer. The five convolutional layers will locate the presence of a weapon and identify the location with a bounding box. In these layers, preceding layers are convolved from feature maps with the help of learnable kernels. In order to form the feature maps of output, the kernel output should pass in between the linear and the nonlinear enabling function. The feature map of each output is produced by convolving the input map with definite kernel that is merged subsequently and the last mapping of output pass through the linear or the nonlinear enabling function. Sub-sampling layer also called as pooling layer executes the less sampled functioning. There is no change in the number of feature mapping of both the input and outputs at any layer. At any instance, there is equal number of output maps for total number of input maps. Sampling function is based on the magnitude of task involved in downsampling; the magnitude of spatial extent decreases for an output

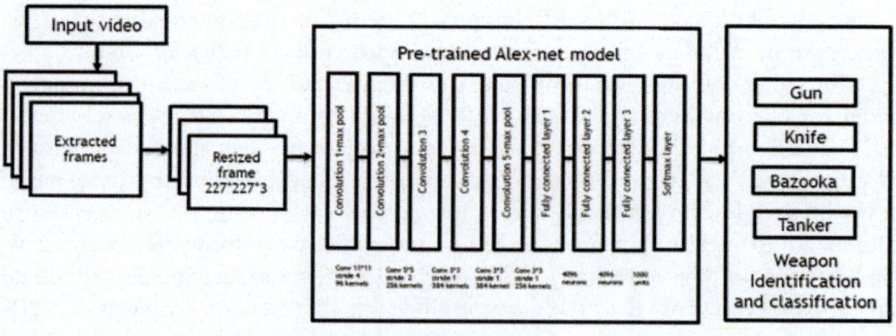

Fig. 2 Proposed weapon detection system using deep learning

map consecutively. The sampling kernel of magnitude 2*2 is applied at the first, second and fifth layers, and there is reduction in size of each output with respect to the input by half. Max-pooling and average pooling functions are then executed.

Convolution layers are followed by three fully connected layers for classification of weapon which will predict the weapon based on the kind of class. These fully connected feed-forward neural network layers enumerate the value of every class for features obtained. The feature map of the last layer is passed to the completely connected layer and is represented as a vector. The scalar value represented in the vector is then used by the softmax layer for categorization. Localization and classification task are carried out in the fully connected softmax layer. The weapons detected are then labeled based on class score obtained.

4 Experimental Results

We have created our own dataset comprising of 13,743 images obtained from ImageNet dataset and Internet sources for training the model. Each weapon image in the dataset has annotation specifying a bounding box followed by a corresponding class label for the target weapon object present in the image. Weak labeling is done since part locations or component labels are not specified. These images are of various sizes and are preprocessed before giving as an input to deep CNN model. Images are split into 70% for training and 30% for testing purpose. The sample images used for training the proposed Alex-net CNN model are shown in Fig. 3.

Performance review is an extremely critical assignment at the end of improvement procedure. It is notable to show that the proposed approach accomplishes an appropriate execution, and it accomplishes a notable improvement over the existing calculations. The proposed approach is executed by utilizing the Python 3.6. The

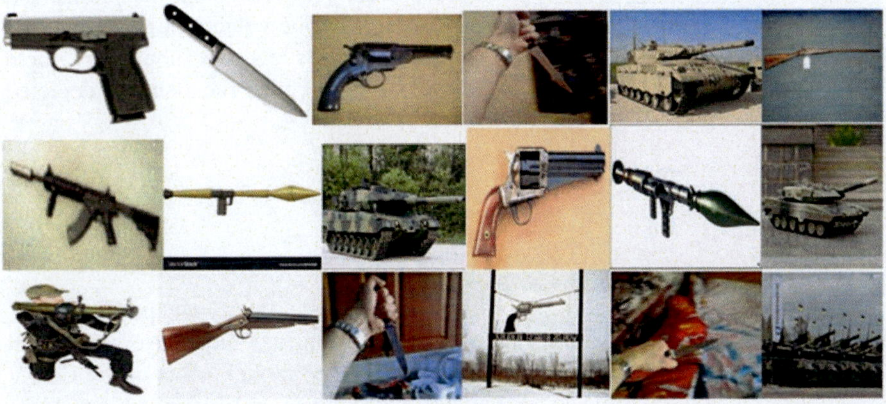

Fig. 3 Representative images from the dataset used for pre-training the Alex-net

usage is running under 64-bit Windows 10 operating system on a PC with Intel Core i5 processor and 8 GB RAM.

In order to estimate the performance of the proposed framework, accuracy, sensitivity and specificity of the proposed model are measured using the rate of true positive (TP), false positive (FP) and precision.

True Positive Rate$_{\text{weapon}}$ $\left(\text{TPR}_{\text{Weapon}}\right)$: It is the assurance of proportion of those pictures containing weapons, which are precisely seen by the framework and conveyed by Eq. (1). It is otherwise called sensitivity or recall in AI.

$$\text{TPR}_{\text{Weapon}} = \frac{\text{TP}}{(\text{TP} + \text{FN})} \tag{1}$$

where FN is false negative which indicates the miss in correct detection and classification of weapons.

False Positive Rate$_{\text{weapon}}$ $\left(\text{FPR}_{\text{Weapon}}\right)$: It is the assurance of proportion of those negative pictures, which are wrongly seen by the system.

$$\text{FPR}_{\text{Weapon}} = \frac{\text{FP}}{(\text{FP} + \text{TN})} \tag{2}$$

where TN is true negative which corresponds to failure in correct detection and classification of weapons.

Accuracy: It is calculated to determine the closeness of the measured value to the actual value for the class results obtained after training.

$$\text{Accuracy} = \frac{(\text{TP} + \text{TN})}{(\text{TP} + \text{TN} + \text{FP} + \text{FN})} \tag{3}$$

In the next stage, frames of the video input are compared with the pre-trained Alex-net model to predict the four types of weapons. The video inputs given to the proposed system are obtained from video sharing Web site—YouTube. The sample frames for the weapon-based videos and the output obtained for them are depicted in Fig. 4 (Table 1).

5 Conclusion

This paper presents a deep learning-oriented approach to identify weapons like gun, knife, bazooka and tanker from videos. Deep learning technique using Alex-net increases the overall recognition rate due to pre-trained model. Most of the CCTV systems generate low-quality and low perseverance images, and our pre-trained Alex-net model can effectively identify the weapons. Experimental outcome shows that the proposed approach attains an accuracy of 95% for gun, 93% for bazooka, 94%

Fig. 4 Results of the proposed weapon detection system. **a–d** Sample input frames taken from various YouTube videos. **e–h** Output frames for the corresponding input frames

Table 1 Classification report

Weapons	Total images	Positive predicted value (in %)	False detected rate (in %)	Accuracy
Gun	3,848	95.19	4.81	95.99
Bazooka	3,280	93.39	6.61	93.32
Knife	3,520	94.36	5.64	94.40
Tanker	3,095	92.92	7.08	92.79

for knife and 92% for tanker detection. Results can be further improved by increasing the volume of the dataset with variants of weapons.

Acknowledgements We wish to thank JSS MahaVidyapeetha, Karnataka, India, for the constant support and encouragement in our work.

References

1. F. Gelana, A. Yadav, Firearm detection from surveillance cameras using image processing and machine learning techniques, in *Smart Innovations in Communication and Computational Sciences* (Springer, Singapore, 2019), pp. 25–34
2. R. Olmos, S. Tabik, F. Herrera, Automatic handgun detection alarm in videos using deep learning. Neurocomputing **275**, 66–72 (2018)
3. M. Singleton, B. Taylor, J. Taylor, Q. Liu, Gun identification using tensorflow, in *International Conference on Machine Learning and Intelligent Communications* (Springer, Cham, 2018), pp. 3–12

4. G.K. Verma, A. Dhillon, A handheld gun detection using faster r-cnn deep learning, in *Proceedings of the 7th International Conference on Computer and Communication Technology* (2017, November), pp. 84–88
5. Y. Elmir, S.A. Laouar, L. Hamdaoui, Deep learning for automatic detection of handguns in video sequences, in *JER* I (2019, April).
6. A. Kaur, L. Kaur, Concealed weapon detection from images using SIFT and SURF, in *2016 Online International Conference on Green Engineering and Technologies (IC-GET)*. IEEE (2016, November), pp. 1–8
7. G. Flitton, T.P. Breckon, N. Megherbi, A comparison of 3D interest point descriptors with application to airport baggage object detection in complex CT imagery. Pattern Recogn. **46**(9), 2420–2436 (2013)
8. R. Gesick, C. Saritac, C.C. Hung, Automatic image analysis process for the detection of concealed weapons, in *Proceedings of the 5th Annual Workshop on Cyber Security and Information Intelligence Research: Cyber Security and Information Intelligence Challenges and Strategies* (2009, April), pp. 1–4
9. R.K. Tiwari, G.K. Verma, A computer vision based framework for visual gun detection using Harris interest point detector. Procedia Comput. Sci. **54**, 703–712 (2015)
10. R.K. Tiwari, G.K. Verma, A computer vision based framework for visual gun detection using SURF, in *2015 International Conference on Electrical, Electronics, Signals, Communication and Optimization (EESCO)* (2015, January), pp. 1–5
11. N.B. Halima, O. Hosam, Bag of words based surveillance system using support vector machines. Int. J. Secur. Appl **10**(4), 331–346 (2016)
12. S. Ren, K. He, R. Girshick, J. Sun, Faster r-cnn: towards real-time object detection with region proposal networks, in *Advances in Neural Information Processing Systems* (2015), pp. 91–99
13. P.F. Felzenszwalb, R.B. Girshick, D. McAllester, D. Ramanan, Object detection with discriminatively trained part-based models. IEEE Trans. Pattern Anal. Mach. Intell. **32**(9), 1627–1645 (2009)
14. P.F. Felzenszwalb, D.P. Huttenlocher, Pictorial structures for object recognition. Int. J. Comput. Vision **61**(1), 55–79 (2005)
15. R. Girshick, J. Donahue, T. Darrell, J. Malik, Rich feature hierarchies for accurate object detection and semantic segmentation, in *Proceedings of the IEEE Conference on Computer Vision and Pattern Recognition* (2014), pp. 580–587
16. J.R. Uijlings, K.E. Van De Sande, T. Gevers, A.W. Smeulders, Selective search for object recognition. Int. J. Comput. Vision **104**(2), 154–171 (2013)

2D Pose Estimation of Subject Body via Deep Neural Networks

K. Seemanthini and S. S. Manjunath

Abstract A large part of a computer vision project's task is to identify an object in a camera and also to identify its position and orientation of a specific system of coordinates. The pose of an object is the combination of its position and orientation. Efficient pose estimation is a much desired trait of any application in the computer vision sector. We propose an industrially efficient model which will effectively estimate the pose of an object in any given image. Our team proposes a new methodology toward the field of pose estimation which employs an object detection algorithm You Look Only Once (YOLOv3) trained to detect human body parts and limbs. These detected regions are then used to obtain a 2D pose of the human in focus. Our approach is to adapt the existing YOLO network supplied with new training images with annotated body parts of humans and changes to the network configuration file. A post-processing step would be to convert the bounding boxes to a few pixels or to obtain the centroid of the bounding boxes to draw a point indicating the position of the limbs and draw lines to connect the limbs together.

Keywords Deep learning approach · Pose estimation · YOLO · Darknet

1 Introduction

Human pose estimation refers to computer vision techniques that detect human figures in images and video, so that one could determine where that person's joints are in the image or video. This technology is not used to recognize the person in an image, i.e., there is no personal identifiable information associated to pose detection. The algorithm is simply estimating where key body joints are.

K. Seemanthini
Dayananda Sagar Academy of Technology and Management, Bangalore, India

S. S. Manjunath (✉)
ATME College of Engineering, Mysore, India

V. Bhateja et al. (eds.), *Data Engineering and Intelligent Computing*,
Advances in Intelligent Systems and Computing 1407,
https://doi.org/10.1007/978-981-16-0171-2_48

The problem statement can be defined as "localization of human joints in images or videos." It is also defined as the search for a specific pose in space of all articulated poses. Depending on the number of people being tracked, pose estimation can be classified into single-person and multi-person pose estimation. The pose of the person is estimated by showing a skeletal figure of the subject with their joints presented as nodes of the graph.

The main stream of work in this field has been motivated mainly by the first challenge, the need to search in the large space of all possible articulated poses. Inferring the pose of multiple people in images, especially socially engaged individuals, presents a unique set of challenges.

Our team proposes a new methodology toward the field of pose estimation which employs an object detection algorithm You Look Only Once (YOLOv3) trained to detect human body parts and limbs. In our scenario, the object happens to be a person, and the task is to identify the position and the orientation of the person in front of the camera. There are two popular approaches to solving this class of problems.

Top-down: The entire person is followed by the limbs of that person.

Bottom-up: The person's limbs and body parts are detected directly by-passing the first step.

These detected regions are then used to obtain a 2D pose of the human in focus. The primary choice of YOLO is the impressive speed at which it processed video frames.

Natively, YOLO network was trained on COCO and ImageNet, and with its current implementation, YOLO cannot be applied to our problem statement. Our approach is to adapt the existing YOLO network supplied with new training images with annotated body parts of humans and changes to the network configuration file. A post-processing step would be to convert the bounding boxes to a few pixels or to obtain the centroid of the bounding boxes to draw a point indicating the position of the limbs and draw lines to connect the limbs together.

2 Related Works

Redmon et al. [1] proposed a state-of-the-art approach to solve the problem of object detection, namely "You Only Look Once" (YOLO). This proposed way of detecting multiple objects in a given image proved to be faster and most efficient than the existing models of the time. This algorithm finds all objects in an image grid simultaneously. When working with the YOLO algorithm, we are not searching for interesting regions in our image that could potentially contain an object. YOLO sees the entire image during training and test time, so it implicitly encodes contextual information about classes as well as their appearances, unlike the sliding window or region-based techniques, thus making less than half the number of background errors compared to Fast R-CNN. It uses features from the entire image to predict each bounding box. It also predicts all bounding boxes across all classes for an image simultaneously and the bounding boxes and the class probabilities for these boxes.

The authors made significant improvements to the original YOLO algorithm by adopting techniques like batch normalization, higher resolution classification, anchor boxes, Darknet-19, etc. This improved version was called the YOLO9000 [2]. The name 9000 comes from the fact that the improved algorithm was used to train over 9000 images across different datasets.

This was followed by the next version, YOLOv3 [3]. This state-of-the-art algorithm included some major improvements such as bounding box prediction, class prediction, feature pyramid network, Darknet-53; which made use of 53 convolution layers in its network.

We have used the YOLO algorithm as the basis of our project where the bounding boxes of the objects detected by the algorithm are the location of the 13 classes (head, torso, hip, right shoulder, right elbow, eight wrist, right knee, right ankle, left shoulder, left elbow, left wrist, left knee and left ankle) in a human body.

3 Methodology

Our methodology proposes the use of the YOLOv3 network to identify the limb locations that make up the subject body. The data which is used to train the YOLOv3 network consists of a mixture of sports images and numerous frames of individuals extracted from a video.

The frames and images are manually annotated using annotation software which highlight various joint locations which are visible in the frame. The annotations for each frame are saved in a text file with the same filename as its respective frame.

The native YOLO network was trained on the COCO dataset which comprises 80 classes of objects. Since our work is inspired from the working of YOLO, with suitable changes to the YOLO architecture, it can be trained on our dataset for limb detection. The changes made to the configuration of the YOLO network are discussed further in this paper. Our dataset is split into training and validation before training our modified YOLO network on these frames.

Once trained on the required number of iterations, we test the trained network on the validation images. We also feed in the complete video from which the frames were extracted, and the network predicts the joint locations for every frame in the video. The output frame is stitched back into a video.

The online version of the volume will be available in LNCS Online. Members of institutes subscribing to the Lecture Notes in Computer Science series have access to all the pdfs of all the online publications. Non-subscribers can only read as far as the abstracts. If they try to go beyond this point, they are automatically asked, whether they would like to order the pdf, and are given instructions as to how to do so (Fig. 1).

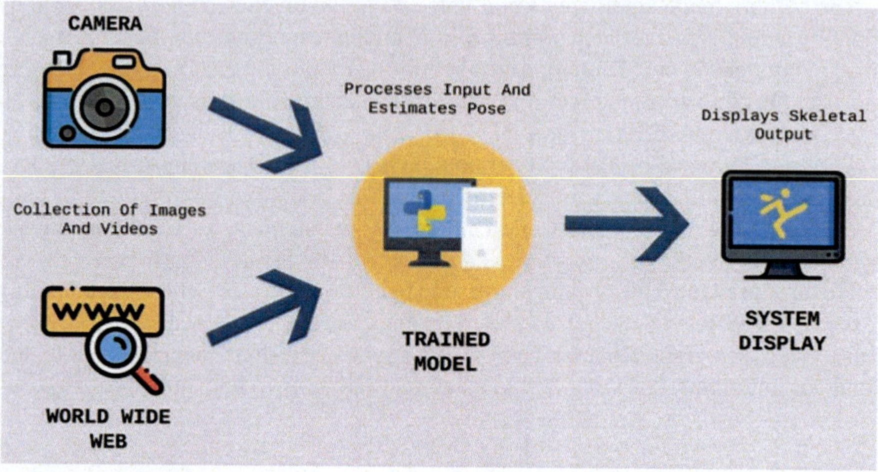

Fig. 1 System architecture

4 Dataset Details

The dataset has been created manually by searching for images of good quality in which each of the joints is fairly visible. To ensure diversity of images, we chose multiple frames from different sporting events. Keeping in mind that the model would learn better when it has a context to learn for each image, we considered multiple videos. These videos contained a single subject, performing a variety of poses. We extracted multiple frames from each video and further expanded our dataset. The dataset has a total of 610 images.

The correct BibTeX entries for the Lecture Notes in Computer Science volumes can be found at the following Website shortly after the publication of the book: https://www.informatik.uni-trier.de/~ley/db/journals/lncs.html.

5 Annotation Details

Annotations describe the coordinates and the size of the rectangle enclosing the object or limb in our case in the input image or frame. To be more specific, annotations for a particular object comprise <height>, <width>, <x>, <y>, <class>. Here, **height** and **width** specify the dimensions of the rectangular annotation enclosing the limb. **x** and **y** are the centroid locations to the rectangle. Class is self-descriptive and identifies the class name of the limb, e.g., right shoulder, torso, hip, etc.

There are many open-sourced annotations software available which create annotations for a set of images. However, we must convert the annotations into a specific format, since we are dependent on Darknet (YOLOv3 framework); we require annotations which are compatible with its format. Darknet requires each input training image to have a text file with the same filename as its image. The text file consists of the annotations of all the objects present in its respective image with the format as specified above.

6 Annotation Details

Anchor boxes are a set of predefined bounding boxes of a certain height and width. These boxes are defined to capture the scale and aspect ratio of specific object classes you want to detect and are typically chosen based on object sizes in your training datasets.

The concept of anchors might be tricky to understand if one does not have prior understanding of the YOLO algorithm. In brief, all the annotations created are plotted on a graph with its respective **x and y** values on a 2D plane. We obtain a bunch of data points, and we then apply k-means clustering to cluster the data points into a specified number of clusters. In our work, we clustered it into five clusters. Anchor values are the centroids of all the clusters.

Thus, by applying k-means algorithm on the input bounding boxes and obtaining the centroids of these clusters help us select ideal predefined size for the boxes which have captured the scale and aspect ratio for our dataset.

7 Training of YOLO Object Detector

The prepared data is used to train the YOLO model. The model is trained to detect all the classes that have been specified in the configuration file. Here, the model is trained to detect 13 classes in total. The network detects the joints by dividing each image into regions and predicts bounding boxes and probabilities for each region. These bounding boxes are weighed by the predicted probabilities. Darknet, the framework that is used to implement YOLO, prints out the objects that it detected, its confidence, and how long it took to find them.

Darknet framework provides a wide-range of choices to select the YOLO network architecture starting from Darknet-19 to Darknet-153. It also provides a YOLO network based on ResNet or DenseNet architecture. Considering the size of our custom dataset and the limitations, we faced in our hardware specifications, and we decided to use Darknet-19 also known as Tiny-YOLO.

Tiny-YOLO model: Tiny-YOLO is a variation of YOLO architecture which is designed to be even faster than YOLO. It is approximately 442% faster than the vanilla YOLO architecture. It achieves upward of 244FPS on a single GPU. The

Layer	Type	Filters	Size/Stride	Input	Output
0	Convolutional	16	3 × 3/1	416 × 416 × 3	416 × 416 × 16
1	Maxpool		2 × 2/2	416 × 416 × 16	208 × 208 × 16
2	Convolutional	32	3 × 3/1	208 × 208 × 16	208 × 208 × 32
3	Maxpool		2 × 2/2	208 × 208 × 32	104 × 104 × 32
4	Convolutional	64	3 × 3/1	104 × 104 × 32	104 × 104 × 64
5	Maxpool		2 × 2/2	104 × 104 × 64	52 × 52 × 64
6	Convolutional	128	3 × 3/1	52 × 52 × 64	52 × 52 × 128
7	Maxpool		2 × 2/2	52 × 52 × 128	26 × 26 × 128
8	Convolutional	256	3 × 3/1	26 × 26 × 128	26 × 26 × 256
9	Maxpool		2 × 2/2	26 × 26 × 256	13 × 13 × 256
10	Convolutional	512	3 × 3/1	13 × 13 × 256	13 × 13 × 512
11	Maxpool		2 × 2/1	13 × 13 × 512	13 × 13 × 512
12	Convolutional	1024	3 × 3/1	13 × 13 × 512	13 × 13 × 1024
13	Convolutional	256	1 × 1/1	13 × 13 × 1024	13 × 13 × 256
14	Convolutional	512	3 × 3/1	13 × 13 × 256	13 × 13 × 512
15	Convolutional	255	1 × 1/1	13 × 13 × 512	13 × 13 × 255
16	YOLO				
17	**Route 13**				
18	Convolutional	128	1 × 1/1	13 × 13 × 256	13 × 13 × 128
19	Up-sampling		2 × 2/1	13 × 13 × 128	26 × 26 × 128
20	**Route 19 8**				
21	Convolutional	256	3 × 3/1	13 × 13 × 384	13 × 13 × 256
22	Convolutional	255	1 × 1/1	13 × 13 × 256	13 × 13 × 256
23	YOLO				

Fig. 2 Tiny-YOLOv3 network structure

model size is less than 50 MB and has a faster inference speed. Tiny-YOLOv3 has 24 layers and implements a Darknet-19 structure (Fig. 2).

The YOLOv3-Tiny creates a feature pyramid with strong semantics at two scales by adopting subsampling layers and a fusion approach. As shown in Figure, the size of the two scales is 13 × 13 and 26 × 26, which are obtained in the detection of ordinary size targets, respectively. Finally, two scales are merged in the end. The architecture is constructed as a feature pyramid, wherein predictions are independently made on each level. The feature pyramid has rich semantics via a top-town pathway and lateral connections. In this way, YOLOv3-Tiny has the ability to detect small targets.

8 Estimation of Pose

Once the model has been trained, the weights of the network are saved in a weights file. A video is fed as input to perform the task of pose estimation. As the model detects the pose of the subject in images, the frames of the video are fed as input to the detector. A Python script is responsible for extracting all the frames from the video and feeding them to the detector in a linear fashion.

The model is loaded with the weights file of the model that was trained on our custom dataset. This model is then presented with each input frame for performing the task of joint detection. Using the network weights, the model predicts these joints by drawing multiple bounding boxes around the joint. Each of these bounding boxes that enclose the joint has a confidence score that has been assigned to it by the model. Based on these confidence scores of all bounding boxes of each predicted joint, the model returns a single bounding box as the final bounding box. The bounding box is presented as a list of its corner points and the box's height and width. A Python script is used to calculate the centroid of each predicted bounding box. This centroid represents the position of that respective joint. The joints are then separated into four different arrays. Here, each array represents a single line of joint connections. For example, line1 contains the joints: head, torso and hip. Using the OpenCV library, these joints are then connected by lines. This results in the construction of a skeletal figure that has been drawn on top of the frame with joints presented as points. This skeletal figure represents the pose of the subject in that particular frame. Similarly, all the frames of the input video are drawn with their respective poses.

Once all the frames have been returned with their respective skeletal figure drawn on them, all the frames are appended in their respective input order. This results in the creation of a video with each frame containing the pose of the respective subject. This output video is then saved locally, which allows for viewing of the output.

9 Data Flow

Data Collection: The process of gathering data was done manually by finding images on the Internet of people in different poses of diverse ethnicity. Mainly, we obtained our data by extracting frames from videos of people dancing which provided us with frames containing various different poses. The videos were around 1 min in length, and every 4–5 frames of the video were extracted using VLC media player (Fig. 3).

Annotation: This is the technique of creating trainable data for a neural network by manually drawing bounding boxes around the region of interest and associating them with the classes that we want to be predicted. The format of the annotation is the class ID of the joint followed by the x and y coordinates of the top-left corner of the bounding box and then the width and height of the box itself which provides enough information about the location and the type of joint within the box to the network. This information is stored in a text file of the same name as that of the image.

Train–Test split: This is an integral part of the process as it splits the existing dataset into two clear groups of training and testing dataset. By default, the split ratio is 20% of the total number of datasets to be used as test data. The data is also shuffled before splitting so as to make the data more diverse.

Anchor Calculation: Anchor boxes are an essential part of prediction as they are responsible for accurately localizing the object or joint to be predicted. It does so by

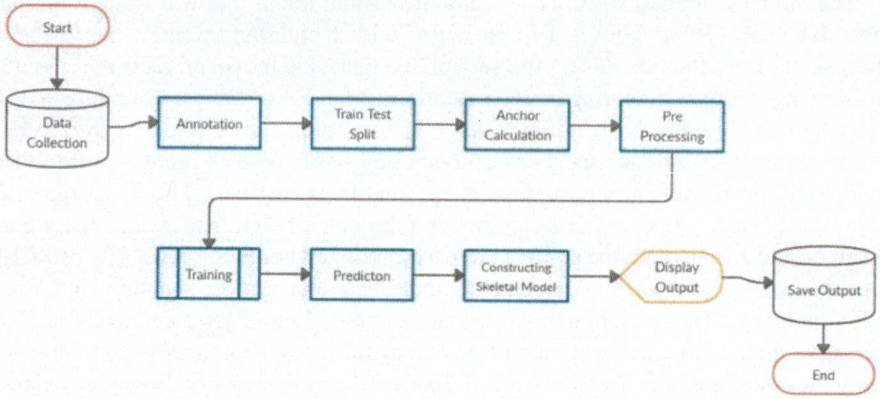

Fig. 3 Data flow diagram

calculating the Intersection Over Union (IOU) of various boxes to find the one with the highest overlap of boxes. Initially calculating the anchors for each bounding box helps the model learn more precisely about each of the joint classes.

Pre-processing: The images that are fed into the network are altered in a way that best fits the training model. The resolution is changed to multiple of 32, and the image saturation and contrast are altered.

Training: Once the dataset is ready, they are used to train the YOLOv3 model. The model uses a Darknet framework and is made up of layers trained on ImageNet and layers of detection networks. The configurations of the network can be altered in the .cfg file. The weights are generated after every 1000 epochs and at the point of stopping the training.

Predictions: Once the model is trained and the weights are obtained, test images are fed to the model to predict the joints of the subject in the image. The resulting output will contain the bounding boxes over the different joints being predicted.

Constructing skeletal model: The resulting bounding boxes predicted on an image can be used to create a skeletal model. The centroids of each bounding box are calculated which acts as keypoints. With a suitable logic, the different keypoints are connected to obtain the skeletal frame of the subject in the image.

Output: The resultant output is a skeletal model overlaid on top of the image. This can also be done for videos where each frame of the video is considered an image to be predicted.

10 Training Details

After initial research, we decided to go ahead and implement the concept of transfer learning. Transfer learning would ensure that our model learned with more diversity and finally increase its performance. Transfer learning also seemed viable as it allowed us to train on more data, while we faced hardware limitations which would be necessary for training the model with large datasets.

At first, we loaded the model with the pre-trained YOLO weights and began training the model on a small section of our dataset. Since athletes are involved in a lot of physical movement, we decided to go with creating a dataset of images of various athletes of different sports. To add more continuity in the poses, we also added frames that we extracted from a dance video. Our initial dataset was made of these images and had 160 frames. We trained the model using this dataset for about 8000 epochs. At this stage, the model showed a loss of 6.47. Having overfitted the model with this data, it performed very well on our dataset and predicted the pose of each frame with fairly high accuracy.

For the next stage of training, we decided to further implement transfer learning and loaded the model with the network weights that we retrieved from the previously trained model. This time, we added frames from two more videos to the original dataset of 160 frames and created a new dataset of 610 images. We trained our model with this dataset and with the loaded weights of the previously trained model and ran it for about 8000 epochs. Along with training the model with a larger dataset, we also managed to bring down the loss of the exhibited by the model down to 4.97.

To train our YOLO network, we were dependent on Google Colab which is a free cloud-based service to train our machine learning models on. Colab provides Tesla K-80 GPUs for a runtime session of 12 h after which the VM loses its memory. Since Google Colab provides us with Tesla K-80 GPUs, we decided to exploit the free service opportunity more by installing CUDA drivers. "Compute Unified Device Architecture." **CUDA** is a parallel computing platform developed by NVIDIA and introduced in 2006. **CUDA** enables developers to speed up compute-intensive applications by harnessing the power of GPUs for the parallelizable part of the computation.

The effect of installing CUDA was witnessed immediately with faster computations and optimized RAM memory utilization. We could also train the network faster which enabled us to see the results faster. We later applied transfer learning on the latest saved weights of a checkpoint.

After every checkpoint, we increased the width and height of input from 416 $\times$ 416 to 608 $\times$ 608. The effect of this is an increase in precision of the detected objects and also increase in batch size from 64 to 128 after every checkpoint. Sometimes, this led to out of memory error.

Table 1 Preliminary results

Average loss	6.847788
Threshold value operating on	0.2
Initial learning rate	0.000
Final learning rate	0.001
Batch size	64

11 Results

The preliminary results were improved by making a few changes to the configuration file. These changes help the model to better train on the dataset available (Table 1).

- The first change done to improve results was to change the network resolution height and width from 416 × 416 to 608 × 608 and 832 × 832 which helped in increasing the precision of detections.
- Adding flip = 0 played a significant role as it allowed the model to distinguish left and right joints as separate classes. The prediction of the joints was more accurately distinguished from their counterparts
- The batch size and subdivisions size were also altered. The batch size was increased from 64 to 128 in one of the iterations of the code which resulted in the model being trained with 128 images in every single epoch. This added a considerable amount of performance overhead.
- Pre-trained model weights added made the training process more efficient due to the fact that the fact that pre-trained weights were a result of training the model on other datasets which already gave the model an idea about the objects it was going to predict.

The resulting model trained with a lower loss at similar epoch check points (Figs. 4 and 5; Table 2).

There are certain cases where the joints are not localized at all or are classified under the wrong class. These situations arise when the joint is in fast motion which causes blurring of that particular joint, or sometimes, the joint is not visible properly, i.e., partially visible or the joint blends with the background of the image.

12 Conclusion

The proposed system deals with an efficient way of estimating the pose of a person in the video, with it being able to join various limbs that are detected by the model and giving us a skeletal figure of the subject in the video. The system employs a trained YOLOv3 model that detects the joints of the subject in the video. This deep learning approach proves to be better than the classical approach which used the pictorial structures framework where an object was represented by a collection of

Fig. 4 Preliminary results

Fig. 5 Final results

Table 2 Current results

Average loss	4.975247
Threshold	0.1
NMS threshold	0.3
Initial learning rate	0.000
Final learning rate	0.001
Batch size	64

"parts" arranged in a deformable configuration. Here, a part is an appearance template which matches in an image and so comes with the limitation of having a pose model not depending on image data. The performance of the system can be improved with the availability of larger datasets of high-quality, filtered data and also with the application of larger computational power. Improved performance will ensure that the model effectively detects joints that might be obstructed by small objects such as clothes and hair and also ensures higher confidence percentages while detecting all joints.

This system can be employed to develop a number of applications that take advantage of pose estimation such as video games or for graphics development. Pose estimation can prove to be an important asset in the field of security where efficient pose estimation might help predict a crime and allow for law enforcement to perform necessary steps to avoid it from happening.

References

1. J. Redmon, S. Divvala, R. Girshick, A. Farhadi, You only look once: unified, real-time object detection, in *Proceedings of the IEEE Conference on Computer Vision and Pattern Recognition* (2016), pp. 779–788
2. J. Redmon, A. Farhadi, YOLO9000: better, faster, stronger, in *Proceedings of the IEEE Conference on Computer Vision and Pattern Recognition* (2017), pp. 7263–7271
3. J. Redmon, A. Farhadi, Yolov3: an incremental improvement. arXiv preprint arXiv:1804.02767 (2018)

Efficient COVID-19 Diagnosis Approach Using Multi-scale Retinex and Convolution Neural Network

P. Sudir, M. C. Hanumantharaju, and V. N. Manjunath Aradhya

Abstract Post coronavirus disease (COVID-19), pandemic has triggered and provided an opportunity to researchers to contribute towards diagnosis and prognosis of coronavirus disease. As it was found from the literature review that computer vision could assist tremendously in solving complex problems related to healthcare and contributing towards fighting against controlling COVID-19. Recent works have motivated the authors to propose a convolution neural networks (CNN)-based solution towards classifying of chest X-ray images as of COVID and non-COVID cases. Prior to the classification, the images undergo pre-processing using popular image enhancement technique, namely retinex algorithm, and the results obtained from the enhancement algorithm are utilized in the classification. Experimental results conducted on 25 images from Kaggle's chest X-ray image (non-COVID case) dataset and 25 images from Dr. Joseph Cohen dataset (COVID case) have shown that proposed method has good accuracy and proves as an efficient algorithm towards diagnosis of COVID-19 pandemic.

Keywords COVID · Convolution neural networks · Machine learning · Image enhancement · VGG16

P. Sudir (✉)
Department of ECE, SJCIT, Chikkaballapur, India

M. C. Hanumantharaju
Department of ECE, BMS Institute of Technology and Management, Bangalore, India
e-mail: mchanumantharaju@bmsit.in

V. N. Manjunath Aradhya
Department of Computer Applications, JSS Science and Technology University, Mysuru, India
e-mail: aradhya@sjce.ac.in

V. Bhateja et al. (eds.), *Data Engineering and Intelligent Computing*,
Advances in Intelligent Systems and Computing 1407,
https://doi.org/10.1007/978-981-16-0171-2_49

1 Introduction

Various sectors in the present age, such as transportation, e-commerce, education, health care, electronic communication, etc., are undergoing rapid changes due to advent of artificial intelligence and machine learning. Health sector in particular has seen and possibly will see a rapid development in the future specifically in the area of prevention [1], diagnosis [2] and curing of diseases. Genomic analysis [3], medical decision support [4, 5], 3D printing of medical equipments, human/animal body parts [6], designer babies [7], ageing research [8], brain computer interface [9], drug design [10], etc., are just few, where tremendous usage of artificial intelligence (AI) and machine learning (ML) has been witnessed.

Advancements in machine learning and data science are well adopted in health-care applications to extreme levels due to huge amount of health data generated by patients such as vaccinations, diagnostic report, X-ray results, CT scan reports, DNA sequences, past medical history and so on. One vital usage of ML approach is in the digital diagnosis. Pattern classification and segmentation algorithms serve as second pair of eye to doctors to extract knowledge about patient's health reports such as X-ray report, CT scan report, ultrasound report, etc. One such popular and extensively researched area is in breast cancer detection [11].

With the advent of COVID-19 pandemic in early 2020, diagnostic testing using AI and ML techniques has taken giant leap as pandemic has spread worldwide, and prompt diagnostic is the present situation demand [12]. In corona infection case, chest X-ray [13, 14] is a discriminating tool, and it decides initially the clinical situation of a patient. The role of radiologist is to facilitate doctors to take decision dealing with COVID-19 infection visible abnormalities like ground glass patterned areas, lesions, consideration and crazy paving patterns in X-ray speak about the percentage of infection; hence, computer-based vision approaches play a vital role in COVID-19 diagnosis and help in understanding of the abnormalities throughout the disease course. Solving problem of recognizing and classifying abnormalities mentioned above involve various iterative approaches to find the best model for classification. Sethy et al. [15] proposed a deep feature extraction-based method to detect COVID-19 which relies on the features extracted from a pre-trained CNN. The deep features are fed to the classifier for training purpose. The deep features obtained from each CNN networks are used by SVM classifier. After that, the classification is performed, and the performance of all classification models are measured. Based on the result obtained, it was found that better machine learning algorithms and neural network architectures can increase the rate of satisfactory result that provided X-ray images are pre-processed, and then, deep features are extracted; the methodology justifying this is explained next.

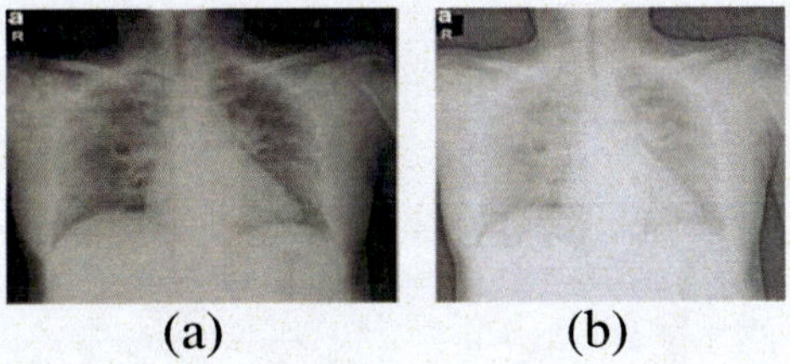

Fig. 1 Chest X-ray image. **a** Before enhancement. **b** After enhancement

1.1 *Multi-scale Retinex-Based Pre-processing Method*

The proposed pre-processing technique in this paper uses the most popular image enhancement algorithm, namely multi-scale retinex algorithm. This algorithm is well known as Multi-scale Retinex (MSR) and was initially proposed by Jobson et al. [16] to enhance the aerial, medical and natural sceneries. X-ray images in Digital Imaging and Communications in Medicine (DICOM) format are first converted into HSV domain since HSV space separates value or intensity from hue and saturation. HSV domain used in the enhancement of X-ray images preserves hue, stretches the saturation part and highlights the value component. Some of the important papers on ratinex for image enhancement can be seen in [17–19]. In this paper, the parameters in the retinex algorithm are adjusted such that it is suited for medical images. The parameters intensify the details of chest X-ray images of COVID-19 patients and assists in accurate classification. The retinex method not only enhance the details of X-ray images but also improves overall appearance of the image. As it is presented in Fig. 1 and Table 1 results of COVID and non-COVID classification that the retinex-based pre-processing renders a better classification.

1.2 *Introduction to CNN*

In machine learning, CNN is complex feed forward neural networks used for object classification and recognition thanks to their high accuracy, visual cortex features used in CNN; CNN's are used extensively from Facebook's photo tagging to self-driving cars [20]. Typical CNN has four major segments.

1. **Convolution layer**: Here, learned features convolves with input data using 2D convolution layer.
2. **RELU layers**: Rectified linear activation transforms the summed weightage input from the node into the activation of the node/output for that input. It is easier to use to train and achieves better performance.
3. **Pooling layers**: This layer combines the outputs of clusters of neurons into a single neuron in the next layer.

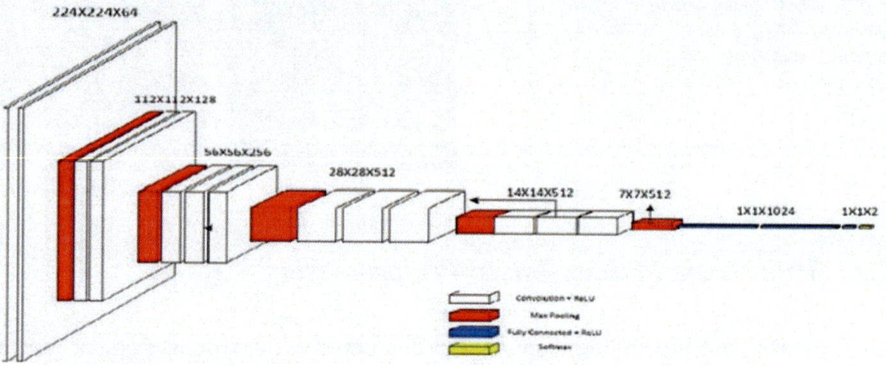

Fig. 2 VGG architecture

4. **Fully connected layer**: The features generated by convolution and pooling are fed into fully connected NN structure, which initiates the decisive classification decision.

Simonyan et al. [21] from the University of Oxford proposed a VGG16 convolutional neural network model proposed by in the paper 'Very Deep Convolutional Networks for Large-Scale Image Recognition'. It was an improvement over Alex Net because it replaced large kernel-sized filters. It has proved to be excellent among various vision model architecture [22]. It has unique characteristic of large number of hyper-parameter with convolution layers of 3×3 filter with stride 1, padding and maxpool layer of 2×2 filter of stride 2. At the end, it incorporates a two fully connected layer followed by a softmax for output. It has 16 weighted layers. Many deep learning image classification problems utilize VGG16, as it is a great building block for learning purpose and easy to implement (Fig. 2).

2 Setting Up of Diagnostic Models

1. 25 Images from Kaggle's chest X-ray image [23] (includes healthy and pneumonic case) dataset were chosen for negative COVID-19 cases.
2. 25 images form Dr. Joseph Cohen [24] Dataset (actual COVID affected case) was chosen for COVID positive cases.

The following steps were involved in creating COVID-19 X-ray image dataset (Fig. 3).

1. Metadata-CSV file was parsed from Dr. Cohen's repository.
2. All rows that are (i) Positive for COVID-19 case. (ii) Posterior anterior (PA) view of lungs

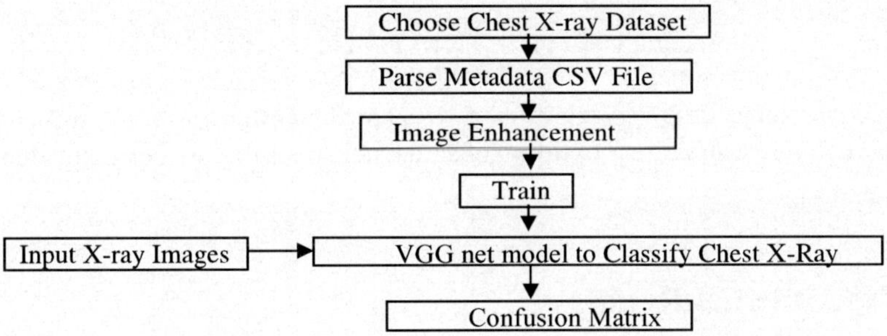

Fig. 3 Flowchart of proposed methodology

Next, COVID-19 training was performed using keras, tensor flow and deep learning. Initial learning rate of 0.00125 epochs and batch size hyper-parameter of 8 was fixed. Total of 100 images were considered for experiment purpose. Images were considered in order, enhanced and resized to 224 × 224 pixels to feed to CNN. Data augmentation is performed by setting the random image rotation setting to 20 degrees clockwise or counter clockwise. Next, VGG net model is set-up for fine tuning. VGG 16 network are instantiated with weights pre-trained on image net. The network is compiled with learning rate decay and Adam optimizer. Confusion matrix to derive the accuracy, sensitivity and specificity is generated.

Next, key metrics are used to measure the efficiency of the model which is derivative measurement metrics that are chosen.

- True positive (TP) number of correctly classified COVID patterns.
- True negative (TN) number of correctly classified patients who are healthy.
- False positive (FP) number of misclassified patients who are healthy.
- False negative (FN) no. of misclassified patients who are affected by COVID.

Based on these, following metrics are defined.

- **Precision**—ratio of correctly classified patients with COVID to the total patients classified as affected by COVID.

$$\text{Precision} = \text{TP}/\text{TP} + \text{FN} \tag{1}$$

- **Recall**—ratio of correctly classified COVID patients to patients who are affected by COVID.

$$\text{Recall} = \text{TP}/\text{TP} + \text{FN} \tag{2}$$

- Accuracy is the fraction of predictions the designed model got right.

$$\text{Accuracy} = \text{Number of correct predictions}/\text{total number of predictions} \tag{3}$$

$$\text{Accuracy} = (\text{TP} + \text{TN})/(\text{TP} + \text{TN} + \text{FP} + \text{FN}) \quad (4)$$

- Sensitivity measures the proportion of actual positives that are correctly identified.
- Specificity measures the proportion of actual negatives that are correctly identified as such.

3 Experimental Results

Experiment was conducted in two ways. First was to check the necessity of enhancement before diagnosis. The 40 COVID and Non-COVID dataset images were fed without enhancement, and model was trained. Next, 10 images (not enhanced) were chosen for testing. Next, the training images were enhanced using multi-scale retinex algorithm, and the model was built. Then, enhanced images (Testing) were fed to model to diagnose, and it was found that that before enhancement, 10 COVID X-ray images were classified as non-COVID X-ray images but after enhancement these reduced to 4 thus increasing the accuracy by 60% and proving the necessity to enhance X-ray Images and then diagnose.

Below table shows that accuracy of diagnosis increases for COVID cases after enhancement of X-ray images.

Tables 1 and 2 shows that enhancement increases precision and recall to a considerable level.

Second was to check the efficiency of proposed model; hence, the work has been compared with Sethy and Behera [15] which proposes the deep learning-based methodology for detection of corona virus infected patient using X-ray images. The support vector machine classfourfies the corona affected X-ray images from others

Table 1 Result of proposed method before enhancement

	Precision (P)	Recall (R)	$F1$-score $= 2PR/(P + R)$
COVID	0.8	0.6	0.68
Normal	1.00	0.8	0.89

Accuracy	Sensitivity	Specificity
0.9	1.0	0.8

Table 2 Result of proposed method after enhancement

	Precision	Recall	$F1$-score
COVID	0.93	0.82	0.85
Normal	1.00	0.8	0.89

Accuracy	Sensitivity	Specificity
0.94	1.0	0.87

Table 3 Result of proposed method and state of art method after enhancement

	Accuracy	Sensitivity	Specificity
Proposed	0.94	1	0.87
Deep feature and SVM based [15]	0.9	1	0.8

using the deep feature. The comparative Table 3 is shown below which shows that the proposed method is superior in all measurements.

4 Conclusions

Growing COVID-19 pandemic has triggered global public health emergency forcing researchers worldwide to focus their expertise in artificial intelligence, machine learning and deep learning in efficient prognosis and diagnosis of COVID-19. In this regard, the proposed method serves as a small step towards utilization of image enhancement technique and deep learning approaches to strengthen the classification rate and accuracy of radiologists and help them make a quick and accurate diagnosis of COVID-19. Results show that enhanced COVID X-ray images are accurately diagnosable, and proposed model has higher detection accuracy compared to state-of-art-techniques. Although smaller dataset is chosen, result shows good accuracy and paves way for further research choosing larger dataset and better training algorithms along with proposed enhancement technique.

References

1. S.T. Leatherdale, J. Lee, Artificial intelligence (AI) and cancer prevention: the potential application of AI in cancer control programming needs to be explored in population laboratories such as COMPASS. Cancer Causes Control **30**(7), 671–675 (2019)
2. D.S. Kermany et al., Identifying medical diagnoses and treatable diseases by image-based deep learning. Cell **172**(5), 1122–1131 (2018)
3. R. Dias, A. Torkamani, Artificial intelligence in clinical and genomic diagnostics. Genome Med. **11**(1), 1–12 (2019)
4. E.B. Sloane, R.J. Silva, Artificial intelligence in medical devices and clinical decision support systems, in *Clinical Engineering Handbook* (Academic Press, Cambridge, 2020), pp. 556–568
5. E.H. Shortliffe, M.J. Sepúlveda, Clinical decision support in the era of artificial intelligence. JAMA **320**(21), 2199–2200 (2018)
6. Y. Zhou, Research on development and problems of 3D printing technology under intelligent background, in *2019 12th International Conference on Intelligent Computation Technology and Automation (ICICTA)* (IEEE, 2019)
7. B. Steinbock, Designer babies: choosing our children's genes. Lancet **372**(9646), 1294–1295 (2008)
8. J. Campisi et al., From discoveries in ageing research to therapeutics for healthy ageing. Nature **571**(7764), 183–192 (2019)

9. C. Guger, B.Z. Allison, N. Mrachacz-Kersting, *Brain-Computer Interface Research: A State-of-the-Art Summary 7* (Springer, Cham, 2019), pp. 1–9
10. H.S. Chan, H. Shan et al., Advancing drug discovery via artificial intelligence. Trends Pharmacol. Sci. **40**(8), 592–604 (2019)
11. A. Rodríguez-Ruiz et al., Detection of breast cancer with mammography: effect of an artificial intelligence support system. Radiology **290**(2), 305–314 (2019)
12. V.M. Aradhya, M. Mahmud, B. Agarwal, D. Guru, M.S. Kaiser, One shot cluster based approach for the detection of COVID-19 from chest X-ray images. Preprints 2020, 2020070656. https://doi.org/10.20944/preprints202007.0656.v1
13. M. Ilyas, H. Rehman, A. Nait-ali, Detection of Covid-19 from chest X ray images using artificial intelligence: an early review. arXiv preprint arXiv:2004.05436 (2020)
14. J. Hang et al., Covid-19 screening on chest x-ray images using deep learning based anomaly detection. arXiv preprint arXiv:2003.12338 (2020)
15. P.K. Sethy, S.K. Behera, Detection of coronavirus disease (covid-19) based on deep features. Preprints 2020030300 (2020)
16. Z. Rahman, D.J. Jobson, G.A. Woodell, Multi-scale retinex for color image enhancement, in *Proceedings of 3rd IEEE International Conference on Image Processing*, vol. 3 (IEEE, 1996)
17. M.C. Hanumantharaju, V.N. Manjunath Aradhya, M. Ravishankar, A. Mamatha, A particle swarm optimization method for tuning the parameters of multiscale retinex based color image enhancement, in *Proceedings of the International Conference on Advances in Computing, Communications and Informatics* (2012), pp. 721–727
18. M.C. Hanumantharaju, M. Ravishankar, D.R. Rameshbabu, M. Aradhya, A new framework for retinex-based colour image enhancement using particle swarm optimization. Int. J. Swarm Intell. **1**(2), 133–155 (2014)
19. M.C. Hanumantharaju, M. Ravishankar, D.R.R. Babu, V.N.M. Aradhya, An efficient metric for evaluating the quality of color image enhancement, in *Indian International Conference on Artificial Intelligence* (2011), pp. 1016–1026
20. S. Han, H. Mao, W.J. Dally, Deep compression: compressing deep neural networks with pruning, trained quantization and Huffman coding. arXiv preprint arXiv:1510.00149 (2015)
21. K. Simonyan, A. Zisserman, Very deep convolutional networks for large-scale image recognition. arXiv preprint arXiv:1409.1556 (2014)
22. https://medium.com/datadriveninvestor/cnn-architecture-series-vgg-16-with-implementation-part-i-bca79e7db415
23. www.kaggle.com
24. J.P. Cohen, P. Morrison, L. Dao, COVID-19 image data collection. arXiv preprint arXiv:2003.11597 (2020)

Snooze Breaker: Driver Drowsiness Detection Using Machine Learning

Sinchana N Rao, N. Sunitha, M S Swathi, V V Shreevaishnavi, and B V Ajay Prakash

Abstract Accidents in roads have become unquestionably a major threat in cities. The prime reason for accidents in road is irresponsible driving. Main cause for accident is driver drowsiness, alcoholism and careless driving. This paper aims to develop a solution which can detect drivers' exhaustion and to issue a timely warning, hence increasing the transportation safety. Snooze breaker (SB) is proposed to detect driver's drowsiness and provides alarm notification based on Haar cascade algorithm for face and dlib's facial landmarks to correctly map the regions like mouth and eye on the face. The measures that can be used for drowsiness detection are eye closures, frequency of yawing, etc. SB is built using Arduino Uno hardware, video camera, open source computer vision library (OpenCV), ultrasonic sensor (UV), gas sensor and corresponding alert system in the form of buzzer.

Keywords Drowsiness · Dlib · OpenCV · Eye Aspect Ratio (EAR) · Yawning

1 Introduction

Passenger safety has been a major concern to all societies in any country in the world. Around 1.5 lakh people die in road accidents in India, some of them due to drowsiness of the driver [12]. Drowsy driving is a key reason for road casualties. Falling asleep at the wheel is a driver's worst nightmare. The danger, risk and the dreadful results of drowsy driving are alarming. Sleepiness may be caused due to alcohol consumption, routine medicine intake by the driver, change of sleeping patterns because of work. Drowsiness makes you pay less attention to the road and affects your ability to make good decisions. This problem can be addressed through Internet of things (IoT) and machine learning. IoT means to connect the unconnected. IoT connects devices which can sense and control the physical entities by making smarter objects. These are connected through intelligent networks. Machine learning (ML) is an part of artificial intelligence (AI) which provides solution that can automatically learn and enhance

S. N. Rao (✉) · N. Sunitha · M. S. Swathi · V. V. Shreevaishnavi · B. V. Ajay Prakash
Department of Computer Science and Engineering, SJB Institute of Technology, Bengaluru, India

V. Bhateja et al. (eds.), *Data Engineering and Intelligent Computing*,
Advances in Intelligent Systems and Computing 1407,
https://doi.org/10.1007/978-981-16-0171-2_50

from past experience without being programmed explicitly. ML approaches focus on the developing programs that can get data and make use for learning themselves. Face recognition owned a significant consideration in the field of image operations. Object detection is a computer technology, which connects the image processing and computer vision. The primary goal of face detection algorithm is to identify faces within the frame. Viola–Jones proposed real-time face detector [12], which is capable of detecting the faces in real time with high accuracy. In this paper, we use behavioral aspects of the driver to determine the drowsy state of the person driving. By combining eye closure rate and yawning state of the driver, drowsiness can be detected. We built the SB and stopped it when the driver was in an uncontrolled state of his drowsiness.

2 Literature Survey

Abtahi et al. [1] propose a method for yawn detection using geometrical features of the mouth. Multiple images employed by the algorithm identify the mouth region, thus enabling the system to detect yawn condition and alert the driver. Assari et al. [2] Driver drowsiness is identified using facial expression. Hardware is implemented based on infrared light that benefits over any environment lighting conditions. Face landmarks obtained by using the background subtraction technique. Alshaqaqi et al. [3] The driver's facial behavior is captured through an IR camera. Hough transform helps in drawing boundaries around the eye region. Anjali et al. [4] check the weariness level of the driver by using eye feature. Implementation of image processing is done in order to detect the blink rate. Rigane et al. [5] describe an approach in determining the driver drowsiness using fuzzy logic simulation. Eye landmarks are extracted using classification algorithms and fed to the fuzzy system for drowsiness detection. Yauri-Machaca et al. [6] examine all facial features of the driver, and images are processed using MATLAB, which indicates whether the person is sleepy or not and warns the driver. Kahlon et al. [7] With image processing tools, images captured by the camera are converted from RGB to grayscale. The ratio of white and black pixels is estimated and verifies the eye's shape. Yashwanth et al. [8] The OpenCV library of Python helps to convert video streams into frames. Both eye aspect ratio (EAR) and mouth aspect ratio (MAR) are estimated to identify whether the subject is drowsy. Deng et al. [9] Human face is tracked by a video stream and sent to the cloud server. Lee et al. [11] OpenCV and Dlib in Python are used to extract facial features and calculate the frequency of eye closure, yawning and head movements.

3 Proposed Methodology

Identification of drowsiness is possible through analysis of eye states, constant yawns, and thus considering these features, system will be able to distinguish between drowsy driver and normal driver. Main steps include identification of facial landmarks such as eye and mouth region, checking for drowsiness, alert the driver and stop the model. The proposed methodology consists of an aggregate of several components that are interlinked to produce the final artifact as illustrated in Fig. 1.

3.1 *Initialization*

Identification of eye and mouth region is our main target. A camera is used to continuously track the facial landmark and movement of eyes and mouth of the driver, which is then converted to frames. Then, we set the counter to have a count on the

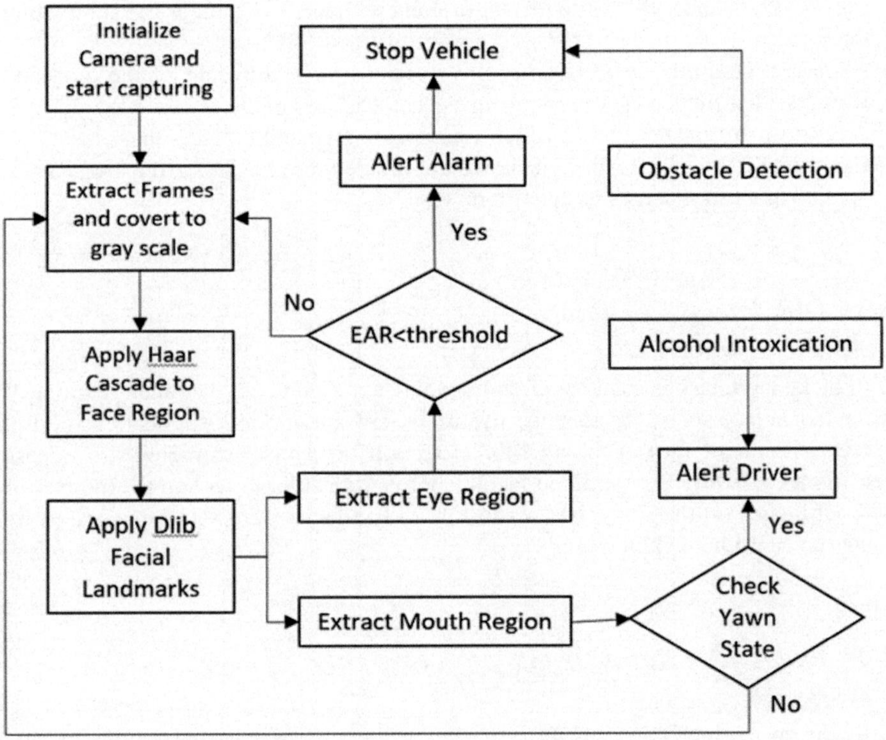

Fig. 1 Proposed methodology for detecting and alerting driver drowsiness

number of consecutive frames of eye closure. The frames will be red–green–blue (RGB) format, we convert it into gray for increasing processing speed.

3.2 *Haar Cascade Classifier*

There are many machine learning object detection algorithms like support vector machines, neural network-based face detection that is used to identify objects in an image or video. Haar cascade classifier is based on the concept proposed by Paul Viola and Michael Jones in 2001 [10]. OpenCV has many pretrained classifiers of which Haar classifiers are one of them. The algorithm has four stages:

- Haar Feature Selection: The algorithm must be trained from a lot of negative images (no faces) and positive (with faces) images and thus would we able to detect objects in other images. Haar features like line, edge and four-rectangle features are extracted from these images.
- Creating Integral Images: Each of the Haar feature is resulted by computing the sum of pixels under the white rectangle along with black rectangle and subtracting these sums.
- Adaboost Training: The classifier must select best feature out of the thousand features identified, i.e., removing irrelevant features. This is achieved through AdaBoost. A threshold is set which classifies to face and no face images.
- Cascading Classifiers: All the features are divided into smaller groups which are applied at various stages of classifiers.

3.3 *Dlib Facial Landmark*

Dlib facial landmark is used for circumscribing and points out the facial features of the person's face such as eyes, nose, eyebrows, jaw and mouth. Applications of dlib are face alignment, face swapping, blink detection, head pose estimation, drowsiness detection, etc. Dlib has pretrained facial landmark detector to reckon the indexes of 68 coordinates on the image shown in Fig. 2. The facial parts can be accessed by using the coordinate values.

3.4 *Eye Aspect Ratio Graph*

After the eye region is perceived from the face, the next step is to determine the eye blink. We use the EAR formula to determine the blink rate of the eye. The formula for eye aspect ratio graph (EAR) is given below.

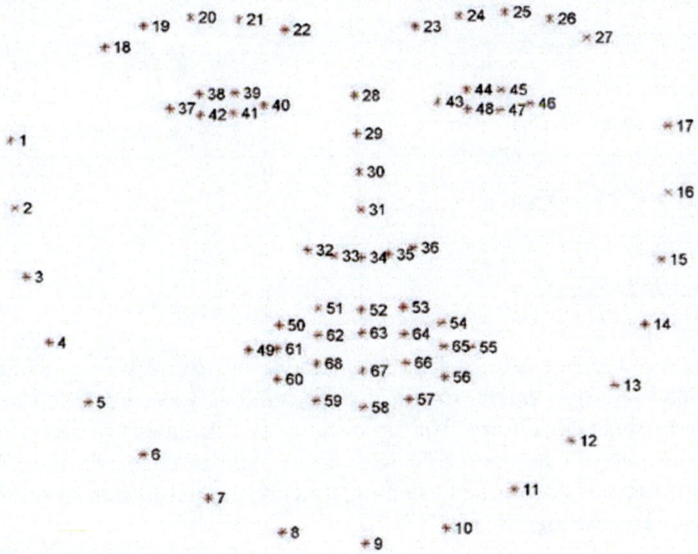

Fig. 2 68 facial landmarks

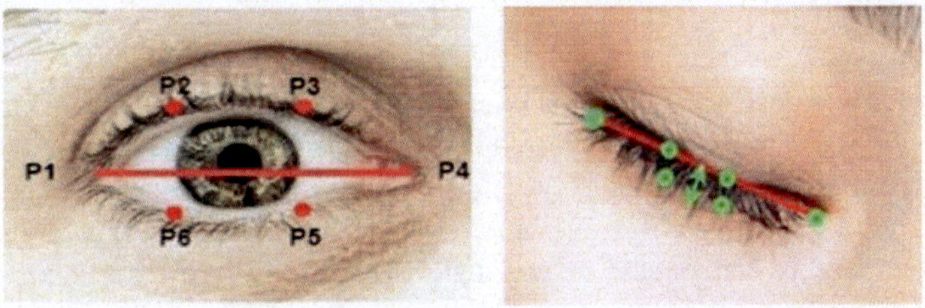

Fig. 3 Eye landmarks

$$EAR = \frac{\|p_2 - p_6\| + \|p_3 - p_5\|}{2\,\|p_1 - p_4\|} \tag{1}$$

In the equation above, the Euclidian is calculated as used in the formula to get the EAR ratio. By observing the ratio, we can judge whether the person is drowsy or not. Consider the below Fig. 3 shown. When the eye is widely open, the EAR would be large and remains constant over time. Whereas when the eye is closed, i.e., when there is a blink, the EAR value falls down dramatically.

The graph shown in Fig. 4 gives the variation in EAR ratio. It remains stable when eye is open and falls rapidly when eye is shut. Then again, it increases indicating a single has taken place.

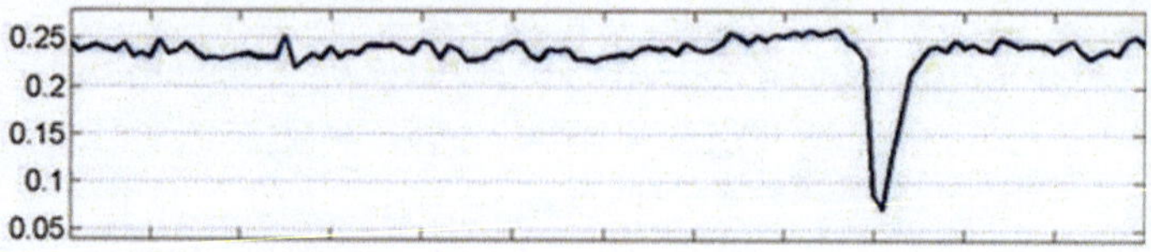

Fig. 4 Eye Aspect Ratio Graph

3.5 *Mouth Detection*

Mouth region of the face is to be drawn from the frame to check for yawning. When mouth opened to larger extent, then it is considered as yawning. Open mouth can be identified using dlib library. The lip distance is calculated by taking the mean distance. Then, it will be compared with a threshold to check drowsiness of the person. If the person is confirmed yawning, an alert sound is made and continued till the person stops yawning.

4 Implementation

First, we initialize necessary libraries like dlib, OpenCV, imutils, NumPy and SciPy. Then, the video stream input is converted into frames. These frames are in RGB color format, we need to convert it into grayscale for increasing the processing speed, since after conversion, the only color available is shades of gray. The conversion might result in decrease in sharpness, so we scale and resize the image to preserve the resolution. Next faces are identified in the image, if found it returns the coordinates as Rect(x, y, w, h). Through Haar cascade algorithm, faces are identified from the frames. Drowsiness is detected by considering the eye and mouth movements. We set threshold for eye blink rate; say we give 0.3 here. This threshold is compared with the EAR obtained in consecutive frames (here 30). Similarly, yawing threshold is set, say 17. If these values obtained go beyond the set respective threshold, it is considered as drowsy situation, and immediately serial communication will be made through Zigbee and the Arduino powered SB, then stops the movement along with creating a buzzer sound. Figure 5 shows the architecture of SB. It has a dual motor. We have used LM293D as the motor driver to drive SB using the microcontroller. Sensors like UV sensor and gas sensor are used to detect any obstacle to the SB and to detect alcoholic intoxication, respectively. When UV sensor detects any obstacles to the SB in motion, it readily stops, thus preventing accident. Major causes for accident are drowsiness due to alcohol consumption. Gas sensor is used to detect the alcoholic consumption through the concentration of the gas. It measures the conductivity of target gas and increases with increase in alcohol concentration. If found alcoholic, a buzzer noise is made to alert the driver.

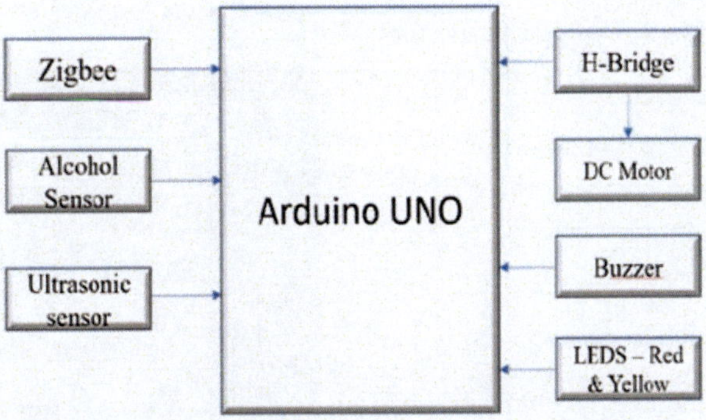

Fig. 5 Hardware architecture of snooze breaker

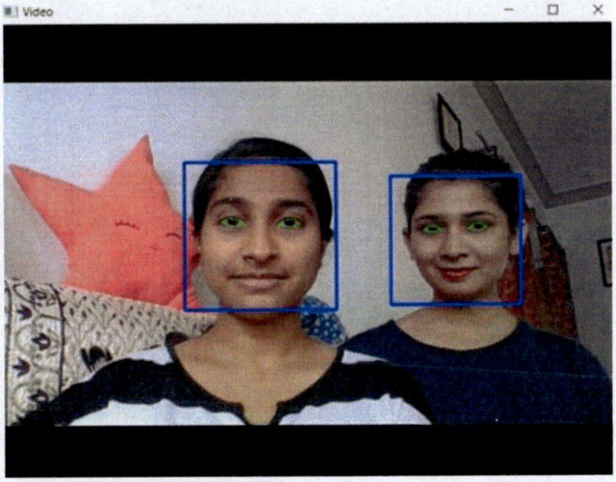

Fig. 6 Face detection

5 Experiment and Results

Using Haar cascade classifier, the faces in the frame can be detected as seen in Fig. 6 and mark the bounding box across the identified faces. Using dlib's 68 facial landmarks, the eye and mouth region is extracted as seen in Fig. 7. Here, we mark out the region and make further computation. The EAR ratio is calculated to check for drowsiness. When the EAR drops below threshold for given set of consecutive frames, then the person is considered drowsy. An alarm is raised, and the snooze breaker speed is driven to left and stopped. Figure 8 notifies driver is drowsy and can observe that the EAR ratio has dropped below the threshold.

Similarly, when yawning is noticed, alert system is enabled as seen in Fig. 9. Yawning is a sign of sleepiness, hence by alerting, the driver could make him awake. The alarm does not stop until the driver stops yawning. The SB has a UV sensor which detects the object (vehicles) approaching by observing the distance to the obstacle. If an obstacle is encountered, alert sound is generated, and then, SB comes to halt, thus avoiding the obstacle. SB senses obstacle with the help of UV sensor.

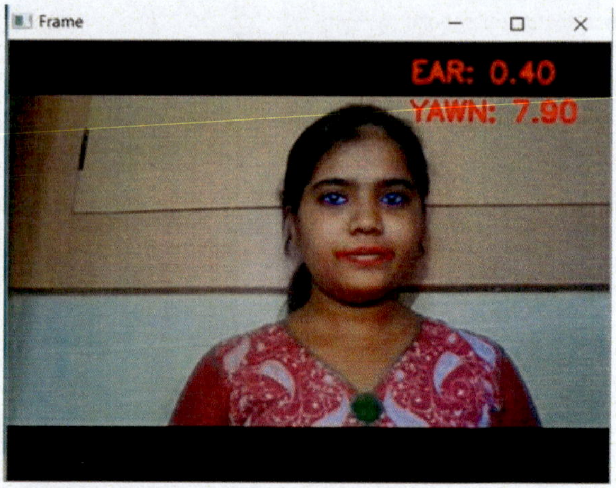

Fig. 7 Facial landmarks for eye and mouth region are extracted

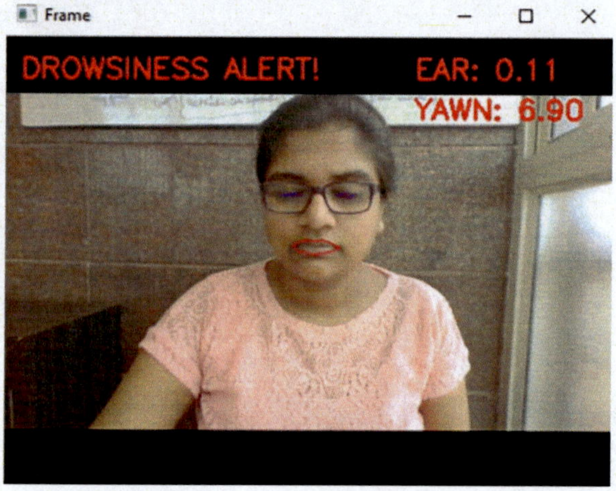

Fig. 8 Drowsiness alert

The alcohol sensor used in SB gives result based on the alcohol concentration in the air. The LED on alcohol sensor is high indicating alcohol is detected. Experimental results of parameter boosting in Haar cascade classifier are given in Table 1.

6 Conclusion

OpenCV provides pretrained classifier like Haar cascade classifier which proves to be effective for face detection. The hyperparameters of Haar cascade classifier such as scale factor and min neighbors are varied with different values to get the better accuracy. Using machine learning algorithms, driver drowsiness detection can be

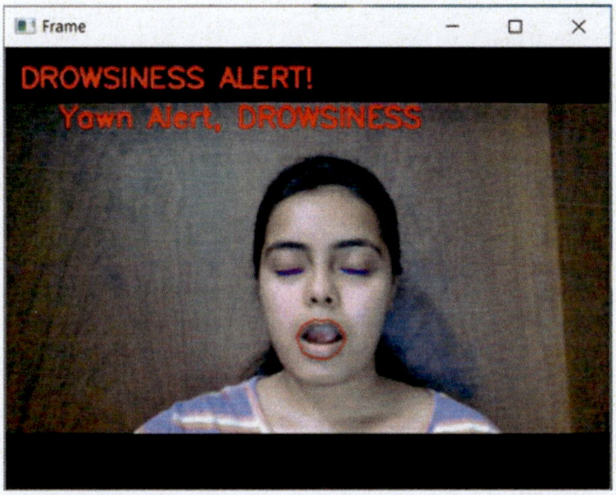

Fig. 9 Yawn and no blink

Table 1 Experimental results of parameter boosting in Haar cascade classifier

scaleFactor	minNeighbors	detecMultiScale: Result
1.0–1.03	3	Unnecessary face detection and drowsiness is not properly identified
1.04	4	Reduction in number of unnecessary face detection
1.05–1.06	5	Time taken to detect drownsiness is 7–8 s
1.07–1.08	5	Time taken to detect drownsiness is 4–5 s
1.09–1.1	6	Drowsiness is detected properly and time taken is 3–4 s

measured accurately. The system will be capable of identifying drowsiness of the driver, by considering the eye blink rate and yawning condition. If the driver is confirmed drowsy, then warning signal will be issued, and the model comes to halt. It also alerts in case of any obstacles and gives alarm in case of alcohol. Snooze breaker, the driver drowsiness detection system, can be used for heavy and commercial vehicles. Accelerometer can be added to the system to monitor the abnormal driving of the driver.

References

1. S. Abtahi, B. Hariri, S. Shirmohammadi, Driver drowsiness monitoring based on yawning detection, in *2011 IEEE International Instrumentation and Measurement Technology Conference, Binjiang*, pp. 1–4 (2011)
2. M.A. Assari, M. Rahmati, Driver drowsiness detection using face expression recognition, in *2011 IEEE International Conference on Signal and Image Processing Applications (ICSIPA), Kuala Lumpur*, pp. 337–341 (2011)

3. B. Alshaqaqi, A.S. Baquhaizel, M.E. Amine Ouis, M. Boumehed, A. Ouamri, M. Keche, Driver drowsiness detection system, in *2013 8th International Workshop on Systems, Signal Processing and their Applications (WoSSPA), Algiers*, pp. 151–155 (2013)
4. K.U. Anjali, A.K. Thampi, A. Vijayaraman, M.F. Francis, N.J. James, B.K. Rajan, Real-time nonintrusive monitoring and detection of eye blinking in view of accident prevention due to drowsiness, in *2016 International Conference on Circuit, Power and Computing Technologies (ICCPCT), Nagercoil*, pp. 1–6 (2016)
5. O. Rigane, K. Abbes, C. Abdelmoula, M. Masmoudi, A fuzzy based method for driver drowsiness detection, in *2017 IEEE/ACS 14th International Conference on Computer Systems and Applications (AICCSA), Hammamet*, pp. 143–147 (2017)
6. M. Yauri-Machaca, B. Meneses-Claudio, N. Vargas-Cuentas, A. Roman-Gonzalez, Design of a vehicle driver drowsiness detection system through image processing using Matlab, in *2018 IEEE 38th Central America and Panama Convention (CONCAPAN XXXVIII), San Salvador*, pp. 1–6 (2018)
7. M. Kahlon, S. Ganesan, Driver drowsiness detection system based on binary eyes image data, in *2018 IEEE International Conference on Electro Information Technology (EIT), Rochester, MI*, pp. 0209–0215 (2018)
8. C. Yashwanth, J.S. Kirar, *"Driver's Drowsiness Detection," TENCON IEEE Region 10 Conference (TENCON)* (Kochi, India, 2019), pp. 1622–1625
9. K.H. Lee, W. Kim, H.K. Choi, B. Tae Jang, A study on feature extraction methods used to estimate a driver's level of drowsiness, in *21st International Conference on Advanced Communication Technology (ICACT), PyeongChang Kwangwoon_Do,Korea(South)*, pp. 710–713 (2019)
10. P. Viola, M. Jones, Rapid object detection using a boosted cascade of simple features, in *Accepted Conference on Computer Vision and Pattern Recognition* (2019)
11. W. Deng, R. Wu, Real-time driver-drowsiness detection system using facial features. IEEE Access **7**, 118727–118738 (2019)
12. Artificial Intelligence for Drowy drivers. https:timesofindia.indiatimes.com/business/india-business/sleepy-at-the-wheel-ai-to-keep-eye-on-drowsy-drivers/articleshow/67098413.cms

Brain Tumor Segmentation in MRI Sequences Using Autoencoders

C. Naveena, V. N. Manjunath Aradhya, S. Poornachandra, and S. B. Rudraswamy

Abstract Manually delineating brain tumor sub-locality in MR images is very time consuming, requires experienced annotators and has high inter-rater variability. Accurate unmanned segmentation of brain cancer helps in better detection and diagnosis of the disease. Further it aids in treatment planning and monitoring. With recent computational advances in graphic processing units (GPUs), drastically reducing the computational cost of neural networks, we are motivated to propose a six-layer deep autoencoder for segmenting brain tumors in MR images. The autoencoder is trained using 18×18 2D patches extracted from multi-sequence MRI T1,T2,T1c & FLAIR. The model is pre-trained using Rmsprop and fine-tuned by exercising stochastic gradient descent (SGD). The proposed model obtained dice score of 0.88, sensitivity of 0.89 and specificity of 0.86 for the whole tumor region.

Keywords MRI · Gliomas · Brain Tumor · Deep autoencoder · SGD

1 Introduction

Gliomas are among the widespread types of brain tumor which are grouped in low-grade(LGG) and high-grade(HGG) brain tumors based on their different diagnostic and prognostic outcomes [1]. The MRI which is a non-invasive and non-ionizing medical imaging modality is employed to discover tumor phenotypes and suggest the appropriate course of treatment such as radiotherapy, chemotherapy and surgical resection. The anatomical MRI modalities used for brain tumor diagnosis and prognosis include T1, T2, T1c & fluid attenuated inversion recovery (FLAIR). The characteristics of brain tumor observed in these MRI sequences

C. Naveena (✉)
SJB Institute of Technology, Affiliated to VTU, Bengaluru, India

V. N. Manjunath Aradhya · S. B. Rudraswamy
JSS Science and Technology University, Mysuru, India

S. Poornachandra
Srinivas University College of Engineering and Technology, Mangaluru, India

V. Bhateja et al. (eds.), *Data Engineering and Intelligent Computing*,
Advances in Intelligent Systems and Computing 1407,
https://doi.org/10.1007/978-981-16-0171-2_51

include non-enhancing/necrotic (hyper-intense in T1), peri-tumoral edema (hyper-intense in FLAIR) and enhancing tumor (hyper-intense in T1c). Manual delineation of these tumor subregions provides useful prognostic information based on volume, shape, texture and intensity. However, manual annotation process demands significant expertise on the part of the annotator and is time consuming. Also, the manual annotation process leads to inter- and intra-rater differences [2]. With these consequences, numerous research groups have proposed semiautomatic and automatic brain tumor partition techniques. During recent years, deep learning techniques have emerged as the preeminent approach in computer vision applications across a variety of problem domains like self-driving cars, natural scene classification, visual question answering, etc. Motivated with these developments in the domain of machine learning and computer vision, we propose brain tumor segmentation model, which is a type of deep learning model.

This paper is organized in the ensuing manner, Section II discusses the different researchers actively contributing in the domain of brain tumor segmentation, Section III presents the methodology for segmenting brain tumors, and Section IV explains the implementation details. Section V focusses on the results obtained, and Section VI discusses the future enhancement and concludes the paper.

2 Related Work

This section encompasses on various researchers actively contributing in the domain of partitioning brain tumors using automatic methods.

Meier et al. [3] focussed on partition of brain tumor by implementing conditional random field. The pairwise CRFs are estimated using SGD of a max-margin learning algorithm. The evaluation of this parameter learning method is compared to the method which employs hand-tuned parameters.

Kamitsas et al. [4] implemented 3D CNN for segmenting brain tumors, the model comprises two parallel convolutional routes that process data at multiple scales to obtain a larger receptive field for final categorization besides maintaining the computational cost lower. Ellawa et al. [5] presented partition of brain tumor using random forest classifier, and the method is employed based on features of gradient, appearance and context.

Karnawat et al. [6] proposed CNN architecture based on radiomics for segmenting brain tumor. The model comprises convolutional & pooling layers, further two fully connected layers and one categorization layer with two pathways. The network is trained using Adam optimizer.

Zhou et al. [7] presented partition of brain tumor cells by utilizing cascaded neural networks. The data are processed using two-stage architecture. In the first stage, 2D slices are applied U-net in order to capture coarse segmentation mask. In the second stage, local details are considered to obtain precise and fine segmentation output. Islam and Ken [8] proposed brain tumor partition by employing convolutional neural network with hypercolumn features. The network employs fully connected layers,

ReLu activation and sparse pixel prediction which are employed during training for efficient mini-batch generation.

Li and Shen [9] proposed segmentation of brain tumors by employing multiview deep learning framework. The model comprises three sub-networks and processes brain images in the direction of x-axis, y-axis and z-axis. The network is trained using Adam optimizer. Amorim et al. [10] proposed 3D U-nets for segmentation of brain tumors. The model processes $64\times64\times64$ patches acquired from MRI modalities as inputs for the segmenting whole tumor and tumor core enhancing tumor.

Hence, these are some of the different researchers actively participating in the field of partition of brain tumors. The ongoing research in this field inspired us to present our work in the field.

3 Methodology

The proposed model focuses on the autoencoders for partition of brain tumor tissues in MRI brain scans. Firstly, the input MRI scans are preprocessed by employing N4ITK to revise the magnetic field bias occurred at the time of acquisition of MR images. Then, the intensities are standardized by computing the zero mean unit variance (Z-score) to align the data. The preprocessed images are provided as input to the model for partitioning the brain tumor regions.

3.1 Preprocessing of MR Images

MRI machines have different magnetic field strengths (1.5T/3T) provided by various vendors in the market. Owing to these differences in magnetic field power, intensity inhomogenity is initiated in the image acquisition process. This bias in the intensity levels affects the forecast of correct labels. In order to meticulate this effect originated due to magnetic field bias, nonparametric, non-uniform intensity normalization technique is applied. An advanced version of N3, N4ITK bias field correction algorithm is solicited to erase the intensity inhomogenities in MR images. And, further the intensities are normalized using the zero mean unit variance(Z-score) method. The ensuing relation depicts the Z-score technique [?,?].

$$X_n = (X - \mu)/\sigma \tag{1}$$

Here X constitutes original image, μ is the mean of X, and σ is the standard deviation of X, respectively. X_n constitutes output image. Figure 1 displays preprocessed images using N4ITK technique and Z-score technique.

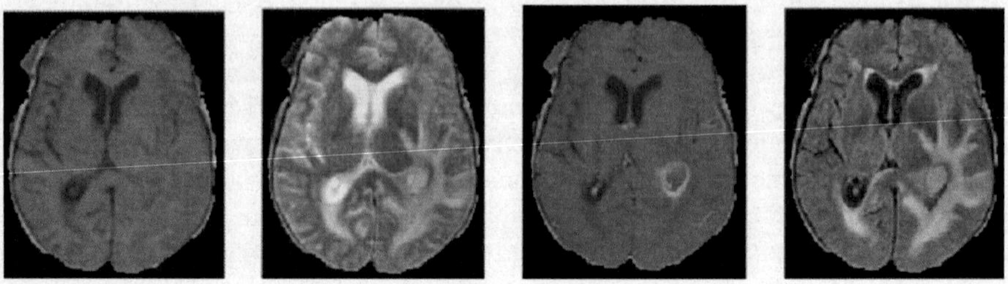

Fig. 1 Preprocessed MR images using N4 bias field correction and Z-score technique (T1, T2, T1c and Flair)

3.2 Network Architecture

In this proposed work, a six-layer deep autoencoder for partitioning brain tumor sub-locality from MR images is employed. Autoencoders are a family of neural network trained to revamp the input data. A deep autoencoder is developed by stacking multiple layers of encoders with each layer pre-trained in an unsupervised manner. And, further fine-tuning is done using labeled data by adding a classification layer to pre-trained encoder. The proposed autoencoder model is trained using 2D patches of size 18×18 extracted from four MRI modalities T1, T2, FLAIR and T1c. The network consists of 1296 units at the input layer and 4000,3500, 2500, 1000, 500 in the hidden layers and at the end five neurons for five output classes. Table 1 depicts the network model.

The network is pre-trained layer-by-layer using the extracted 2D patches from the region of interest(ROI) for 30 epochs. In each layer, the weights and biases are initialized using Xavier initialization and trained using mean-squared error as error function with RmsProp as an optimizer. In the fine-tuning phase, the weights and biases of the penultimate layer are initialized to zeros. The weights of the network are learnt by minimizing the negative log likelihood with SGD as an optimizer with a dropout of 20% and learning rate of 0.005. The network is pre-trained for 180 HGG subjects and fine-tuned for 40 LGG subjects.

Table 1 Autoencoder network model for the proposed work

Input Layer	Hidden	Hidden	Hidden	Hidden	Hidden	output
1296	4000	3500	2500	1000	500	5

4 Experimental Setup

4.1 Dataset

In this work, the presented model for the segmenting brain tumor is accomplished on BraTS 2015 dataset. Each patient in the dataset consists of four MR image sequences, namely T1, T1c, T2 & fluid attenuated inversion recovery(FLAIR). The MRI sequences of individual patient are organized to T1 modality with 1 mm $\times$ 1 mm $\times$ 1 mm isotropic resolution. The MR images are skull sacked with a resolution of 240$\times$240$\times$155. The BraTS dataset totally embodies 274 patient's data comprising of 220 high-grade cases and 54 low-grade cases. The ground truth has four intra-tumoral subregions specifically necrosis—the dead part of the tumor, enhancing tumor, non-enhancing tumor and edema the swelling region around the tumor core. The necrosis has a tag value 1, the edema has a tag value 2, the non-enhancing has a tag value 3, and the enhancing has a tag value 4. The partition results are computed for complete tumor, the tumor core(non-enhancing and necrosis) and the enhancing tumor.

4.2 Assessment Metrics

The partition output for implemented architecture is computed for three regions in the brain tumor area, the complete tumor (all tumor classes), tumor core(non-enhancing tumor and necrosis) and enhancing tumor. For all the three cancerous regions, the partition results are evaluated on the basis of three assessment metrics: dice similarity coefficient, sensitivity (true positive rate or positive predictive value) and specificity (true negative rate). The following relation represents the dice similarity coefficient.

$$D(P, T) = 2 \times \frac{(|P_1 \wedge T_1|)}{(|P_1| + |T_1|)} \tag{2}$$

where P_1 is the predicted lesion area, and T_1 is the true lesion area. |.| is the count of voxels in the true lesion area or the predicted lesion area. The sensitivity is calculated by commissioning the ensuing relation.

$$Sensitivity(P, T) = \frac{|P_1 \wedge T_1|}{|T_1|} \tag{3}$$

The true negative rate or specificity is computed by commissioning the ensuing equation.

$$Specificity(P, T) = \frac{|P_0 \wedge T_0|}{|T_0|} \tag{4}$$

where P_0 is the predicted normal area, and T_0 is the true normal area.

4.3 Implementation Details

In this work, the brain tumor partition model is implemented using Theano. Theano is a powerful Python library for optimizing and evaluating complex mathematical expressions involving multi-dimensional arrays. Theano is integrated with cuDNN for effective execution of deep learning primitives which contributes for maximized throughput and memory usage [11].

The training of the autoencoder consists of an unsupervised phase known as pre-training and a supervised phase called fine-tuning. Patches for pre-training and fine-tuning are extracted within the vicinity of the brain tumor region. For pre-training, the weights and biases in each of the layer are initialized by employing Xavier initialization [12], and RmsProp [13] is used as an optimizer. Fine-tuning the previous layer and the decision layer is initialized with zeros. And, the weights of network are learnt by reducing the negative log likelihood function using SGD [14]. To prevent overfitting, all layers used dropout [15] of 20%. The network is trained using a mix of HGG and LGG data. The model is trained on NVIDIA GTX 970 graphics card.

5 Results and Discussion

This section focusses on partition results acquired for proposed autoencoder model. The preprocessing steps helped the model in prediction of brain tumors. The 2D patches extracted from the four MRI modalities are used for training and testing purposes. The BraTS 2015 dataset is compartmentalized into 80% for training and 20% for testing. In the pre-training, the learning rate is 0.001 and the network is optimized using RmsProp technique. The fine-tuning phase has a learning rate of 0.005 with SGD as an optimizer with a dropout of 20%. The partition results are assessed for three tumoral regions *complete tumor*, *tumor core* and the *enhancing tumor*. The proposed model has obtained promising results for partitioning the intra-cancerous subregions in MR images. The experimental results produced a dice score of 0.88, sensitivity of 0.89 and specificity 0.86 for complete tumor. The dice score of 0.87, sensitivity of 0.84 and specificity of 0.82 are yielded for tumor core. The dice score of 0.86, sensitivity of 0.82 and specificity of 0.83 are yielded for enhancing tumor, respectively. To overcome the limitation of this model and enhance the performance to yield more accurate results, as a future work, we plan to present two additional nodes at the output layer to delineate both the HGG and LGG tumors more accurately. The evaluated results are compared with different models proposed by different researchers actively contributing in the area of partition of brain tumors. Table 2 depicts the comparison of dice score, sensitivity and specificity obtained for the complete tumor. Figure 2 shows segmentation results for the proposed model.

Table 2 Comparison of dice scores, sensitivity and specificity for complete tumor

Model	Dice score	Sensitivity	Specificity
Proposed Model	0.88	0.89	0.86
Havaei et al. [16]	0.88	0.89	0.73
Tustison et al. [17]	0.87	0.85	0.89
Abdelrahman et al. [5]	0.80	0.98	0.84
Kamnitsas et al. [4]	0.89	0.89	0.72
Karnawat et al. [6]	0.89	0.99	0.89
Periera et al. [18]	0.87	0.89	0.86
Maeier et al. [3]	0.83	0.85	0.83

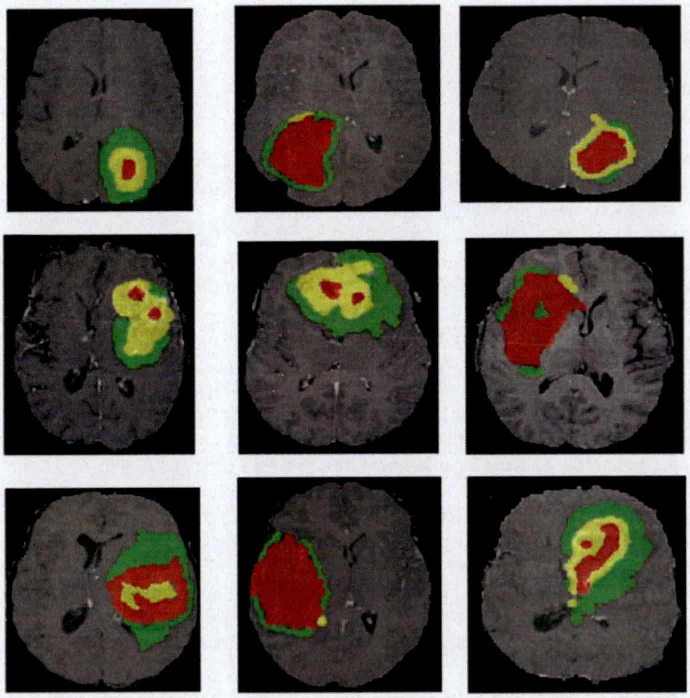

Fig. 2 Segmented results of BRATS 2015 test data—green edema,yellow-enhancing tumor and red necrotic core and non-enhancing

6 Conclusion

In this work, the deep autoencoder network model is proposed for the commissioning of segmentation of brain tumor in MRI scans. Basically, the preprocessing of MR images yielded better results in the segmentation task. The 2D patches extracted from the vicinity of the tumor region are fed into the deep encoder model, thus reducing the class imbalance problem. The network is pre-trained using the RmsProp optimizer and then fine-tuned using the SGD as an optimization technique. Thus, the model proposed yielded comparatively good results in the segmentation task. As a future proposal, we plan to add two more nodes in the output for HGG tumor and the LGG tumors, thereby which helps in accurate delineation of tumor subregions.

References

1. M. Lisa, DeAngelis, Brain tumors. New England J. Med. (2001)
2. B.H. Menze et al., The multimodal brain tumor image segmentation benchmark (BRATS). IEEE Trans. Med. Imaging **34**(10), 1993–2024 (2015). Oct
3. M. Raphael, K. Venetia, H. Simon, W. Roland, M. Reyes, Parameter learning for CRF-based tissue segmentation of brain tumors, in *Proceedings of MICCAI-BraTS*. Munich (2015)
4. K. Kamnitsas et al., DeepMedic on brain tumor segmentation, in *Proceedings of MICCAI-BraTs*, pp. 18–22 (2016)
5. A. Ellawa et al., Brain tumor segmantation using random forest trained on iterative selected patients, in *Proceedings of MICCAI-BraTs*, pp 18–22 (2016)
6. A. Karnawat et al., Radiomics-based Convolutional Neural Network (RadCNN) for brain tumor segmentation on multi-parametric MRI, in *Proceedings of MICCAI-BraTs*, pp. 147-153 (2017)
7. C. Zhou et al., Brain tumor segmentation with cascaded convolutional neural networks, in *Proceedings of MICCAI-BraTS Conference*, pp. 328-333 (2017)
8. M. Islam et al., Fully convolutional network with hypercolumn features for brain tumor segmentation, in *Proceedings of MICCAI-BraTs Conference*, pp. 108-115 (2017)
9. Y. Li et al., MvNet: multi-view deep learning framework for multimodal brain tumor segmentation, in *Proceedings of MICCAI-BraTS*, pp. 169-174 (2017)
10. P.H.A. Amorim, et al., 3D U-nets for brain tumor segmentation in MICCAI 2017 BraTS challenge, in *Proceedings of MICCAI-BraTS Conference*, pp. 09-14 (2017)
11. Theano Development Team. Theano: A Python Framework for Fast Computation of Mathematical Expressions (2016). arXiv:1605.02688v1
12. X. Glorot et al., Understanding the difficulty of training deep feedforward neural networks, in *International Conference on Artificial Intelligence and Statistics*, p. 249256 (2010)
13. G. Hinton et al., Lecture 6.5–RmsProp: Divide the Gradient by a Running Average of its Recent Magnitude (2012)
14. P. Vincent et al., Stacked denoising autoencoders: learning useful representations in a deep network with a local denoising criterion. J. Mach. Learn. Res. 3371-3408 (2010)
15. N. Srivastava, G.E. Hinton, A. Krizhevsky, I. Sutskever, R. Salakhutdinov, Dropout: a simple way to prevent neural networks from overfitting. J. Mach. Learn. Res. **15**, 1929–1958 (2014)
16. M. Havaei et al., Brain Tumor Segmentation with Deep Neural Networks. arXiv:1505.03540 (2016)
17. T. Nick, W. Max, D. Chris, B. Avants, ANTs and arboles, in P*Proceedings of MICCAI-BraTs*. Nagoya, Japan (2013)
18. S. Pereira, A. Pinto, V. Alves, C.A. Silva, Brain tumor segmentation using convolutional neural networks in MRI images. IEEE Trans. Med. Imaging **35**(5), 1240–1251 (2016)

An Investigation into Smart Contract Deployment on Ethereum Platform Using Web3.js and Solidity Using Blockchain

Sandeep Kumar Panda and Suresh Chandra Satapathy

Abstract Blockchain is a decentralized distributed platform, where multiple peers can exchange digital assets securely all over the globe without third party based on trustiness. In blockchain, blocks are connected by using hash that maintains the order of transaction. In this paper, we have proposed a novel approach of understanding proof of work and proof of stake which are considered as consensus algorithms that are used to manage and control blocks on distributed platform. Here, the paper also presents the Ethereum, a public blockchain platform which is used to implement decentralized application over the Internet using a contract (agreement) in digital form written by using solidity language that defines the way to exchange digital assets is called smart contact. Lastly, the paper concludes with implementation methodology for creation, deployment on Ethereum blockchain, and interaction of smart contract using Node.js, Web3 library, and Infura API.

Keywords Blockchain · Ethereum · Smart contracts

1 Introduction

A Blockchain [1] is a decentralized and distributed database system which comprises of various records bundled to form a block. All these blocks are chained together to from a blockchain. Here, the data structure linked list is used widely. One of the well-known applications of blockchain technology is the cryptocurrencies like bitcoin, ether, etc.

The blockchain is not controlled by single entity but it is controlled or managed by the group of people who are using that blockchain. For example, if a person wants to

S. K. Panda (✉)
Department of Computer Science and Engineering, Icfai Tech (Faculty of Science and Technology), ICFAI Foundation for Higher Education (Deemed to be University), Hyderabad, Telangana, India
e-mail: sandeeppanda@ifheindia.org

S. C. Satapathy
School of Computer Engineering, KIIT Deemed to be University, Bhubaneswar, Odisha, India

V. Bhateja et al. (eds.), *Data Engineering and Intelligent Computing*,
Advances in Intelligent Systems and Computing 1407,
https://doi.org/10.1007/978-981-16-0171-2_52

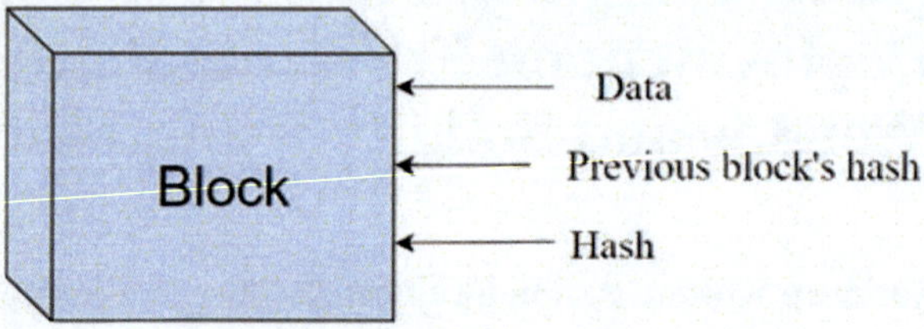

Fig. 1 Structure of a block

transfer money to another person, then there is a need to trust a third party like banks or financial institutions. However, with the use of blockchain, a person can directly transfer money to another person without the need of a third party. The drawback with a centralized entity is that they are singular in number, and all it requires is one person with administrative privileges in an organization to manipulate the data, intentionally or unintentionally.

On the other hand, the blockchain being a distributed system no person has an authority over another person. A blockchain is hosted by a group of people on the network from all over the world. It is built on top of P2P network [1, 2], which means anyone can be a part of network. Each person or node in the network has the entire copy of a blockchain which makes it difficult to tamper with the data without being noticed by another person or node in that blockchain network. This aspect is further emphasized in the paper.

A block comprises of data that is to be stored (type of data to be stored depends on type of blockchain), hash of the previous block and hash of the current block. Figure 1 shows the structure of a block.

A hash [3] is generated by the combination of all the data in block by using some hashing techniques [2]. Hash represents the current state of the block. It is similar to a digital fingerprint which makes the block more secure and almost impossible to tamper with. Apart from hashing, blockchain technology adopts asymmetric key mechanism. In asymmetric key encryption, every user has a pair of keys, namely public and private key [3]. Private key is to be stored securely with the creator; public key can be forked to anyone. While doing a transaction/activity in blockchain the sender account utilizes the public key of the receiver account to encrypt the data and the receiver uses his own private key to decrypt the data. By this mechanism, secrecy and anonymity of sender account are maintained.

The blockchain provides a mechanism to detect the tampered or invalid blocks. In blockchain, the starting block is called genesis block. In Fig. 2, it is depicted that every block holds the hash of the previous block; by this, if an attacker tries to modify the contents of the block 2, then its hash will be modified since a hash is generated by the combination of all the data in the block, i.e., if the data in the block changes, then the hash also will change which results in an error. In the block 3, the hash of previous block is old one and makes all the subsequent blocks as invalid as recursively. On top of that the blockchain provides another security mechanism known as proof of work [4, 10] which consume a lot of time and computational power. If the attacker wants to make the tampered block valid, then he has to redo the proof of work of

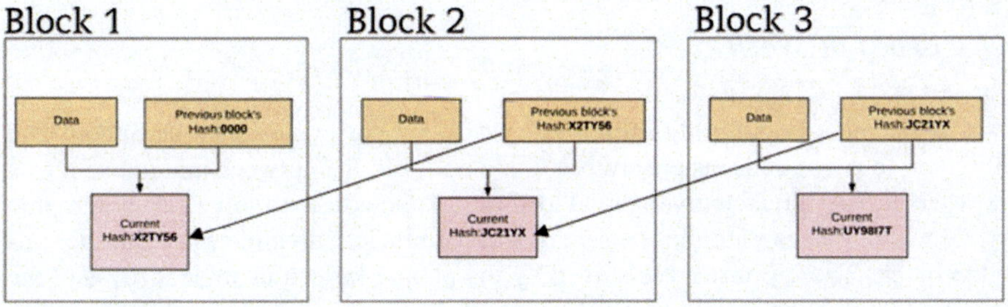

Fig. 2 Blockchain structure

the current block and all the subsequent blocks in the blockchain and has to take control of more than 50% of the blockchain network without being noticed to make everyone in the network accept the block which is highly impossible.

Whenever a new block is generated in the blockchain, it is broadcasted or notified to all of the nodes in the network. The block will be verified and added to the blockchain. The process of verification and adding a block to the blockchain is known as proof of work [4, 10] which is a consensus algorithm.

Motivation and Organization

Nowadays, digital documents are available everywhere in many forms in Internet. However, there is no standard copyright protocol available till date for copyright management of these documents. Two major challenges in digital content management are: the users cannot authenticate the digital documents as the content sources are often unknown, and the content providers have no methods or tools to track usage or identify possible copyright violations of their digital documents. In this paper, we investigate how blockchain technology provides a potential solution to these challenges.

The organization of this paper is as follows. Section 2 gives information about various consensus algorithms. The details of the different blockchain platforms are described in Sect. 3. Section 4 explains the deployment and interaction methods of smart contracts in blockchain technology. Conclusion and future research directions are discussed in Sect. 5.

2 Consensus Algorithms

Blockchain being a distributed system, it is required for everyone in the network to reach an agreement on the authenticity of the block. To achieve distributed consensus, numerous different algorithms were developed. Of these, the proof of work [4, 10] and the proof of stake are the prominent algorithms used to achieve consensus.

2.1 Proof of Work

This is the most popular algorithm being used by currencies such as bitcoin and Ethereum to reach consensus between many nodes in the network and to use it as a way to secure the blockchain. In this algorithm, all nodes have to solve a cryptographic puzzle which comes with the transactions in the block and validate the transactions in the block. These puzzles are solved by the people known as miners [6], and the first one to solve the puzzle will get a reward. For bitcoin, the reward is 12.5 BTC [7], and for ether is 5 ETH [8].

The puzzle is may be like "*Given data X, find a number n such as that the hash of n appended to X results is a number less than Y.*" To solve a puzzle, a miner needs to have a lot of computational power, resources like electricity, cooling fans, and it consumes a lot of time. The time taken to solve the puzzle and validate the transactions is known as block time [8]. The higher the computational power, the higher chance a miner gets the reward. To increase the chance of getting the reward, a group of miners can come together to combine the computational power to mine a block which is called pool mining [6] and distribute the rewards evenly across everyone in the pool.

According to digiconomist [5], from the Fig. 3, it is inferred that bitcoin miners alone uses 73 TWh of electricity which is enough to power 5 million households in US or even power the entire country of New Zealand or Hungary.

Due to this pool mining, the blockchain is becoming centralized as it supposed to be decentralized. To overcome these problems, a new algorithm is proposed known as proof of stake [4, 10].

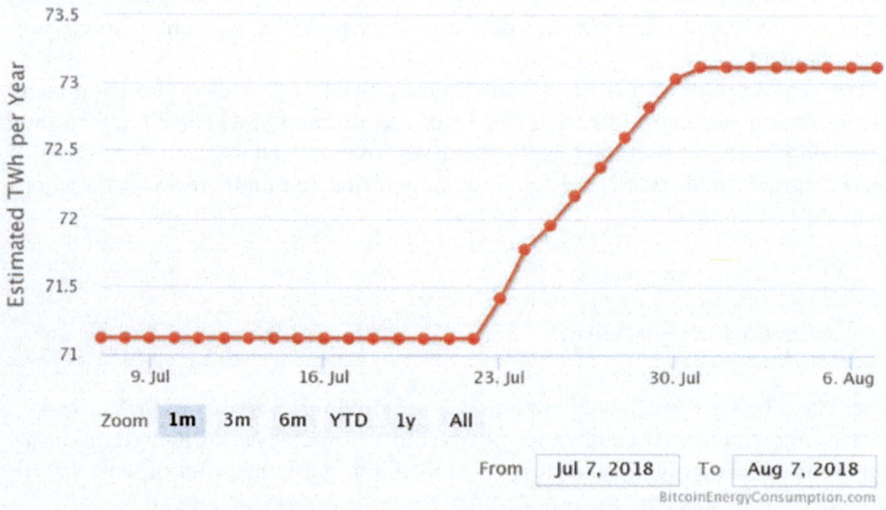

Fig. 3 Bitcoin energy consumption index chart

2.2 *Proof of Stake*

It was first suggested on the bitcointalk forum back in 2011 [9], but the first digital currency to use this method was Peercoin [12] in 2012. Unlike the proof of work, the blocks that are verified and validated will be known as forged [4, 10] or minted not mined, and the users who validate transactions and create new blocks are known as forgers or validators [10].

In this algorithm, the creator of a new block is chosen in a pseudo-random way, depending on the user's wealth, also defined as "stake." To become a forger, the person must put their own coins into the network as stake. It is one kind of security deposit. The higher the stake, the higher are the chances for a validator to be chosen to forge the new block. The validator or forger receives the transaction fees as a reward for forging a block. With the proof of stake algorithm, no new coins are generated. If a validator approves a fraudulent transaction, then they will lose a part of their stake or even lose all the stake. Proof of stake makes a 51% attack [11, 12] virtually impossible, i.e., he has to control at least 51% of all the digital currency in existence which would make it very expensive.

In proof of stake, there is no need for large computational power and resources because the forgers are selected pseudo randomly. Thus, the proof of stake solves the major problems of proof of work algorithm.

3 Platforms of Blockchain Technology

Due to the advantages and security of blockchain technology, many platforms have emerged, namely Ethereum, hyperledgerfabric, ripple, IBM bluemix blockchain, etc. This paper mainly focuses on Ethereum platform and smart contracts.

3.1 *Ethereum*

Ethereum [12, 13] is an opensource platform built on top of blockchain. Ethereum is a public blockchain that allows decentralized apps to be deployed and run inside it. Key features of Ethereum is its flexibility, scalability, lower transaction fees. Ethereum is termed by lot of communities as a "World computer" [14]. The fundamental innovation of Ethereum is Ethereum virtual machine [15] which can run program of any language, given enough time and memory. EVM is the primary consensus mechanism for Ethereum. Every node in the Ethereum network implements EVM. EVM is a runtime environment for smart contract and is secluded from main Ethereum network, which catalyzes testing and enhancing smart contracts.

Ethertweet [16] is a decentralized app built on Ethereum. No company/central authority is not controlling this application. It is censorship resistant, i.e., the published message can only be deleted by the publisher no intermediate person cannot access it/modify it.

3.2 Smart Contract

The term smart contract [13] was coined by NickSczabo in 1997 [14]· Smart contracts are analogs to general contracts in the real world but they are in digital form and are self-executing. The modern economy is driven by contracts. A contract between employer and an organization. A contract between a shareholder and a company, a contract between tenant and landlord, etc. In all the above cases, both the parties trust a third party which is controlled by government, banks, or centralized authorities. Smart contracts are built with ideology of trust in digital world/logic/code rather than relying on trust of a third party. Smart contracts strengthen the idea of "Code is Dictator."

Smart contract is a piece of code with embedded business logic, rules of an agreement and stores asset (anything of important value)/currency. The sole purpose of the smart contract is to move assets/currency, represent ownership that are stored in it based on conditions met. The creator of the smart contract imbibes a set of rules/logic to handle the state of the asset stored in it. Smart contracts are stored inside the blocks in the blockchain network, due to which they inherit the properties of the blockchain technology, i.e., Immutable and distributed. So, once a smart contract is written, it is almost impossible to tamper it.

Smart contracts dwell in Ethereum blockchain. Besides tracking the transactions made, smart contracts programs the transactions too. Smart contracts are written by solidity programming language. Solidity [11, 12] is a contract oriented, high-level language for forging the smart contracts. It is a statically typed language. Solidity is compiled to bytecode that is executable on the EVM. Syntax of solidity is similar to ECMAScript.

From Fig. 4, it is inferred that execution of mathematical computations of smart contracts incurs "Gas Cost."

The author of smart contract must specify the gas limit so that the validators can decide to mine it before validating the smart contract. If the reward of the smart

	Value	Mnemonic	Gas Used	Subset	Removed from stack	Added to stack	Notes
2	0x00	STOP	0	zero	0	0	Halts execution.
3	0x01	ADD	3	verylow	2	1	Addition operation
4	0x02	MUL	5	low	2	1	Multiplication operation.
5	0x03	SUB	3	verylow	2	1	Subtraction operation.
6	0x04	DIV	5	low	2	1	Integer division operation.
7	0x05	SDIV	5	low	2	1	Signed integer division operation
8	0x06	MOD	5	low	2	1	Modulo remainder operation

Fig. 4 Gas cost for computation of smart contract

contract is satisfactory, then miner chooses to invoke smart contract function call in their next block. Every smart contract has its own address and is useful while sending or receiving the ether. Smart contracts can identify the person who calls it; by this, we can impose the privileged access to contract. Remix is a browser-based solidity IDE for coding, deploying, testing the smart contracts. Various execution environments like JavaScriptVM, Injected Web3, Web3 provider are listed in remix.

4 Deployment and Interaction Methods

VSCode has the support for the solidity language. Figure 5 represents that the smart contract, when compiled gives two objects, namely bytecode and ABI. On deploying a smart contract to Ethereum network, the bytecode of smart contract will be stored in the blockchain, and it returns the address of the smart contract called contract address. Just with the help of bytecode, it is highly impossible to know the functionality of a smart contract; hence, ABI plays a crucial role. ABI is abbreviated as application binary interface. ABI is required to interact with the deployed smart contract using contracts address. If any account wants to invoke a function of smart contract, then it utilizes ABI to hash the function and build EVM bytecode to invoke the function. It is a JavaScript formatted object (JSON) that depicts various functionalities of contract like input/output parameters, name of functions, payability (whether ether can be sent or not), state mutability of functions. It facilitates the users to create appropriate messages for contract.

In this paper, for creating and deploying the contract on to Ethereum network, Node.js and its modules are used. For compiling a solidity file in node.js, a node module "solc" has to be installed on the machine. Using this module, the node.js will compile the solidity file which gives the bytecode and the ABI. To interact with

Fig. 5 Compilation of smart contract

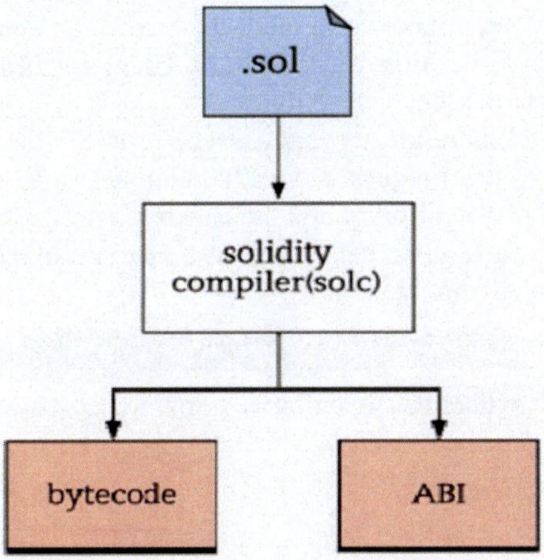

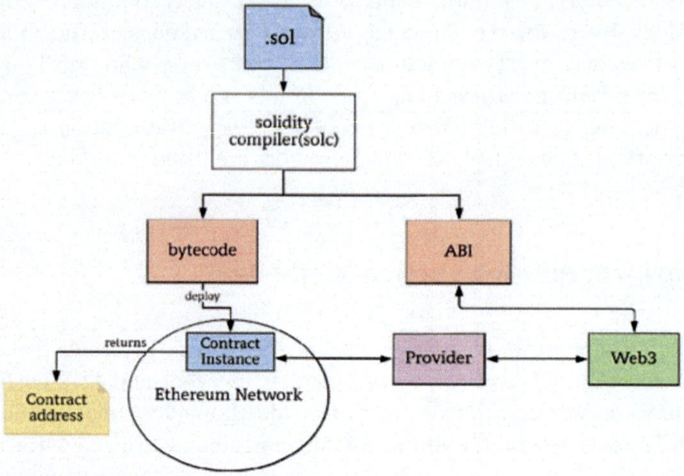

Fig. 6 Deploying methodology

contract on the blockchain, Web3 library is used. The Web3 library is used to connect to the Ethereum network from an application. The ABI is given to the Web3 library, which it uses to give a programmatic access to the contract deployed on the Ethereum network.

To utilize Web3, an object of Web3 class must be instantiated.

```
const Web3 = require('web3');
const web3 = new Web3(); //Instance
```

Multiple instances of Web3 library can be created, and each instance is used to connect to a different Ethereum network. To connect or communicate with an Ethereum network, the instance of Web3 requires a communication layer known as provider illustrated in Fig. 6. There are different types of provider, and each can be used to reach out different.

Ethereum networks, every provider has a set of identical methods to send or receive a request from Ethereum network. The provider acts as a medium between the Web3 library and the Ethereum network; just like when two persons are talking to each other, the air acts as a medium to transfer the sound waves from one person to another person.

```
const web3 = new Web3(provider);
```

where the **provider = ganache.provider();**//Local Test Network.

The ganache is a local Ethereum test network which can be installed through the **"ganache-cli"** npm module. It has to be required or imported into the project to create a local Ethereum test network.

```
const ganache = require('ganche-cli');
web3 = new Web3( ganache.provider() );
```

When the contract is deployed in a local network, there is no need for security and public and private keys to unlock the accounts that are provided by test network (ganache). All the accounts are unlocked, and there is no requirement for ether. However, in case of an Ethereum Main network or Ethereum public test networks, the public and private keys are required to unlock the accounts, and the accounts should have some ether (wei) to deploy the contract. The main network uses real ether for deployment and manipulation or interaction with contract. Hence, the Ethereum test networks (Rinkeby, Kovan, Ropsten) which use a fake ether, for development and testing purposes before deploying the contract to the main network. The process for deploying the contract to the test network and main network is same.

4.1 *Deployment of Smart Contract to Rinkeby Test Network*

To deploy a smart contract, an account in Ethereum client like Metamask [14] or mist browser is required with some amount of ether. In the paper, metamask is used as Ethereum client. Metamask is a web browser extension. When an account is created in the metamask, a 12-word mnemonic is displayed. A 12-word mnemonic is 12 random words which are used to generate (uses BIP39 algorithm) a series of accounts, each with a public key, private key, and an account address. The rinkeby faucet (https://faucet.rinkeby.io) can be used to get some ether into the metamask account (Fig. 7).

The mnemonic is passed as an argument to the provider to unlock the accounts in the metamask. To deploy the contract in public networks, a connection with a node on the public network is required. To become a node in Ethereum network, one needs to configure their machine, which is a little tedious job and requires a lot of time and effort. So, instead a service called Infura is used to connect to a node already configured and existing on the network. Infura is a public API which gives us the ability to access a node on the network. Figure 8 depicts the relation between Infura, provider, and mnemonic. The Infura has the nodes on all the Ethereum networks including the main network as shown in Fig. 7. Get the api key and the link to connect to the rinkeby network from https://infura.io.

As said above, a provider is needed to communicate with the Ethereum test network. The provider has to be setup manually specifying the network to connect and the account mnemonic to unlock the accounts. For this purpose, a node module "truffle-hdwallet-provider" should be installed and imported into the project.

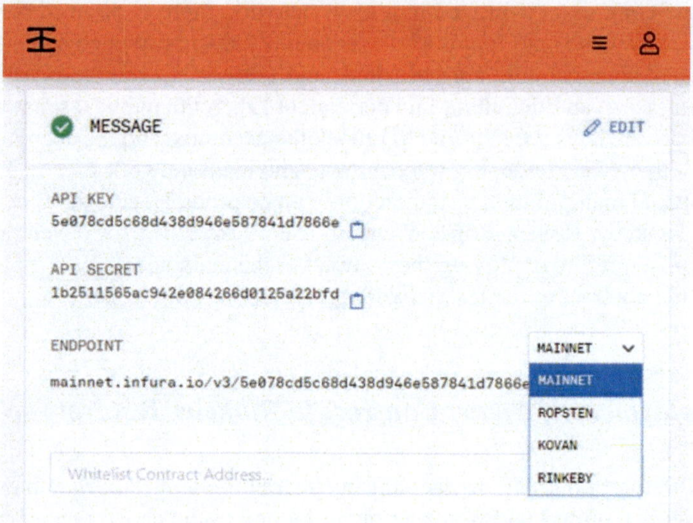

Fig. 7 Infura interface

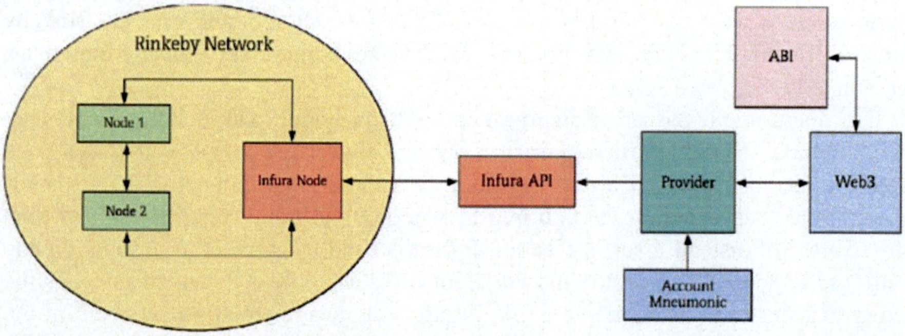

Fig. 8 Connection through Infura API

```
const HDWalletProvider = require('truffle-hdwallet-provider');
const Web3 = require('web3');
const provider = new HDWalletProvider (
  'inject begin slice burger pen midnight calm never latin face left retire',
//Mnemonic
  'https://rinkeby.infura.io/v3/5e078cd5c68d438d946e587841d7866e'        //Infura-
Rinkeby-link
);
const web3 = new Web3(provider);
```

As mentioned, when a smart contract is compiled by a solidity compiler, it generates a bytecode and ABI, which are required in the deployment of the contract on to the rinkeby network. To interact with the Ethereum network, the Web3 instance is used. For example, the script to get access to the accounts in the metamask which is given below.

```
const accounts = await web3.eth.getAccounts();
```

The Web3 library has many modules inside it which are used to work with different types of networks. Here, "eth" is an Ethereum module in Web3 which has a method "getAccounts()." The await key word is used since it is an asynchronous method.

After getting the accounts, the contract can be deployed using one of the accounts. The contract deployment script,

```
const result = await new web3.eth.Contract(JSON.parse(interface))
  .deploy( { data: '0x'+bytecode, arguments: [' ']} )
  .send( {gas: '1000000', from: accounts[0] } ); //gas limit
```

The "Contract" in the above code is a constructor in the Ethereum module for creating a new contract, which has "interface (ABI)" as an argument. The "deploy ()" is a method in contract which has "bytecode and arguments: [15]" as the arguments. The arguments array contains all initial values or messages required for the contract. The "send ()" method takes two arguments: One is the gas limit, and the other is the account address from which the contract is deployed. The send method will actually send this transaction to the network and the return the address with many other parameters to the result variable. The contract is deployed to rinkeby Test network and the address value can be known from result.options.address.

Now, the contract is deployed, a front-end app or a webpage should be created which can interact with the deployed contract. For this, the contract address and the ABI should be imported to the app. Whenever the metamask is running inside the browser, it will automatically inject an instance of Web3 library into the current active page. This Web3 also has a provider which is connected to the public networks available in the metamask. Figure 9 interprets the mechanism of front-end and the deployed contract. Using this Web3 with provider, contract address, and the ABI form the contract, the app will be able to interact with the contract on the rinkeby network.

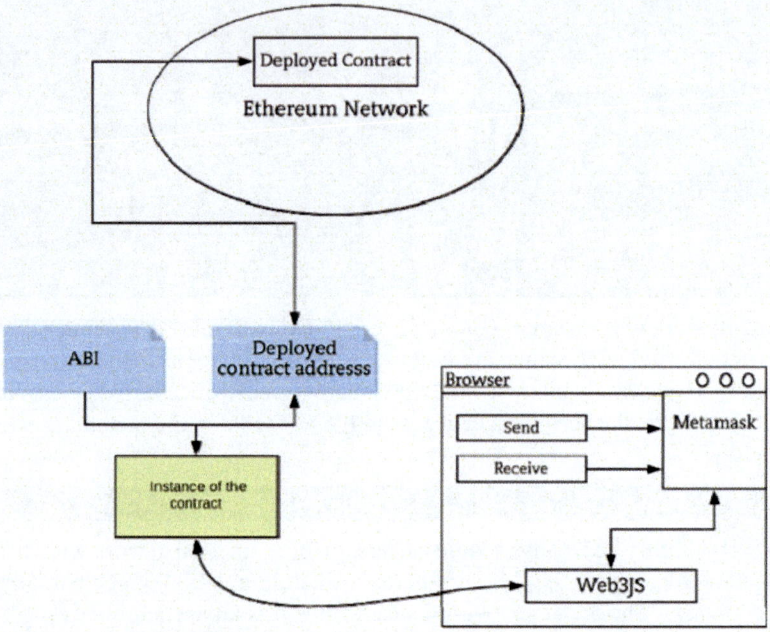

Fig. 9 Interaction with front end

5 Conclusions and Future Scope

The main goal of our research work is to develop an approach for addressing the fundamentals of the visionary blockchain technology which is pertinent in revolutionizing current centralized economy to a decentralized one. We have also discussed consensus algorithms in which flaws are identified, and it paved the path for new consensus algorithms like Casper, which are aimed at reinforcing the current algorithms. We have also proposed a novel approach of deploying a smart contract and interacting with it. Our work can be extended in various directions. In this paper, we have tried to explore some of the unexplored areas related to Ethereum, smart contract, Web3 library. But, still, several unexplored points and open issues related to research reported in this paper, both theoretical and practical needs further investigation. Now, we briefly outline the following possible extensions to our work:

1. Smart contracts are only as smart as their creator.
2. Bug bounty programs on smart contracts have sprung to audit and improve them.
3. Ethereum and smart contracts are presumed to be blockchain 2.0, which in turn evolved at breakneck speeds and let to blockchain 3.0, where Cardano, Zilliqa, and Aion are few examples.
4. Our methodology can be extended to address the problems of kickstarter, e-voting, secure land records maintenance, and other applications, etc.

References

1. S. Nakamoto, Bitcoin: A Peer-to-Peer Electronic Cash System. (2008).
2. E.K. Lua, J. Crowcroft, A survey and comparison of peer-to-peer overlay network schemes. IEEE Commun. Surv. Tutorials **7**(2), 72–93 (2005)
3. J.S. Coron, What is cryptography? IEEE Secur Privacy J **12**(8) (2006)
4. BitFury Group. Proof of Stake versus Proof of Work White Paper (2015)
5. Bitcoin Energy Consumption Index. https://digiconomist.net/bitcoin-energy-consumption. Last accessed 2018/07/20
6. Matt Springer. Is Bitcoin currently experiencing a selfish miner attack? https://scienceblogs.com/builtonfacts/2014/01/11/is-bitcoin-currently-experiencing-a-selfish-miner-attack/. Last accessed 2018/07/23
7. How Bitcoin Mining/Block Rewards Work. https://www.anythingcrypto.com/guides/bitcoin-mining-block-rewards-2018. Last accessed 2018/07/24
8. https://bitinfocharts.com/ethereum/. Last accessed 2018/07/23
9. Proof of Stake Instead of Proof of Work. https://bitcointalk.org/index.php?topic=27787.0. 2018/07/25
10. Marko Vukolic. The quest for scalable blockchain fabric: Proof-of-work vs. bft replication, in *Proceedings of the IFIP WG 11.4 Workshop iNetSec 2015, Open Problems in Network Security*, LNCS, vol. 9591 (Springer, Zurich, Switzerland), pp. 112–125 (2015)
11. V. Buterin, Proof of stake: How I learned to love weak subjectivity, in *Ethereum Blog*. https://blog.ethereum.org/2014/11/25/proof-stake-learned-love-weak-subjectivity/. Last accessed 2018/07/26
12. V. Buterin, *A Next-Generation Smart Contract and Decentralized Application Platform*. https://github.com/ethereum/wiki/wiki/White-Paper/. Last accessed 2018/07/27
13. Ethereum Docs. https://www.ethereum.org/. Last accessed 2018/07/27
14. Travis Patron. What's the big idea behind Ethereum World Computer? https://www.coindesk.com/whats-big-idea-behind-ethereums-world-computer/.Last accessed 2018/07/28
15. https://github.com/ethereum/wiki/wiki/Ethereum-Virtual-Machine-(EVM)-Awesome-List. Last accessed 2018/07/28
16. EtherTweet: Microblogging on the Ethereum Blockchain. https://ethertweet.net/. Last accessed 2018/07/27

An Investigation into the Usability of Blockchain Technology in Internet of Things

Sandeep Kumar Panda, D. Chandrasekhar Rao, and Suresh Chandra Satapathy

Abstract Internet of Things (IoT) has been the primary way for assuring the usable devices becoming smart and intelligent. The main challenge in IoT is to identify effective IoT issues. There are many challenges and difficulties have to face with this technology in the area of security and storage. In this regard, it is the responsibility to assure the user that devices are trustworthy. Blockchain technology is the key technology to provide every kind of security to Internet of Things using the decentralized architecture and peer-to-peer network connection. Cloud storage technology is very much reliable about the data storage. In this paper, we proposed a novel model for smart home application which focuses about the challenges while integrating blockchain technology with Internet of Things.

Keywords Internet of Things · Blockchain · Decentralized architecture

1 Introduction

Process of making your home, village, cities, devices into smart home, smart village, smart city, and smart device, respectively, by connecting it to the Internet is Internet of Things (IoT). Nowadays, IoT plays a vital role to make thing smart, reduce man power, saves time in every field. In this cutting edge technology, things are able to share their data with each other easily through Internet. Devices and objects are

S. K. Panda (✉)
Department of Computer Science and Engineering, Icfai Tech (Faculty of Science and Technology), ICFAI Foundation for Higher Education (Deemed to be University), Hyderabad, Telangana, India
e-mail: sandeeppanda@ifheindia.org

D. C. Rao
Department of Information Technology, Veer Surendra Sai University of Technology, Burla, Odisha, India
e-mail: dcrao_it@vssut.ac.in

S. C. Satapathy
School of Computer Engineering, KIIT Deemed to be University, Bhubaneswar, Odisha, India

V. Bhateja et al. (eds.), *Data Engineering and Intelligent Computing*,
Advances in Intelligent Systems and Computing 1407,
https://doi.org/10.1007/978-981-16-0171-2_53

Fig. 1 A clear vision about IoT

built with sensors shares their data to the hub which is later integrate from different devices and applies analytics to fulfill the specific needs of the user while fulfilling the needs of the user. Every industry welcomes this technology with a huge success, and this amazing concept of IoT is introduced by Kew Ashton, co-founder of MIT's Auto-ID centers in 1999 by presenting radio frequency identification as essential to the Internet of Things [1].

Figure 1 gives us a clear vision about IoT works which change the physical world into virtual world with the help of Internet is called nothing but Internet of Things.

By integrating blockchain technology with IoT, blockchain helps in improving the security cause, storage cause of IoT. The process of making a transaction in a valid and transparent way without involve of third-party authentication is called blockchain technology. It will be not a mistake if one can say that blockchain technology is the technology of trust. As an example, to understand what blockchain technology is, assume A wants to send $100 to B but, A can't send the money without transaction fees and a third-party guarantor. But, in blockchain technology, A can send the money not only to B but also to many people at the same time.

In this paper, the challenges, issues, and benefits are discussed when blockchain is integrated with IoT and analyze the performance when different blocks are connected with IoT device.

The rest of the paper is structured as mentioned. In Sect. 2 describes the introduction of blockchain technology. Next, benefits and challenges of blockchain technology with Internet of Things are presented in Sect. 3. In Sect. 4, the different platforms are used for blockchain technology for Internet of Things which is discussed. In Sect. 5, the proposed model for a blockchain-based smart home application along with its architecture is discussed. Finally, the conclusions and future scope are extensible described in Sect. 6.

2 Basic Concepts

There are two types of blockchain, i.e., public blockchain and private blockchain [2]. In public blockchain, anyone can read and write data into it, whereas in private blockchain technology, only known entities can change data into it. Transaction takes place in blockchain technology between two users provided with two keys i.e. public key and private key. Digital signature cryptography is also one of a media used for recognition of transaction in two phases, i.e., signing phase and verification phase. In signing phase, sender sends the encrypt data with private key and receiver receives the signal and decrypt with his public key and recognize the transaction that receiver receives the transaction from correct user. Sometimes, conflicting of transaction is possible. Consensus algorithm [3] is used to select which transaction is going to take place. By the longest chain subsequence rule, the block which provides the longest chain will participate in blockchain. To validate whether the transaction is correct or not, there are some miners who solve a cryptographic puzzle and generate a number, called mining technology. Miners are nothing but computer user who solves the puzzle first shares the solution to all other computers on the network so that the miner is awarded with 12.5 bitcoin that is of worth $200,000. This is called proof of work (PoW) [4]. To participate in a blockchain, every block should have validate and unique hash to connect with one another. Every block has a parent block with whom every block is connected. The first block of the blockchain is called genesis block. The technology in which hash is generated is called hashing technology. Hashing is the technology in which it can take an input and give the output of a fixed length character, i.e., of 256-bit length. Every block provides the unique number solved by the minors along with data as input and by using algorithm called SHA-256 [5] to create the unique hash of the given block. When the hashing of one block is completed, it participates in blockchain.

3 Literature Review

The Internet of Things and blockchain technology are two trending technology discussed among top industries. Combining these two technologies by producing blockchain-based IoT solution is giving more and more impact on business development, financial growth, and achieving significant cost efficiencies. From many points of view, IoT can hugely benefited from the services provided by blockchain technology, and the two most important benefit are uniformity and security. Apart from this, cloud computing has contributing tremendous effort on IoT technology to a huge success [6]. Basically, cloud computing works on centralized architecture in which all data are stored in one system so that every network does not have a clear vision on the data and how these data are processed, whereas blockchain works on decentralized architecture, where every system is connected with each other so that each and every system have a clear vision on data processing. For this architecture, now data should be reliable and traceable.

Therefore, blockchain is what a cloud is supposed to be. An interconnected and transparent network that removes the need for centralized authority, thereby solving the security risks of cloud computing.

4 Blockchain Platform for Internet of Things

Bitcoin was the first blockchain platform invented by anonymous person Satoshi Nakamoto. It is implemented as a secure payment system in every field. In IoT domain, devices accept this technology to make micro payments. As every technology set some instruction, bitcoin also set instruction or condition including scripting language while transaction.

Ethereum is also one of the most trusted platforms in blockchain technology [4]. Ethereum is nothing but a virtual machine (Ethereum Virtual KMachine) consisting of blockchain technology with including scripting language. For its community, it is the most popular platform to build block-based IoT application.

Hyperledger [5] is an open-source platform which has had really a good impact on which blockchain-based application is developed. This faster is usually faster than other platforms.

Quorum [6] is another blockchain platform that is widely accepted for its services in financial sector. Through consensus algorithm, it receives data and store securely. To create link between physical assets and blockchain, the quorum platform has been used by chronicled [6].

5 Proposed Model—Smart Home

Who else would not be wanted to operate his own home smartly at his fingertips? IoT-based smart home does exactly the same thing for you. Whether you are in vacation, or in work or at somewhere, then do not be worried about your smart TV which is whether on or off, your refrigerator, your entertainment system all are at your fingertips. You can control them at anywhere through your smart phone or tablets. A decade before exactly the same thing is happened through X10 approach. In this architecture, devices are connected to electrical line through which data are transmitted from one device to another device. But sometimes, when there is huge data to transform sometime wire maybe noisy and traffic situation occurs. So, this is the biggest drawback of the system. To avoid such type of disturbance, ZigBee and ZWave architecture came to the picture. In ZigBee architecture, there are more ways to communicate with other device, whereas ZWare works on source routing algorithm.

Figure 2 shows how an IoT-based application works. In IoT-based application, device should connect with sensor, and for security reason, a blockchain technology should be applied. In the blockchain technology for IoT-based application, a policy header is there which allows the user to control all the transaction happened in his smart home. If owner permits by providing them by a key, then device can connect with each other, and this algorithm is called Diffie Hellman algorithm. Every smart device stores their data in cloud storage from where the service provider collects the data and after analysis provide smart instruction to the user. The data stored in

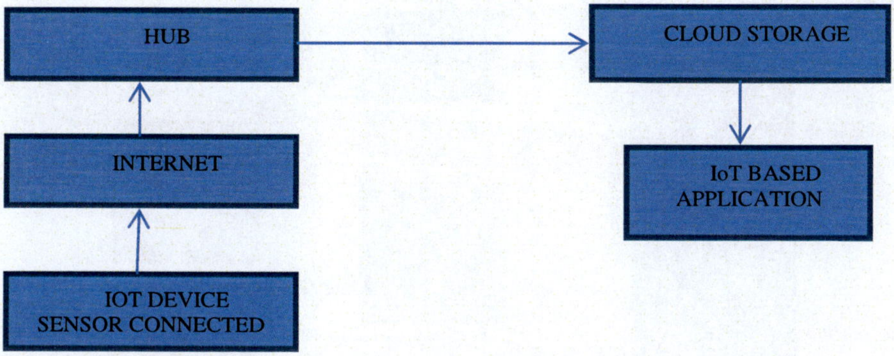

Fig. 2 Working of IoT

cloud storage should have a unique block number and hash number which gives the authentication to the user. In cloud storage, data stored in first come first out service. After completion of one block, next block is ready to encrypt to the cloud storage. As we know every block has unique hash number so there is no chance for data collision. Here, the user can decide to store the data in which cloud storage. In the next section, we will briefly discuss about the block-based IoT architecture of application.

5.1 Architecture

1. Device sends data to store in cloud storage
2. To generate block number and hash number
3. Different block number and hash number generated
4. Send the generated block number and hash number to cloud storage
5. Sends back with a unique block number
6. Stores the transaction.

Figure 3 shows the whole architecture of block-based IoT application. In the above figure, it shows how the device stores the data in cloud storage and stores the details of transaction in blockchain.

User has to create an account to upload data in cloud storage. When device wants to store the data in cloud storage, it sends the data to miner first like in the above figure the refrigerator and oven send their data to their respective miner. Miner applies the blockchain concept and receives a block no. and hash no., and the miner sends the data to cloud storage. Cloud storage store the data and for configuration, sends a unique block no. to miner and miner sends the block no. to blockchain to store the details of the transaction. In IoT application, it is not necessary to store every devices data in single cloud storage. As an example, in a smart home, one user does not want to share his data with other person, and then, he can store his data separately in other

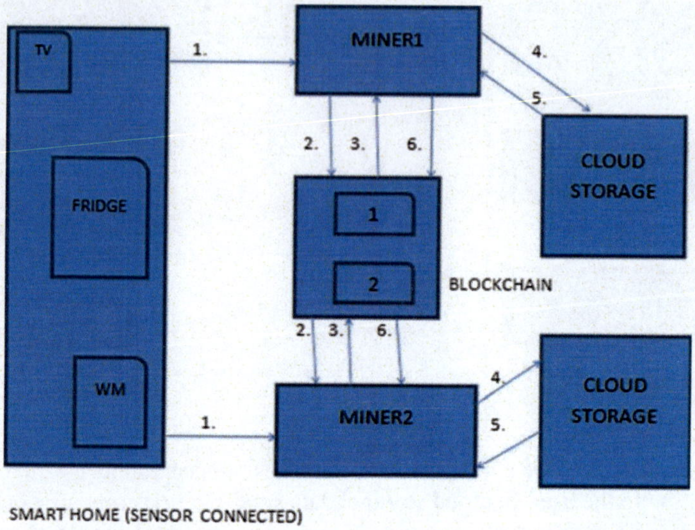

Fig. 3 Blockchain-based IoT architecture

cloud storage. In this figure, the refrigerator and oven store their data in different cloud storage.

Some example of smart devices and their work—

- Lock and unlock the door whenever you want
- Get notification on your mobile from refrigerator and kitchen when ingredients are going to finish
- Warm the water, bedroom automatically whenever you want.
- Get notification about your daily routine
- Get recipe list from your kitchen.

5.2 *Monitoring*

Monitoring phase tells us about how user operates their smart device at their fingertips at any time.

1. Send me the current availability in refrigerator
2. Has refrigerator has a access to chain
3. Yes, it has access to chain
4. Send me the current availability of the device
5. Data sent
6. Data sent.

In Fig. 4, the user wants to know the availability of food in his refrigerator. User asks for data to the miner. When miner gets the instruction, immediately it

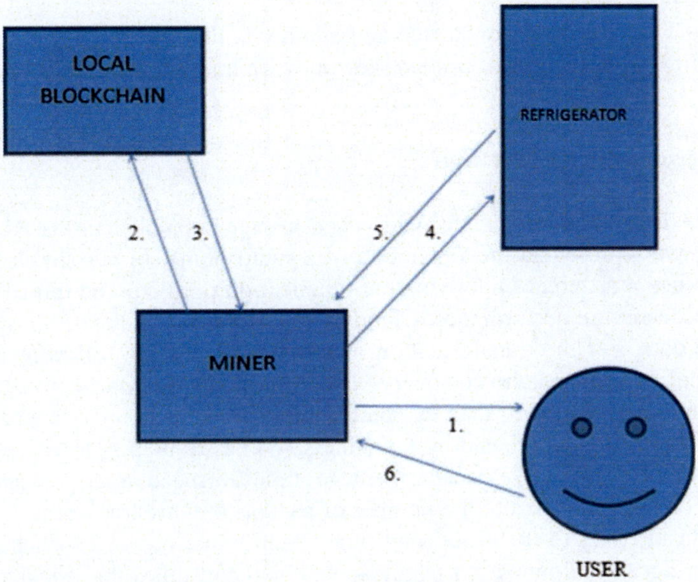

Fig. 4 Process of monitoring

asks permission to local blockchain. After getting the permission by the transaction holder, it asks the refrigerator for data, and after the refrigerator sends the data to the miner, miner sends the desired data to the user.

5.3 Evaluation and Analysis

In this section, we discussed security analysis and performance evaluation related to our proposed work.

5.3.1 Security Analysis

For security analysis, every smart device needs to be concerned about three things, i.e., confidentiality, integrity, and availability. Confidentiality describes about identity, whereas integrity refers to check whether the message is delivered or not, and availability means the data availability. To increase the number of smart home availability, the smart home is to be protected from different attackers. This can only be done by reducing the number of transactions stored in blockchain in every second, so the processing delay should decrease as compared to the existing smart home. To increase the effectiveness of the smart home, we should take care of two security attacks. One is distributed denial service (DDOS) attack, where attacker uses the infected IoT device over new device and another is linking attack, where attacker creates a link with the help of primary key between transaction and data layer.

5.3.2 Performance Evaluation

To analyze the evaluation, transformation and storage in blockchain-based architecture, we have represented the simulation of a smart home. In our simulated smart home, we use a different underlying communication protocol and three simulated sensors to detect the performance. Cloud storage collects the data from miner and stores the data. For the transformation of transaction, we are following two flow patterns, i.e., periodic pattern and query-based pattern. In periodic approach, devices will continuously send their data to cloud storage, whereas in query based, devices will response to the query send by the miners. While simulating these tasks, more evaluation will be done like packet overhead, time overhead, energy consumption, etc. Packet overhead refers to the number of packets that are transformed, whereas time overhead refers to the processing time taken by the miner. Blockchain-based IoT design doesn't allow packet overhead and also consumes the valuable time of the user.

6 Discussion

If we are going to accept a good thing, then we must have to face some challenges. Similarly, there are a number of challenges takes place while connecting blockchain technology with Internet of Things.

Uniformity

Blockchain technology is a trustworthy technology for IoT devices and application [7]. Implementing blockchain technology in IoT device can help users to identify the device and data provided by the device.

Storage

After a lot of success from every field, still blockchain has storage issue that still not solved. Combining IoT application with blockchain technology is a great challenge. Blockchain technology is made for transaction process not for storing a huge data produced by blockchain-based IoT application. So, it stores the data in cloud storage by implementing blockchain technology. When one user wants to keep the data in cloud storage, it sends the data along with the unique number solved from cryptographic puzzle by miner, and with the help of hashing technology, it produces a hash to match with the previous data stored by the block. After confirming that there is some place in cloud storage to store some data, it matches the hash code with the previous one and after matching the data stored in cloud storage. This is how block-based IoT application stores data in cloud storage. Now, using the data mining technology, data are compressed from bigger size to smaller size, so that storage of high volume of data is possible.

Security

In IoT technology, devices are operated under some defined mechanism and protocol. The most secured standard protocol is opted with the application of blockchain technology in IoT. IoT connect so many devices through Internet and come across a huge data which should be secured from fraud attack, hackers, and some unacceptable things [8]. Security is a major concern in any application. The increasing number of hackers, frauds, is nowadays gives more challenging tasks to user. Basically, IoT devices are work on sensor. Data travels through the sensor to the system, as a result data misplace chances are higher than the other system. So, blockchain technology is a key technology to provide security improvements in IoT application. The main question arises after integration of blockchain with IoT is data reliability. As blockchain technology works on decentralized architecture, data travels from one to another in a transparent way. So, any unusual things can't happen with the data. Sometimes due to the failure of devices, the sensors are damaged to work properly. In this situation, many hackers are try their best to capture the data. This could not be happened if the device is properly checked before use. This is the reason the IoT device are properly checked before their integration with blockchain technology.

7 Conclusions

In our research work, we have proposed the issues faced during implementation of blockchain technology with Internet of Things integrated with home automation system. The main aim of our research work is to highlight the issues related to implementation of smart things and providing the solution by using blockchain technology. We recovered the issues by considering consensus algorithm and smart contract. In this paper, we explored some areas of blockchain technology which were yet to be unexplored. Here, we describe the virtual things with Internet through blockchain technology. In our future work, we can provide the reliability and security of smart devices by using blockchain technology.

References

1. K. Ashton, That "Internet of Things". RFID J. **1** (2009)
2. M. Bob, *Blockchain: The Invisible Technology that's Changing the World*. USA (2017)
3. N. Giang-Troung, K. Kyungbaek, A Survey About Consensus Algorithms used in Blockchain **14**, 101–128 (2018)
4. A. Biryukov, D. Khovratovich, I. Pustogarov, Deanonymisation of clients in Bitcoin P2P network, in *ACM Conference on Computer and Communication Security (CCS)*, pp. 15–29 (2014)

5. P. Rahul, *Optimising the SHA256 Hashing Algorithm for Faster and More Efficient Bitcoin Mining* (2013)
6. M. Aazam, E. Huh, Fog computing and smart gateway-based communication for cloud things, in *Proceeding of the 2nd International Conference on Future Internet of Things and Cloud*, pp. 27–29 (2014)
7. S. Gan, An IoT Simulator in NS3 and a Key-Based Authentication Architecture for IoT Devices using blockchain, pp. 234–345 (2016)
8. D. Ali, S. Salil, J. Raja, Blockchain in Internet of Things: Challenges and Solution. The University of New South Wales (UNSW), Introduction to Cryptography. Springer Science & Business Media (2013)

Optimization of Block Query Response Using Evolutionary Algorithm

Sandeep Kumar Panda, Siba Prasad Dash, and Ajay Kumar Jena

Abstract The term blockchain can be practically viewed as a database as storage, and retrieval of information can be performed within blocks. But database cannot be termed as blockchain. In the simplest definition, it is the time stamped series of data along with heterogeneity which is managed by a cluster of systems with security implementing the cryptographic applications. Practically, it uniformly stores the information in blocks, where each block contains the hashed information from the previous block to provide cryptographic security. This implementation mechanism is absolutely decentralized supporting peer-to-peer storage of data to all the access points. Accordingly, it captures the prerequisite and relevant information logically structured in blocks. Each block is attached to its preceding blocks by applying the cryptographic generic function, where the details of the initial block are required to be focused. Based upon the non-recursive processing of data, it cannot repeat the data already validated in the block. In case of decentralization, rather having faith on single entity, it is essential to focus on other linked entities and to maintain integrity of data during the application. After that as soon as the data is associated with the blockchain, the information can be stored permanently, and the particular block will be associated with the succeeding blocks to update the information. In this paper, it is intended to optimize the response of queries blocks implementing evolutionary algorithms approach.

Keywords Block · Evolutionary · Query · Optimization · Cluster

---

S. K. Panda (✉)
Computer Science and Engineering, IcfaiTech (Faculty of Science and Technology), ICFAI Foundation for Higher Education (Deemed to be University), Hyderabad, India
e-mail: sandeeppanda@ifheindia.org

S. P. Dash
National Informatics Centre, Odisha State Centre, Bhubaneswar, Odisha, India

A. K. Jena
School of Computer Engineering, KIIT Deemed to be University, Bhubaneswar, Odisha, India

V. Bhateja et al. (eds.), *Data Engineering and Intelligent Computing*,
Advances in Intelligent Systems and Computing 1407,
https://doi.org/10.1007/978-981-16-0171-2_54

1 Introduction

Considering the chaining mechanism, it is understood that the new block usually maintains the procedure from fetching and transacts the data focusing on the list of unused data of specific address. The indexing technique in this case maintains all the block information performing the search expansion and responsible towards evaluation of queries. It is also clear that the blockchain technology supports the processing of queries efficiently and maintains the integrity of data. Somehow, the implementation mechanism in some specific applications is bit challengeable. Many times, the protocols linked with this technique seeks the identification of space with objects during processing. As such the mechanism with sequential approach is not so effective to obtain optimality during query processing. In such case, the cryptographic approach can be adopted during propitiating the application with traceable data. Also, it will be essential to concentrate on the scale of system implementation as well as support of query system. Sometimes, linking with coded keys as well as data generation tends to location of the queries and processed data. So, the main challenge here is the frequent updates of the data values. As the authenticated data value captures the entire relevancy during processing, managing the complete transaction of data is really a great challenge. The technology linked with these transactions will initiate from the initial block and update the position up to final block. Categorizing on multiple databases with heterogeneity and focusing on the specified range of queries, it is observed that the optimized cost of queries depends on the recursive traversal of queries within the range as well as relational operators applied to them. Accordingly, the resultant queries transact linking with the blocks within the specified timestamp.

2 Review of Literature

Casino et al. [1] in their work have focused on blockchain technology and discussed various impacts associated with decentralization, immutability and transparency capabilities. They have also in their work discussed the developmental issues linked with crypto currencies.

Draskovic et al. [2] during their research have cited the examples of cryptocurrencies and observed the significance of permitting individual entity with data from sensors during communication as well as enabling the cryptocurrencies towards blockchain applications.

Hartgerink et al. [3] in their study observed the dissimilarities of blockchains linked with scarcity as well as market principles. In this regard, they have also impacted towards principle and proper implementation of blockchain techniques.

Tennant et al. [4] in their research practically considered the timestamp linked with institutional framework on specific domain. They have focused on basic rules and parameters associated with datasets, hypothesis, etc., and imparting the same with specified framework.

Anwar et al. [5] in their work have signified the distributed technology approach linked with data and applications with blockchain. They have also observed the implementation mechanisms of distributed framework as well as synchronization of transactions. They have also found similarities with data in blocks linked chronologically during cryptographically applications.

In [6] many applications linked with blockchain have been discussed. While focusing on timestamps, it is observed that specific data can be accumulated and can be reliably processed concurrently by applying crypto currencies.

Deshpande et al. [7] in their work have prioritized the application of blockchains. As per their survey, the enhancement as well as accessibility of blockchain applications is the major factors towards designing and implementing new redundant interfaces.

In [8], the specific set of rules like ocean protocol is being responsible towards incentivizing the re-usage of data implementing the blockchain technology in the system which can be used to link the data with smart real-type applications.

Dannen et al. [9] in their work have focused on smart contracts linked to counteract vulnerabilities. During their research, they have come through the possibilities of implementation of commercial service providers focusing on runtime verification with security.

In [10], the specific concept named as coalition of automated legal applications. Intellectual property is introduced which is also community-driven protocol and responsible to form a consistent framework and to eliminate the dependence linked with central servers phase.

3 Application of Particle Swarm Optimization Towards Query Response as Well as Streaming the Blocks

In general, the particle swarm optimization is otherwise termed as metaheuristic approach and is linked with autonomous components. In this case, the particle searches for optimality in the process of application. Somehow, the implementation of the algorithm is based on the particles in the form of data packets. Each data packet will be associated with other relevant information along with the position of the corresponding particle while initiating the search mechanism. Accordingly, the implementation of specific tasks, as well as clusters can be efficiently carried out implementing the technique.

3.1 *Algorithm 1: Listing the Query Response and Streaming the Blocks*

Step 1: Obtain the query sets and fix the recursion parameter, recn. Initially, recn = null.

Step 2: Link the query sets with other databases updating the recursion parameter, recn.

Step 3: Obtain the join index value as well as point of split of queries.

Step 4: Perform sorting operation to the queries based on join index value as well as query performance index.

Step 5: Assign the recursion parameter, recn and join index value to blocks.

Step 6: Obtain the updated performance index of queries along with join index values.

Step 7: Simulate the blocks containing updated recn and join index values to obtain optimality.

3.2 *Algorithm 2: Linking Blocks Towards Generalization Based on Abstraction*

Step 1: Obtain the size of blocks.

Step 2: Update the query performance index linked with each block and evaluate the query execution time during query transact.

Step 3: Find the query specifying nodes with processing elements assigned with each block.

Step 4: Obtain the optimality of each block.

Step 5: Based on query terms and recursion parameter of each block obtain the query transact response and simulate each block.

Step 6: Generate the updated query evaluation performance index of each block and search for other unconnected blocks to obtain optimized result.

Step 7: Reconfigure again all the linked blocks and apply the aggregation mechanisms.

Experimental analysis

Considering the Fig. 1 as well as Table 1, it is clearly understood that during process of execution, the query terms are directly dependent upon the number of blocks associated with the system.

Similarly, considering the Fig. 2 as well as Table 2, categorically time of execution of queries depends on the databases linked within the system.

As cited in Fig. 3 and Table 3, the consistency of query execution plans is maintained up to certain level. Considering fixed size queries, the efficiency of query plans is enhanced with linkage of more servers in the system.

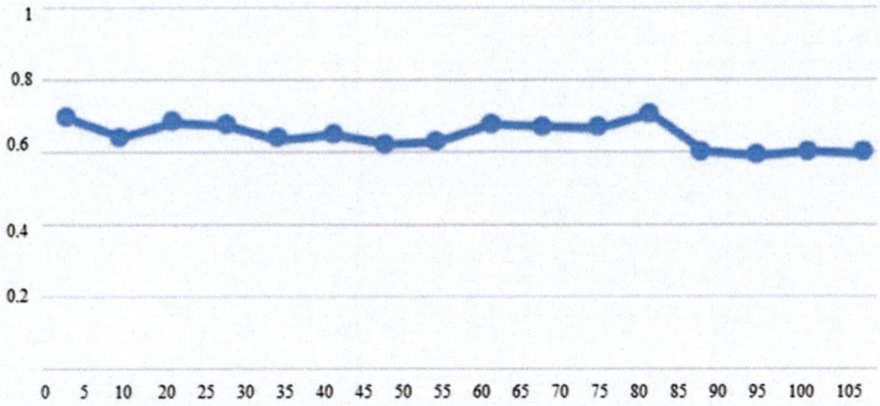

Fig. 1 No. of blocks with performance of query terms in execution process

Table 1 Performance evaluation of blocks with query term execution

Sl. no	No. of blocks	No. of query terms per block	Cost of query terms during execution process (ms)
1	28	55	0.65
2	38	67	0.62
3	68	82	0.64
4	80	91	0.68

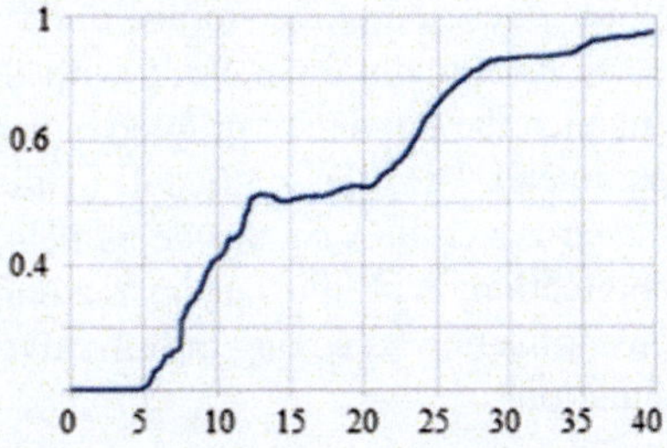

Fig. 2 No. of databases with execution of queries

Table 2 Association of databases with query execution

Sl. no	No. of associated databases	Cost of execution of queries (ms)
1	18	0.52
2	30	0.83
3	38	0.90
4	41	0.97

Fig. 3 No. of servers with query execution plans

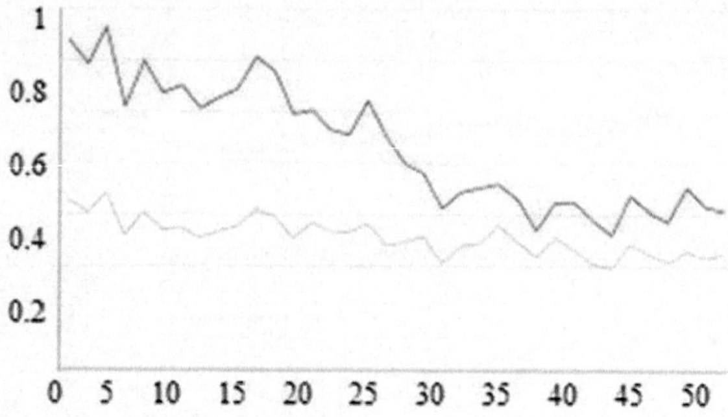

Table 3 Association of servers with query execution plans

Sl. no	No. of database servers	Queries associated with each database	Cost of query execution plans
1	20	90	0.82
2	30	90	0.65
3	40	90	0.57
4	50	90	0.47

4 Discussion and Future Direction

While considering the storage of data as well as query processing mechanisms, it is essential to maintain integrity with all the associated blocks, and also it is required to enhance the efficiency of queries updating the query evaluation performance index parameter. The rules applicable in this case during linking the entities and particularly the strong entities should be authenticated with proper identification of keys in the application. With the varying size of queries, the experimental results should focus on application achieving the effectiveness and obtaining optimal solution in practical situations.

5 Conclusion

Practically, the blockchain technology is provisioned with appropriate solution to enhance the device to device communication security issues, and as such it is a great challenge in present situation. The application in such cases should address the issues related to security measures towards Internet of Things. Also, the blockchain technology is deployed to concentrate on the issue linked with decentralized systems. To authenticate and judge each cluster within the blocks as well as to stream each block, the metaheuristic technique in this case is applied. It is also observed that achieving maximum security measures, the blockchain technology provides the centralized authentication mechanism prioritizing the individual local blockchains.

References

1. F. Casino, T.K. Dasaklis, C. Patsakis, *A Systematic Literature Review of Blockchain-Based Applications: Current Status, Classification and Open Issues* (Telemat, Informat, 2018).
2. D. Draskovic, G. Saleh, *Datapace—Decentralized Data Marketplace Based on Blockchain.* Available online at: https://www.datapace.io/datapace_whitepaper.pdf (2017)
3. C. Hartgerink, *Concerns About Blockchain for Science. Medium.* Available online at: https://bit.ly/2kloBDB. Google Scholar (2018)
4. J.P. Tennant, The state of the art in peer review. FEMS Microbiol. Lett. **365**, 1–10 (2018). https://doi.org/10.1093/femsle/fny204,PubMedAbstract|CrossRefFullText|GoogleScholar
5. Anwar H. *Blockchain vs. Distributed Ledger Technology.* Available online at: https://bit.ly/2SFTRZ0, Google Scholar, 2019.
6. K. Werbach, Trust, but verify WhyThe blockchain needs the law. Berkeley Tech. Law J. **33**, 487–550 (2018). https://doi.org/10.15779/Z38H41JM9N,CrossRefFullText|GoogleScholar
7. A. Deshpande, K. Stewart, L. Lepetit, S. Gunashekar, Distributed Ledger Technologies/Blockchain: Challenges, Opportunities and the Prospects for Standards. Technical report, The British Standards Institution (BSI) (2017)
8. Ocean Protocol. *The Ocean Protocol- A Decentralized Data Exchange Protocol to Unlock Data for AI.* Available online at: https://oceanprotocol.com/tech-whitepaper.pdf (2019)
9. C. Dannen, *Introducing Ethereum and Solidity: Foundations of Cryptocurrency and Blockchain Programming for Beginners*, 1st edn. Apress, Berkeley, CA. doi: https://doi.org/10.1007/978-1-4842-2535-6 CrossRef Full Text|Google Scholar (2017)
10. Vaultitude. *Vaultitude—Intellectual Property Blockchain Platform.* Available online at: https://bit.ly/3208YBO (2018)

Optimized Trust and Reputation-Based Routing Protocol in Delay-Tolerant Networks

Pradosh Kumar Gantayat and Satyabrata Das

Abstract The delay-tolerant network (DTN) has captured the interests of research scholars for its increasing importance in the demanding environment. Routing of data among two nodes when they are not within the communication range is a tedious work in DTNs as the end-to-end routing path cannot be guaranteed. In our proposed work, we propose a trust and reputation-based routing protocol in the DTNs to provide secure transmission of data between the destination and source. For the transfer of data, a node nearer to the source must be normal and should possess increased trust and reputation. The proposed trust and reputation-based protocol consists of acknowledgments, node lists, and aging steps to make effective communication. The forwarding node is chosen using reputation and trust-based context-aware routing (RCAR) algorithm. Upon handling a message, the dispatcher waits for acknowledgment from receiver, with which the sender increases the trust and reputation of the forwarding node. After the receiving of the message, the node itself is added to the node list. The proposed trust and reputation-based routing protocol produced less delay and high packet delivery ratio (PDR) of 0.0427 and 0.9917, respectively.

Keywords Routing · Delay-tolerant networks · Trust · Reputation · Acknowledgment

1 Introduction

DTNs [1] are the sparse mobile networks where there exists no end-to-end path between sender and receiver so it is based on store-and-forward mechanism to transfer and finding out the next relay node toward destination [2]. That is, the DTNs utilize the store-and-forward approach unlike conventional networks where the packets are

P. K. Gantayat (✉)
Department of Computer Science and Engineering, Veer Surendra Sai University of Technology, Burla, Sambalpur, Odisha, India

S. Das
Department of Information Technology, Veer Surendra Sai University of Technology, Burla, Sambalpur, Odisha, India

V. Bhateja et al. (eds.), *Data Engineering and Intelligent Computing*,
Advances in Intelligent Systems and Computing 1407,
https://doi.org/10.1007/978-981-16-0171-2_55

forwarded over the fixed links [3, 4]. In the present era, there is an increased interest in the routing of DTN [5], but not more attention is given to misbehavior that arises in routing [6]. In DTNs, a node can misbehave by drop of the packets, which may be done intentionally, and these characteristics cause misbehaviors in routing [7]. Routing in DTN is not a simple work as it follows packet replication policy of the store–carry–forward technique. Replication may act as an effective problem, when the replicated packets exceed certain limits [8]. Incentive mechanisms [9] are used for the increase of the participation rate of the network nodes and are developed using trust-based reputation systems [10]. The selfish nodes are either incentivized with the trust for enhancing participation or avoided through isolation in passing the message. The trust-reputation-based routing works on only non-collaborative misbehaving nodes, which are unwilling to act.

The trust is utilized as the utility in obtaining the vector of exchange among the nodes that were encountered. The context-based routing needs the identification of nodes to perform routing. Thus, the privacy of node identity is very important in case of context-based routing, in addition to the preservation of message [11]. The challenges of routing in DTNs are: (i) DTNs do not have any previously available wireless or any wired-based networks. Hence, the conventional MANET protocols, such as AODV [12] and DSR, cannot be directly applied to DTN due to the need of fully connected graph to transfer a message. (ii) DTN nodes preserve the messages till finding an intermediate node to transfer the message to the sink node. Selecting the relay node is difficult as it deteriorates the performance that leads to loss of message. Due to the presence of communication delay, it is tedious to design an optimal forwarding criterion for the routing process [13]. DTNs comprise its routing protocols, and the conventional DTN routing protocols, like PROPHET [14], have adopted the pure opportunistic or social-based relay selection [2].

2 Literature Review

The literature survey is performed with four papers as follows: Gianluca and Angelica [15] developed the reputation-based protocol for contrasting the black holes that outperformed the conventional T-ProPHET with respect to the metrics, like average delay and delivery ratio, but when there occurred a loss of message, the node cannot know whether it is due to black hole or any other situation had occurred. Reina et al. [16] designed the multi-objective-based optimization framework that permitted the designer of the network to contain a complete discovery of the data dissemination algorithm with outputs. It produced worst performance in terms of consumption of energy. Miao et al. [17] modeled the protocol of privacy preserving probabilistic prediction-based routing (4PR) which provided enhanced preservation of privacy, but reduced the latency of delivery. Bucur and Iacca [18] developed two algorithms, where the number of generations needed is small and consistent, but cannot be applied to variable-sized individual's problems.

2.1 Challenges

The major challenges involved are: The provenance-oriented trust models in the sensor networks did not consider the behaviors of the attack [19]. The trust management schemes [20] addressed the selfish characteristics and the collaborative attacks present in the DTNs. However, these solutions were not able to deal the routing characteristics explicitly [7]. Solutions, like COCOWA [21] and CWS [22], aimed on avoiding the selfish characteristics and do not deal other misbehaviors [7]. An approach based on confidence factor to estimate the indirect trust was designed in [13] that produced a high trust value in case of indirect trust estimation and led to colluding attacks [7].

3 Proposed Reputation and Trust-Based Context-Aware Routing Protocol

The proposed trust and reputation-based routing protocol aims to develop a routing protocol in DTN using optimization technique. The trust and reputation-based mechanisms are efficient in providing security in terms of access control and trust of a node. When the reputation is less, the chance of a node to be a black hole is higher. The chances of choosing a forwarding node depend on its reputation, and thus a node with high reputation is selected as the forwarding node. The problem of lowering the effect of black holes in a DTN is proposed in this paper on the basis of trust and reputation. The proposed trust and reputation-based protocol consists of three main steps, namely acknowledgments, node lists, and aging, which make efficient communication and adjust to the variation in the operating criteria of the DTN.

3.1 Delay-Tolerant Networks

The DTNs act as the networks with decentralized features, where the nodes communicate with cooperative manner to transfer the data from sender to receiver. The topology of DTN is dynamic, as the nodes possessing high mobility create disconnections among the nodes, network partitioning, and increased end-to-end delays. However, as per the mobile ad hoc network, the DTNs mobility acts as a way to forward the information in an indirect way in an opportunistic manner.

3.1.1 Attacks Due to Malicious Nodes in Delay-Tolerant Networks

If a node acts selfish, it sends messages only if it is a friend of a source, or the sink node, whereas a well-behaved node acts unselfishly without considering the social ties. The malicious nodes normally aim to disturb the basic routing characteristics of the DTN. A malicious node may create the trust-oriented attacks.

3.1.2 Estimation of Trust and Reputation in Delay-Tolerant Networks

The trust and reputation-based methods for the transmission of information in DTN are effective in offering security in terms of trustworthiness and the access control of a node in the network.

$$\text{Trust and Reputation based Utility function, } \mathrm{TR}_{xy} = \left(G_{xy} + J_{xy} + K_{xy}\right) * M_{xy} \tag{1}$$

where G_{xy} indicates the direct trust, J_{xy} indicates the indirect trust, K_{xy} is the history trust, and M_{xy} is the utility function.

$$M_{xy} = P(E/A) \tag{2}$$

This equation is the condition of probability of event of message delivery to the event of attack.

$$P(E/A) = \frac{1}{\sqrt{2\Pi\mu^2}} \exp\left(\frac{-(W-\alpha)^2}{2\mu^2}\right) \tag{3}$$

4 Results and Discussion

See Fig. 1.

4.1 Performance Metrics

The metrics used in the analysis of the proposed trust and reputation-based routing protocol are delay and PDR.

4.2 Performance Analysis

See Fig. 2.

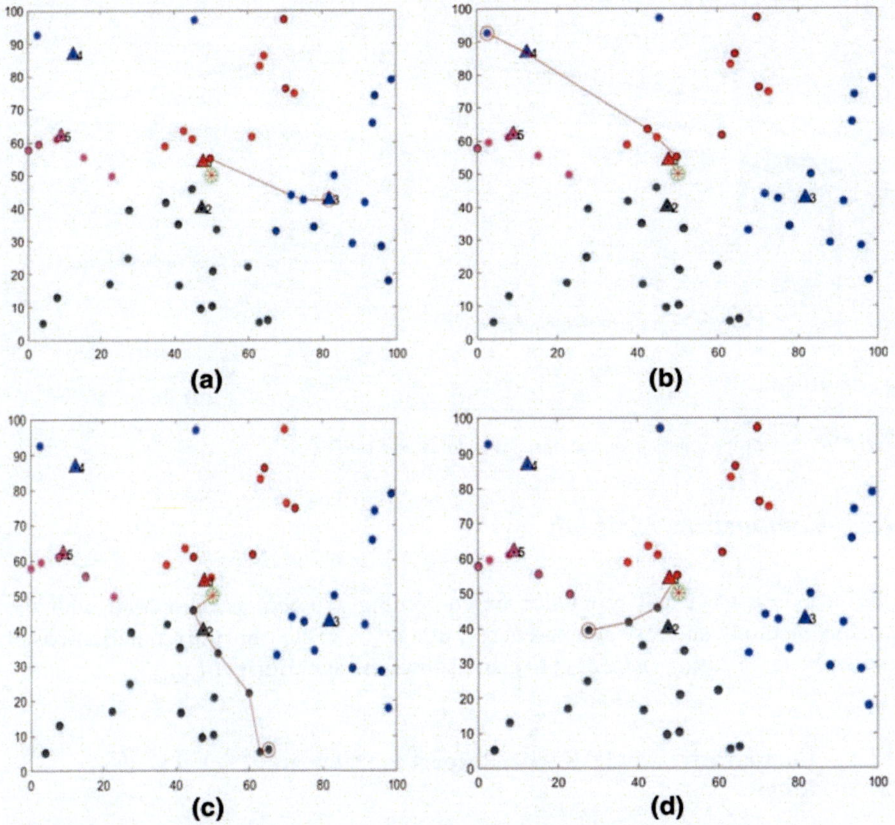

Fig. 1 Simulation result of proposed method: **a** path from cluster head 3 to the destination, **b** path through cluster head 4 to the sink node, **c** path through cluster head 2 to the destination, and **d** path from a node to the sink node

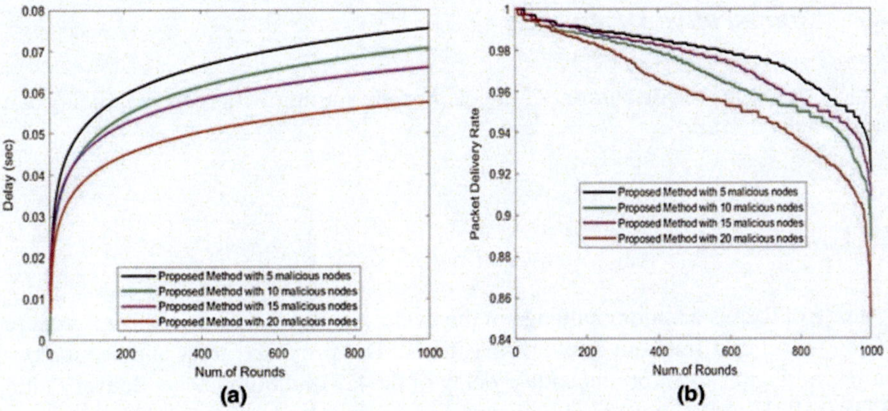

Fig. 2 Performance analysis based on **a** delay and **b** PDR

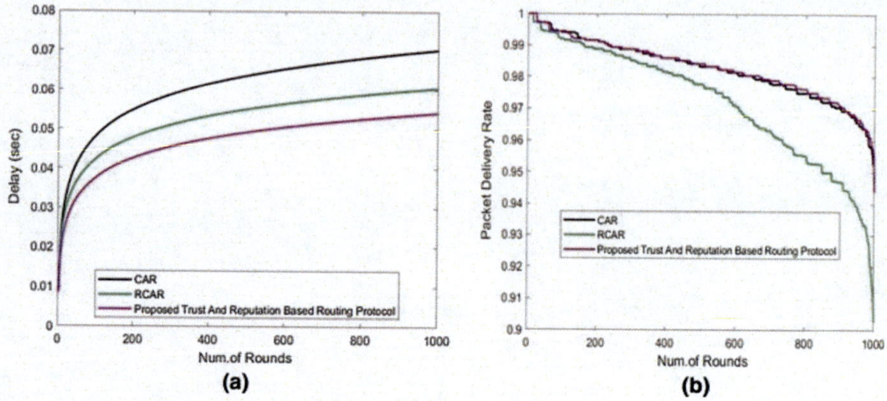

Fig. 3 Comparative analysis on the basis of **a** delay and **b** PDR

4.3 *Comparison Methods*

The proposed trust and reputation-based routing protocol is compared with the existing methods, such as context-aware routing (CAR) [15] and the reputation-based context-aware routing (RCAR) [15], and shows the superiority of it.

4.3.1 Comparative Analysis with Respect to Delay and Packet Delivery Ratio

See Fig. 3.

4.4 *Comparative Discussion*

Table 1 describes the discussion of the methods in routing with respect to delay and packet delivery ratio.

5 Conclusion

Routing of data is a major challenge in the delay-tolerant networks (DTNs) because of long-term and frequent network partitions. The proposed trust and reputation-based routing protocol produced less delay of 0.0427 and high packet delivery ratio (PDR) 0.9917 that shows its effectiveness in routing. In the future, DTN gateways will be used to lower the consumption of energy during the estimation of trust and reputation.

Table 1 Comparative analysis of the methods with respect to delay and packet delivery ratio

Number of rounds	Metrics	Methods		
		CAR	RCAR	Proposed trust and reputation-based routing protocol
200	Delay	0.0549	0.0479	0.0427
	PDR	0.9817	0.9889	0.9917
400	Delay	0.0615	0.0533	0.0476
	PDR	0.9761	0.9816	0.9861
600	Delay	0.0654	0.0565	0.0505
	PDR	0.9813	0.9711	0.9816
800	Delay	0.0681	0.0587	0.0526
	PDR	0.9750	0.9553	0.9763
1000	Delay	0.0702	0.0604	0.0542
	PDR	0.9369	0.9029	0.9438

References

1. P.K. Gantayat, S. Das, Award-reward protocol based on clustering and trust level for routing in delay tolerant network. J. Appl. Secur. Res. **14**, 270–287 (2019)
2. C.C. Sobin, V. Raychoudhury, G. Marfia, A. Singla, A survey of routing and data dissemination in delay tolerant networks. J. Netw. Comput. Appl. **67**, 128–146 (2016)
3. Z. Gao, H. Zhu, S. Du, C. Xiao, R. Lu, PMDS: a probabilistic misbehavior detection scheme in DTN, in *Proceedings of IEEE International Conference on Communications (ICC)* (2012), pp. 4970–4974
4. Y. Cao, Z. Sun, N. Wang, H. Cruickshank, N. Ahmad, A reliable and efficient geographic routing scheme for delay/disruption tolerant networks. IEEE Wirel. Commun. Lett. **2**, 603–606 (2013)
5. Y. Cao, Z. Sun, N. Wang, M. Riaz, H. Cruickshank, X. Liu, Geographic-based spray-and-relay (gsar): an efficient routing scheme for dtns. IEEE Trans. Veh. Technol. **64**, 1548–1564 (2015)
6. P.K. Gantayat, S. Das, Secure trust level routing in delay tolerant network with fuzzy empowered node categorization technique. J. Comput. Theor. Nano Sci. **16**, 2294–2298 (2019)
7. P. Asuquo, H. Cruickshank, C.P. Anyigor Ogah, A. Lei, Z. Sun, A distributed trust management scheme for data forwarding in satellite DTN emergency communications. IEEE J. Sel. Areas Commun. **6**(2), 246–256 (2018)
8. A.K. Singh, R. Pamula, IRS: incentive based routing strategy for socially aware delay tolerant networks, in *Proceedings of 5th International Conference on Signal Processing and Integrated Networks (SPIN)* (2018)
9. X. Jiang, X.Y. Bai, A survey on incentive mechanism of delay tolerant networks, in *Proceedings of 10th International Conference on Wavelet Active Media Technology Information Process* (2013), pp. 191–197
10. W. Sherchan, S. Nepal, C. Paris, A survey of trust in social networks. ACM Comput. Surv. **45**(4) (2013)
11. S.K. Dhurandher, A. Kumar, M.S. Obaidat, Cryptography-based misbehavior detection and trust control mechanism for opportunistic network systems. IEEE Syst. J. **12**(4), 3191–3202 (2018)
12. C. Perkins, E. Belding-Royer, S. Das, *Ad Hoc On-Demand Distance Vector (AODV) Routing*. IETF Network Working Group (2003)
13. A.M. Shabut, K.P. Dahal, S.K. Bista, I.U. Awan, Recommendation based trust model with an effective defence scheme for manets. IEEE Trans. Mob. Comput. **14**, 2101–2115 (2015)

14. A. Lindgren, A. Doria, O. Schelen, Probabilistic routing in intermittently connected networks. ACM SIGMOBILE Mob. Comput. Commun. Rev. **7**(3), 19–20 (2003)
15. D. Gianluca, D. Angelica, Towards a reputation-based routing protocol to contrast blackholes in a delay tolerant network. Ad Hoc Netw. **10**, 1167–1178 (2012)
16. D.G. Reina, R. Ciobanu, S. Toral, C. Dobre, A multi-objective optimization of data dissemination in delay tolerant networks. Expert Syst. Appl. **57**, 178–191 (2016)
17. J. Miao, O. Hasan, S. Mokhtar, L. Brunie, A. Hasan, 4PR: privacy preserving routing in mobile delay tolerant networks. Comput. Netw. **111**, 17–28 (2016)
18. D. Bucur, G. Iacca, Improved search methods for assessing delay-tolerant networks vulnerability to colluding strong heterogeneous attacks. Expert Syst. Appl. **80**, 311–322 (2017)
19. J.-H. Cho, I.-R. Chen, PROVEST: provenance-based trust model for delay tolerant networks. IEEE Trans. Dependable Secure Comput. **15**(1), 151–165 (2018)
20. I.-R. Chen, F. Bao, M.J. Chang, J.-H. Cho, Dynamic trust management for delay tolerant networks and its application to secure routing. IEEE Trans. Parallel Distrib. Syst. **25**(5), 1200–1210 (2014)
21. E. Hernandez-Orallo, M. Serrat Olmos, J.-C. Cano, C. Calafate, P. Manzoni, CoCoWa: a collaborative contact-based watchdog for detecting selfish nodes. IEEE Trans. Mobile Comput. **14**, 1162–1175 (2015)
22. J. Dias, J. Rodrigues, C. Mavromoustakis, F. Xia, A cooperative watchdog system to detect misbehavior nodes in vehicular delay-tolerant networks. IEEE Trans. Ind. Electron. (99), 1 (2015)

Liver Disease Prediction Using Machine Learning Algorithm

Sambit Mohanty, Pradosh Kumar Gantayat, Sachikanta Dash, Bhabani P. Mishra, and Shiba Ch. Barik

Abstract Over the last few decades, liver diseases are one of the reasons for a large number of death cases worldwide and have emerged as a life-threatening disease also. By WHO report, around 59% of the mortality and 46% of global diseases are because of chronic diseases and around 35 millions of people worldwide die due to chronic diseases. Now, we are living in the era of information where millions of data are generating from various sources every day. We can use these data to improve our healthcare services or proper identification of diseases. We have collected patient data from open source platform and applied various kinds of data analysis techniques and machine learning (ML) approaches applied to see the pattern of the data sets. Then, a performance comparison between these models is made to get highly accurate model for predicting liver disease.

Keywords Liver disease · Machine learning · KNN · Random forest

1 Introduction

The liver is the most essential and vital part of the human body; it weights around 1500 g and is located on the right upper abdominal quadrant. In human body, liver plays a major role in metabolism. Liver produces digestive fluid called bile which helps in digestion and breaks down of fat or lipids. Bile is very much essential for intestinal absorption of fat-soluble vitamins like A, D, K, E and cholesterol [1]. It also neutralizes the gastric acid that enters to intestine from stomach. Bile also serves as

S. Mohanty (✉)
CTSC, Hyderabad, India

P. K. Gantayat · S. Dash
DRIEMS Autonomous Engineering College, Cuttack, India

B. P. Mishra
GATE, Berhampur, Odisha, India

S. Ch. Barik
DRIEMS Polytechnic, Cuttack 754022, India

V. Bhateja et al. (eds.), *Data Engineering and Intelligent Computing*,
Advances in Intelligent Systems and Computing 1407,
https://doi.org/10.1007/978-981-16-0171-2_56

a vehicle for excretion of bile pigments, many drugs, heavy metals like copper, zinc, etc. Bile also helps to remove cholesterol from the body and send it through waste materials [2]. It also serves as a detergent for the body to eliminate waste particles from our bodies. Nowadays, lots of people are suffering from the liver disease because of various reasons like changing lifestyle, smoking, liquor consumption, infection, and so on. There exist various types of liver disease like hepatitis A, hepatitis B, hepatitis c, autoimmune hepatitis, primary biliary cholangitis, liver cancer, bile duct cancer, hemochromatosis, cirrhosis, etc., so the early detection of these diseases can prevent patient from death; for this purpose, proper diagnosis is required which can help the doctors to detect these kinds of serious liver disease [3].

We are living in the age of information where data play a very vital role in decision making. Every day, massive amount of data is generating from various sources like social media, shopping malls, hospital, laboratories, and so, if we can use these data in a meaningful way, then we can make our life easier. In this paper, we have shown how data can be useful to make decision whether a patient is suffering from liver disease or not. This will assist doctors more in making their decision while diagnosing patients.

2 Algorithm Used

We have collected these patient data from UCI Machine Learning Repository, which is free and easily accessible. UCI is a large database which consists of a large number of data set for research purpose, and these data sets can be accessed by anyone. As per UCI, these data sets contain patient information which is collected from the northeast part of Andhra Pradesh. In this data set, there are total of 583 patient records present on which 416 are patients suffering from liver disease, and 167 people are not with liver disease. This data set contains 442 female and 141 male patients' information. This data set has 11 attributes or columns or features like:

Age: Patient's age
Gender: Gender of the patients
Direct Bilirubin: Conjugated bilirubin in mg/dL
Total Bilirubin: Total bilirubin in mg/dL
Alkaline Phosphotase: Value of ALP in IU/L
Alanine Aminotransferase: Value of ALT in IU/L
Aspartate Aminotransferase: Value of AST in IU/L
Total Proteins: Total proteins in g/dL
Albumin: Total albumin in g/dL
Albumin and Globulin Ratio: The A/G ratio
Result: It gives the presence or absence of liver disease.

We have used Jupyter Notebook, which is freely available software for performing machine learning operations. For machine learning purpose, we need to import the Sklearn module, which contains all the essential algorithms and functions. We need to import Python NumPy module and Pandas module, for data analysis purposes. To plot different graphs, we need to import Matplotlib module, which contains all the methods related to plotting graphs. We stored our data in a CSV file, and through Python Pandas module, the data are imported to the notebook as a data frame. After importing the data, we can apply various data analysis and machine learning algorithms for classification and prediction.

Before applying any machine learning algorithm, we need to check properly our data set so that we can be aware of what data are present in our data set. For this kind of operation, we need to perform an exploratory data analysis technique that gives a complete view of the data set with its properties and some known and unknown things about the data.

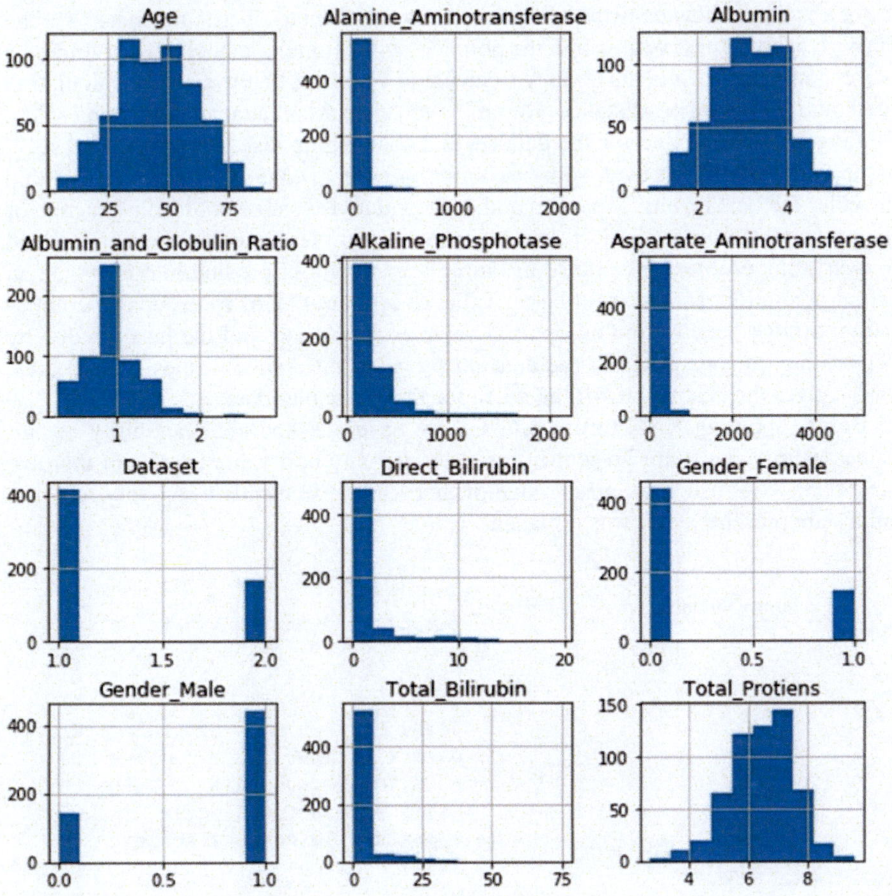

Fig. 1 Various features of data set

Before applying our machine learning algorithms on the data set, we can see how the features of our data set look like in plots in Fig. 1. For this, we can use the Matplotlib module.

2.1 K-Nearest Neighbor Algorithm

KNN is a nonparametric supervised learning algorithm that creates the model structure from the given data set. In KNN, *k* is a number that decides how many nearest neighbors are used to provide the best result for the data set. Let '*a*' is point whose label is to be predicted then first, we have to find the *K* closest point near to '*a*'. Then make a voting of how many nearest data points represent to a particular class [4]. The class gets the highest voting, and the point '*a*' will belong to that class. For finding a closet point near to our data point, we can use any of these below distance calculation methods like Euclidean distance, Hamming distance, Manhattan distance, and so on. Finding the best *k* value for the data set is a challenging task. For every data set, *k* value is different. However, some assumption says *k* value is an odd number that provides the best results. We can say that the value of *k* is the controlling factor for the problem. Research shows that if we take *k* value a less number, then it may lead to overfitting problem; i.e., noise has more impact on our prediction rather than an actual prediction. If we take a larger value of *k*, then it leads to expensive computation. So, we need to find an optimal value of *k*. We can find the best *k* value by performing our operation on the data set by using different *k* values. The *k* value which gives the best result will be taken for that particular data set.

Before applying KNN to the data set, we have checked the availability of null values as in (Fig. 2) are present in the dataset if any null value is present then we have replaced it using the mean value of that feature. In our data set, very few null values are present, and it was replaced.

Fig. 2 Features with null value

```
data.isnull().sum()

Age                           0
Gender                        0
Total_Bilirubin               0
Direct_Bilirubin              0
Alkaline_Phosphotase          0
Alamine_Aminotransferase      0
Aspartate_Aminotransferase    0
Total_Protiens                0
Albumin                       0
Albumin_and_Globulin_Ratio    4
Dataset                       0
dtype: int64
```

In our data set, one categorical value is present, which cannot be given as input to the machine learning algorithm as a machine learning algorithm only takes numerical values as input, so this 'gender' column needs to be converted into a numerical value. For this, we used a function from Sklearn module, which converts categorical values into numeric form. We divided our data set into two parts: X stores all the features, and y stores all the labels or outputs. For applying KNN on the data set, we need a K value, but we do not know which k value is good for our data set. To find the best suitable k value for our data set, we have taken K value in a range of (1, 21) and find the best score for all the K values, for which K value it will give the best score then that is the best suitable value for our model. To know the best K value, we have plotted a graph using Matplotlib.

From Fig. 3, it clearly shows that at $K = 10$ the accuracy is more in comparison with all other k values in the range $(1, 21)$. The accuracy value for $K = 10$ is 0.708455.

```
knn_classifier = KNeighborsClassifier(n_neighbors = 10)#k=10 highest
score=cross_val_score(knn_classifier,X,y,cv=10)
```

```
score.mean()
```

```
0.7084556071651952
```

2.2 *Random Forest Classifier*

It is a type of supervised learning algorithm. As the name says forest basically, it contains trees, i.e., decision trees. Random forest gives its prediction value basing upon the prediction of decision trees used. In random forest, many decision trees are present, and from the data set, few features and labels are given to each decision tree on random basis [5]. Then, these decision trees perform their operation on the supplied data points and predict their score. As multiple decision trees are present, random forest makes voting for classification-type problems. The majority of decision trees belong to which class result of random forest is that. Random forest can also be used as a repressor. In this case, the output is a continuous value, which is the mean or median of all predicted values of the decision tree [6].

Random forest provides better results in comparison with decision trees as decision tree suffers from low bias and high variance problem, so it may not give accurate prediction values. So to overcome such type of problem in random forest, many number of decision trees are used so that the high variance problem can be reduced and it will provide more accurate prediction [7]. In fact, RF is one of the most popular ML algorithms used for both regression and classification problem. In real life, it is used for design of recommendation system, credit card fraud detection, classification [8] of loyal loan applicants, disease prediction, and so on. Random forest may be slow in terms of prediction time as it uses many decision trees.

```
knn_classifier = KNeighborsClassifier(n_neighbors = 10)#k=10 highest
score=cross_val_score(knn_classifier,X,y,cv=10)
```

```
score.mean()
```

```
0.7084556071651952
```

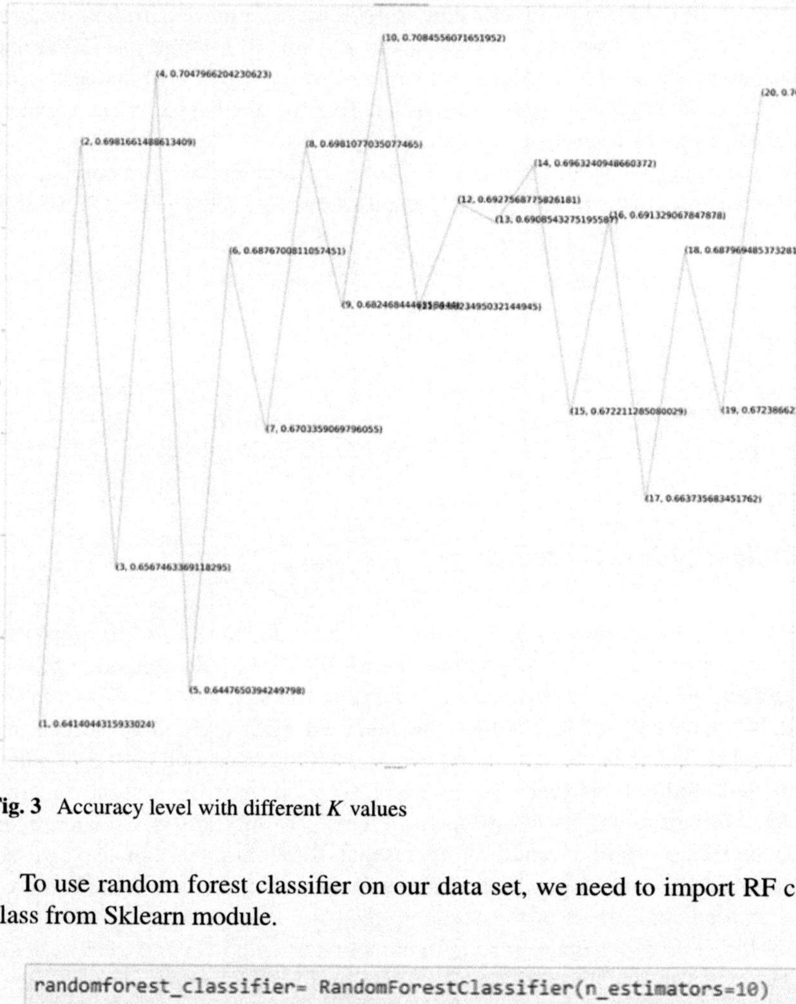

Fig. 3 Accuracy level with different *K* values

To use random forest classifier on our data set, we need to import RF classifier class from Sklearn module.

```
randomforest_classifier= RandomForestClassifier(n_estimators=10)

score=cross_val_score(randomforest_classifier,X,y,cv=10)
```

```
score.mean()
```

```
0.6878474678806894
```

Here, we have used a parameter n_estimator which is a controlling parameter of random forest algorithm, which indicates how many number of decision trees are

Fig. 4 Prediction value for different ML approaches

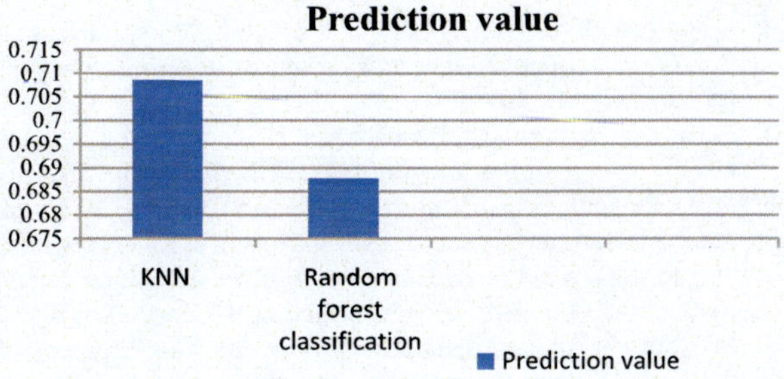

there inside the RF classifier. More the number of value of n_estimator i.e. more number of decision tree more accuracy we can get.

3 Comparison

We have used 2 classification algorithms on our data set to get the prediction for liver disease. For KNN with k value $= 10$, we get accuracy value 0.7084; for random forest classification with 10 decision trees, we get accuracy value 0.6878. However in Fig. 4, it is clearly seen that KNN model predicts more accurate in our case.

.

4 Conclusion and Future Scope

In this paper, we have applied only 2 machine learning algorithms on the data set to predict the accuracy value but there exist other machine learning algorithms which can be applied on the data set which may perform better. We can also improve the performance of these algorithms by adjusting tuning parameters and other adjustment parameters.

References

1. A. Rairikar, V. Kulkarni, V. Sabale, H. Kale, A. Lamgunde, Liver disease prediction using data mining techniques, in *IEEE International Conference on Intelligent Computing and Control (I2C2)*, 1–8 June 2017
2. H. Almarabeh, E. Amer, A study of data mining techniques accuracy for healthcare. Int. J. Comput. Appl. **168**(3), 12–17 (2017)
3. Y. Khourdifi, B. Mohamed, Liver disease prediction and classification using machine learning algorithms optimized by particle swarm optimization and ant colony optimization. Int. J. Intell. Eng. Syst. **12**(1) (2019)

4. J. Santhana Krishnan, S. Geetha, Prediction of liver disease using machine learning algorithms, in *1st International Conference on Innovations in Information and Communication Technology (ICIICT)* (IEEE, 2019)
5. A.H. Chen, S.Y. Huang, P.S. Hong, C.H. Cheng, E.J. Lin, HDPS: heart disease prediction system, in *2011 Computing in Cardiology* (IEEE, 2011), pp. 557–560
6. V. Krishnaiah, G. Narsimha, N. Subhash Chandra, Heart disease prediction system using data mining technique by fuzzy K-NN approach, in *Emerging ICT for Bridging the Future—Proceedings of the 49th Annual Convention of the Computer Society of India (CSI)*, vol. 1 (Springer International Publishing, Switzerland, 2015), pp. 371–384
7. M.J. Sorich, An intelligent model for liverdisease diagnosis. Artif. Intell. Med. **47**, 53–62 (2009)
8. S. Dash, R.K. Das, An implementation of neural network approach for recognition of handwritten Odia text, in *Advances in Intelligent Computing and Communication. Lecture Notes in Networks and Systems*, vol. 109, ed. by M. Mohanty, S. Das (Springer, Singapore, 2020)

Hybrid Cryptography for Data Security in Wireless Sensor Network

Alakananda Tripathy, Sateesh Kumar Pradhan, Ajit Kumar Nayak, and Alok Ranjan Tripathy

Abstract Wireless sensor network plays a major role in collecting data in different fields. To achieve authentication, confidentiality of data and to transmit over the network, cryptography is needed. The paper proposes a new model known as hybrid encryption technique (HET) which discusses the attainment of data security in sensor network using single-key and two-key encryption algorithms. The proposed method uses the advance encryption standard for encrypting the data using the key which is encrypted using elliptic curve cryptographic algorithm for encrypting the message, and then Lempel–Ziv–Welch (LZW) compression technique is used to compress the data, thus reducing the size of the ciphertext. The proposed algorithm provides high security of data by maintaining security primitive like confidentiality, authentication and integrity. The algorithm has less encryption and decryption time as compared to the existing algorithm.

Keywords Advance encryption standard · Elliptic curve cryptography · LZW · Throughput · Ciphertext

1 Introduction

Although the fast intrigue and research in the fields of WSN have occurred, utilization of sensors for specific administrations is not new. A sensor system can be expected to have numerous spatially conveyed self-governing detecting gadgets which course

A. Tripathy
Department of CSE, S'O'A, Deemed to be University, Bhubaneswar, Odisha, India

S. K. Pradhan
Department of Computer Science and Application, Utkal University, Vani Vihar, Bhubaneswar, Odisha, India

A. K. Nayak
Department of CSIT, S'O'A, Deemed to be University, Bhubaneswar, Odisha, India

A. R. Tripathy (✉)
Department of Computer Science, Ravenshaw University, Cuttack, Odisha, India

V. Bhateja et al. (eds.), *Data Engineering and Intelligent Computing*,
Advances in Intelligent Systems and Computing 1407,
https://doi.org/10.1007/978-981-16-0171-2_57

the data to a hub and which can make the best utilization of the obtained data. The use of WSN is expanding exponentially because of the highlights, for example, scalability, adaptability, convenience, mobility, accessibility, minimal effort and so on [1]. Wireless sensor network (WSN) arranges applications commonly including the perception of some physical wonder through examining nature. To observe and record conditions at different areas, sensor network is used which is a collection of particular transducers. Normally, the parameters that are observed are temperature, mugginess, weight, wind speed, power of vibration, sound, control line voltage, level of poison and imperative body work [2]. A sensor network comprises sensor nodes at a different location, and every node is little, lightweight and convenient to use.

In wireless sensor network, different types of attacks are performed on data that is to be transmitted in the network [3]. To ensure authentication, integrity, confidentiality, availability and non-repudiation cryptography [4] play an important role. As sensor network has limited battery life, small memory traditional algorithm could not deal with this constraint. The size of memory should be minimum that is required to store the encoded message, and the size of key used for encoding should be minimum [5].

The different requirements of security like data authentication, confidentiality and integrity can be achieved using the number of cryptographic algorithms. To prevent unauthorized access to data is known as authentication. Securing the information from unauthorized parties is known as data confidentiality. Not alternating the data in transit by an adversary integrity is achieved. For achieving the different security goals, data needs to be encoded in such a manner that it cannot be read by the adversary, and the process of encoding data is known as encryption. The most commonly used encryption techniques are private key and public key encryption.

An algorithm proposed in this paper is based on Advanced Encryption Standard (AES) [6], elliptic curve cryptosystem (ECC) and Lempel–Ziv–Welch (LZW) compression technique. It is used to achieve data security and authenticity and LZW compression which reduces the encrypted data size. This algorithm provides faster encryption with minimum key and also reduces the size of the encrypted message.

In this paper, related work is discussed in Sect. 2 and outline of the proposed work is mentioned in Sect. 3. Section 4 presents the result, and the conclusion is discussed in Sect. 5.

2 Related Work

The plaintext content is encoded using elliptic curve cryptography to determine the cipher is the basic objective of the security protocol architecture proposed by Subasree and Sakthivel [7], and the generated encoded message is delivered through the secure path to the receiver. The hash value is determined for then identical plaintext using MD5 at the same time, which has been modified by ECC into ciphertext. Two-stage hybrid cryptographic algorithm (THCA) was a synthesis of both symmetric and asymmetric algorithms in parallel proposed by Rizk and Yasmin [8].

Zhu [9] suggested Security Protocol Architecture, where the symmetric algorithm is needed for the generation of the ciphertext, and to secure the key asymmetric algorithm is used, and the coded data is verified by the digital signature technique. The key is used only once to improve the key management, and also security of the data is enhanced. After the verification of signature, the recipient acquires the data. DES algorithm is used for the transmission of data, and the key used to encrypt the data is generated by RSA public key cryptosystem as discussed by Ren and Zhiqian [10]. The random number generator is used to generate the session key only once to encrypt the data by DES algorithm, and the plaintext is encrypted to have the ciphertext. The public key management centre generates the public key that is used to encrypt the session key, on the other hand, using RSA public key algorithm.

Tripathy et al. [11] proposed a hybrid model where the key used for encrypting data using the RC4 algorithm is generated using elliptic curve cryptography; the model needs less time for data encryption and decryption.

A lightweight three-factor authentication and key agreement protocol were discussed by Jiang et al. [12] which provides end-to-end secure communication. This protocol is secure against all the attacks.

Dynamic authentication for secure communication for protection of privacy in the e-health system with limited computing capacity and memory size was proposed by Zhang et al. [13].

3 Proposed Work

An algorithm is proposed in this section using both public key and private key algorithms. The proposed model achieves high security levels at less time of execution. In this method, encryption is done to provide high security and compression in order to reduce the length of encrypted data needed for communication. The proposed method first uses advanced encryption standard to encrypt the data and then followed by the Lempel–Ziv–Welch (LZW) compression technique to compress the ciphertext by reducing the length of the message. The key used for encryption is encrypted using elliptic curve cryptography to provide high security and then compresses message communicated to the receiver through the secure link.

3.1 Proposed Encryption Algorithm

The data is encrypted using the AES [14] symmetric key algorithm in the first phase of encryption. The keys used in the AES algorithm can be 128, 192 or 256 bits. The keys used in the proposed method are generated using elliptic curve cryptography [15]. Elliptic curve cryptography [15] is based on algebraic structure of an elliptic curve over a finite field.

An elliptic curve over a finite field is defined as in Eq. (1):

$$y^2 = x^3 + (i \times x) + y \quad (1)$$

Here, (x, y) are the points on the curve and (i, j) are the coefficients that must satisfy Eq. (2):

$$(4i^3 + 27j^2) \bmod p \neq 0 \quad (2)$$

where p is the prime number and the values of the variable are from 1 to $(p - 1)$. The key generation algorithm for elliptic curve is given in Algorithm 1:

Algorithm 1 Generating keys using elliptic curve
INPUT: Parameter domain of elliptic curve with prime number up to order n
OUTPUT: Public key Pub and private key Pri

1. Select the private key Pri from interval Pri $\in [1, n - 1]$
2. Compute public key Pub = PriX where X is the point on the curve
3. Return public key Pub and Private key Pri
4. Stop

In this algorithm, Pri is the private key and Pub is the public key. This model is more secure as it uses both elliptic curve and AES algorithm. During the encryption phase, the input is first XORed with 128-bit round key with 10 different rounds of operation like SubBytes, ShiftRow and MixColumn and then adds round key for each round. The number of rounds can be 10, 12 or 14. Once the data is encrypted, it is then compressed using the Lempel–Ziv–Welch (LZW) [16] technique. In this technique, a dictionary is maintained for the input data to keep correspondence between the longest encounter words and a rundown of code values. The words are then replaced by their comparing codes, thus by compressing the input file. Then, the obtained ciphertext is transmitted in the network. The encrypted text along with the encrypted key used for encryption is transmitted to the receiving node. The encryption algorithm for the proposed model is discussed in Algorithm 2.

Algorithm 2 Encryption algorithm
INPUT: Original message (P)
OUTPUT: Encrypted message (C)

1. Initialize the state array S of 16 bytes with the plaintext
2. The plaintext is XORed with the round key
3. Split the plaintext into bytes
4. To obtain the ciphertext step 5 to step 8 is performed for 9 rounds

5. In SubBytes, each byte of plaintext is converted to different values by passing through the substitution box
6. ShiftRow is 4×4 matrix where each cell of the state contains a byte. Except first row, each row is shifted to left by 1 byte, 2 bytes and 3 bytes.
7. Each column of the input array is multiplied with constant matrix to get new column.
8. Result of MixColumn is XORed with the round key
9. In the 10th round, SubBytes and ShiftRow operations are performed which is again XORed with 10th round key to obtain the ciphertext
10. Maintain a table with the encrypted string
11. Initialize X with the first character
12. Repeat steps 12 to 17 until end of string
13. Read next character Y from the table
14. Are X and Y in the string then
15. New string will be X + Y
16. Otherwise output the code
17. Add entry to the table for X + Y
18. Ciphertext C is obtained
19. Stop

3.2 *Proposed Decryption Algorithm*

Decryption is the opposite of encryption. The LZW technique is used to decompress the encoded data, and the encoded data is decrypted using AES decryption algorithm to obtained the plaintext. The steps of the algorithm for decryption are explained in Algorithm 3.

Algorithm 3 Decryption algorithm
INPUT: Encrypted message (C)
OUTPUT: Original message (P)

1. The deciphered text is maintained in a table of single characters
2. Set Z = first input character
3. Output the translation of Z
4. Repeat steps 6 to 15 not the end of the input string
5. Set N = next character
6. If N is not in the table then
7. Set S = translation of Z
8. Set S = S + C
9. Otherwise

10. S = translation of N
11. Output S
12. Set C = first character of S
13. Add Z + C to the table
14. Assign Z = N
15. end while
    //After decompressing the text AES is performed in reverse direction to undo the steps.
16. Inverse the final round of AES in encryption by performing Add roundkey, ShiftRow and SubBytes.
17. Inverse the Main 9 round of encryption by performing Add roundkey, MixColumn, ShiftRow and SubBytes.
18. Inverse the initial round by performing Add roundkey.
19. Stop

4 Result and Discussion

The proposed algorithm is evaluated based on the time needed for encryption and decryption throughput and the ciphertext length. The existing algorithms as discussed in Subasree and Sakthivel [7], Rizk and Yasmin [8], Zhu [9], Ren and Zhiqian [10] and HCA [17] are compared with the proposed hybrid model in the context of encryption and decryption time, throughput and the size of the ciphertext. The encryption time of the proposed model (HET) is compared with the other algorithms which show that the proposed algorithm is better as compared to the existing ones. Encryption time is the time consumed by the proposed model to encode the data to maintain the confidentiality of the information, and the time needed to decode the ciphered data to obtain the original message is known as time of decryption. Similarly, the decryption time of the proposed model is minimum as compared to the existing algorithms. The time of encryption and decryption for the various sizes of plaintext individually for different algorithms is explained in Tables 1 and 2. The ciphertext length specifies the

Table 1 Encryption time of different hybrid algorithms (in ns)

Plaintext (in bytes)	Subashree	Ren	Zhu	THCA	HCA	HET
609	2063	1432	998	998	548	450
25,615	3683	1490	1022	1022	990	452
35,080	3465	1468	1059	1059	1037	474
61,386	5198	3019	3143	3143	2345	480
184,162	6892	4970	3814	3814	3187	486
209,777	10,575	6460	4836	4836	4177	938

Table 2 Decryption time of different hybrid algorithms (in ns)

Plaintext (in bytes)	Subashree	Ren	Zhu	THCA	HCA	HET
609	1078	756	562	562	221	206
25,615	1085	821	713	713	636	388
35,080	1082	953	824	824	762	390
61,386	1197	864	891	891	795	398
184,162	1306	1075	907	907	856	408
209,777	2391	1896	1620	1620	1492	796

Table 3 Ciphertext size of different hybrid algorithms

Plaintext (in bytes)	Subashree	Ren	Zhu	THCA	HCA	HET
609	609	601	609	641	648	489
25,615	25,615	25,610	25,615	25,648	25,646	20,540
35,080	35,080	35,070	35,080	35,114	35,117	28,170
61,386	61,386	61,369	61,386	61,418	61,420	49,389
184,162	184,162	184,143	184,162	184,196	184,204	147,873
209,777	209,777	209,753	209,777	209,844	209,850	168,413

size needed for the encrypted data to store in the memory. The size of the ciphertext for different hybrid algorithms is discussed in Table 3.

The proposed model required minimum space to store the encrypted message irrespective of the size of the plaintext.

Throughput is measured as the rate at which an algorithm is processed. The maximum is the throughput lesser is the power consumption in the wireless sensor network. As shown in Fig. 1, it is clear that the proposed method has maximum throughput than the existing model. So, the proposed method consumes less power than the existing model. The result of HET along with other algorithms for the time of encryption and decryption is shown in Figs. 2 and 3.

5 Conclusion

To have better security of the message by maintaining the minimum memory and time for sensor networks, a hybrid model is proposed in this paper. The algorithm provides double encryption by using symmetric algorithm AES and then asymmetric algorithm known as elliptic curve cryptography (ECC). It also reduces the size of the ciphertext by using the LZW compression technique. The proposed algorithm first uses Advance Encryption Standard to encrypt the message and then compresses it, and then the compressed data is again encrypted using elliptic curve cryptography (ECC) which provides high security. The encryption and decryption time of the

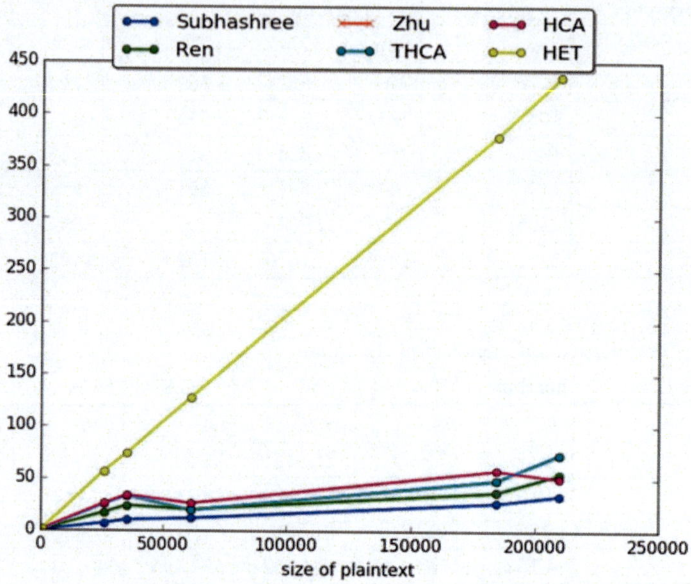

Fig. 1 Throughput of different hybrid algorithms

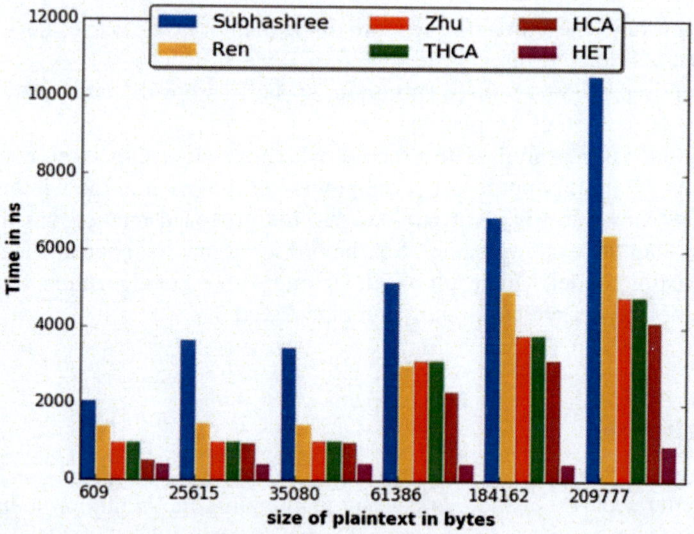

Fig. 2 Encryption of different hybrid algorithms (in ns)

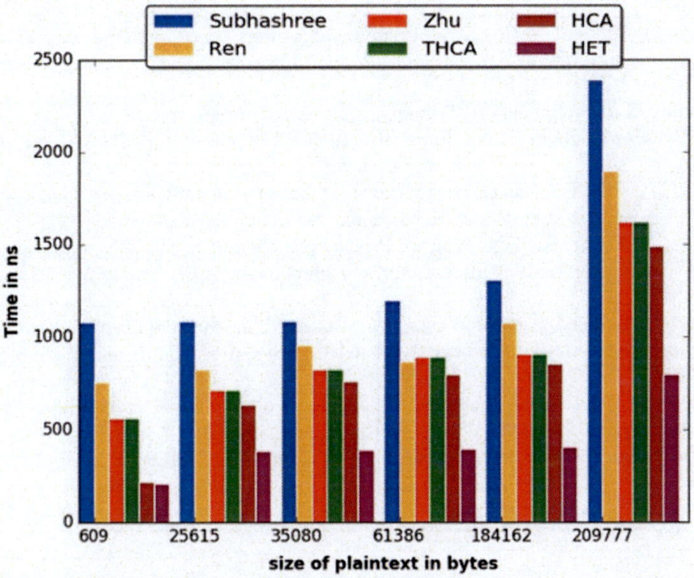

Fig. 3 Decryption of different hybrid algorithms (in ns)

proposed algorithm is compared with the existing hybrid algorithms and shows that it requires less time as compared to other algorithms. The algorithm provides high security, but the key used for encryption is not made secure. In future work, emphasis can be given to make the key more secure by enhancing the security of the algorithm.

References

1. A. Faquih, P. Kadam, Z. Saquib, Cryptographic techniques for wireless sensor networks: a survey, in *IEEE Bombay Section Symposium (IBSS)*, Mumbai (2015), pp. 1–6
2. A. Perrig, J. Stankovic, D. Wagner, Security in wireless sensor networks (2004), pp. 53–57
3. C.K. Marigowda, M. Shingadi, Security vulnerability issues in wireless sensor networks: a short survey. Int. J. Adv. Res. Comput. Commun. Eng. **2**, 2765–2770 (2013)
4. E.A.M. Anita, R. Geetha, E. Kannan, A novel hybrid key management scheme for establishing secure communication in wireless sensor networks. Wirel. Pers. Commun. (3), 1419–1433 (2015)
5. S. Faye, F.M. Jean, Secure and energy-efficient geocast protocol for wireless sensor networks based on a hierarchical clustered structure. IJ Netw. Secur. **15**(3), 151–160 (2013)
6. W.E. Burr, Selecting the advanced encryption standard. IEEE Secur. Priv. **1**(2), 43–52 (2003)
7. S. Subasree, N.K. Sakthivel, Design of a new security protocol using hybrid cryptography algorithms. IJRRAS **2**(2), 95–103 (2010)
8. R. Rizk, A. Yasmin, Two-phase hybrid cryptography algorithm for wireless sensor networks. J. Electr. Syst. Inf. Technol. **2**(3), 296–313 (2015)
9. S. Zhu, Research of hybrid cipher algorithm application to hydraulic information transmission, in *International Conference on Electronics, Communications and Control (ICECC)* (IEEE, 2011)

10. W. Ren, M. Zhiqian, A hybrid encryption algorithm based on DES and RSA in bluetooth communication, in *Second International Conference on Modeling, Simulation and Visualization Methods* (IEEE, 2010)
11. A. Tripathy, S.K. Pradhan, A.R. Tripathy, A.K. Nayak, A new hybrid cryptography technique in wireless sensor network. Int. J. Innovative Technol. Exploring Eng. (IJITEE) **8**(10), 121–131 (2019)
12. Q. Jiang, S. Zeadally, J. Ma, D. He, Lightweight three-factor authentication and key agreement protocol for internet-integrated wireless sensor networks. IEEE Access **5**, 3376–3392 (2017)
13. L. Zhang, Y. Zhang, S. Tang, H. Luo, Privacy protection for e-health systems by means of dynamic authentication and three-factor key agreement. IEEE Trans. Ind. Electron. **65**(3), 2795–2805 (2017)
14. V. Rijmen, J. Daemen, Advanced encryption standard, in *Proceedings of Federal Information Processing Standards Publications* (National Institute of Standards and Technology, 2001), pp. 19–22
15. N. Koblitz, A. Menezes, S. Vanstone, The state of elliptic curve cryptography. Des. Codes Crypt. **19**(2), 173–193 (2000)
16. H.N. Dheemanth, LZW data compression. Am. J. Eng. Res. (AJER) **3**(2), 22–26 (2014)
17. K.M. Abdullah, E.H. Houssein, H.H. Zayed, New security protocol using hybrid cryptography algorithm for WSN, in *1st International Conference on Computer Applications & Information Security (ICCAIS)* (IEEE, 2018)

Superimposing Periodic Subgraph Mining in Dynamic Social Network

S. Vairachilai

Abstract Nowadays, complex network has been inspired by the observation of many real-world network patterns such as pattern of mobile phone usage, gene expression, e-mail habits, protein interactions in networks, regional interaction in archaeological contexts and co-authorship in research publications. The detection of interaction patterns that occur at fixed time periods can be mined in a complex network. This article proposes an algorithm to predict an interaction pattern hidden behind these networks, namely periodic subgraphs embedded in a complex network. The primary data structures used in tracking the periodic subgraphs are lists, tables, data frames, matrices and vectors. The algorithm can be implemented using analytical tools that do not require advanced data structures such as pattern trees and indexing techniques such as hash maps. In this paper, our algorithm processes a dynamic network in two stages. The first stage of the algorithm extracts all maximal subgraphs which appear between successive time steps. In second stage, it maintains the periodicities of the maximal subgraphs identified and flushes out those graphs whose periodicities ceased to exist. The algorithm is tested by generating a dynamic network using random graph generating function and by superimposing periodic subgraphs on the network at different time interval periods.

Keywords Dynamic network · Subgraph mining · Maximal subgraph · R programming

1 Introduction

Dynamic networks are very powerful model for representing time-varying structure in the network. The interaction among these networks varies with respect to time. In dynamic network analysis, frequent and periodic substructures are the basic patterns which can be discovered in a temporal collection of graphs. The several methods and

---

S. Vairachilai (✉)
Department of Computer Science and Engineering, CVR College Of Engineering, Hyderabad 501510, India
e-mail: vairachilai2676@gmail.com

V. Bhateja et al. (eds.), *Data Engineering and Intelligent Computing*,
Advances in Intelligent Systems and Computing 1407,
https://doi.org/10.1007/978-981-16-0171-2_58

algorithms have been developed to analyse and discover interesting patterns in graph data applications [1]. Mining the communities on temporal graphs has more attention due to the several real-time applications [2, 3]. Elfeky et al. [4] developed an algorithm to find the periodic patterns in time series database. They found the periodic patterns of unknown periods without affecting the time complexity in database. Cao et al. [5] proposed a solution to mine periodic patterns in spatiotemporal data. They proposed an efficient algorithm for retrieving maximal periodic patterns. Lahiri and Berger-Wolf [6, 7] introduced periodic patterns mining in dynamic networks. They proposed PSEMiner algorithm that finds periodic patterns in polynomial time. Tanbeer et al. [8] introduced periodic–frequent patterns that the patterns occur frequently and regularly in a database. Apostolico et al. [9] introduced ListMiner algorithm to solve unessential tree node creation problem and speed up PSEMiner algorithm.

The periodic and frequent interactions among the elements in a data set can be modelled as a graph. In graph, the set of vertices are represented as individual elements and the set of edges are represented as interactions between them. The interactions between the individual elements are noted over a period of time. The time periods are divided into discrete time steps of equal duration. These data formulate a dynamic network. The size and the complexity of the network increase from social network analysis (SNA) to dynamic network analysis (DNA). The algorithm is also difficult to predict and hard to detect patterns hidden behind these networks. In the last few years, the discovery of graph structures that occur a significant number of times in a dynamic network has been an active research. It focuses on dynamic state and dynamic event networks [10]. In dynamic state networks, links have a tendency to endure. These applications are found in protein network, social network and collaboration networks. In dynamic state networks, dynamic event networks have links which represent brief events which may not endure for a prolonged period. The examples of such networks include e-mail and telephone call graphs. The proposed superimposing periodic subgraph mining time complexity is not increased exponentially.

The rest of this paper is organized as follows. In Sect. 2, periodic subgraph mining has been discussed. In Sect. 3, the simulated network and the periodic subgraphs superimposed at different time intervals are discussed. The proposed algorithms are discussed in Sect. 4. In Sects. 5 and 6, the time complexity and experimental results are discussed. The conclusion and the direction of future work are discussed in Sect. 7.

2 Related Work

2.1 Preliminaries

Dynamic Network: In dynamic network, <G1,...,GT> consisting of simple graphs Gt = <Vt,Et> over a unique labelled set of vertices {Vt} and a minimum support threshold which represents the minimum number of time steps in which a periodic subgraph appears, the algorithm to detect periodic subgraphs will list all periodic subgraphs having different periods appearing at different time steps of the dynamic network.

Periodic Subgraph: A subgraph is periodic if its support (occurrence frequency) in a given dynamic network is not less than a minimum support threshold and it occurs in a regular time interval (step). Support is the number of subgraphs that contain the pattern (occurrence frequency). The minimum support values (σ) lie between $1 \leq \sigma \leq t$, where 't' is the number of time steps in the dynamic network.

Maximal Common Subgraph (MCS): The MCS is obtained by taking the intersection of their set representations. The MCS in graph is unique and could possibly be the empty graph with no vertices or edges.

2.2 *Periodic Subgraph*

A sequence of subgraphs uniquely labelled vertices in a dynamic network, the periodic subgraph is a graph which reappears at regular intervals. Apostolico et al. [11] discussed the periodic subgraph mining problem that consists of discovering maximal subgraphs that repeat at regular intervals. When a series of observations were captured as temporal graphs, detecting periodically recurrent sequence of subgraphs having uniquely labelled edges demands efficient algorithms.

Lahiri and Berger-Wolf [6] have proposed an algorithm to detect interaction patterns hidden behind such dynamic networks by identifying subgraphs that occur periodically in the network. The primary data structures used in tracking the subgraphs are pattern tree, subgraph hash map and an auxiliary data structure known as a descriptor to generate support set. The algorithm was implemented using a pattern tree that maintains information about all patterns that are either currently periodic or could become periodic at a future time step. As each time step of the dynamic network is read, the pattern tree is traversed and updated by flushing out the patterns that are no longer periodic, and also identifying new possible periodic subgraphs. In this article, an attempt was made to develop a similar algorithm using the concepts of data sets and data structures such as lists, tables, data frames, matrices and vectors in R and SAS programming languages. The developed algorithm is tested using randomly generated graphs and by superimposing periodic subgraphs that appear over time periods. The algorithm was implemented using R environment which provides an appropriate data structure to handle graph databases and also provides functions to generate random graphs.

3 Simulated Network

One can use existing graphs captured at a sequence of time steps from real-world problems and detect maximal common subgraphs that occur either frequently or periodically. However, gathering such data is extremely time-consuming. Alternatively, one can choose to use graph generators to create and simulate real-world problems. In this article, an attempt was made to generate a time series of random graphs as a dynamic network and superimpose periodic subgraphs that occur with different periodicities during different time intervals. Many of the studies of random graphs have their origin back to the work of Erdős and Rényi [12–14] whose fundamental work has laid the foundation for the theory of random graph generation. There are two standard models for Erdős random graph generation [14], in which two parameters were used one to control the number of nodes while generating the random graph and the other to control the density or a number of edges in the network. For example, the random graph model $G(N, M)$ assigns a uniform probability to all graphs with N nodes and M number of edges, while the other graph model $G(N, p)$ generates a graph with N number of nodes and the edges were chosen randomly with probability p. The later model is used in this research to generate random graphs at each time step of the dynamic network using Erdos–Renyi graph function. To maintain a unique set of labels for the vertices over different time periods, we have assigned unique numbers to the vertices after generating the graphs at each time step. In real-world applications such as social media networks, all nodes are not likely to be active simultaneously. To achieve this condition, we tested the model with different probability values and fixed a low value for p and eliminating nodes with zero vertex degrees.

To superimpose periodic subgraphs with different periodicities, we have considered a couple of standard connected graphs such as star and ring. We superimpose such graphs having a small number of nodes and edges to the randomly generated graphs over different time periods of the network. The entire sequence of graphs generated was stored as data sets, and an index file was maintained to retrieve any graph at any selected time step randomly. As an example, we have shown the graphs at time step 10 before and after superimposing subgraphs in Figs. 1, 2 and 3. Figure 1 shows the randomly generated graph at time step 10. This randomly generated graph consists of 50 nodes. Figure 2 depicts periodic subgraphs superimposed at different time intervals. In this figure, the periodic subgraphs with periods 1 and 2 are superimposed between steps 2–8 and 4–10, respectively. Figure 3 shows the random graph along with the period's graphs embedded at time step 10. The standard connected star graph with 6 nodes is generated.

4 Proposed Algorithm

The algorithm proposed in this article was implemented using R environment. The advantage of using R is that it gives an additional leverage to use statistical computing

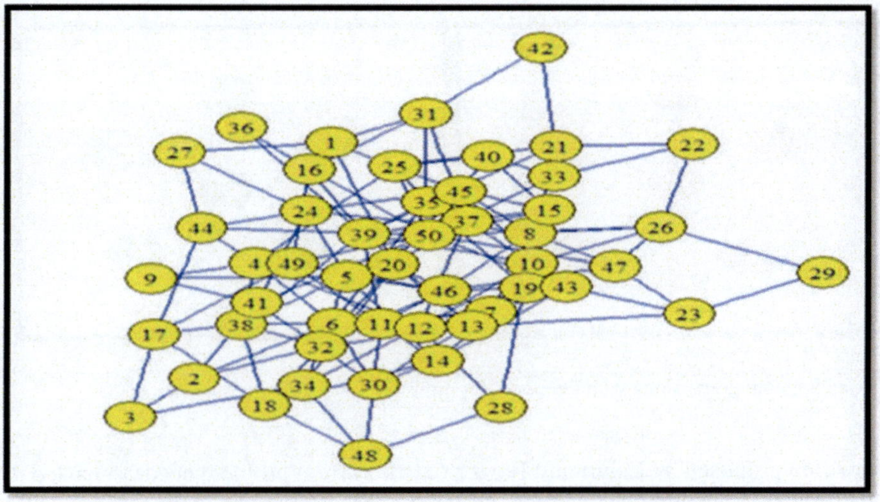

Fig. 1 Random graph at time step 10

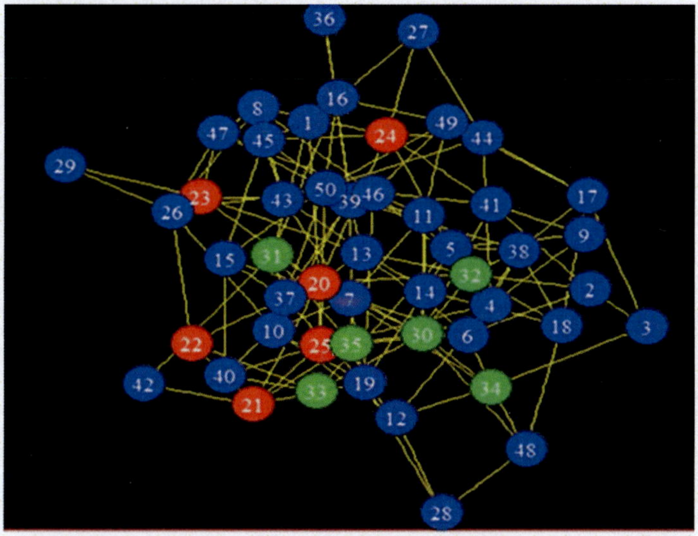

Fig. 2 Periodic subgraphs superimposed at different time intervals

features and also allows us to plot the graphs resulted at various stages of computations. The built-in functions allow us to intersect and decompose the subgraphs and maintain in appropriate lists. To implement the algorithm, we need to have appropriate data structures to store the temporal series of graphs and the intermediate subgraphs resulting from intersections of graphs between various time steps. The

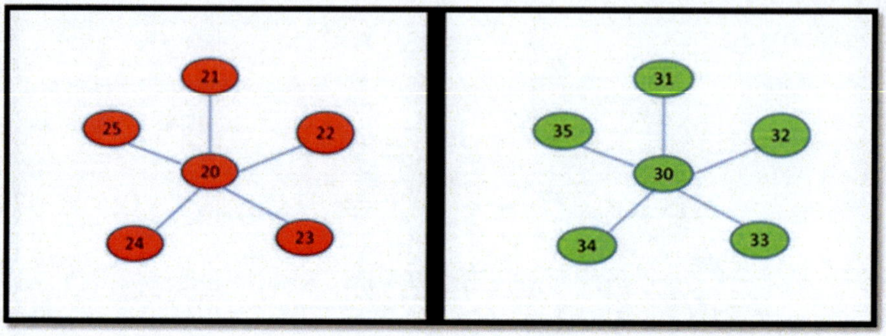

Fig. 3 Random graph along with the period's graphs embedded at time step 10

algorithm proposed by Lahiri and Berger for the current problem needs to have data structures such as link lists, pattern trees and appropriate indexing techniques such hash map. Data analysis tools such as R allow any application to use five common data structures: vectors, matrices and arrays, factors, lists and data frames. We have made use some of these data structures to implement the proposed algorithm.

The first phase of the algorithm reads the input stream of graphs sequentially. The graph at every time step is intersected with the rest of the graphs for all possible time periods. Maximal subgraphs were extracted from the intersected graphs and eliminating isolated graphs. The set of maximal subgraphs is stored in a vector for each one of the periods as a list of graphs. As we have considered randomly generated graphs at each time step, the maximal subgraphs for each time period were observed reasonably low.

4.1 Algorithm for the Phase-I

Input: Time Series of N graphs

Output: Vector of list of maximal subgraphs

Algorithm

1. for (i = 1, i < n, i++) //i denotes the time step for the first graph to be intersected
2. {
3. for (j = i, 1, i − 1) //j denotes the time step for the second graph to be intersected
4. {
5. firstGraph<-read (Gi)
6. secondGraph<-read (Gj)
7. intersectedGraph<-graph. intersection (firstGraph, secondGraph)
8. intersectedGraph<-removeIsolatedVertices(intersectedGraph)
9. maximalSubGraph[j]<-extractmaximalSubGraph(intersectedGraph)

Table 1 Output at the end of the phase-I of the algorithm

	F	S	P	N	From	To
[1]	2	1	1	3	1	6
[2]	3	2	1	4	7	12
[3]	3	1	2	5	13	18
[4]	4	3	1	5	19	26
[5]	4	2	2	6	27	30
[6]	4	1	3	7	31	34
[7]	5	4	1	6	35	41
[8]	5	3	2	7	42	51
[9]	5	2	3	8	52	57
[10]	5	1	4	9	58	65
…	…	…	…	…	…	…
[185]	20	6	14	34	1290	1296
[186]	20	5	15	35	1297	1305
[187]	20	4	16	36	1306	1309
[188]	20	3	17	37	1310	1317
[189]	20	2	18	38	1318	1326
[190]	20	1	19	39	1327	1333

```
table<-(I, k, p, nm) //p is the time period between the intersected graphs and
nm is the number of maximal subgraphs stored in the List
}
}
```

The output at the end of phase-I of the algorithm is shown in Table 1.

The output of the phase-I maintains a table for each intersected graph the following information:

F	Time step of the first intersected graph.
S	Time step of the second intersected graph.
P	Time period at which the maximal subgraph appeared.
N	Next time the subgraph expected to appear to maintain its period (P).
From and To	The first and the last maximal subgraph stored in the list at time step S.

4.2 Algorithm for the Phase-II

Input: maximalSubGraphs [] //Vector of Lists of Maximal Subgraphs

Table [F, S, P, N, From, To] //Table of information for each maximal subgraph with minimum support threshold of 3-time steps

Output: Periodic graphs with different periods and their associated start and end time steps

Algorithm

```
for (p = 1, p < ((N/2) − 1), p++)
{ // p is the periodicity of the maximal subgraph
matrix<-Table [P==p]
for (row = 1, row <= maxRow, row = row + p)
{ //maxRow is the last row of the matrix
for(i = From(row), i < To(row), i++)
{ // i is the graph index at the current row of the matrix
currentGraph<-g[i]
for (j = 1, j <= row + p, j++) // j is the graph index at the next row of the
    matrix
nextGraph<-g[[gindex_2]]
intersectedGraph<-graph. intersection (currentGraph, nextGraph)
if (intersectedGraph! = ø) (replace the intersected graph as the current graph
    for future intersection and break the loop)
}
if (intersectedGraph==ø) (flush out the currentGraph and remove the same
    from the list)
}
}
```

During the second phase of the algorithm, the maximal subgraphs stored in the vector of a list of graphs were further processed. The aim of this phase is to keep track whether the already appeared periodic maximal subgraph with minimum support threshold of three time steps continues to maintain its periodicity at the future time steps. If the subgraph ceased to maintain its periodicity, it is flushed out in the output stream and removed from the list of graphs maintained in the vector. Sub-tables were formed from the table resulted in phase-I to subset the information for each of the periodicities. The maximal subgraphs resulted from the phase-1 are further intersected with the subgraphs in the rest of the steps. If the intersection results in the empty graph at any point of time, it denotes that its periodicity ceased to exist and hence the graph is routed. If the intersection is not empty, upgrade the intersected graph as the current graph for the next step.

5 Complexity Analysis of Proposed Algorithm

The complexity of an algorithm indicates that how fast or slow the particular algorithm performs. The time complexity is generally calculated by counting the numbers of basic operations carried out by the algorithm. The time complexity for phase-I algorithm and phase-II algorithm (Tables 2 and 3) is tested in two ways: (i) the

Table 2 Analysis of the time complexity for phase-I and phase-II algorithm for fixed number of edges

Number of edges	Number of time steps	Phase-I		Phase-II	
		Execution time (in seconds)	Maximal subgraphs	Execution time (in seconds)	Periodic subgraphs
40	20	4.23	961	14.71	15
40	30	9.24	2318	34.89	38
40	40	22.50	4179	51.34	50
40	50	31.30	6279	64.72	57
40	60	45.58	9313	82.67	72

Table 3 Analysis of the time complexity for phase-I and phase-II algorithm for fixed number time steps

Number of edges	Number of time steps	Phase-I		Phase-II	
		Execution time (in seconds)	Maximal subgraphs	Execution time (in seconds)	Periodic subgraphs
40	20	4.23	961	14.71	15
50	20	6.15	1399	33.87	33
60	20	4.66	1740	60.95	33
70	20	5.48	2017	74.80	54
80	20	5.18	2342	108.06	78

numbers of edges are fixed, and the numbers of time steps are increased by 10. (ii) The numbers of time steps are fixed, and the numbers of edges are increased by 10. In both cases, the execution time is not increased linearly. At the same time, the execution time increase is not increased exponentially also. The time taken to intersect the two graphs and to find the maximal subgraphs depends on the R package. The structure of the maximal subgraphs depends on the number of subgraphs present in the dynamic network.

6 Experimental Results

The algorithm developed to process the dynamic network was tested on randomly generated time series of graphs. We have generated a network of different sizes and superimposed periodic subgraphs at different time intervals. A randomly generated network at twenty time steps with fifty nodes and randomly generated edges and two periodic subgraphs were superimposed with periods of one and two between ten time steps, and the phase-I algorithm has generated 1327 maximal subgraphs

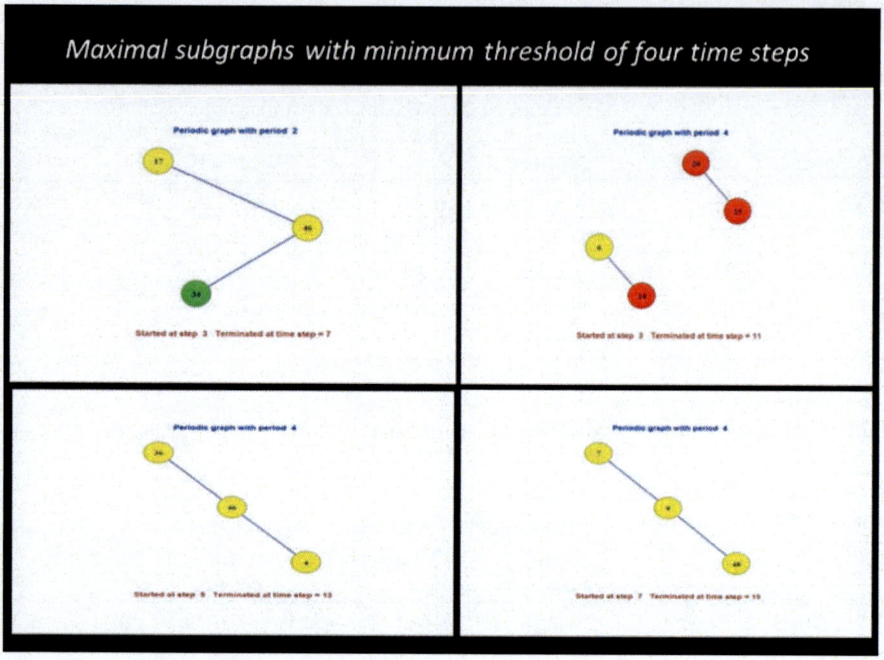

Fig. 4 Maximal subgraphs with minimum thresholds of four-time steps

with minimum support threshold of three time steps. Figure 4 shows the maximal subgraphs with minimum thresholds of four time steps. On further processing, these subgraphs, the phase-II of the algorithm flushed out 89 maximal subgraphs for more than four time steps as minimum support threshold limit. Out of these 89 graphs, more than 46 graphs had single edge graphs.

We recollect that subgraphs shown in Fig. 2 were superimposed as periodic subgraphs with periods of 1 and 2, respectively, between time steps 1 and 10. The algorithm was able to detect these graphs as periodic subgraphs with maximum periods of 4 time steps as shown in Fig. 5a, b. The graphs shown in Fig. 5c, d are the other maximal periodic subgraphs detected with periods of 4 and 2 time steps, respectively. Though the superimposed graph with period 1 remained periodic with a maximum period of 4 time steps between time steps 2 and 10, it appeared as a subgraph to another maximal subgraph with period 4 between time steps 1 and 9. Similarly, though the superimposed graph with period 2 remained periodic with a maximum period of 4 time steps between time steps 2 and 10, it appeared as a subgraph to another maximal subgraph with period 2 between time steps 6 and 10. The workflow to extract the maximal subgraph is shown in Fig. 6.

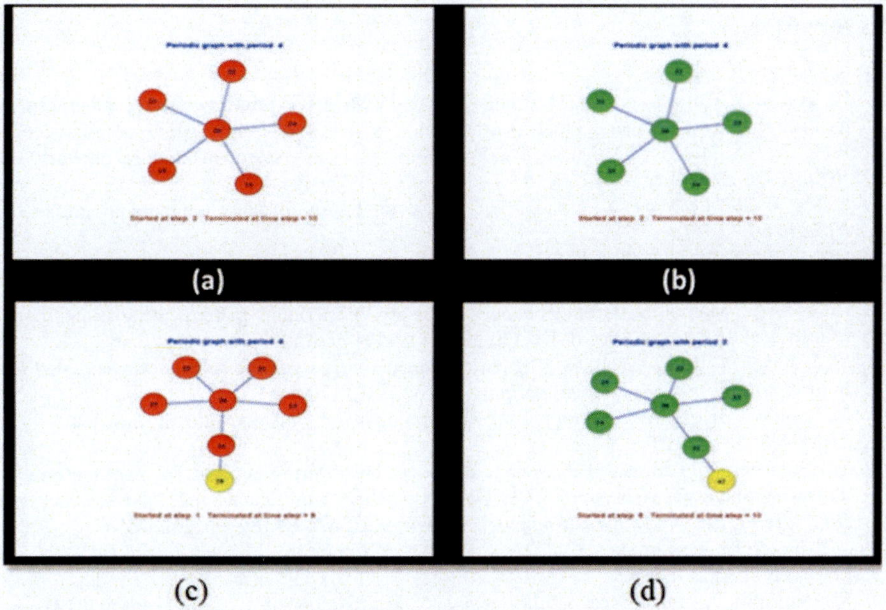

Fig. 5 Superimposed graphs detected as maximal periodic subgraphs with different periodicities

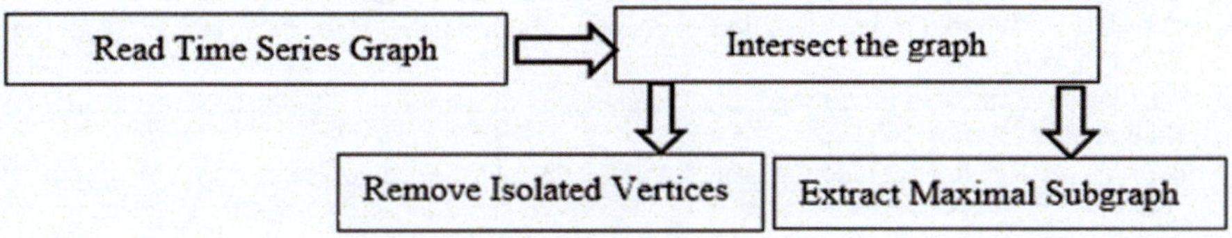

Fig. 6 Workflow to extract the maximal subgraph

7 Conclusions and Perspectives

The proposed algorithm was implemented using igraph package under R environment and tested using randomly selected streaming graphs generated by the Erdos–Renyi graph function. The implemented algorithm predicts the subgraphs superimposed on the dynamic network simulated along with other closed periodic subgraphs. The complexity of the algorithm is established as the algorithm depends on various data structures built in R environment. The developed algorithm was also tested to detect periodic subgraphs based on attributes attached to the vertices and edges of the graphs. In future, this algorithm can be tested with the social network application domains.

References

1. B. Boden, S. Gunnemann, H. Hoffmann, T. Seidl, MiMAG: mining coherent subgraphs in multi-layer graphs with edge labels. Knowl. Inf. Syst. **50**(2), 417–446 (2017)
2. R.-H. Li, J. Su, L. Qin, J.X. Yu, Q. Dai, Persistent community search in temporal networks, in *IEEE ICDE* (2005)
3. S. Ma, R. Hu, L. Wang, X. Lin, J. Huai, Fast computation of dense temporal subgraphs, in *IEEE ICDE* (2017)
4. M.G. Elfeky, W.G. Aref, A.K. Elmagarmid, Periodicity detection in time series databases. IEEE Trans. Knowl. Data Eng. **17**(7), 875–887 (2005)
5. H. Cao, N. Mamoulis, D.W. Cheung, Discovery of periodic patterns in spatiotemporal sequences. IEEE Trans. Knowl. Data Eng. **19**(4), 453–467 (2007)
6. M. Lahiri, T.Y. Berger-Wolf, Mining periodic behavior in dynamic social networks, in *IEEE International Conference on Data Mining*, pp. 373–382 (2008)
7. M. Lahiri, T. Berger-Wolf, Periodic subgraph mining in dynamic networks. J. Knowl. Inf. Syst. **24**(3), 467–497 (2010)
8. S.K. Tanbeer, C.F. Ahmed, B.S. Jeong, Y.K. Lee, Discovering periodic frequent patterns in transactional databases, in *Pacific–Asia Conference on Advances in Knowledge Discovery and Data Mining. Lecture Notes in Computer Science*, vol. 5476 (2009), pp. 242–253
9. A. Apostolico, M. Barbares, C. Pizzi, Speedup for a periodic subgraph miner. Inf. Process. Lett. **111**(11), 521–523 (2011)
10. K. Macropol, A. Singh, Reachability analysis and modeling of dynamic event networks, in *Machine Learning and Knowledge Discovery in Databases. Lecture Notes in Computer Science*, vol. 7523 (2012), pp. 442–457
11. A. Apostolico, P.L. Erdős, E. Győri, Z. Lipták, C. Pizzi, Efficient algorithms for the periodic subgraphs mining problem. J. Discrete Algorithms **17**, 24–30 (2012)
12. P. Erdős, A. Rényi, On the evolution of random graphs. Publ. Math. Inst. Hung. Acad. Sci. 17–61 (1960)
13. P. Erdős, A. Rényi, On the strength of connectedness of random graphs. Acta Math. Acad. Sci. Hungar. 267 (1961)
14. P. Erdős, A. Rényi, On random graphs I. Publicationes Math. Debrecen **6**, 290–297 (1959)

Hate Tweet Extraction from Social Media Text Using Autoencoder Wrapped Multinomial Naive Bayes Classifier

Ashutosh Bhoi and Rakesh Chandra Balabantaray

Abstract Social media is one of the biggest source for gathering real-time data, and information spreading in day-to-day life. In the name of freedom of speech, many people are using derogatory words and offensive languages which create a lot of problem for the concern person, organization, society, and government. To reduce the dehumanizing activities arises due to these hate speeches, we need to filter them from the social media text. Here, we propose a system which analyzes the micro-blog text to classify them as hate or non-hate speech and filter out the hate speeches. Our presented prototype system also handles non-English text as the social media platforms are not confined to English only. Various classifiers along with several learning techniques are used to exhibit their effectiveness for an accomplished solution. An autoencoder wrapped Multinomial Naive Bayes (MNB) classifier is incorporated in the system for a significant gain in the performance. The simplicity in its implementation makes our system superior to other state-of-the-art systems available in the literature. Finally, the prototype system is evaluated on various publicly available hate-related datasets to showcase its effectiveness on this challenging task.

Keywords Hate speech · Tweets analysis · Autoencoder · Classification · Feature selection.

1 Introduction

Social media platforms like Twitter, WhatsApp, and Facebook have evolved as a prime mode of communication in last decade. These online social networking platforms are generating huge volume of real-time data. The users who are connected through these social platforms come from various locations, genders, cultures, mind-

A. Bhoi (✉)
Department of Computer Science & Engineering, GCEK Bhawanipatna, Bhawanipatna, India

R. Chandra Balabantaray
Department of Computer Science & Engineering, IIIT Bhubaneswar, Bhubaneswar, India
e-mail: rakesh@iiit-bh.ac.in

V. Bhateja et al. (eds.), *Data Engineering and Intelligent Computing*,
Advances in Intelligent Systems and Computing 1407,
https://doi.org/10.1007/978-981-16-0171-2_59

set, and educational backgrounds. Many a times, difference in belief, judgment, or opinions lead to verbal attacks like abuse, insult, teasing, and cursing, etc. In democratic countries like India, these micro-blog users are misusing their freedom of speech and repeatedly using offensive languages, derogatory terms, and hate speeches in their posts. These kind of posts incite violence or express prejudice against a person or particular group especially on the basis of race, color, disability, religion, physical trait, national origin, or sexual orientation [14]. This can lower the self-esteem of people, leading to mental illness and negative impact on the society as a whole. Toxic languages may further lead to cyberbullying, which is coming as a major reason behind many dehumanizing activities or even suicides [3]. In many recent incidents like Delhi riots in the year 2020, it is observed that these kind of hatred posts are working as a catalyst which is a major concern for the society and government. To overcome this issue, an automatic hate text classification system is the need of the hour. Here, we propose a social media text classification system which automatically categorizes hate text from non-hate text. Automatic detection of hate speech may reduce its spreading and further minimizes any type of antisocial activities. We consider tweets dataset to evaluate the effectiveness of our proposed technique.

The related work about the task is explored in Sect. 2. In Sect. 3, it is discussed in details about the proposed methodology. The complete specifications about the case study and dataset are presented in Sect. 4. The results and discussion are elaborated in Sect. 5. At last, Sect. 6 concludes our work along with the findings and further scope for improvement.

2 Related Work

Social media text is one of the major sources of real-time unstructured text. Social scientists are analyzing this data to extract information from it, which are applied in numerous applications [4]. However, many people are using these platforms for wrong purposes and spreading hate speeches in the society. Due to the presence of these kind of ill content in micro-blogs, they are playing a larger role in hate crime. Because of communal concern and how hate speeches are propagating through Internet, we are strongly motivated to analyze automatic detection of hate speech. Researchers have attempted this challenging task in different dimensions in the near future [16].

The main issue in detecting hate speech is defining hate speech and lack of any generalized system [11]. Hate speech identification with the help of comment embedding is proposed where paragraph2vec is used for finding the distributed representation in joint space [6]. [17] proposed a character n-gram model along with some extra linguistic features to categorize the hate tweets. A logistic regression-based SVM is proposed for hate speech and offensive language identification where logistic regression is used for dimensionality regression and SVM performs the classification task [5]. Multiple deep learning architecture-based hate tweets categorization system is proposed using semantic word embedding [2]. In [18], the authors suggested a pattern-based unigram approach to find hateful and offensive expressions present on tweets. A supervised algorithm-based cascading voting and AdaBoost classifier is suggested to detect the cyberbullying in social commentary [13].

3 Proposed Methodology

The proposed system classifies the tweets into predefined categories like hate, non-hate, offensive, racism, sexism, insulting, non-insulting, non-aggressive, overtly aggressive, covertly aggressive, etc. in various publicly available hate speech datasets. The architecture of the proposed prototype system is illustrated in Fig. 1.

3.1 Crawling and Preprocessing of Tweets

The first thing that the system does is crawling of tweets from Twitter streaming using available APIs. Once the tweets are crawled, we preprocess each tweet in the elementary steps. We removed user handles, email IDs, retweet symbols, HTML tags, hashtags, etc. from the tweets. All delimiters like extra spaces, colon, semicolon, periods, and comma are also removed from the original text.

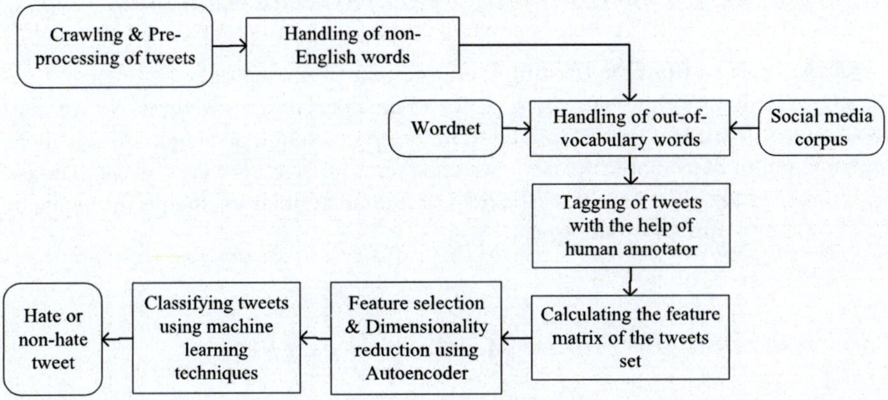

Fig. 1 Proposed architecture

3.2 Handling Non-english Tweets

Although English is the most widely accepted language by these social platform users, still many people are using their regional languages for communication. To build any generic system for these kind of code-mix tweets, first thing that we have to do is to translate all text into a single language. Here, we add a language translation component which converts the non-English texts to their equivalent English texts. *Google Trans* is used here for the language translation task.

3.3 Handling of Out-of-Vocabulary Words

The biggest challenge for handling unstructured text like social media text is due to the presence of spelling errors, frequently used acronyms and abbreviations. Here, we mainly focus on those out-of-vocabulary (OOV) or misspelled words to make them syntactically correct. To handle the slangs, acronyms, and abbreviations, a secondary social media corpus (e.g., btw "by the way," fb "Facebook," 4m "from," c "see," b4 "before," pls"please," 2day "today," and lv "love") is also used to improve the accuracy. All syntactically incorrect words are substituted with their corresponding normalized words.

3.4 Tagging of Tweets with the Help of Human Annotator

It is also observed from the existing literature that if labeled data are present, then the supervised algorithms may be superior to unsupervised algorithms. So we may tag the tweets based on the interpretation of the engaged human annotators. As there are some publicly available datasets, we use them directly. However, if we increase the number of tagged tweets with the help of human annotators, the performance of the system may further be improved.

3.5 Generating the Feature Matrix of Tweets Set

Once tagging of normalized tweets is done successfully, stemming and stopwords removal from the tweets set are done. TF-IDF is used here for generating the feature matrix. To compare the effectiveness of our proposed model with popular deep learning-based classifiers, we also tried the word embedding-based feature vector generation. Word vectors are generated using GloVe pretrained model.

3.6 Classifying the Tweets Using Autoencoder Wrapped MNB Classifier

To perform the hate tweet classification task, the feature matrix is input into different state-of-the-art statistical classifiers like Multinomial Naive Bayes (MNB) [8], SVM [15], and random forest [1] to categorize the tweets based on the pre-specified classes of hate and non-hate tweets. Deep learning models like CNN [7] and LSTM [12] are also used previously for the categorization task. CNN takes advantages in choosing the important features, and LSTM networks have ascertained good abilities for learning sequential data. It is reported in the literature that both approaches provide improved results in the natural language processing applications. A combined autoencoder wrapped MNB model is applied here to perform the challenging, short, and noisy text classification task.

Feature selection using autoencoder

In all statistical machine learning-based classification models, feature selection and dimensionality reduction are one of the most vital tasks. The feature matrix generated using TF-IDF technique is generally sparse matrix, as there are only few words present in the tweet. Here, each distinct word in the tweet set is treated as a feature. There may be many non-contributing features which are conflicting to each other and further degrading the performance of the classification system. From the literature, it is found that autoencoder is very effective for feature subset selection and dimensionality reduction. Here, the autoencoder is wrapped with the MNB classifier, so that it selects the most optimal set of features which give the maximum classification performance.

Autoencoder

Autoencoder is a special type of neural network which is used to learn efficient data representation. Like any other neural network, it has one input, one output, and one or more hidden layer(s). Due to its unsupervised learning mechanism, it may be a powerful tool in many natural language processing (NLP) applications as there are huge scarcity of labeled data. There are two sides of an autoencoder such as reduction side (encoder) and reconstruction side (decoder). The precision to reconstruct the original input from the reduced representation makes it powerful for dimensionality reduction tasks.

The encoding and decoding steps are represented as follows:

$$\sigma : X \rightarrow F \tag{1}$$

$$\tau : F \rightarrow X \tag{2}$$

$$\sigma, \tau = \arg\min_{\sigma,\tau} \parallel X - (\tau \circ \sigma)X \parallel^2 \tag{3}$$

Here, we use the denoising autoencoder to achieve better representation by altering the decoding criterion [10]. The training steps of denoising autoencoder performs as mention below:

Step-1. The initial input X is noised or corrupted into X' through isotropic Gaussian mapping

$$X' \sim q_D(X' \mid X) \tag{4}$$

Step-2. The noised or corrupted input X' is then mapped to a intermediate encoding with the same process as that of an autoencoder

$$i = f_{\varphi}(X') = s(WX' + b) \tag{5}$$

Step-3. From the intermediate encoding, the model decodes

$$z = g_{\varphi'}(i) \tag{6}$$

Here, autoencoder works as a wrapper for the MNB classifier to extract the best performance. It selects the optimal set of features with respect to their performance.

Multinomial Naive Bayes Classifier(MNB)

The most contributing features in any text documents are the words or tokens. The words or phrases in a sentence, tweet or document have maximum dependencies or relationships to their adjacent words or phrases. These sequence of words follow the multinomial distribution that motivates us to use the Multinomial Naive Bayes for our categorization problem. In numerous NLP applications, Multinomial Naive Bayes has validated itself to be an accurate, fast, and reliable classifier. In this MNB classifier, feature vector of each tweet depicts the Tf-Idf scores of each word present in the tweet. The multinomial distribution works well for both integer and fractional feature counts. Here, $sklearn \cdot naive\_bayes \cdot MultinomialNB$ is used for the implementation task.

4 Description of Case Study and Dataset

Twitter is mostly used for the case study. In one case, we have used Facebook as a case study. The complete description about all the datasets used for the validation of our proposed system is mentioned in this section. Apart from publicly available datasets, we may also built our own dataset. However, this human annotation process is a tedious and challenging task due to agreeing mismatch and no universal definition of hate speech. During any riot, political propaganda or any emergency situation hate speeches may be very critical. Our automation system filters such hate tweets from tweet stream. Few publicly available datasets are used to validate our model [9, 17] Table 1.

Table 1 Details about used datasets

Sl. No.	Dataset	Year	Source	Language (s)	No. of Labels	Labels
1	TRAC (Twitter)	2018	Tweets	English & Hindi	3	Non-aggressive, overly aggressive, covertly aggressive
2	TRAC (Facebook)	2018	Facebook posts	English & Hindi	3	Non-aggressive, overly aggressive, covertly aggressive
3	Kaggle	2018	Tweets	English	2	Insulting, non-insulting

5 Results and Discussion

All experimentation that are carried out to evaluate the efficiency of our proposed system is presented in this section. Here, the model performance is compared with many state-of-the-art classifiers used for this task. In all datasets, we have divided the total samples into four ratios of training and testing. The ratios are ranging from 60:40 to 90:10 and their mean results are shown in the table in five different runs. We have also shown the fivefold and tenfold cross-validation results of our model. Here, we have used precision, recall, f-measure, and accuracy as performance measures. TensorFlow 2.0 machine learning library is used to implement autoencoder, LSTM, and CNN-based models. All hyperparameters of the models are set based on the literature and they are fine-tuned accordingly. Word embeddings were initialized using publicly available 200-dimensional GloVe word vectors trained on a 2 billion tweets and 27 billion tokens of the Twitter data corpus.

The classification results of various statistical machine learning and deep learning models for hate tweet detection are given in Table 2. From all results, it is observed that LSTM-based classifier outperforming others in all aspects. CNN-based classifier is close to the result of LSTM. Among all statistical machine learning-based models, Multinomial Naive Bayes is outperforming the other two. It may be noticed that among all datasets, the models are working better in TRAC-Facebook dataset. The reason may be due to size of Facebok posts, which are larger in size than tweets. Due to larger size, the context may be learned better. The other reason may be due to larger number of samples present in the dataset. The sample size may also be the reason in case of TRAC-Twitter and Kaggle dataset (Table 3).

Table 2 Performance without any feature selection

Dataset	Classifier	Avg. results of 4 different ratios				fivefold cross valiation				tenfold cross valiation			
		A	P	R	F1	A	P	R	F1	A	P	R	F1
TRAC (Twitter)	SVM	74.81	79.54	71.63	75.39	74.76	79.51	71.57	75.33	74.64	79.43	71.45	75.23
	Multinomial Naive Bayes	76.73	80.57	72.62	76.39	76.68	80.52	72.55	76.33	76.59	80.45	72.42	76.22
	Random Forest	75.64	79.69	71.92	75.61	75.55	79.65	71.83	75.54	75.43	79.51	71.72	75.41
	CNN	78.58	82.13	74.81	78.30	78.52	82.04	74.75	78.23	78.45	81.91	74.66	78.12
	LSTM	79.04	82.98	75.04	78.81	78.96	82.95	74.99	78.77	79.81	82.83	74.86	78.64
TRAC (Facebook)	SVM	76.21	80.67	72.12	76.16	76.14	80.60	72.05	76.09	76.02	80.49	71.96	75.99
	Multinomial Naive Bayes	78.03	81.76	73.34	77.32	77.98	81.71	73.29	77.27	77.89	81.58	73.18	77.15
	Random Forest	76.87	80.83	72.73	76.57	76.79	80.75	72.66	76.49	76.65	80.62	72.52	76.36
	CNN	79.94	83.06	75.52	79.11	79.91	83.03	75.46	79.06	79.82	82.91	75.35	78.95
	LSTM	80.74	83.75	75.97	79.67	80.70	83.70	75.91	79.61	80.58	83.59	75.78	79.49
Kaggle	SVM	74.91	79.63	71.69	75.45	74.85	79.58	71.63	75.40	74.73	79.47	71.49	75.27
	Multinomial Naive Bayes	76.88	80.68	72.71	76.49	76.81	80.65	72.64	76.44	76.68	80.53	72.55	76.33
	Random Forest	75.74	79.73	71.98	75.66	75.68	79.67	71.91	75.59	75.54	79.56	71.79	75.48
	CNN	78.69	82.25	74.92	78.41	78.62	82.19	73.86	77.80	78.48	82.08	74.77	78.25
	LSTM	79.15	83.09	75.17	78.93	79.03	83.05	75.11	78.88	78.92	82.91	75.02	78.77

Table 3 Performance of all models using autoencoder for feature selection

Dataset	Classifier	Avg. results of 4 different ratios				fivefold cross valiation				tenfold cross valiation			
		A	P	R	F1	A	P	R	F1	A	P	R	F1
TRAC (Twitter)	SVM	77.42	81.55	74.37	77.79	77.38	81.51	74.34	77.76	77.28	81.39	74.27	77.67
	Multinomial Naive Bayes	78.93	82.91	74.98	78.75	78.87	82.87	74.93	78.70	78.78	82.73	74.84	78.59
	Random Forest	77.89	81.95	74.68	78.15	77.84	81.90	74.64	78.10	77.72	81.78	74.56	78.00
	CNN	78.58	82.13	74.81	78.30	78.52	82.04	74.75	78.23	78.45	81.91	74.66	78.12
	LSTM	79.04	82.98	75.04	78.81	78.96	82.95	74.99	78.77	79.81	82.83	74.86	78.64
TRAC (Facebook)	SVM	78.57	82.59	75.41	78.84	78.52	82.51	75.35	78.76	78.41	82.39	75.24	78.65
	Multinomial Naive Bayes	80.03	83.93	75.97	79.75	79.98	83.81	75.93	79.68	79.89	83.68	75.78	79.53
	Random Forest	78.97	82.99	75.71	79.18	78.90	82.93	75.64	79.12	78.75	82.82	75.52	79.00
	CNN	79.67	83.18	75.85	79.35	79.91	83.03	75.46	79.06	79.82	82.91	75.35	78.95
	LSTM	80.09	84.01	76.09	79.85	80.70	83.70	75.91	79.61	80.58	83.59	75.78	79.49
Kaggle	SVM	77.49	81.62	74.45	77.87	77.45	81.57	74.40	77.82	77.32	81.43	74.26	77.68
	Multinomial Naive Bayes	79.04	82.98	75.06	78.82	78.99	82.95	75.00	78.77	78.87	82.86	74.94	78.70
	Random Forest	77.97	82.04	74.75	78.23	77.91	81.98	74.69	78.17	77.80	81.86	74.55	78.03
	CNN	78.69	82.25	74.92	78.41	78.62	82.19	73.86	77.80	78.48	82.08	74.77	78.25
	LSTM	79.15	83.09	75.17	78.93	79.03	83.05	75.11	78.88	78.92	82.91	75.02	78.77

The results which we got after using autoencoder for feature selection are represented in Table 2. The results show a significant improvement in the performance of all statistical machine learning-based classifiers. It may be observed that if feature representation and subset selection are done properly, then machine learning-based classification systems are at par with other deep learning-based models. Among all statistical machine learning-based classifiers, our suggested Multinomial Naive Bayes (MNB) is giving better results than the other two. The performance gap between MNB and LSTM is marginal which may be considered a significant contribution as deep learning-based models are resource and data hungry. If we may get similar results with statistical classifiers, then it will be very useful as they do not require that much amount of resources. Looking at the results of all the datasets, it may also noticed that in TRAC-Facebook performance is better than the other two. Finally, it is also found that the speed of computation in autoencoder-based MNB is quite fast in comparison to LSTM and CNN which is due to the dimensionality reduction of autoencoder.

6 Conclusion and Future Work

In this research, we have proposed an autoencoder wrapped Multinomial Naive Bayes classification model for hate speech identification from social media text. We also tried few other statistical classifiers along with LSTM and CNN, two popular deep learning-based models. With limited resources unlike LSTM and CNN, our proposed system giving results at par with them. We validated the effectiveness of our system with three hate speech social media datasets on tweets and Facebok posts. All the models are giving better results on TRAC-Facebook dataset, may be due to the content size of Facebook posts and sample size of the dataset. The computational speed is also faster in our proposed model as compared to LSTM and CNN. In our future work, we may improve the performance by reducing the misclassification rates.

References

1. I. Alfina, R. Mulia, M.I. Fanany, Y. Ekanata, Hate speech detection in the Indonesian language: a dataset and preliminary study, in *2017 International Conference on Advanced Computer Science and Information Systems (ICACSIS)*, pp 233–238, IEEE (2017)
2. P. Badjatiya, S. Gupta, M. Gupta, V. Varma, Deep learning for hate speech detection in tweets, in *Proceedings of the 26th International Conference on World Wide Web Companion*, pp 759–760 (2017)
3. C.P. Barlett, J.B. Heath, C.S. Madison, C.C. DeWitt, S.M. Kirkpatrick, You're not anonymous online: the development and validation of a new cyberbullying intervention curriculum. Psychol. Popular Media Culture (2019)
4. A. Bhoi, R.C. Balabantaray, Named entity recognition from social media text: a comparative study. Int. J. Control Theor. Appl. **10**(19), 9–15 (2017)

5. T. Davidson, D. Warmsley, M. Macy, I. Weber, Automated hate speech detection and the problem of offensive language, in *Eleventh International AAAI Conference on Web and Social Media* (2017)
6. N. Djuric, J. Zhou, R. Morris, M. Grbovic, V. Radosavljevic, N. Bhamidipati, Hate speech detection with comment embeddings, in *Proceedings of the 24th International Conference on World Wide Web*. pp. 29–30 (2015)
7. B. Gambäck, U.K. Sikdar, Using convolutional neural networks to classify hate-speech, in *Proceedings of the First Workshop on Abusive Language Online*, pp. 85–90 (2017)
8. K.K. Kiilu, G. Okeyo, R. Rimiru, K. Ogada, Using naïve Bayes algorithm in detection of hate tweets. Int. J. Sci. Rese. Publ. **8**(3) (2018)
9. R. Kumar, A.K. Ojha, S. Malmasi, M. Zampieri, Benchmarking aggression identification in social media, in *Proceedings of the First Workshop on Trolling, Aggression and Cyberbullying (TRAC-2018)*, pp. 1–11 (2018)
10. M. Leyli-Abadi, L. Labiod, M. Nadif, Denoising autoencoder as an effective dimensionality reduction and clustering of text data, in *Pacific-Asia Conference on Knowledge Discovery and Data Mining*, pp. 801–813. Springer (2017)
11. S. MacAvaney, H.R. Yao, E. Yang, K. Russell, N. Goharian, O. Frieder, Hate speech detection: challenges and solutions. PloS one **14**(8) (2019)
12. G.L. De la Pena Sarracén, R.G. Pons, C.E.M. Cuza, P. Rosso, Hate speech detection using attention-based lstm. EVALITA Eval. NLP Speech Tools Italian **12**, 235 (2018)
13. M.O. Raza, M. Memon, S. Bhatti, R. Bux, Detecting cyberbullying in social commentary using supervised machine learning, in *Future of Information and Communication Conference*, pp. 621–630. Springer (2020)
14. K. Relia, Z. Li, S.H. Cook, R. Chunara, Race, ethnicity and national origin-based discrimination in social media and hate crimes across 100 us cities. Proc. Int. AAAI Conf. Web Soc. Media **13**, 417–427 (2019)
15. D. Robinson, Z. Zhang, J. Tepper, Hate speech detection on twitter: feature engineering vs feature selection, in *European Semantic Web Conference*, pp. 46–49. Springer (2018)
16. A. Schmidt, M. Wiegand, A survey on hate speech detection using natural language processing, in *Proceedings of the Fifth International Workshop on Natural Language Processing for Social Media*, pp. 1–10 (2017)
17. Z. Waseem, D. Hovy, Hateful symbols or hateful people? Predictive features for hate speech detection on twitter, in *Proceedings of the NAACL Student Research Workshop*, pp. 88–93 (2016)
18. H. Watanabe, M. Bouazizi, T. Ohtsuki, Hate speech on twitter: a pragmatic approach to collect hateful and offensive expressions and perform hate speech detection. IEEE Access **6**, 13825–13835 (2018)

A Reduced Feature Representation Scheme for Offline Handwritten Character Recognition

Raghunah Dey and Rakesh Chandra Balabantaray

Abstract Recognizing handwritten characters (HCR) is a significant and challenging area of study in the field of computer vision and pattern recognition. Here, a sliding window-based platform for effective visualization of handwritten character images is designed. Three specific feature extraction strategies are suggested to recognize the characters. Those are an angular movement of shape-dependent characteristics, center-to-boundary of the text line, and center-to-boundary of text's edge feature. The feature selection method is implemented after producing the features to reduce the dimension of the feature matrix. So that, the implementation time can be reduced, somehow. The accuracy from the reduced feature set has also been computed to assess the effectiveness. The feature selection is performed using an extra tree classifier. The characters in this study are recognized using the random forest classifier. The suggested method achieves quite better accuracy with the limited feature set.

Keywords Handwritten character · Random forest · Edge detection · Extra tree classifier · Feature importance.

1 Introduction

In the previous few years, the identification of handwritten English alphabets was widely researched. Digital paper processing, such as information from data entry, credit cards, tax, medical insurance forms, is the HCR's applications. Several organizations, companies, health, and legal institutions hold many paper reports such as printouts of replies, reviews, datasheets. These records are often different and require

R. Dey (✉) · R. Chandra Balabantaray
Cross Lingual Information Access Laboratory Department of Computer Science and Engineering, International Institute of Information Technology, Bhubaneswar, India
e-mail: c118003@iiit-bh.ac.in

R. Chandra Balabantaray
e-mail: rakesh@iiit-bh.ac.in

V. Bhateja et al. (eds.), *Data Engineering and Intelligent Computing*, Advances in Intelligent Systems and Computing 1407,
https://doi.org/10.1007/978-981-16-0171-2_60

time to handle simultaneously. When these documents are older, they may get corrupted with passing time. To avoid all these issues, there is a need to develop an effective offline handwriting recognition program that converts all these documents into digital text form.

Section 2 discusses the survey documents. The used dataset is listed in Sect. 3. Section 4 defines preprocessing including the extraction of features. Section 5 outlines the recognition process and also the features selection. The experimental results are shown in Sect. 6. Finally, Sect. 7 ends the paper.

2 Literature Survey

The random forest algorithm for IC inspection system has been built for an embedded character recognition program [1]. The device was tested depending on the form of feature extraction and in terms of classification accuracy. A learning system with the usage of a deep CNN for handwritten words was proposed in [2]. The approach is to identify through word spotting. This requires just a dimensions of 32, of the feature vector. The paper [3] used a variety of features and classifiers, such as nearest neighbors, decision trees, and random forest for the recognition. It focused on the influences of features on the quality of classification. Two types of features by means of parabola curve and power curve fitting were suggested for Gurmukhi characters [4]. The efficiency of the method was contrasted with other feature extraction systems. Four types of features: zoning, projection profile, freeman chain code, and histograms are used for handwritten Latin vowel recognition [5]. It was employed using K-NN and fuzzy min-max methods of classification. In the [6], the handwritten Shui character recognition was presented using convolutional neural network. One of the feature extraction methods, namely structural, horizontal, and vertical histogram, was analyzed in [7]. Two more projections using these feature vectors were created and called as mean H histogram and mean V histogram. Handwritten numerals from EMNIST dataset were identified from geometric and gradients using NN [8]. The features are selected with minimum and maximal relevance redundancy (mRMR), specific shared knowledge feature selection, and relief feature selection techniques. A features extraction technique that combines the unique features of the geometric, hybrid zone based, and gradient are proposed [9]. The MLP with backpropagation, Levenberg-Marquardt algorithm, and CNN was used here. The paper [10] presented for Amharic handwritten character and Chars74K number detection using HOG, LBP, and graphical features. LDA has been used to reduce the total features.

3 Dataset Description

For experiment, the Chars74k handwritten Latin alphabets are used. The Chars74k dataset contains three distinct sub-dataset categories (all of which are 62 classes), i.e., *EnglishHnd*, *EnglishFnt*, and *EnglishImg* [14]. Database Chars74k consists of png format images. These images are grouped into 62 classes [0–9], [A–Z], and [a–z]. For computational purposes, the handwritten Latin alphabets from the *EnglishHnd* dataset are included here. The data for *EnglishHnd* was analyzed by a tablet computer, and 3410 number of character samples written by 55 authors. These are all black and white samples. Images are 900 * 1200 pixels and are positioned in unparalleled positions.

4 Preprocessing

In the preprocessing segment, we have concentrated only on three tasks. These are binarization, thinning, and edge detection. Followed by this, the feature extraction and selection procedures are performed.

Binarization is one of the main activities of preprocessing. Having got the gray picture, a global thresholding technique from Otsu [19] was implemented for the binarized version of the same input sample. We apply the method of thinning [20] after binarization. The stroke width may not be consistent everywhere since it is handwritten documentation. This is why it is important to create a thinned picture. Here is the "Canny's edge detection algorithm" [18] with the Gaussian gradient used to detect the edge. The probability of detecting true edge points will be maximized in the Canny edge detection technique.

4.1 Feature Extraction

4.1.1 "Angular Movement of Shape-Based Feature" (AMSF) in Tabular Form

The AMSF feature in string form is explained in [21]. The AMSF algorithm positions the current text pixel in the center of the window and checks its neighboring location, as explained in paper [16]. There are nine cells on the sliding window. Since the current pixel occupies a centered cell, then the other eight cells help the feature vector to be created. We have 256, that is 2^8, number of possibilities of different distinctive feature values. The codes are created only in one direction, from left to right of the sliding window. That is why the optimal combination number of characters is $256/2 = 128$. The sliding window scans the image into two sections: one in the lower section and one in the upper section. Therefore, the total combination of characters is 128 * 2= 256. Creating the feature matrix is shown in Fig. 1.

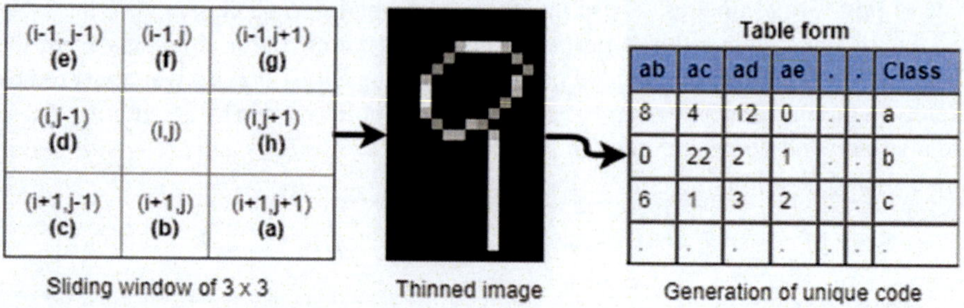

Fig. 1 Unique feature generation for every character sample

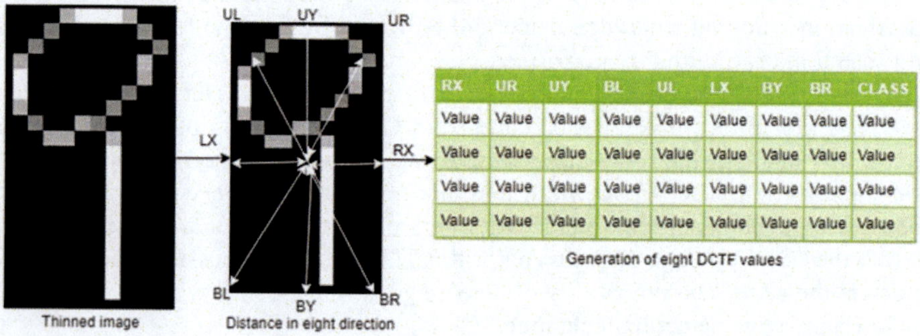

Fig. 2 DCTF-type feature generation

4.1.2 "Distance from the Center to Thinned Text Feature" Generation (DCTF)

Every character has different shapes. The distance from the center of these different text images is different. Based on this definition, the distance is computed from the central. The size from the middle location to its thin text outline is determined in eight directions. The main concept is illustrated in Fig. 2. Points RX, UR, UY, BL, UL, LX, BY, and BR are the locations on a binarized image, whereby in each of eight directions, the algorithm seeks the first white pixel. This length is Euclidean's distnace of the angular position and is a floating value. The DCTF feature matrix will store these values.

4.1.3 "Distance from the Center to the Outer Edge of Text Feature" Generation (DCEF)

In certain cases, because of the unclear, blurred, and haziness of the image sample, the binarized thinned version of the image is not correctly created. A new type of feature called DCEF is introduced, based on the distance from the outer edge to the

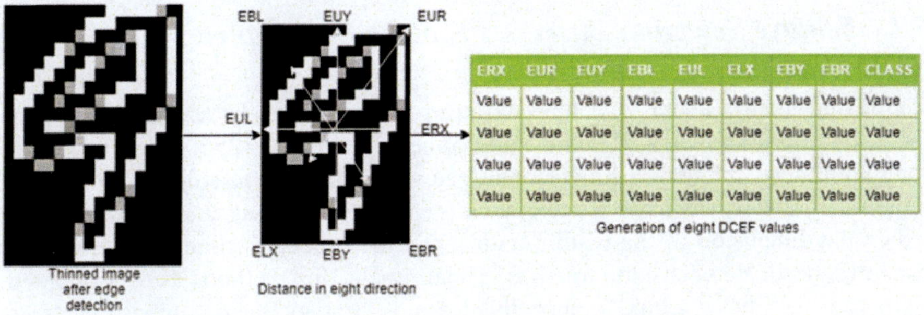

Fig. 3 DCEF-type feature generation

middle of the character image. The mechanism is illustrated in Fig. 3. Each of ERX, EUR, EUY, EUL, ELX, EBL, EBY, and EBR eight angular distances from the center is computed according to the cell positions to generate the DCEF feature matrix. There are two thin lines for an edged version of the image sample. So, always the outer edge will be encountered first. The distance will be from the outer edge to its center.

5 Proposed Character Recognition

A random forest classification scheme for handwritten pictures is proposed here. In Fig. 4, all stages of the proposed structure are shown. It consists primarily of three stages. These are preprocessing, extracting of features, and finally the recognizing. To build the recognition system, all the three kinds of feature sets are used. To reduce the time required for the execution of the recognition system, a reduced feature matrix is needed. That is why there a feature selection technique is implemented based on the feature importance.

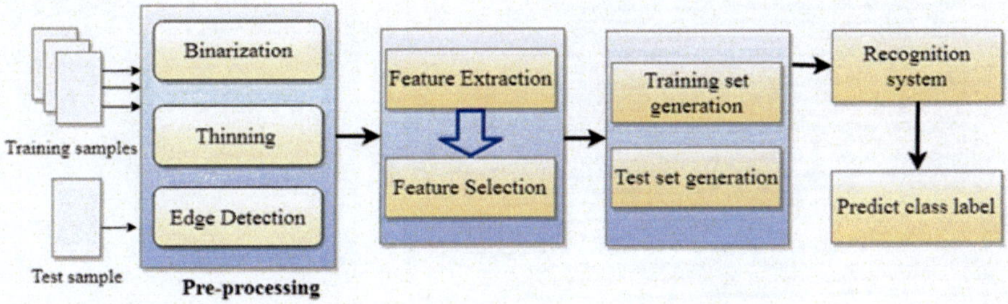

Fig. 4 Proposed offline character recognition process

5.1 Feature Selection Based on Feature Importance

The feature matrix consists of three types of features, namely AMSF, DCTF, and DCEF. The feature set will possess the dimension of 272 (256 + 8 + 8). So, to train with these number of dimensional feature matrix, the classifier will take a bit more time. That is why a feature selections technique is necessary. By implementing which the dimension of the feature matrix and the execution time of the classifier can be reduced. Here, an extra tree-based feature selection method has been adopted. Extra tree classifier is a kind of ensemble learning strategy that combines the effects of the different decision trees gathered in a "forest" to generate the effects of its classification [17]. After applying feature selection, 100 numbers of best features are considered based on their feature importance. For better understanding, the best 30 features are shown graphically in Fig. 5.

5.2 Recognition Using Random Forest Classifier

In random forests [11], the common error in the tree classification system relies on the frequency and similarity between specific trees in the system. The random collection of features to distinguish each node results in error levels which are more stable in terms of noise than Adaboost. Internal estimates track and display the reaction to the increased number of features used in splitting fault, power, and correlation. The procedure incorporates the "bagging" principle of Breiman [12] with a random range of features to create a series of decision trees with managed variance. There are a variety of decision trees containing random forests. Every tree is graded, and we conclude for the same class the tree is "voting." The forest picks the most vote category, including all forest trees.

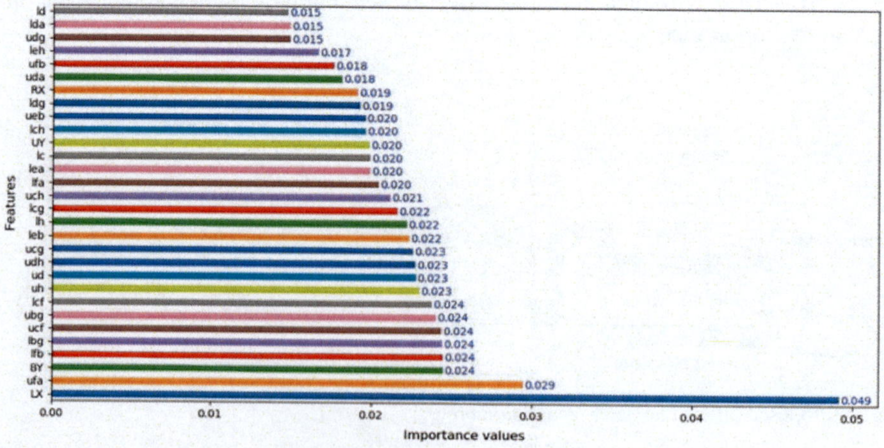

Fig. 5 Best 30 selected features based on their importance value

6 Results and Discussion

For the conducted experimental programs, we have used Python 3.7.0 programming language. The standard *ExtraTreesClassifier()* and *RandomForestClassifier()* are used from the *sklearn* package for our feature selection and character recognition task, respectively. These all tasks are performed by a system having an intel core i3 processor of 2.30 GHz with 6GB of RAM.

Two experiments are conducted for the proposed model. One is showing the difference of their recognition accuracy based on the complete feature set (272 dimensions) and the selected best 100 features. The outcomes are shown in Table 1. From which, we can see that there is no such huge difference in the achieved accuracy. But we can reduce the required execution time about 45% by means of reduced number of features. Second is the comparison among the result of proposed framework, and the result existing is shown in Table 2. The proposed offline character recognition framework can be seen to outperform the existing systems.

Table 1 Recognition accuracy with the execution time for the dataset

Dataset samples	No. of class	Dimension	Acc(%)	Exc. time (in sec)	Dimension	Acc(%)	Exc. time (in sec)
(0–9)	10	272	91.45	244	100	91.23	125
(a–z)	26	272	84.71	302	100	84.43	165
(A–Z)	26	272	90.65	310	100	90.41	174
(0–9) + (a–z)	36	272	82.73	345	100	82.59	186
(0–9) + (A–Z)	36	272	83.46	352	100	83.16	193
(a–z) + (A–Z)	52	272	77.19	403	100	76.86	273
All	62	272	68.23	480	100	68.05	296

Table 2 Comparing with the various feature generation methodologies

Dataset samples	No. of class	Methods exists	Dimension	Accuracy	
				Existing	Proposed (%)
(0–9)	10	Relief [8]	150	88.81%	91.23
(0–9)	10	JMI [8]	180	88.59%	
(0–9)	10	mRMR [8]	190	88.37%	
(A–Z)	26	HMM [13]	64	73.46%	90.41
(A–Z)	26	12 Direction [15]	12	90.00%	
(A–Z)	26	geometric, zone-based, gradient [9]	145	88.46%	

7 Conclusion

In this paper, a feature representation paradigm for offline handwritten character recognition is presented. It uses a random forest algorithm for identification of character and an extra tree classifier for choosing the best features from the complete feature range. The suggested model is applied to a series of handwritten character pictures. The model was also evaluated with performance metrics including identification accuracy and the dimension of the feature set. The experimental findings focused on just 100 features, collected from the images. These indicate that this model's efficiency is substantially better than some other models.

References

1. C.W. Jian et al., Embedded character recognition system using random forest algorithm for IC inspection system.J. Telecommunication Electronic Comput. Eng. (JTEC) **10**(1–3), 121–125 (2018)
2. P. Krishnan, C. V. Jawahar. Hwnet v2: An efficient word image representation for handwritten documents. Int. J. Document Anal. Recogn. (IJDAR) **22**(4), 387–405 (2019)
3. W. Homenda, W. Lesinski, Features selection in character recognition with random forest classifier, in *International Conference on Computational Collective Intelligence* (Springer, Berlin, 2011)
4. M. Kumar, R.K. Sharma, M.K. Jindal, Efficient feature extraction techniques for offline handwritten Gurmukhi character recognition. Natl. Acad. Sci. Lette. **37**(4), 381–391 (2014)
5. S. Benchaou, O. El Melhaoui, Features extraction for offline handwritten character recognition, in *Europe and MENA Cooperation Advances in Information and Communication Technologies* (Springer, Cham, 2017), pp. 209–217
6. Y. Weng, C. Xia, A new deep learning-based handwritten character recognition system on mobile computing devices. Mobile Netw. Appl. (2019). https://doi.org/10.1007/s11036-019-01243-5
7. V. Prasad, Y. Jayanta Singh, A Study on Structural Method of Feature Extraction for Handwritten Character Recognition
8. S. Kalita, D. Gautam, A. Kumar Sahoo, R.A. Kumar, Combined approach of feature selection and machine learning technique for handwritten character recognition (June 24, 2019). Int. J. Adv. Stud. Sci. Res. **4**(4) (2019). Available at SSRN: https://ssrn.com/abstract=3408997
9. S.A. Ayyadevara, P.N.V. Sai Ram Teja, M. Rajesh Kumar, Handwritten character recognition using unique feature extraction technique, in *3rd IEEE International Conference on Recent Trends in Electronics* (IEEE, Information & Communication Technology (RTEICT), 2018), p. 2018
10. B.Y. Reta, D. Rana, G.V. Bhalerao, Amharic handwritten character recognition using combined features and support vector machine, in *2018 2nd International Conference on Trends in Electronics and Informatics (ICOEI)*, Tirunelveli (2018), pp. 265–270
11. L. Breiman, Random forests. Mach. Learn. **45**, 5–32 (2001). https://doi.org/10.1023/A:1010933404324
12. L. Breiman, Bagging predictors. Mach. Learn. **26**(2), 123–40 (1996)
13. T. Saurabh, Stochastically optimized handwritten character recognition system using Hidden Markov Model. J. Electronics Commun. Eng. **5**(1), 1–6 (2016)
14. The Chars74K dataset Character Recognition in Natural Images. http://www.ee.surrey.ac.uk/CVSSP/demos/chars74k/

15. J. Supardi, I.A. Hapsari, M.M. Siraj, Handwritten alphabets recognition using twelve directional feature extraction and self organizing maps, in *2014 International Conference on Computer, Control, Informatics and Its Applications (IC3INA)* (IEEE, 2014)
16. R. Dey, R.C. Balabantaray, A novel sliding window approach for offline handwritten character recognition, in *2019 International Conference on Information Technology (ICIT)* (IEEE, 2019)
17. J. Piri, P. Mohapatra, Exploring fetal health status using an association based classification approach, in *2019 International Conference on Information Technology (ICIT)* (IEEE, 2019), pp. 166–171
18. J. Canny, A computational approach to edge detection. IEEE Trans. Pattern Anal. Mach. Intell. PAMI-8(6), 679–698 (1986)
19. N. Otsu, A threshold selection method from gray-level histograms. IEEE Trans. Syst. Man Cybern. **9**(1), 62–66 (1979)
20. Orit Baruch, Line thinning by line following. Pattern Recogn. Lett. **8**(4), 271–276 (1988)
21. R. Dey, R.C. Balabantaray, J. Piri, A robust handwritten digit recognition system based on sliding window with edit distance, in *2020 IEEE International Conference on Electronics, Computing and Communication Technologies (CONECCT)*, Bangalore, India (2020), pp. 1–6, https://doi.org/10.1109/CONECCT50063.2020.9198385

Correction to: Data Engineering and Intelligent Computing

Vikrant Bhateja, Suresh Chandra Satapathy, Carlos M. Travieso-González, and V. N. Manjunath Aradhya

Correction to:
V. Bhateja et al. (eds.),
***Data Engineering and Intelligent Computing*, Advances in Intelligent Systems and Computing 1,**
https://doi.org/10.1007/978-981-16-0171-2

The original version of the book was inadvertently published with incorrect volume number, and this has been corrected as 1407. This has now been amended throughout the book to the correct volume number.

The updated version of the book can be found at
https://doi.org/10.1007/978-981-16-0171-2

V. Bhateja et al. (eds.), *Data Engineering and Intelligent Computing*,
Advances in Intelligent Systems and Computing 1407,
https://doi.org/10.1007/978-981-16-0171-2_61

Author Index

V. Bhateja et al. (eds.), *Data Engineering and Intelligent Computing*,
Advances in Intelligent Systems and Computing 1407,
https://doi.org/10.1007/978-981-16-0171-2